Learning Scientific Programming with Python

Second Edition

Learn to master basic programming tasks from scratch with real-life, scientifically relevant examples and solutions drawn from both science and engineering. Students and researchers at all levels are increasingly turning to the powerful Python programming language as an alternative to commercial packages and this fast-paced introduction moves from the basics to advanced concepts in one complete volume, enabling readers to gain proficiency quickly.

Beginning with general programming concepts such as loops and functions within the core Python 3 language, and moving on to the NumPy, SciPy and Matplotlib libraries for numerical programming and data visualization, this textbook also discusses the use of Jupyter Notebooks to build rich-media, shareable documents for scientific analysis. The second edition features a new chapter on data analysis with the pandas library and comprehensive updates, new exercises and examples. A final chapter introduces more advanced topics such as floating-point precision and algorithm stability, and extensive online resources support further study. This textbook represents a targeted package for students requiring a solid foundation in Python programming.

Christian Hill is a physicist and physical chemist currently working at the International Atomic Energy Agency. He has over 25 years' experience of programming in the physical sciences and has been programming in Python for 15 years. His research uses Python to produce, analyze, process, curate and visualize large data sets in the area of spectroscopy, plasma physics and material science.

Learning Scientific Programming with Python

Second Edition

CHRISTIAN HILL

CAMBRIDGE
UNIVERSITY PRESS

University Printing House, Cambridge CB2 8BS, United Kingdom

One Liberty Plaza, 20th Floor, New York, NY 10006, USA

477 Williamstown Road, Port Melbourne, VIC 3207, Australia

314–321, 3rd Floor, Plot 3, Splendor Forum, Jasola District Centre, New Delhi – 110025, India

79 Anson Road, #06–04/06, Singapore 079906

Cambridge University Press is part of the University of Cambridge.

It furthers the University's mission by disseminating knowledge in the
pursuit of education, learning, and research at the highest international levels of excellence.

www.cambridge.org
Information on this title: www.cambridge.org/9781108745918
DOI: 10.1017/9781108778039

First published 2015
Second edition 2020

A catalogue record for this publication is available from the British Library.

ISBN 978-1-108-74591-8 Paperback

Additional resources for this publication at www.cambridge.org/hill2 and https://scipython.com/

Contents

v

Acknowledgments

For Emma, Charlotte and Laurence

Many people have helped directly or indirectly in the preparation of this book, in partic-ular Jonathan Tennyson at UCL, and Laurence Rothman and Iouli Gordon for hosting my sabbatical year at the Harvard-Smithsonian Center for Astrophysics.

Many of the errors and omissions in the first edition of this book were pointed out by just a few people who were helpful enough to get in touch, notably Stafford Baines, Matthew Gillman and Stuart Anderson. Those that remain are, of course, entirely my own fault.

Special thanks are also due to Helen Reynolds, Chris Pickard, Alison Whiteley, James Elliott, Lianna Ishihara and Milo Shaffer. As ever, I owe much to the support, encouragement and friendship of Natalie Haynes.

Code Listings

1 Introduction

1.1 About This Book

This book is intended to help scientists and engineers learn version 3 of the Python programming language and its associated libraries: NumPy, SciPy, Matplotlib and pandas. No prior programming experience or scientific knowledge in any particular field is assumed. However, familiarity with some mathematical concepts such as trigonometry, complex numbers and basic calculus is helpful to follow the examples and exercises.

Python is a powerful language with many advanced features and supplementary packages; while the basic syntax of the language is straightforward to learn, it would be impossible to teach it in depth in a book of this size. Therefore, we aim for a balanced, broad introduction to the central features of the language and its important libraries. The text is interspersed with examples relevant to scientific research, and at the end of most sections there are questions (short problems designed to test knowledge) and exercises (longer problems that usually require a short computer program to solve). Although it is not necessary to complete all of the exercises, readers will find it useful to attempt at least some of them. Where a section, example or exercise contains more advanced material that may be skipped on first reading, this is indicated with the symbol ◊.

In Chapter 2 of this book, the basic syntax, data structures and flow control of a Python program are introduced. Chapter 3 is a short interlude on the use of the `pyplot` library for making graphical plots of data: this is useful to visualize the output of programs in subsequent chapters. Chapter 4 provides more advanced coverage of the core Python language and a brief introduction to object-oriented programming. There follows another short chapter introducing the popular IPython and Jupyter Notebook environments, before chapters on scientific programming with NumPy, Matplotlib, SciPy and pandas. The final chapter covers more general topics in scientific programming, including floating-point arithmetic, algorithm stability and programming style.

Readers who are already familiar with the Python programming language may wish to skim Chapters 2 and 4.

Code examples and exercise solutions may be downloaded from the book's website at https://scipython.com. Note that while comments have been included in these downloadable programs, they are not so extensive in the printed version of this book: instead, the code is explained in the text itself through numbered annotations (such as ❶). Readers typing in these programs for themselves may wish to add their own explanatory comments to the code.

1

1.2   About Python

Python is a powerful, general-purpose programming language devised by Guido van Rossum in 1989.[1] It is classified as a high-level programming language in that it automatically handles the most fundamental operations (such as memory management) carried out at the processor level ("machine code"). It is considered a higher-level language than, for example, C, because of its expressive syntax (which is close to natural language in some cases) and rich variety of native data structures such as lists, tuples, sets and dictionaries. For example, consider the following Python program which outputs a list of names on separate lines.

Listing 1.1 Outputing a list of names using a program written in Python

```
# eg1-names.py: output three names to the console.

names = ['Isaac Newton', 'Marie Curie', 'Albert Einstein']
for name in names:
    print(name)
```

Output:

```
Isaac Newton
Marie Curie
Albert Einstein
```

Now compare this with the equivalent program in C.

Listing 1.2 Outputing a list of names using a program written in C

```
/* eg1-names.c: output a list of names to the console. */
#include <stdio.h>
#include <stdlib.h>

const char *names[] = {"Isaac Newton", "Marie Curie", "Albert Einstein"};

int main(void)
{
    int i;

    for (i = 0; i < (sizeof(names) / sizeof(*names)); i++) {
        printf("%s\n", names[i]);
    }

    return EXIT_SUCCESS;
}
```

Even if you are not familiar with the C language, you can see there is quite a lot of overhead involved in coding even this simple task in C: two includes of libraries not loaded by default, explicit declarations of variables to hold the list ("array", in C) of names, names, a counter, i, and explicit indexing of this array in a for loop; you even need to add the line endings ("\n" is the "newline" character). This source code

---

[1] Until recently, Python's "benevolent dictator for life" (BDFL).

then has to be *compiled* – converted into the machine code that the computer processor understands – before it can be run (*executed*). Furthermore, there is plenty of scope for errors (bugs): trying to print the name stored in name[10] will likely cause junk to be output: the C compiler won't stop you from accessing this non-existent name.

The same program written in three lines of Python is clean and expressive: we do not have to explicitly declare that names is a list of strings, there is no need for a loop counter like i and there are no separate libraries to include (import in Python). To run the Python program, one simply needs to type python eg1-names.py which will automatically invoke the Python "interpreter" to compile and then run the resulting "bytecode" (a kind of intermediate representation of the program between its source and the ultimate machine code that Python dispatches to the processor).

Python's syntax aims to ensure that "There should be one – and preferably only one – obvious way to do it." This differs from some other popular high-level languages such as Ruby and Perl, which take the opposite approach, encapsulated by the mantra "there's more than one way to do it." For example, there are (at least) four obvious ways to output the same list in Perl:[2]

Listing 1.3 Different ways to output a list of names using a program written in Perl

```perl
@names = ("Isaac Newton", "Marie Curie", "Albert Einstein");
# Method 1
print "$_\n" for @names;

# Method 2
print join "\n", @names;
print "\n";

# Method 3
print map { "$_\n" } @names;

# Method 4
$" = "\n";
print "@names\n";
```

(Note also Perl's famously concise but somewhat opaque syntax.)

1.2.1   Advantages and Disadvantages of Python

Here are some of the main advantages of the Python programming language and why you might want to use it:

- Its clean and simple syntax makes writing Python programs fast and generally minimizes opportunities for bugs to creep in. When done right, the result is high-quality software that is easy to maintain and extend.
- It's free – Python and its associated libraries are free of cost and open source, unlike commercial offerings such as Mathematica and MATLAB.

---

[2] Well, obvious to Perl programmers.

- Cross-platform support: Python is available for every commonly available computer system, including Windows, Unix, Linux and macOS. Although platform-specific extensions exist, it is possible to write code that will run on any platform without modification.
- Python has a large library of modules and packages that extend its functionality. Many of these are available as part of the "Standard Library" provided with the Python interpreter itself. Others, including the NumPy, SciPy, Matplotlib and pandas libraries used in scientific computing, can be downloaded separately at no cost.
- Python is relatively easy to learn. The syntax and idioms used for basic operations are applied consistently in more advanced usage of the language. Error messages are generally meaningful assessments of what went wrong rather than the generic "crashes" that can occur in compiled lower-level languages such as C.
- Python is flexible: it is often described as a "multi-paradigm" language that contains the best features from the procedural, object-oriented and functional programming paradigms. There is little need for the work-arounds required in some languages when a problem can only be solved cleanly with one of these approaches.

So where's the catch? Well, Python does have some disadvantages and isn't suitable for every application.

- The speed of execution of a Python program is not as fast as some other, fully compiled languages such as C and Fortran. For heavily numerical work, the NumPy and SciPy libraries alleviate this to some extent by using compiled-C code "under the hood," but at the expense of some reduced flexibility. For many, many applications, however, the speed difference is not noticeable and the reduced speed of execution is more than offset by a much faster speed of *development*. That is, it takes much less time to write and debug a Python program than to do the same in C, C++ or Java.
- It is hard to hide or obfuscate the source code of a Python program to prevent others from copying or modifying it. However, this doesn't mean that successful commercial Python programs don't exist.
- A common complaint about Python has historically been that its rapid development has led to compatibility issues between versions. Certainly there are important differences between Python 2 and Python 3 (described in the next section and Appendix B), but the complaint stems from the fact that within the Python 2 series there were major improvements and additions to the language that meant that code written in a later version (say, 2.7) would not run on an earlier version of Python (e.g. 2.6), although code written for an earlier version of Python will always run on a later version (within the same branch, 2 or 3). If you use the latest version of Python (see Section 1.3) you probably won't run into a problem, but some operating systems that come with Python are rather conservative and install by default only an older version.

1.2.2   Python 2 or Python 3?

On 1 January 2020, Python 2 reached its "end of life": it will receive no further updates or official support, and it is the newer Python 3 version that is being actively maintained and developed. Although the differences between the two versions may seem minor, code written in Python 3 will not run under Python 2 and vice versa: Python 3 is not *backward-compatible* with its predecessor. **This book teaches Python 3**.

Since its release in 2009, the number of users and extent of library support for Python 3 has grown to the point that new users would find little benefit in learning Python 2 except to maintain legacy code.

There are several reasons for major change between versions (breaking your users' existing code is not something to be undertaken lightly): Python 3 fixes some ugly quirks and inconsistencies in the language and provides *Unicode support* for all strings (eliminating a lot of the confusion that is created in dealing with Unicode and non-Unicode strings in Python 2). Unicode is an international standard for the representation of text in most of the writing systems in the world.

It is anticipated that most users of this book will not have trouble converting their own code between the two versions of Python if necessary. The major differences are listed and more information is given in Appendix B.

1.3   Installing Python

The official website of Python is www.python.org, and it contains full and easy-to-follow instructions for downloading Python. However, there are several full distributions which include the NumPy, SciPy and Matplotlib libraries (the "SciPy Stack") to save you from having to download and install these yourself:

- *Anaconda* is available for free (including for commerical use) from www.anaconda.com/distribution. It installs both Python 2 and Python 3, but the default version can be selected either before downloading as indicated on this web page, or subsequently using the "conda" command.
- *Enthought Deployment Manager (EDM)* is a similar distribution with a free version and various tiers of paid-for versions including technical support and development software. It can be downloaded from https://assets.enthought.com/downloads/.

In most cases, one of these distributions should be all you need. We provide some platform-specific notes below.

The source code (and binaries for some platforms) for the NumPy, SciPy, Matplotlib and IPython packages are available separately at:

- NumPy: https://github.com/numpy/numpy
- SciPy: https://github.com/scipy/scipy
- Matplotlib: https://matplotlib.org/users/installing.html
- IPython: https://github.com/ipython/ipython

- Jupyter Notebook and JupyterLab: https://jupyter.org/

Windows

Windows users have a couple of further options for installing the full SciPy stack: *Python(x,y)* (https://python-xy.github.io) and *WinPython* (https://winpython.github.io/). Both are free.

macOS

macOS (formerly Mac OS X), being based on Unix, comes with Python, but it is usually an older version of Python 2. You must not delete or modify this installation (it's needed by the operating system), but you can follow the instructions above for obtaining Python 3 and the SciPy stack. macOS does not have a native *package manager* (an application for managing and installing software), but the two popular third-party package managers, Homebrew (https://brew.sh/) and MacPorts (www.macports.org), can both supply Python 3 and its packages if you prefer this option.

Linux

Almost all Linux distributions come with Python 2, but usually not Python 3, so you may need to install it from the links above: the Anaconda and Enthought distributions both have versions for Linux. Most Linux distributions come with their own software package managers (e.g. apt in Debian and rpm for RedHat). These can be used to install Python 3 and its libraries, though finding the necessary package repositories may take some research on the Internet. Be careful not to replace or modify your system installation as other applications may depend on it.

1.4   The Command Line

Most of the code examples in this book are written as stand-alone programs which can be run from the *command line* (or from within an *integrated development environment* (IDE) if you use one: see Section 10.3.2). To access the command-line interface (also known as a console or terminal) on different platforms, follow the instructions below.

- Windows 7 and earlier: *Start > All Programs > Command Prompt*; alternatively, type cmd in the *Start > Run* input box.
- Windows 8: *Preview* (lower left of screen) > *Windows System: All apps*; alternatively type cmd in the search box pulled down the top right corner of the screen.
- Windows 10: From the Start Menu (Windows icon, lower left of screen) > *Windows System > Command Prompt*; alternatively type cmd in the search box accessed from the bottom-left corner of the screen, next to the Windows icon.
- Mac OS X and macOS: *Finder > Applications > Utilities > Terminal*
- Linux: if you are not using a graphical interface you are already at the command line; if you are, then locate the Terminal application (distributions vary, but it is usually found within a *System Utilities* or *System Tools* subfolder).

Commands typed at the command line are interpreted by an application called a *shell*, which allows the user to navigate the file system and is able to start other applications. For example, the command

```
python myprog.py
```

instructs the shell to invoke the Python interpreter, sending it the file `myprog.py` as the script to execute. Output from the program is then returned to the shell and displayed in your console.

2 The Core Python Language I

2.1 The Python Shell

This chapter introduces the syntax, structure and data types of the Python programming language. The first few sections do not involve writing much beyond a few statements of Python code and so can be followed using the Python *shell*. This is an interactive environment: the user enters Python statements that are executed immediately after the *Enter* key is pressed.

The steps for accessing the "native" Python shell differ by operating system. To start it from the command line, first open a terminal using the instructions from Section 1.4 and type python.

To exit the Python shell, type exit().

When you start the Python shell, you will be greeted by a message (which will vary depending on your operating system and precise Python version). On my system, the message reads:

```
Python 3.7.5 (default, Oct 25 2019, 10:52:18)
[Clang 4.0.1 (tags/RELEASE_401/final)] :: Anaconda, Inc. on darwin
Type "help", "copyright", "credits" or "license" for more information.
>>>
```

The three chevrons (>>>) are the *prompt*, which is where you will enter your Python commands. Note that this book is concerned with Python 3, so you should check that the Python version number reported on the first line is Python 3.X.Y where the precise values of the minor version numbers X and Y should not be important.

Many Python distributions come with a slightly more advanced shell called *IDLE*, which features tab-completion, and *syntax highlighting* (Python keywords are colored specially when you type them). We will pass over the use of this application in favor of the newer and more advanced *IPython* environment, discussed in Chapter 5.

It is also possible for many installations (especially on Windows) to start a Python shell directly from an application installed when you install the Python interpreter itself. Some installations even add a shortcut icon to your desktop which will open a Python shell when you click on it.

2.2   Numbers, Variables, Comparisons and Logic

2.2.1   Types of Numbers

Among the most basic Python objects are the numbers, which come in three *types*: integers (type: `int`), floating-point numbers (type: `float`) and complex numbers (type: `complex`).

Integers

Integers are whole numbers such as 1, 8, −72 and 3847298893721407. In Python 3, there is no limit to their magnitude (apart from the availability of your computer's memory). Integer arithmetic is exact.

For clarity, it is possible to separate any pair of digits by an underscore character, "_." For example, `299_792_458` is interpreted as the same number as `299792458`.

Floating-Point Numbers

Floating-point numbers are the representation of real numbers such as 1.2, −0.36 and 1.67263×10^{-7}. They do not, in general, have the exact value of the real number they represent, but are stored in binary to a certain precision (on most systems, to the equivalent of 15–16 decimal places),[1] as explained in Section 10.1. For example, the number $\frac{4}{3}$ is stored as the binary equivalent of $1.3333333333333325931846502\ldots$, which is nearly (but not quite) the same as the infinitely repeating decimal representation of $\frac{4}{3} = 1.3333\cdots$. Moreover, even numbers that do have an exact *decimal* representation may not have an exact *binary* representation: for example 1/10 is represented by the binary number equivalent to $0.1000000000000000555111512\ldots$ Because of this finite precision, floating-point arithmetic is not exact but, with care, it is "good enough" for most scientific applications.

Any single number containing a period (".") is considered by Python to specify a floating-point number. Scientific notation is supported using "e" or "E" to separate the significand (mantissa) from the exponent: for example, `1.67263e-7` represents the number 1.67263×10^{-7}.

As with integers, pairs of digits may be separated by an underscore. For example, `1.602_176_634e-34`.

Complex Numbers

Complex numbers such as $4+3j$ consist of a real and an imaginary part (denoted by j in Python), each of which is itself represented as a floating-point number (even if specified without a period). Complex number arithmetic is therefore not exact but subject to the same finite precision considerations as `floats`.

A complex number may be specified either by "adding" a real number to an imaginary one (denoted by the j suffix), as in `2.3 + 1.2j` or by separating the real and imaginary parts in a call to `complex`, as in `complex(2.3, 1.2)`.

---

[1] This corresponds to the implementation of the IEEE-754 double-precision standard.

Example E2.1   Typing a number at the Python shell prompt simply echoes the number back to you:

```
>>> 5
5
>>> 5.
5.0
>>> 0.10
❶ 0.1
>>> 0.0001
0.0001
>>> 0.0000999
❷ 9.99e-05
```

Note that the Python interpreter displays numbers in a standard way. For example:
❶ The internal representation of 0.1 discussed earlier is rounded to "0.1," which is the shortest number with this representation.
❷ Numbers smaller in magnitude than 0.0001 are displayed in scientific notation.
   A number of one type can be created from a number of another type with the relevant *constructor*:

```
>>> float(5)
5.0
>>> int(5.2)
5
>>> int(5.9)
❶ 5
>>> complex(3.)
❷ (3+0j)
❸ >>> complex(0., 3.)
3j
```

❶ Note that a positive floating-point number is *rounded down* in casting it into an integer; more generally, int *rounds towards zero*: int(-1.4) would yield -1.
❷ Constructing a complex object from a float generates a complex number with the imaginary part equal to zero.
❸ To generate a pure imaginary number, you have to explicitly pass two numbers to complex with the first, real part, equal to zero.

2.2.2   Using the Python Shell as a Calculator

Basic Arithmetic

With the three basic number types described earlier, it is possible to use the Python shell as a simple calculator using the operators given in Table 2.1. These are *binary* operators in that they act on two numbers (the *operands*) to produce a third (e.g. 2**3 evaluates to 8).

   Python 3 has two types of division: floating-point division (/) always returns a floating-point (or complex) number result, even if it acts on integers. Integer division

Table 2.1 Basic Python
arithmetic operators

+	Addition
–	Subtraction
*	Multiplication
/	Floating-point division
//	Integer division
%	Modulus (remainder)
**	Exponentiation

(//) *always rounds down* the result to the nearest smaller integer ("*floor division*"); the type of the resulting number is an int only if both of its operands are ints; otherwise it returns a float. Some examples should make this clearer:

Regular ("*true*") floating-point division with (/):

```
>>> 2.7 / 2
1.35
>>> 9 / 2
4.5
>>> 8 / 4
2.0
```

The last operation returns a float even though both operands are ints.

Integer division with (//):

```
>>> 8 // 4
2
>>> 9 // 2
4
>>> 2.7 // 2
1.0
```

Note that // can perform integer arithmetic (rounding down) on floating-point numbers. The modulus operator gives the remainder of an integer division:

```
>>> 9 % 2
1
>>> 4.5 % 3
1.5
```

Again, the number returned is an int only if both of the operands are ints.

Operator Precedence

Arithmetic operations can be strung together in a sequence, which naturally raises the question of *precedence*: for example, does 2 + 4 * 3 evaluate to 14 (as 2 + 12) or 18 (as 6 * 3)? Table 2.2 shows that the answer is 14: multiplication has a higher precedence than addition and is evaluated first. These precedence rules are overridden by the use of *parentheses*: for example, (2 + 4) * 3 = 18.

Operators of equal precedence are evaluated left to right with the exception of exponentiation (**), which is evaluated right to left (that is, "top down" when written using the conventional superscript notation). For example,

Table 2.2 Python arithmetic operator precedence

**	(highest precedence)
*, /, //, %	
+, -	(lowest precedence)

```
>>> 6 / 2 / 4          # the same as 3 / 4
0.75
>>> 6 / (2 / 4)        # the same as 6 / 0.5
12.0
>>> 2**2**3            # the same as 2**(2**3) == 2**8
256
>>> (2**2)**3          # the same as 4**3
64
```

In examples such as these, the text following the hash symbol, #, is a comment that is ignored by the interpreter. We shall sometimes use comments in this to explain more about a statement, but it is not necessary to type it in if you try out the code.

Methods and Attributes of Numbers

Python numbers are *objects* (in fact, everything in Python is an object) and have certain *attributes*, accessed using the "dot" notation: <object>.<attribute> (this use of the period has nothing to do with the decimal point appearing in a floating-point number). Some attributes are simple values: for example, complex number objects have the attributes real and imag, which are the real and imaginary (floating-point) parts of the number:

```
>>> (4 + 5j).real
4.0
>>> (4 + 5j).imag
5.0
```

Other attributes are *methods*: callable functions that act on their object in some way.[2] For example, complex numbers have a method, conjugate, which returns the complex conjugate:

```
>>> (4 + 5j).conjugate()
(4-5j)
```

Here, the empty parentheses indicate that the method is to be *called*, that is, the function to calculate the complex conjugate is to be run on the number $4 + 5j$; if we omit them, as in (4 + 5j).conjugate, we are referring to the method itself (without calling it) – this method is itself an object!

Integers and floating-point numbers don't actually have very many attributes that it makes sense to use in this way, but if you're curious you can find out how many bits an integer takes up in memory by calling its bit_length method. For example,

---

[2] In this book, we will use the terms *method* and *function* interchangeably. In Python, everything is an object and the distinction is not as meaningful as it is in some other languages.

```
>>> (3847298893721407).bit_length()
52
```

Note that Python allocates as much memory as is necessary to exactly represent the integer.

Mathematical Functions

Two of the mathematical functions that are provided "by default" as so-called *built-ins* are abs and round.

abs returns the absolute value of a number as follows:

```
>>> abs(-5.2)
5.2
>>> abs(-2)
2
>>> abs(3 + 4j)
5.0
```

This is an example of *polymorphism*: the same function, abs, does different things to different objects. If passed a real number, x, it returns $|x|$, the nonnegative magnitude of that number, without regard to sign; if passed a complex number, $z = x + iy$, it returns the modulus, $|z| = \sqrt{x^2 + y^2}$.

The round function (with one argument) rounds a floating-point number to the nearest integer, employing *Banker's rounding*:[3]

```
>>> round(-9.62)
-10
>>> round(7.5)
8
>>> round(4.5)
4
```

One can also specify the number of digits of precision after the decimal point as a second argument to round():

```
>>> round(3.141592653589793, 3)
3.142
>>> round(96485.33289, -2)
96500.0
```

Python is a very modular language: functionality is available in packages and modules that are *imported* if they are needed but are not loaded by default: this keeps the memory required to run a Python program to a minimum and improves performance. For example, many useful mathematical functions are provided by the math module, which is imported with the statement

```
>>> import math
```

The math module concerns itself with floating-point and integer operations (for functions of complex numbers, there is another module, called cmath). These are called by passing one (or sometimes more than one) number to them inside parentheses (the numbers are said to act as *arguments* to the function being called). For example,

---

[3] In Banker's rounding, half-integers are rounded to the nearest *even* integer.

Table 2.3 Some functions provided by the `math` module. *Angular arguments are assumed to be in radians.*

`math.sqrt(x)`	$\sqrt{x}$
`math.exp(x)`	e^x
`math.log(x)`	$\ln x$
`math.log(x, b)`	$\log_b x$
`math.log10(x)`	$\log_{10} x$
`math.sin(x)`	$\sin(x)$
`math.cos(x)`	$\cos(x)$
`math.tan(x)`	$\tan(x)$
`math.asin(x)`	$\arcsin(x)$
`math.acos(x)`	$\arccos(x)$
`math.atan(x)`	$\arctan(x)$
`math.sinh(x)`	$\sinh(x)$
`math.cosh(x)`	$\cosh(x)$
`math.tanh(x)`	$\tanh(x)$
`math.asinh(x)`	$\text{arsinh}(x)$
`math.acosh(x)`	$\text{arcosh}(x)$
`math.atanh(x)`	$\text{artanh}(x)$
`math.hypot(x, y)`	The Euclidean norm, $\sqrt{x^2 + y^2}$
`math.factorial(x)`	$x!$
`math.erf(x)`	The error function at x
`math.gamma(x)`	The gamma function at x, $\Gamma(x)$
`math.degrees(x)`	Converts x from radians to degrees
`math.radians(x)`	Converts x from degrees to radians
`math.isclose(a, b)`	Test if a and b are equal to within some tolerance

```
>>> import math
>>> math.exp(-1.5)
0.22313016014842982
>>> math.cos(0)
1.0
>>> math.sqrt(16)
4.0
```

A complete list of the mathematical functions provided by the `math` module is available in the online documentation;[4] the more commonly used ones are listed in Table 2.3.

The `math` module also provides two very useful nonfunction attributes: `math.pi` and `math.e` give the values of π and e, the base of the natural logarithm, respectively.

It is possible to import the math module with "`from math import *`" and access its functions directly:

```
>>> from math import *
>>> cos(pi)
-1.0
```

---

[4] https://docs.python.org/3/library/math.html.

However, although this may be convenient for interacting with the Python shell, it is not recommended in Python programs. There is a danger of name conflicts (particularly if many modules are imported in this way), and it makes it difficult to know which function comes from which module. Importing with `import math` keeps the functions bound to their module's *namespace*: thus, even though `math.cos` requires more typing it makes for code that is much easier to understand and maintain.

Example E2.2   As might be expected, mathematical functions can be strung together in a single expression:

```
>>> import math
>>> math.sin(math.pi/2)
1.0
>>> math.degrees(math.acos(math.sqrt(3)/2))
30.000000000000004
```

Note the finite precision here: the exact answer is arccos($\sqrt{3}/2$) = 30°.

The fact that the `int` function rounds down in casting a positive floating-point number to an integer can be used to find the number of digits a positive integer has:

```
>>> int(math.log10(9999)) + 1
4
>>> int(math.log10(10000)) + 1
5
```

2.2.3   Variables

What Is a Variable?

When an object, such as a `float`, is created in a Python program or using the Python shell, memory is allocated for it: the location of this memory within the computer's architecture is called its *address*. The actual value of an object's address isn't actually very useful in Python, but if you're curious you can find it out by calling the `id` built-in method:

```
>>> id(20.1)
4297273888                 # for example
```

This number refers to a specific location in memory that has been allocated to hold the `float` object with the value `20.1`.

For anything beyond the most basic usage, it is necessary to store the objects that are involved in a calculation or algorithm and to be able to refer to them by some convenient and meaningful name (rather than an address in memory). This is what *variables* are for.[5] A variable name can be assigned ("bound") to any object and used to identify that object in future calculations. For example,

```
>>> a = 3
```

---

[5] In Python, it is arguably better to talk of object *identifiers* or *identifier names* rather than variables, but we will not be too strict about this.

```
>>> b = -0.5
>>> a * b
-1.5
```

In this snippet, we create the `int` object with the value 3 and assign the variable name a to it. We then create the `float` object with the value -0.5 and assign b to it. Finally, the calculation a * b is carried out: the values of a and b are multiplied together and the result returned. This result isn't assigned to any variable, so after being output to the screen it is thrown away. That is, the memory required to store the result, a `float` with the value -1.5, is allocated for long enough for it to be displayed to the user, but then it is gone.[6] If we need the result for some subsequent calculation, we should assign it to another variable:

```
>>> c = a * b
>>> c
-1.5
```

Note that we did not have to *declare* the variables before we assign them (tell Python that the variable name a is to refer to an integer, b is to refer to a floating-point number, etc.), as is necessary in some computer languages. Python is a *dynamically typed* language and the necessary object type is inferred from its definition: in the absence of a decimal point, the number 3 is assumed to be an `int`; -0.5 looks like a floating-point number and so Python defines b to be a `float`.[7]

Variable Names

There are some rules about what makes a valid variable name:

- Variable names are *case-sensitive*: a and A are different variables;
- Variable names can contain any letter, the underscore character ("_") and any digit (0–9) ...
- ... but must not *start with* a digit;
- A variable name must not be the same as one of the *reserved keywords* given in Table 2.4;
- The built-in constant names `True`, `False` and `None` cannot be assigned as variable names.

Most of the reserved keywords are pretty unlikely choices for variable names, with the exception of `lambda`. Python programmers often use `lam` if they need to use it. A good text editor will highlight the keywords as you type your program, so this rarely causes confusion.

It is possible to give a variable the same name as a built-in function (e.g. abs and round), but that built-in function will no longer be available after such an assignment,

---

[6] Actually in an interactive Python session the result of the last calculation is stored in the special variable called _ (the underscore), so it isn't really thrown away until overwritten by the *next* calculation.

[7] This is sometimes called *duck-typing* after the phrase attributed to James Whitcomb Riley: "When I see a bird that walks like a duck and swims like a duck and quacks like a duck, I call that bird a duck."

Table 2.4 Python 3 reserved keywords

and	as	assert	async	await	break
class	continue	def	del	elif	else
except	finally	for	from	global	if
import	in	is	lambda	nonlocal	not
or	pass	raise	return	try	while
with	yield	False	True	None	

so this is probably best avoided – luckily, most have names that are unlikely to be chosen in practice.[8]

In addition to the rules mentioned earlier, there are certain style considerations that dictate good practice in naming variables:

- Variable names should be meaningful (area is better than a) ...
- ... but not too long (the_area_of_the_triangle is unwieldy);
- Generally, don't use I (upper-case i), l (lower-case L) or the upper-case letter O: they look too much like the digits 1 and 0;
- The variable names i, j and k are usually used as integer counters;
- Use lower-case names, with words separated by underscores rather than "CamelCase": for example, mean_height and not MeanHeight.[9]

These and many other rules and conventions are codified in a style guide called PEP8 which forms part of the Python documentation[10] (see also Section 10.3.1).

Breaking these style rules will not result in your program failing to run, but it might make it harder to maintain and debug – the person you help might be yourself!

Example E2.3 *Heron's formula* gives the area, A, of a triangle with sides a, b, c as

$$A = \sqrt{s(s-a)(s-b)(s-c)} \text{ where } s = \tfrac{1}{2}(a+b+c).$$

For example,

```
>>> a = 4.503
>>> b = 2.377
>>> c = 3.902
>>> s = (a + b + c) / 2
>>> area = math.sqrt(s * (s - a) * (s - b) * (s - c))
>>> area
4.63511081571606
```

❶ Don't forget to import math if you haven't already in this Python session.

---

[8] For a complete list of built-in function names, see https://docs.python.org/3/library/functions.html.
[9] CamelCase in Python is usually reserved for class names: see Section 4.6.2.
[10] https://legacy.python.org/dev/peps/pep-0008/.

Table 2.5 Python comparison
operators

==	Equal to
!=	Not equal to
>	Greater than
<	Less than
>=	Greater than or equal to
<=	Less than or equal to

Example E2.4 The data type and memory address of the object referred to by a variable name can be found with the built-ins type and id:

```
>>> type(a)
<class 'float'>
>>> id(area)
4298539728      # for example
```

2.2.4   Comparisons and Logic

Operators

The main comparison operators that are used in Python to compare objects (such as numbers) are given in Table 2.5.

The result of a comparison is a *boolean* object (of type bool) which has exactly one of two values: True or False. These are built-in constant keywords and cannot be reassigned to other values. For example,

```
>>> 7 == 8
False
>>> 4 >= 3.14
True
```

Python is able, as far as possible without ambiguity, to compare objects of different types: the integer 4 is promoted to a float for comparison with the number 3.14.

Note the importance of the difference between == and =. The single equals sign is an *assignment*, which does not return a value: the statement a = 7 assigns the variable name a to the integer object 7 and that is all, whereas the expression a == 7 is a test: it returns True or False depending on the value of a.[11]

Care should be taken in comparing floating-point numbers for equality. Since they are not stored exactly, calculations involving them frequently lead to a loss of precision and this can give unexpected results to the unwary. For example,

```
>>> a = 0.01
>>> b = 0.1**2
>>> a == b
False
```

[11] In some languages, such as C, assignment returns the value of whatever is being assigned, which can lead to some nasty and hard-to-find bugs when = is mistakenly used as a comparison operator.

In this example, 0.01 cannot be represented exactly as a floating-point number but is (on my system) stored as a binary number equivalent to 0.010000000000000000208; the result of squaring the floating-point representation of 0.1 on the other hand is 0.010000000000000000194, and these two numbers are not the same. See Section 10.1 for more information.

Since Python 3.5, the math library has provided a function, isclose, to test whether two floating-point numbers are equal to within some absolute or relative tolerance:

```
>>> math.isclose(0.1**2, 0.01)
True
```

The relative tolerance can be set with the rel_tol argument, which defaults to 1.e-9: this is the maximum allowed difference between the two numbers relative to the larger absolute value of them; for example, to test if a and b are within 5% of the larger of them:

```
>>> a = 9.5
>>> b = 10
>>> math.isclose(a, b, rel_tol=0.05)
True
```

This kind of relative comparison is problematic when one of the numbers is zero,[12] in which case it can be helpful to test against an absolute tolerance, set by the abs_tol argument (which defaults to 0):

```
>>> math.isclose(0, 1.e-12)    # relative tolerance comparison fails if rel_tol < 1
False
>>> math.isclose(0, 1.e-12, abs_tol=1.e-10)
True
```

Table 2.6 Truth table for the not operator

P	not P
True	False
False	True

Logic Operators

Comparisons can be modified and strung together with the logic operator keywords and, not and or. See Tables 2.6, 2.7 and 2.8. For example,

```
>>> 7.0 > 4 and -1 >= 0     # equivalent to True and False
False
>>> 5 < 4 or 1 != 2         # equivalent to False or True
True
```

---

[12] The relative difference between any number a and 0 is $(|a| - 0)/|a|$ which is certainly bigger than rel_tol if rel_tol is less than 1.

Table 2.7 Truth table for the and operator

P	Q	P and Q
True	True	True
False	True	False
True	False	False
False	False	False

Table 2.8 Truth table for the or operator

P	Q	P or Q
True	True	True
False	True	True
True	False	True
False	False	False

In compound expressions such as these, the comparison operators are evaluated first, and then the logic operators in order of precedence: not, and, or. This precedence is overridden with parentheses, as for arithmetic. Thus,

```
>>> not 7.5 < 0.9 or 4 == 4
True
>>> not (7.5 < 0.9 or 4 == 4)
False
```

The truth tables for the logic operators are given below; note that, in common with most languages or in Python is the *inclusive or* variant for which A or B is True if both A and B are True, rather than the *exclusive or* operator (A xor B is True only if one but not both of A and B are True[13]).

◇ **Boolean Equivalents and Conditional Assignment**

In a logic test expression, it is not always necessary to make an explicit comparison to obtain a boolean value: Python will try to convert an object to a bool type if needed. For numerical objects, 0 evaluates to False and any nonzero value is True:

```
>>> a = 0
>>> a or 4 < 3        # same as: False or 4 < 3
False
>>>
>>> not a + 1         # same as: not True
False
```

---

[13] This xor built-in operator does not exist in Python, but it can be imported as a function with from operator import xor. The call: xor(a, b) returns True or False.

In this last example, addition has higher precedence than the logic operator not, so a + 1 is evaluated first to give 1. This corresponds to boolean True, and so the whole expression is equivalent to not True. To explicitly convert an object to a boolean object, use the bool constructor:

```
>>> bool(-1)
True
>>> bool(0.0)
False
```

In fact, the and and or operators always return one of their operands and not just its bool equivalent. So, for example:

```
>>> a = 0
❶ >>> a - 2 or a
-2
❷ >>> 4 > 3 and a - 2
-2
❸ >>> 4 > 3 and a
0
```

Logic expressions are evaluated left to right, and those involving and or or are *short-circuited*: the second expression is only evaluated if necessary to decide the truth value of the whole expression. The three examples presented here can be analyzed as follows:

❶ In the first example, a − 2 is evaluated first: this is equal to −2, which is equivalent to True, so the or condition is fulfilled and the operand evaluating to True is returned immediately: −2.

❷ 4 > 3 is True, so the second expression must be evaluated to establish the truth of the and condition. a − 2 is equal to −2, which is also equivalent to True, so the and condition is fulfilled and −2 (as the most recently evaluated expression) is returned.

❸ In the last case, a is 0 which is equivalent to False: the and condition evaluates to False because of this, and so the return value is 0.

Python's Special Value, None

Python defines a single value, None, of the special type, NoneType. It is used to represent the absence of a defined value, for example, where no value is possible or relevant. This is particularly helpful in avoiding arbitrary default values (such as 0, -1 or −99) for bad or missing data.

In a boolean comparison, None evaluates to False, but to test whether or not a variable, x, is equal to None, use

```
if x is None
```

and

```
if x is not None
```

rather than the shortcuts if x and if not x.[14]

---

[14] Note that not x also evaluates to True if x is any of 0, False or an empty data structure such as an empty list, [], or string, ''; it is, therefore, not a very reliable way to test specifically if x is not set to None.

Example E2.5   A common Python idiom is to assign a variable using the return value of a logic expression:

```
>>> a = 0
>>> b = a or -1
>>> b
-1
```

That is (for a understood to be an integer): "set b equal to the value of a unless a $==$ 0, in which case set b equal to -1."

2.2.5   Immutability and Identity

The objects presented so far, such as integers and booleans, are *immutable*. Immutable objects do not change after they are created, though a variable name may be reassigned to refer to a different object from the one it was originally assigned to. For example, consider the assignments:

```
>>> a = 8
>>> b = a
```

The first line creates the integer object with value 8 in memory, and assigns the name a to it. The second line assigns the name b to the same object. You can see this by inspecting the address of the object referred to by each name:

```
>>> id(a)
4297273504
>>> id(b)
4297273504
```

Thus, a and b are references to the same integer object. Now suppose a is reassigned to a new number object:

```
>>> a = 3.14
>>> a
3.14
>>> b
8
>>> id(a)
4298630152
>>> id(b)
4297273504
```

Note that the value of b has not changed: this variable still refers to the original 8. The variable a now refers to a new, float object with the value 3.14 located at a new address. This is what is meant by immutability: it is not the "variable" that cannot change but the immutable object itself – see Figure 2.1.

A more convenient way to establish if two variables refer to the same object is to use the is operator, which determines object *identity*:

```
>>> a = 2432
>>> b = a
```

```
>>> a is b
True
>>> c = 2432
>>> c is a
False
>>> c == a
True
```

Here, the assignment c = 2432 creates an entirely new integer object so c is a evaluates as False, even though a and c have the same *value*. That is, the two variables refer to different objects with the same value.

It is often necessary to change the value of a variable in some way, such as

```
>>> a = 800
>>> a = a + 1
>>> a
801
```

The integers 800 and 801 are immutable: the line a = a + 1 creates a *new* integer object with the value 801 (the right-hand side is evaluated first) and assigns it to the variable name a (the old 800 is forgotten[15] unless some other variable refers to it). That is, a points to a different address before and after this statement.

This reassignment of a variable by an arithmetic operation on its value is so common that there is a useful shorthand notation: the *augmented assignment* a += 5 is the same as a = a + 5. The operators -=, *=, /=, //=, %= work in the same way. C-style increment and decrement operations such as a++ for a += 1 are *not* supported in Python, however.[16]

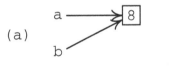

(a)

(b)

Figure 2.1 (a) Two variables referring to the same integer; (b) after reassigning the value of a.

---

[15] That is, the memory assigned for it by Python is reclaimed ("garbage-collected") for general use.

[16] Assignment and augmented assignment in Python are statements not expressions and so do not return a value and cannot be chained together.

Example E2.6   Python provides the operator is not: it is more natural to write c is not a than not c is a.

```
>>> a = 8
>>> b = a
>>> b is a
True
>>> b /= 2
>>> b is not a
True
```

◇ **Example E2.7**   Given the previous discussion, it might come as a surprise to find that

```
>>> a = 256
>>> b = 256
>>> a is b
True
```

This happens because Python keeps a *cache* of commonly used, small integer objects (on my system, the numbers −5 to 256). To improve performance, the assignment a = 256 attaches the variable name a to the existing integer object without having to allocate new memory for it. Because the same thing happens with b, the two variables in this case do, in fact, point to the same object. By contrast,

```
>>> a = 257
>>> b = 257
>>> a is b
False
```

2.2.6   Exercises

Questions

Q2.2.1   Predict the result of the following expressions and check them using the Python shell.

(a)   `2.7 / 2`
(b)   `2 / 4 - 1`
(c)   `2 // 4 - 1`
(d)   `(2 + 5) % 3`
(e)   `2 + 5 % 3`
(f)   `3 * 4 // 6`
(g)   `3 * (4 // 6)`
(h)   `3 * 2 ** 2`
(i)   `3 ** 2 * 2`

Q2.2.2   The operators listed in Table 2.1 are all *binary* operators: they take two operands (numbers) and return a single value. The symbol − is also used as a *unary* operator, which returns the negative value of the single operand on which it acts. For example,

```
>>> a = 4
>>> b = -a
>>> b
-4
```

Note that the expression b = -a (which sets the variable b to the negative value of a) is different from the expression b -= a (which subtracts a from b and stores the result in b). The unary – operator has a higher precedence than *, / and % but a lower precedence than exponentiation (**), so that, for example -2 ** 4 is -16 (i.e. $-(2^4)$, not $(-2)^4$).

Predict the result of the following expressions and check them using the Python shell.

(a)   -2 ** 2
(b)   2 ** -2
(c)   -2 ** -2
(d)   2 ** 2 ** 3
(e)   2 ** 3 ** 2
(f)   -2 ** 3 ** 2
(g)   (-2) ** 3 ** 2
(h)   (-2) ** 2 ** 3

Q2.2.3 Predict and explain the results of the following statements.

(a)   9 + 6j / 2
(b)   complex(4, 5).conjugate().imag
(c)   complex(0, 3j)
(d)   round(2.5)
(e)   round(-2.5)
(f)   abs(complex(5, -4)) == math.hypot(4,5)

Q2.2.4 Determine the value of i^i as a real number, where $i = \sqrt{-1}$.

Q2.2.5 Explain the (surprising?) behavior of the following short code:

```
>>> d = 8
>>> e = 2
>>> from math import *
>>> sqrt(d ** e)
16.88210319127114
```

Q2.2.6 Formally, the integer division a // b is defined as the *floor* of a/b (sometimes written $\lfloor \frac{a}{b} \rfloor$) – that is, the largest integer less than or equal to a / b. The modulus or remainder, a % b (also written $a \bmod b$), is then

$$a \bmod b = a - b \left\lfloor \frac{a}{b} \right\rfloor.$$

Use these definitions to predict the result of the following expressions and check them using the Python shell.

(a)   7 // 4
(b)   7 % 4
(c)   -7 // 4
(d)   -7 % 4
(e)   7 // -4
(f)   7 % -4
(g)   -7 // -4
(h)   -7 % -4

Q2.2.7 If two adjacent sides of a regular, six-sided die have the values a and b when viewed side-on and read left to right, the value on the top of the die is given by $3(a^3b - ab^3) \bmod 7$.

Determine the value on the top of the die if (a) $a = 2, b = 6$, (b) $a = 3, b = 5$.

Q2.2.8 How many times must a sheet of paper (thickness, $t = 0.1$ mm but otherwise any size required) be folded to reach the Moon (distance from Earth, $d = 384\,400$ km)?

Q2.2.9 Predict the results of the following expressions and check them using the Python shell.

(a)   not 1 < 2 or 4 > 2
(b)   not (1 < 2 or 4 > 2)
(c)   1 < 2 or 4 > 2
(d)   4 > 2 or 10/0 == 0
(e)   not 0 < 1
(f)   1 and 2
(g)   0 and 1
(h)   1 or 0
(i)   type(complex(2, 3).real) is int

Q2.2.10 Explain why the following expression does not evaluate to 100.

```
>>> 10^2
8
```

Hint: refer to the Python documentation for *bitwise* operators.

Problems

P2.2.1 There is no exclusive-or operator provided "out of the box" by Python, but one can be constructed from the existing operators. Devise two different ways of doing this. The truth table for the xor operator is given in Table 2.9.

P2.2.2 Some fun with the math module:

(a)   What is special about the numbers $\sin\left(2017\sqrt[5]{2}\right)$ and $(\pi + 20)^i$?
(b)   What happens if you try to evaluate an expression, such as e^{1000}, which generates a number larger than the largest floating-point number that can be represented

in the default double precision? What if you restrict your calculation to integer arithmetic (e.g. by evaluating 1000!)?

(c)   What happens if you try to perform an undefined mathematical operation such as division by zero?

(d)   The maximum representable floating-point number in IEEE-754 double precision is about 1.8×10^{308}. Calculate the length of the hypotenuse of a right-angled triangle with opposite and adjacent sides 1.5×10^{200} and 3.5×10^{201} (i) using the `math.hypot()` function directly and (ii) without using this function.

P2.2.3   Some languages provide a `sign(a)` function which returns -1 if its argument, a, is negative and 1 otherwise. *Python does not provide such a function*, but the `math` module does include a function `math.copysign(x, y)`, which returns the absolute value of x with the sign of y. How would you use this function in the same way as the missing `sign(a)` function?

P2.2.4   The World Geodetic System is a set of international standards for describing the shape of the Earth. In the latest WGS-84 revision, the Earth's *geoid* is approximated to a reference ellipsoid that takes the form of an oblate spheroid with semi-major and semi-minor axes $a = 6\,378\,137.0$ m and $c = 6\,356\,752.314245$ m respectively.

Use the formula for the surface area of an oblate spheroid,

$$S_{\text{obl}} = 2\pi a^2 \left(1 + \frac{1 - e^2}{e} \operatorname{atanh}(e) \right), \quad \text{where } e^2 = 1 - \frac{c^2}{a^2},$$

to calculate the surface area of this reference ellipsoid and compare it with the surface area of the Earth assumed to be a sphere with radius 6371 km.

2.3   Python Objects I: Strings

2.3.1   Defining a String Object

A Python string object (of type `str`) is an ordered, immutable sequence of characters. To define a variable containing some constant text (a *string literal*), enclose the text in either single or double quotes:

```
>>> greeting = "Hello, Sir!"
>>> bye = 'À bientôt'
```

Table 2.9 Truth table for the xor operator

P	Q	P xor Q
True	True	False
False	True	True
True	False	True
False	False	False

Strings can be concatenated using either the + operator or by placing them next to each other on the same line:

```
>>> 'abc' + 'def'
'abcdef'
>>> 'one ' 'two' ' three'
'one two three'
```

Python doesn't place any restriction on the length of a line, so a string literal can be defined in a single, quoted block of text. However, for ease of reading, it is usually a good idea to keep the lines of your program to a fixed maximum length (79 characters is recommended). To break up a string over two or more lines of code, use the line continuation character, "\" or (better) enclose the string literal in parentheses:

```
>>> long_string = 'We hold these truths to be self-evident,'\
...                ' that all men are created equal...'

>>> long_string = ('We hold these truths to be self-evident,'
...                ' that all men are created equal...')
```

This defines the variable long_string to hold a single line of text (with no carriage returns). The concatenation does not insert spaces so they need to be included explicitly if they are wanted. The spaces lining up the opening quotes in this example are optional but make the code easier to read.

If your string consists of a repetition of one or more characters, the * operator can be used to concatenate them the required number of times:

```
>>> 'a' * 4
'aaaa'
>>> '-o-' * 5
'-o--o--o--o--o-'
```

The *empty string* is defined simply as s = '' (two single quotes) or s = "".

Finally, the built-in function, str converts an object passed as its argument into a string according to a set of rules defined by the object itself:

```
>>> str(42)
'42'
>>> str(3.4e5)
'340000.0'
>>> str(3.4e20)
'3.4e+20'
```

For finer control over the formatting of the string representation of numbers, see Section 2.3.7.

Example E2.8 Strings concatenated with the "+" operator can repeated with "*," but only if enclosed in parentheses:

```
>>> ('a'*4 + 'B') * 3
'aaaaBaaaaBaaaaB'
```

2.3.2   Escape Sequences

The choice of quotes for strings allows one to include the quote character itself inside a string literal – just define it using the other quote:

```
>>> verse = 'Quoth the Raven "Nevermore."'
```

But what if you need to include both quotes in a string? Or to include more than one line in the string? This case is handled by special *escape sequences* indicated by a backslash, \. The most commonly used escape sequences are listed in Table 2.10. For example,

```
>>> sentence = "He said, \"This parrot's dead.\""
```
❶
```
>>> sentence
'He said, "This parrot\'s dead."'
```
❷
```
>>> print(sentence)
He said, "This parrot's dead."
>>> subjects = 'Physics\nChemistry\nGeology\nBiology'
>>> subjects
'Physics\nChemistry\nGeology\nBiology'
>>> print(subjects)
Physics
Chemistry
Geology
Biology
```

❶   Note that just typing a variable's name at the Python shell prompt simply echoes its literal value back to you (in quotes).

❷   To produce the desired string including the proper interpretation of special characters, pass the variable to the print built-in function (see Section 2.3.6).

On the other hand, if you want to define a string to include character sequences such as "\n" *without them being escaped*, define a *raw string* prefixed with r:

```
>>> rawstring = r'The escape sequence for a new line is \n.'
>>> rawstring
'The escape sequence for a new line is \\n.'
>>> print(rawstring)
The escape sequence for a new line is \n.
```

Table 2.10 Common Python escape sequences

Escape sequence	Meaning
\'	Single quote (')
\"	Double quote (")
\n	Linefeed (LF)
\r	Carriage return (CR)
\t	Horizontal tab
\b	Backspace
\\	The backslash character itself
\u, \U, \N{}	Unicode character (see Section 2.3.3)
\x	Hex-encoded byte

When defining a block of text including several line endings it is often inconvenient to use \n repeatedly. This can be avoided by using *triple-quoted strings*: new lines defined within strings delimited by """ and ''' are preserved in the string:[17]

```
a = """one
two
three"""
>>> print(a)
one
two
three
```

This is often used to create "docstrings" which document blocks of code in a program (see Section 2.7.1).

Example E2.9   The \x escape denotes a character encoded by the single-byte hex value given by the subsequent two characters. For example, the capital letter "N" is encoded by the value 78, which is 4e in hex. Hence,

```
>>> '\x4e'
'N'
```

The *backspace* "character" is encoded as hex 08, which is why '\b' is equivalent to '\x08':

```
>>> 'hello\b\b\b\b\bgoodbye'
'hello\x08\x08\x08\x08\x08goodbye'
```

Sending this string to the print() function outputs the string formed by the sequence of characters in this string literal:

```
>>> print('hello\b\b\b\b\bgoodbye')
goodbye
```

2.3.3   Unicode

Python 3 strings are composed of *Unicode* characters. unicode is a standard describing the representation of more than 100 000 characters in just about every human language, as well as many other specialist characters such as scientific symbols. It does this by assigning a number (*code point*) to every character; the numbers that make up a string are then encoded as a sequence of bytes.[18] For a long time, there was no agreed encoding standard, but the *UTF-8* encoding, which is used by Python 3 by default, has emerged as the most widely used today.[19] If your editor will not allow you to enter a character directly into a string literal, you can use its 16- or 32-bit hex value or its Unicode character name as an escape sequence:

```
>>> '\u00E9' # 16-bit hex value
```

---

[17] It is generally considered better to use three *double* quotes, """, for this purpose.

[18] For a list of code points, see the official Unicode website's code charts at www.unicode.org/charts/.

[19] UTF-8 encoded Unicode encompasses the venerable 8-bit encoding of the ASCII character set (e.g. A = 65).

```
'é'
>>> '\u000000E9' # 32-bit hex value
'é'
>>> '\N{LATIN SMALL LETTER E WITH ACUTE}' # by name
'é'
```

Example E2.10 Providing your editor or terminal allows it, and you can type them at your keyboard or paste them from elsewhere (e.g. a web browser or word processor), Unicode characters can be entered directly into string literals:

```
>>> creams = 'Crème fraîche, crème brûlée, crème pâtissière'
```

Python even supports Unicode variable names, so identifiers can use non-ASCII characters:

```
>>> Σ = 4
>>> crème = 'anglaise'
```

Needless to say, because of the potential difficulty in entering non-ASCII characters from a standard keyboard and because many distinct characters look very similar, this is not a good idea.

2.3.4   Indexing and Slicing Strings

Indexing (or "subscripting") a string returns a single character at a given location. Like all sequences in Python, strings are indexed with the first character having the index 0; this means that the final character in a string consisting of n characters is indexed at $n - 1$. For example,

```
>>> a = "Knight"
>>> a[0]
'K'
>>> a[3]
'g'
```

The character is returned in a str object of length 1. A nonnegative index counts forward from the start of the string; there is a handy notation for the index of a string counting backward: a negative index, starting at -1 (for the final character) is used. So,

```
>>> a = "Knight"
>>> a[-1]
't'
>>> a[-4]
'i'
```

It is an error to attempt to index a string outside its length (here, with index greater than 5 or less than −6); Python raises an IndexError:

```
>>> a[6]
Traceback (most recent call last):
  File "<stdin>", line 1, in <module>
IndexError: string index out of range
```

Slicing a string, s[i:j], produces a substring of a string between the characters at two indexes, *including* the first (i) but *excluding* the second (j). If the first index is omitted, 0 is assumed; if the second is omitted, the string is sliced to its end. For example,

```
>>> a = "Knight"
>>> a[1:3]
'ni'
>>> a[:3]
'Kni'
>>> a[3:]
'ght'
>>> a[:]
'Knight'
```

This can seem confusing at first, but it ensures that the length of a substring returned as s[i:j] has length j−i (for positive i, j) and that s[:i] + s[i:] == s. Unlike indexing, slicing a string outside its bounds does not raise an error:

```
>>> a = "Knight"
>>> a[3:10]
'ght'
>>> a[10:]
''
```

To test if a string contains a given substring, use the in operator:

```
>>> 'Kni' in 'Knight':
True
>>> 'kni' in 'Knight':
False
```

Example E2.11 Because of the nature of slicing, s[m:n], n−m is always the length of the substring. In other words, to return r characters starting at index m, use s[m:m+r]. For example,

```
>>> s = 'whitechocolatespaceegg'
>>> s[:5]
'white'
>>> s[5:14]
'chocolate'
>>> s[14:19]
'space'
>>> s[19:]
'egg'
```

Example E2.12 The optional third number in a slice specifies the *stride*. If omitted, the default is 1: return every character in the requested range. To return every kth letter, set the stride to k. Negative values of k reverse the string. For example,

```
>>> s = 'King Arthur'
>>> s[::2]
'Kn rhr'
>>> s[1::2]
'igAtu'
>>> s[-1:4:-1]
'ruhtrA'
```

This last slice can be explained as a selection of characters from the last (index -1) down to (but not including) character at index 4, with stride -1 (select every character, in the reverse direction).

A convenient way of reversing a string is to slice between default limits (by omitting the first and last indexes) with a stride of -1:

```
>>> s[::-1]
'ruhtrA gniK'
```

2.3.5   String Methods

Python strings are *immutable* objects, and so it is not possible to change a string by assignment – for example, the following is an error:

```
>>> a = 'Knight'
>>> a[0] = 'k'
Traceback (most recent call last):
  File "<stdin>", line 1, in <module>
TypeError: 'str' object does not support item assignment
```

New strings *can* be constructed from existing strings, but only as new objects. For example,

```
>>> a += ' Templar'
>>> print(a)
Knight Templar
>>> b = 'Black ' + a[:6]
>>> print(b)
Black Knight
```

To find the number of characters a string contains, use the len built-in method:

```
>>> a = 'Earth'
>>> len(a)
5
```

String objects come with a large number of methods for manipulating and transforming them. These are accessed using the usual dot notation we have met already – some of the more useful ones are listed in Table 2.11. In this and similar tables, text in *italics* is intended to be replaced by a specific value appropriate to the use of the method; italic text in [*square brackets*] denotes an optional argument.

Because these methods each return a new string, they can be chained together:

```
>>> s = '-+-Python Wrangling for Beginners'
>>> s.lower().replace('wrangling', 'programming').lstrip('+-')
'python programming for beginners'
```

Example E2.13   Here are some possible manipulations using string methods:

```
>>> a = 'java python c++ fortran '
>>> a.isalpha()
❶ False
>>> b = a.title()
```

Table 2.11 Some common string methods

Method	Description
center(*width*)	Return the string centered in a string with total number of characters *width*.
endswith(*suffix*)	Return True if the string ends with the substring *suffix*.
startswith(*prefix*)	Return True if the string starts with the substring *prefix*.
index(*substring*)	Return the lowest index in the string containing *substring*.
lstrip([*chars*])	Return a copy of the string with any of the leading characters specified by [*chars*] removed. If [*chars*] is omitted, any leading whitespace is removed.
rstrip([*chars*])	Return a copy of the string with any of the trailing characters specified by [*chars*] removed. If [*chars*] is omitted, any trailing whitespace is removed.
strip([*chars*])	Return a copy of the string with leading and trailing characters specified by [*chars*] removed. If [*chars*] is omitted, any leading and trailing whitespace is removed.
upper()	Return a copy of the string with all characters in upper case.
lower()	Return a copy of the string with all characters in lower case.
title()	Return a copy of the string with all words starting with capitals and other characters in lower case.
replace(*old*, *new*)	Return a copy of the string with each substring *old* replaced with *new*.
split([*sep*])	Return a list (see Section 2.4.1) of substrings from the original string which are separated by the string *sep*. If *sep* is not specified, the separator is taken to be any amount of whitespace.
join([*list*])	Use the string as a separator in joining a list of strings.
isalpha()	Return True if all characters in the string are alphabetic and the string is not empty; otherwise return False.
isdigit()	Return True if all characters in the string are digits and the string is not empty; otherwise return False.

```
>>> b
'Java Python C++ Fortran '
>>> c = b.replace(' ', '!\n')
>>> c
'Java!\nPython!\nC++!\nFortran!'
>>> print(c)
Java!
Python!
C++!
Fortran!
>>> c.index('Python')
```
❷ `6`
```
>>> c[6:].startswith('Py')
True
>>> c[6:12].isalpha()
True
```

❶ a.isalpha() is False because of the spaces and '++'.
❷ Note that \n is a single character.

2.3.6   The print Function

In Python 3, print is a built-in *function* (just like the others we have met such as len and round). It takes a list of objects to be output, and the optional arguments end and sep, that specify which characters should end the string and which characters should be used to separate the printed objects respectively. Omitting these additional arguments results in output in which the object fields are separated by a *single space* and the line is ended with a *newline* character.[20] For example,

```
>>> ans = 6
>>> print('Solve:', 2, 'x =', ans, 'for x')
Solve: 2 x = 6 for x
>>> print('Solve: ', 2, 'x = ', ans, ' for x', sep='', end='!\n')
Solve: 2x = 6 for x!
>>> print()

>>> print('Answer: x =', ans/2)
Answer: x = 3.0
```

❶

❶ Note that print() with no arguments just prints the default newline end character.

To suppress the new line at the end of a printed string, specify end to be the empty string: end='':

```
>>> print('A line with no newline character', end='')
A line with no newline character>>>
```

The chevrons, >>>, at the end of this line form the prompt for the next Python command to be entered.

Example E2.14   print can be used to create simple text tables:

```
>>> heading = '| Index of Dutch Tulip Prices |'
>>> line = '+' + '-'*16 + '-'*13 + '+'
>>> print(line, heading, line,
...         '|    Nov 23 1636 |         100 |',
...         '|    Nov 25 1636 |         673 |',
...         '|    Feb  1 1637 |        1366 |', line, sep='\n')

+---------------------------+
| Index of Dutch Tulip Prices |
+---------------------------+
|    Nov 23 1636 |         100 |
|    Nov 25 1636 |         673 |
|    Feb  1 1637 |        1366 |
+---------------------------+
```

---

[20] The specific newline character used depends on the operating system: for example, on a Mac it is "\n" (the "linefeed" character), on Windows it is two characters: "\r\n" ("carriage return" + "line feed").

2.3.7   String Formatting

Introduction to Python 3 String Formatting

In its simplest form, it is possible to use a string's `format` method to insert objects into it. The most basic syntax is

```
>>> '{} plus {} equals {}'.format(2, 3, 'five')
2 plus 3 equals five
```

Here, the `format` method is called on the string literal with the arguments 2, 3 and `'five'` which are interpolated, in order, into the locations of the *replacement fields*, indicated by braces, {}. Replacement fields can also be numbered or named, which helps with longer strings and allows the same value to be interpolated more than once:

```
>>> '{1} plus {0} equals {2}'.format(2, 3, 'five')
'3 plus 2 equals five'
>>> '{num1} plus {num2} equals {answer}'.format(num1=2, num2=3, answer='five')
'2 plus 3 equals five'
>>> '{0} plus {0} equals {1}'.format(2, 2+2)
'2 plus 2 equals 4'
```

Note that numbered fields can appear in any order and are indexed starting at 0.

Replacement fields can be given a minimum size within the string by the inclusion of an integer length after a colon as follows:

```
>>> '=== {0:12} ==='.format('Python')
'=== Python       ==='
```

If the string is too long for the minimum size, it will take up as many characters as needed (overriding the replacement field size specified):

```
>>> 'A number: <{0:2}>'.format(-20)
'A number: <-20>'        # -20 won't fit into 2 characters: 3 are used anyway
```

By default, the interpolated string is aligned to the left; this can be modified to align to the right or to center the string. The single characters <, > and ^ control the alignment:

```
>>> '=== {0:<12} ==='.format('Python')
'=== Python       ==='
>>> '=== {0:>12} ==='.format('Python')
'===       Python ==='
>>> '=== {0:^12} ==='.format('Python')
'===    Python    ==='
```

In these examples, the field is padded with spaces, but this fill character can also be specified. For example, to pad with hyphens in the last example, specify

```
>>> '=== {0:-^12} ==='.format('Python')
'=== ---Python--- ==='
```

It is even possible to pass the minimum field size as a parameter to be interpolated. Just replace the field size with a reference in braces as follows:

```
>>> a = 15
>>> 'This field has {0} characters: ==={1:>{2}}===.'.format(a, 'the field', a)
'This field has 15 characters: ===      the field===.'
```

Or with named interpolation:

```
>>> 'This field has {w} characters: ==={0:>{w}}===.'.format('the field', w=a)
'This field has 15 characters: ===      the field===.'
```

In each case, the second format specifier here has been taken to be `:>15`.

To insert the brace characters themselves into a formatted string, they must be doubled up: use "{{" and "}}".

Formatting Numbers

The Python 3 string `format` method provides a powerful way to format numbers.

The specifiers "d", "b", "o", "x"/"X" indicate a decimal, binary, octal and lower-case/upper-case hex *integer* respectively:

```
>>> a = 254
>>> 'a = {0:5d}'.format(a)        # decimal
'a =   254'
>>> 'a = {0:10b}'.format(a)       # binary
'a =   11111110'
>>> 'a = {0:5o}'.format(a)        # octal
'a =   364'
>>> 'a = {0:5x}'.format(a)        # hex (lower-case)
'a =    fe'
>>> 'a = {0:5X}'.format(a)        # hex (upper-case)
'a =    FE'
```

Numbers can be padded with zeros to fill out the specified field size by prefixing the minimum width with a `0`:

```
>>> a = 254
>>> 'a = {a:05d}'.format(a=a)
'a = 00254'
```

By default, the sign of a number is only output if it is negative. This behavior can also be customized by specifying, before the minimum width:

- "+": always output a sign;
- "-": only output a negative sign, the default; or
- " ": output a leading space only if the number is positive.

This last option enables columns of positive and negative numbers to be lined up nicely:

```
>>> print('{0: 5d}\n{1: 5d}\n{2: 5d}'.format(-4510, 1001, -3026))
-4510
 1001
-3026
>>> a = -25
>>> b = 12
>>> s = '{0:+5d}\n{1:+5d}\n= {2:+3d}'.format(a, b, a+b)
>>> print(s)
  -25
  +12
= -13
```

There are also format specifiers for floating-point numbers, which can be output to a chosen precision if desired. The most useful options are "f": fixed-point notation, "e"/"E": exponent (i.e. "scientific" notation), and "g"/"G": a general format which uses scientific notation for very large and very small numbers.[21] The desired precision (number of decimal places) is specified as ".p" after the minimum field width. Some examples:

```
>>> a = 1.464e-10
>>> '{0:g}'.format(a)
'1.464e-10'
>>> '{0:10.2E}'.format(a)
'  1.46E-10'
>>> '{0:15.13f}'.format(a)
'0.0000000001464'
>>> '{0:10f}'.format(a)
'  0.000000'
```

❶

❶ Note that Python will not protect you from this kind of rounding to zero if not enough space is provided for a fixed-point number.

Formatted String Literals (*f-strings*)

Since version 3.6, Python has supported a further way of interpolating values into strings: a string literal denoted with a f before the quotes can evaluate *expressions* placed within braces, including references to variables, function calls and comparisons. This provides an expressive and concise way to define string objects; for example, given the variables

```
>>> h = 6.62607015e-34
>>> h_units = 'J.s'
```

instead of using the format function:

```
>>> 'h = {h:.3e} {h_units}'.format(h=h, h_units=h_units)
'h = 6.626e-34 J.s'
```

one can simply write:

```
>>> f'h = {h:.3e} {h_units}'
'h = 6.626e-34 J.s'
```

This means that there is no need for the awkward repetition in the format call (h=h, h_units=h_units) and for longer strings with many interpolations it is easier to read. It is also generally faster to execute because the syntax is part of Python's fundamental grammar and no explicit function call is required.

Although it wouldn't generally be a good idea to put a complex expression in an f-string replacement field, it is common to call functions or make comparisons:

```
>>> name = 'Elizabeth'
>>> f'The name {name} has {len(name)} letters and {name.lower().count("e")} "e"s.'
'The name Elizabeth has 9 letters and 2 "e"s.'
```

or even:

```
>>> letter = 'k'
>>> f'{name} has {len(name)} letters and {name.lower().count(letter)} "{letter}"s.'
'Elizabeth has 9 letters and 0 "k"s.'
```

---

[21] More specifically, the g/G specifier acts like f/F for numbers between 10^{-4} and 10^{p} where p is the desired precision (which defaults to 6), and acts like e/E otherwise.

There are few minor things to bear in mind: the quotes used inside an f-string expression should not conflict with those used to delimit the string literal itself (note the use of " above to avoid the clash with the outer f'...' quotes). Also, because f-strings are evaluated once, at runtime, it is not possible to define a reuseable "template":

```
>>> radius = 2.5
>>> s = f'The radius is {radius} m.'
>>> print(s)

The radius is 2.5 m.

>>> radius = 10.3
>>> print(s)

The radius is 2.5 m.
```

For this use-case, traditional `format` string interpolation is better:

```
>>> radius = 2.5
>>> t = 'The radius is {} m.'
>>> print(t.format(radius))

The radius is 2.5 m.

>>> radius = 10.3
>>> print(t.format(radius))

The radius is 10.3 m.
```

In this book we will use both traditional `format` string interpolation and f-strings.

Older C-style Formatting

Python 3 also supports the less powerful, C-style format specifiers that are still in widespread use. In this formulation the replacement fields are specified with the minimum width and precision specifiers following a % sign. The objects whose values are to be interpolated are then given after the end of the string, following another % sign. They must be enclosed in parentheses if there is more than one of them. The same letters for the different output types are used as earlier; strings must be specified explicitly with "%s". For example,

```
>>> kB = 1.380649e-23
>>> 'Here\'s a number: %10.2e' % kB
"Here's a number:    1.38e-23"
>>> 'The same number formatted differently: %7.1e and %12.6e' % (kB, kB)
'The same number formatted differently: 1.4e-23 and 1.380649e-23'
>>> '%s is %g J/K' % ("Boltzmann's constant", kB)
"Boltzmann's constant is 1.38065e-23 J/K"
```

Example E2.15 Python can produce string representations of numbers for which thousands are separated by commas:

```
>>> '{:11,d}'.format(1000000)
'  1,000,000'
>>> '{:11,.1f}'.format(1000000.)
'1,000,000.0'
```

Here is another table, produced using several different string methods:

```
title = '|' + '{:^51}'.format('Cereal Yields (kg/ha)') + '|'
line = '+' + '-'*15 + '+' + ('-'*8 + '+')*4
row = '| {:<13} |' + ' {:6,d} |'*4
header = '| {:^13s} |'.format('Country') + (' {:^6d} |'*4).format(1980, 1990,
                                                                  2000, 2010)
print('+' + '-'*(len(title)-2) + '+',
      title,
      line,
      header,
      line,
      row.format('China', 2937, 4321, 4752, 5527),
      row.format('Germany', 4225, 5411, 6453, 6718),
      row.format('United States', 3772, 4755, 5854, 6988),
      line,
      sep='\n')
```

```
+---------------------------------------------------+
|                Cereal Yields (kg/ha)              |
+---------------+--------+--------+--------+--------+
|    Country    |  1980  |  1990  |  2000  |  2010  |
+---------------+--------+--------+--------+--------+
| China         |  2,937 |  4,321 |  4,752 |  5,527 |
| Germany       |  4,225 |  5,411 |  6,453 |  6,718 |
| United States |  3,772 |  4,755 |  5,854 |  6,988 |
+---------------+--------+--------+--------+--------+
```

2.3.8   Exercises

Questions

Q2.3.1   Slice the string s = 'seehemewe' to produce the following substrings:

(a)   'see'
(b)   'he'
(c)   'me'
(d)   'we'
(e)   'hem'
(f)   'meh'
(g)   'wee'

Q2.3.2   Write a single-line expression for determining if a string is a palindrome (reads the same forward as backward).

Q2.3.3   Predict the results of the following statements and check them using the Python shell.

```
>>> days = 'Sun Mon Tues Weds Thurs Fri Sat'
```

(a)   `print(days[days.index('M'):])`
(b)   `print(days[days.index('M'):days.index('Sa')].rstrip())`
(c)   `print(days[6:3:-1].lower()*3)`
(d)   `print(days.replace('rs', '').replace('s ', ' ')[::4])`
(e)   `print(' -*- '.join(days.split()))`

Q2.3.4 What is the output of the following code? How does it work?

```
>>> suff = 'thstndrdththththththth'
>>> n = 1
>>> print('{:d}{:s}'.format(n, suff[n*2:n*2+2]))
>>> n = 3
>>> print('{:d}{:s}'.format(n, suff[n*2:n*2+2]))
>>> n = 5
>>> print('{:d}{:s}'.format(n, suff[n*2:n*2+2]))
```

Q2.3.5 Consider the following (incorrect) tests to see if the string 's' has one of two values. Explain how these statements are interpreted by Python and give a correct alternative.

```
>>> s = 'eggs'
>>> s == ('eggs' or 'ham')
True

>>> s == ('ham' or 'eggs')
False
```

Problems

P2.3.1

(a)   Given a string representing a base-pair sequence (i.e. containing only the letters A, G, C and T), determine the fraction of G and C bases in the sequence.
(*Hint*: strings have a count method, returning the number of occurrences of a substring.)

(b)   Using only string methods, devise a way to determine if a nucleotide sequence is a palindrome in the sense that it is equal to its own complementary sequence read backward. For example, the sequence TGGATCCA is palindromic because its complement is ACCTAGGT, which is the same as the original sequence backward. The complementary base pairs are (A, T) and (C, G).

P2.3.2 The table that follows gives the names, symbols, values, uncertainties and units of some physical constants.

Defining variables of the form

```
G = 6.6743e-11        # J/K
G_unc = 1.5e-15       # uncertainty
G_units = 'Nm^2/kg^2'
```

use the string object's format method to produce the following output:

(a)   `kB = 1.381e-23 J/K`

Name	Symbol	Value	Uncertainty	Units
Boltzmann constant	k_B	1.380649×10^{-23}	(def)	$J\,K^{-1}$
Speed of light	c	2.99792458×10^8	(def)	$m\,s^{-1}$
Planck constant	h	$6.62607015 \times 10^{-34}$	(def)	$J\,s$
Avogadro constant	N_A	$6.02214076 \times 10^{23}$	(def)	mol^{-1}
Electron magnetic moment	μ_e	$-9.28476377 \times 10^{-24}$	2.3×10^{-31}	$J\,T^{-1}$
Gravitational constant	G	6.67430×10^{-11}	1.5×10^{-15}	$N\,m^2\,kg^{-2}$

(b)
```
G = 0.0000000000667430 Nm^2/kg^2
```

(c) Using the same format specifier for each line,
```
kB   =   1.3807e-23 J/K
mu_e = -9.2848e-24 J/T
N_A  =   6.0221e+23 mol-1
c    =   2.9979e+08 m/s
```

(d) Again, using the same format specifier for each line,
```
===    G = +6.67E-11 [Nm^2/kg^2]    ===
===   µe = -9.28E-24 [      J/T]    ===
```

Hint: the Unicode codepoint for the lower-case Greek letter mu is U+03BC.

(e) (Harder). Produce the output below, in which the uncertainty (one standard deviation) in the value of each constant is expressed as a number in parentheses relative the preceding digits: that is, $6.67430(15) \times 10^{-11}$ means $6.67430 \times 10^{-11} \pm 1.5 \times 10^{-15}$.
```
G = 6.67430(15)e-11 Nm^2/kg^2
mu_e = -9.28476377(23)e-24 J/T
```

P2.3.3 Given the elements of a 3×3 matrix as the nine variables a11, a12, ..., a33, produce a string representation of the matrix using formatting methods, (a) assuming the matrix elements are (possibly negative) real numbers to be given to one decimal place; (b) assuming the matrix is a permutation matrix with integer entries taking the values 0 or 1 only. For example,
```
>>> print(s_a)
[  0.0   3.4  -1.2 ]
[ -1.1   0.5  -0.2 ]
[  2.3  -1.4  -0.7 ]
>>> print(s_b)
[ 0 0 1 ]
[ 0 1 0 ]
[ 1 0 0 ]
```

P2.3.4 Find the Unicode code points for the planet symbols listed on the NASA website (https://solarsystem.nasa.gov/resources/680/solar-system-symbols/) which mostly fall within the hex range 2600–26FF: Miscellaneous Symbols (https://www.unicode.org/charts/PDF/U2600.pdf) and output a list of planet names and symbols.

2.4   Python Objects II: Lists, Tuples and Loops

2.4.1   Lists

Initializing and Indexing Lists

Python provides data structures for holding an ordered list of objects. In some other languages (e.g. C and Fortran) such a data structure is called an *array* and can hold only one type of data (e.g. an array of integers); the core array structures in Python, however, can hold a mixture of data types.

A Python *list* is an ordered, *mutable* array of objects. A list is constructed by specifying the objects, separated by commas, between square brackets, []. For example,

```
>>> list1 = [1, 'two', 3.14, 0]
>>> list1
[1, 'two', 3.14, 0]
>>> a = 4
>>> list2 = [2, a, -0.1, list1, True]
>>> list2
[2, 4, -0.1, [1, 'two', 3.14, 0], True]
```

Note that a Python list can contain references to any type of object: strings, the various types of numbers, built-in constants such as the boolean value True, and even other lists. It is not necessary to declare the size of a list in advance of using it. An empty list can be created with list0 = [] or list0 = list().

An item can be retrieved from the list by indexing it (remember Python indexes start at 0):

```
>>> list1[2]
3.14
>>> list2[-1]
True
>>> list2[3][1]
'two'
```

This last example retrieves the second (index: 1) item of the fourth (index: 3) item of list2. This is valid because the item list2[3] happens to be a list (the one also identified by the variable name list1), and list1[1] is the string 'two'. In fact, since strings can also be indexed:

```
>>> list2[3][1][1]
'w'
```

To test for membership of a list, the operator in is used, as for strings:

```
>>> 1 in list1
True
>>> 'two' in list2:
False
```

This last expression evaluates to False because list2 does not contain the string literal 'two' even though it contains list1 which does: the in operator does not recurse into lists-of-lists when it tests for membership.

Lists and Mutability

Python lists are the first *mutable* object we have encountered. Unlike strings, which cannot be altered once defined, the items of a list can be reassigned:

```
>>> list1
[1, 'two', 3.14, 0]
>>> list1[2] = 2.72
>>> list1
[1, 'two', 2.72, 0]
>>> list2
[2, 4, -0.1, [1, 'two', 2.72, 0], True]
```

Note that not only has list1 been changed, but list2 (which contains list1 as an item) *has also changed.*[22] This behavior catches a lot of people out to begin with, particularly if a list needs to be copied to a different variable.

```
>>> q1 = [1, 2, 3]
>>> q2 = q1
>>> q1[2] = 'oops'
>>> q1
[1, 2, 'oops']
>>> q2
[1, 2, 'oops']
```

Here, the variables q1 and q2 refer to the *same list*, stored in the same memory location, and because lists are mutable, the line q1[2] = 'oops' actually changes one of the stored values at that location; q2 still points to the same location and so it appears to have changed as well. In fact, there is only one list (referred to by two variable names) and it is changed once. In contrast, integers are *immutable*, so the following does not change the value of q[2]:

```
>>> a = 3
>>> q = [1, 2, a]
>>> a = 4
>>> q
[1, 2, 3]
```

The assignment a = 4 creates a whole new integer object, quite independent of the original 3 that ended up in the list q. This original integer object isn't changed by the assignment (integers are immutable) and so the list is unchanged. This distinction is illustrated by Figures 2.2, 2.3 and 2.4.

Lists can be *sliced* in the same way as string sequences:

```
>>> q1 = [0., 0.1, 0.2, 0.3, 0.4, 0.5]
>>> q1[1:4]
[0.1, 0.2, 0.3]
>>> q1[::-1]          # return a reversed copy of the list
[0.5, 0.4, 0.3, 0.2, 0.1, 0.0]
>>> q1[1::2]          # striding: returns elements at 1, 3, 5
[0.1, 0.3, 0.5]
```

---

[22] Actually, it hasn't changed: it only ever contained a series of references to objects: the reference to list1 is the same, even though the references within list1 have changed.

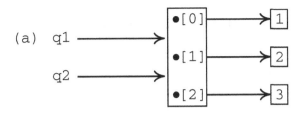

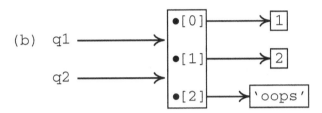

Figure 2.2 Two variables referring to the same list: (a) on initialization and (b) after setting q1[2] = 'oops'.

Taking a slice *copies the data* to a new list. Hence,

```
>>> q2 = q1[1:4]
>>> q2[1] = 99        # only affects q2
>>> q2
[0.1, 99, 0.3]
>>> q1
[0.0, 0.1, 0.2, 0.3, 0.4, 0.5]
```

List Methods

Just as for strings, Python lists come with a large number of useful methods, summarized in Table 2.12. Because list objects are mutable, they can grow or shrink *in place*, that is, without having to copy the contents to a new object, as we had to do with strings. The relevant methods are

- append: add an item to the end of the list;
- extend: add one or more objects by copying them from another list;[23]
- insert: insert an item at a specified index;
- remove: remove a specified item from the list.

---

[23] Actually, any Python object that forms a sequence that can be iterated over (e.g. a string) can be used as the argument to extend

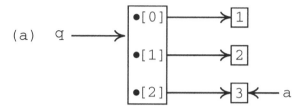

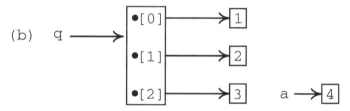

Figure 2.3 A list defined with q = [1, 2, a] where a = 3: (a) on initialization and (b) after changing the value of a with a = 4.

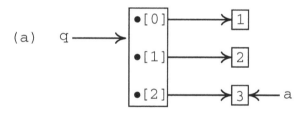

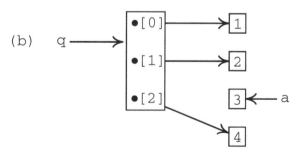

Figure 2.4 A list defined with q = [1, 2, a] where a = 3: (a) on initialization and (b) after changing the value of q with q[2] = 4.

Table 2.12 Some common list methods

Method	Description
append(*element*)	Append *element* to the end of the list
extend(*list2*)	Extend the list with the elements from *list2*
index(*element*)	Return the lowest index of the list containing *element*
insert(*index*, *element*)	Insert *element* at index *index*
pop()	Remove and return the last element from the list
reverse()	Reverse the list in place
remove(*element*)	Remove the first occurrence of *element* from the list
sort()	Sort the list in place
copy()	Return a copy of the list
count(*element*)	Return the number of elements equal to *element* in the list

```
>>> q = []
>>> q.append(4)
>>> q
[4]
>>> q.extend([6, 7, 8])
>>> q
[4, 6, 7, 8]
>>> q.insert(1, 5)  # insert 5 at index 1
>>> q
[4, 5, 6, 7, 8]
>>> q.remove(7)
>>> q
[4, 5, 6, 8]
>>> q.index(8)
3   # the item 8 appears at index 3
```

Two useful list methods are sort and reverse, which sort and reverse the list *in place*. That is, they change the list object, but *do not return a value*:

```
>>> q = [2, 0, 4, 3, 1]
>>> q.sort()
>>> q
[0, 1, 2, 3, 4]
>>> q.reverse()
>>> q
[4, 3, 2, 1, 0]
```

If you do want a sorted *copy* of the list, leaving it unchanged, you can use the sorted built-in function:

```
>>> q = ['a', 'e', 'A', 'c', 'b']
>>> sorted(q)
['A', 'a', 'b', 'c', 'e']   # returns a new list
>>> q
['a', 'e', 'A', 'c', 'b']   # the old list is unchanged
```

By default, sort() and sorted() order the items in an array in *ascending order*. Set the optional argument reverse=True to return the items in descending order:

```
>>> q = [10, 5, 5, 2, 6, 1, 67]
```

```
>>> sorted(q, reverse=True)
[67, 10, 6, 5, 5, 2, 1]
```

Python 3 does not allow direct comparisons between strings and numbers, so it is an error to attempt to sort a list containing a mixture of such types:

```
>>> q = [5, '4', 2, 8]
>>> q.sort()
TypeError: unorderable types: str() < int()
```

Example E2.16 The methods append and pop make it very easy to use a list to implement the data structure known as a *stack*:

```
>>> stack = []
>>> stack.append(1)
>>> stack.append(2)
>>> stack.append(3)
>>> stack.append(4)
>>> print(stack)
[1, 2, 3, 4]
>>> stack.pop()
4
>>> print(stack)
[1, 2, 3]
```

The end of the list is the top of the stack from which items may be added or removed ("last in, first out" (LIFO): think of a stack of dinner plates).

Example E2.17 The string method, split, generates a list of substrings from a given string, split on a specified separator:

```
>>> s = 'Jan Feb Mar Apr May Jun'
>>> s.split()        # By default, splits on whitespace
['Jan', 'Feb', 'Mar', 'Apr', 'May', 'Jun']
>>> s = "J. M. Brown AND B. Mencken AND R. P. van't Rooden"
>>> s.split(' AND ')
['J. M. Brown', 'B. Mencken', "R. P. van't Rooden"]
```

2.4.2   Tuples

The tuple Object

A tuple may be thought of as an immutable list. Tuples are constructed by placing the items inside parentheses:

```
>>> t = (1, 'two', 3.)
>>> t
(1, 'two', 3.0)
```

Tuples can be indexed and sliced in the same way as lists but, being immutable, they cannot be appended to, extended or have elements removed from them:

```
>>> t = (1, 'two', 3.)
>>> t[1]
'two'
>>> t[2] = 4
Traceback (most recent call last):
  File "<stdin>", line 1, in <module>
TypeError: 'tuple' object does not support item assignment
```

Although a tuple itself is immutable, it may *contain* references to mutable objects such as lists. Hence,

```
>>> t = (1, ['a', 'b', 'd'], 0)
>>> t[1][2] = 'c'   # OK to change the list within the tuple
>>> t
(1, ['a', 'b', 'c'], 0)
```

An empty tuple is created with empty parentheses: t0 = (). To create a tuple containing only one item (a *singleton*), however, it is not sufficient to enclose the item in parentheses (which could be confused with other syntactical uses of parentheses); instead, the lone item is given a trailing comma: t = ('one',).

Uses of tuples

In some circumstances, particularly for simple assignments such as those in the previous section, the parentheses around a tuple's items are not required:

```
>>> t = 1, 2, 3
>>> t
(1, 2, 3)
```

This usage is an example of *tuple packing*. The reverse, *tuple unpacking*, is a common way of assigning multiple variables in one line:

```
>>> a, b, c = 97, 98, 99
>>> b
98
```

This method of assigning multiple variables is commonly used in preference to separate assignment statements either on different lines or (very un-Pythonically) on a single line, separated by semicolons:

```
a = 97; b = 98; c = 99     # Don't do this!
```

Tuples are useful where a sequence of items cannot or should not be altered. In the previous example, the tuple object only exists in order to assign the variables a, b and c. The values to be assigned: 97, 98 and 99 are packed into a tuple for the purpose of this statement (to be unpacked into the variables), but once this has happened, the tuple object itself is destroyed. As another example, a function (Section 2.7) may return more than one object: these objects are returned packed into a tuple. If you need any further persuading, tuples are slightly faster for many uses than lists.

Example E2.18 In an assignment using the "=" operator the right-hand side expression is evaluated first. This provides a convenient way to swap the values of two variables using tuples:

```
a, b = b, a
```

Here, the right-hand side is packed into a tuple object, which is then unpacked into the variables assigned on the left-hand side. This is more convenient than using a temporary variable:

```
t = a
a = b
b = t
```

2.4.3   Iterable Objects

Examples of Iterable Objects

Strings, lists and tuples are all examples of data structures that are *iterable* objects: they are ordered sequences of items (characters in the case of strings, or arbitrary objects in the case of lists and tuples) which can be taken one at a time. One way of seeing this is to use the alternative method of initializing a list (or tuple) using the built-in constructor methods list() and tuple(). These take any iterable object and generate a list and a tuple, respectively, from its sequence of items. For example,

```
>>> list('hello')
['h', 'e', 'l', 'l', 'o']
>>> tuple([1, 'two', 3])
(1, 'two', 3)
```

Because the data elements are *copied* in the construction of a new object using these constructor methods, list is another way of creating an independent list object from another:

```
>>> a = [5, 4, 3, 2, 1]
>>> b = a          # b and a refer to the same list object
>>> b is a
True
>>> b = list(a)    # create an entirely new list object with the same contents as a
>>> b is a
False
```

Because slices also return a copy of the object references from a sequence, the idiom b = a[:] is often used in preference to b = list(a).

any and all

The built-in function any tests whether any of the items in an iterable object are equivalent to True; all tests whether all of them are. For example,

```
>>> a = [1, 0, 0, 2, 3]
>>> any(a), all(a)
(True, False)      # some (but not all) of a's items are equivalent to True
>>> b = [[], False, 0.]
>>> any(b), all(b)
(False, False)     # none of b's items is equivalent to True
```

◊ *** Syntax**

It is sometimes necessary to call a function with arguments taken from a list or other sequence. The * syntax used in a function call unpacks such a sequence into positional arguments to the function (see also Section 2.7). For example, the math.hypot function takes two arguments, a and b, and returns the quantity $\sqrt{a^2 + b^2}$. If the arguments you wish to use are in a list or tuple, the following will fail:

```
>>> t = [3, 4]
>>> math.hypot(t)
Traceback (most recent call last):
  File "<stdin>", line 1, in <module>
TypeError: hypot expected 2 arguments, got 1
```

We tried to call math.hypot() with a single argument (the list object t), which is an error. We could index the list explicitly to retrieve the two values we need:

```
>>> t = [3, 4]
>>> math.hypot(t[0], t[1])
5.0
```

but a more elegant method is to *unpack* the list into arguments to the function with *t:

```
>>> math.hypot(*t)
5.0
```

for Loops

It is often necessary to take the items in an iterable object one by one and do something with each in turn. Other languages, such as C, require this type of *loop* to refer to each item in turn by its integer index. In Python this is possible, but the more natural and convenient way is with the idiom:

```
    for item in iterable object:
```

which yields each element of the iterable object in turn to be processed by the subsequent block of code. For example,

```
>>> fruit_list = ['apple', 'melon', 'banana', 'orange']
>>> for fruit in fruit_list:
...     print(fruit)
...
apple
melon
banana
orange
```

Each item in the list object fruit_list is taken in turn and assigned to the variable fruit for the block of statements following the ":" – each statement in this block must be indented by the same amount of whitespace. Any number of spaces or tab characters could be used, but it is **strongly recommended to use four spaces** to indent code.[24]

---

[24] The use of whitespace as part of the syntax of Python is one of its most contentious aspects. Some people used to languages such as C and Java, which delimit code blocks with braces ({...}), find it an anathema; others take a more relaxed view and note that code is almost always indented consistently to make it readable even when this isn't enforced by the grammar of the language.

Loops can be nested – the inner loop block needs to be indented by the same amount of whitespace again as the outer loop (i.e. eight spaces):

```
>>> fruit_list = ['apple', 'melon', 'banana', 'orange']
>>> for fruit in fruit_list:
...     for letter in fruit:
...         print(letter, end='.')
...     print()
...
a.p.p.l.e.
m.e.l.o.n.
b.a.n.a.n.a.
o.r.a.n.g.e.
```

In this example, we iterate over the string items in `fruit_list` one by one, and for each string (fruit name), iterate over its letters. Each letter is printed followed by a full stop (the body of the inner loop). The last statement of the outer loop, `print()`, forces a new line after each fruit.

Example E2.19   We have already briefly met the string method `join`, which takes a sequence of string objects and joins them together in a single string:

```
>>> ', '.join( ('one', 'two', 'three') )
'one, two, three'
>>> print('\n'.join(reversed(['one', 'two', 'three'])))
three
two
one
```

The `reversed` built-in iterates over a sequence backwards, with the advantage (for long sequences) that it does not create a new object or modify the original.

Recall that strings are themselves iterable sequences, and so can be passed to the `join` method. For example, to join the letters of `'hello'` with a single space:

```
>>> ' '.join('hello')
'h e l l o'
```

The range Type

Python provides an efficient method of referring to a sequence of numbers that forms a simple arithmetic progression: $a_n = a_0 + nd$ for $n = 0, 1, 2, \ldots$ In such a sequence, each term is spaced by a constant value, the *stride*, *d*. In the simplest case, one simply needs an integer counter, which runs in steps of one from an initial value of zero: $0, 1, 2, \ldots, N - 1$. It would be possible to create a list to hold each of the values, but for most purposes this is wasteful of memory: it is easy to generate the next number in the sequence without having to store all of the numbers at the same time.

Representing such arithmetic progressions for iterating over is the purpose of the range type. A range object can be constructed with up to three arguments defining the first integer, the integer to stop at and the stride (which can be negative).

```
range([a0=0], n, [stride=1])
```

The notation describing the `range` constructor here means that if the initial value, a0, is not given it is taken to be 0; `stride` is also optional and if it is not given it is taken to be 1. Some examples:

```
>>> a = range(5)          # 0, 1, 2, 3, 4
>>> b = range(1, 6)       # 1, 2, 3, 4, 5
>>> c = range(0, 6, 2)    # 0, 2, 4
>>> d = range(10, 0, -2)  # 10, 8, 6, 4, 2
```

In Python 3, the object created by `range` *is not a list*. Rather, it is an iterable object that can produce integers on demand: range objects can be indexed, cast into lists and tuples, and iterated over:

```
>>> c[1]        # i.e. the second element of 0, 2, 4
2
>>> c[0]
0
>>> list(d)     # make a list from the range
[10, 8, 6, 4, 2]
>>> for x in range(5):
...     print(x)
0
1
2
3
4
```

Example E2.20   The *Fibonacci sequence* is the sequence of numbers generated by applying the rules:

$$a_1 = a_2 = 1, \quad a_i = a_{i-1} + a_{i-2}.$$

That is, the ith Fibonacci number is the sum of the previous two: $1, 1, 2, 3, 5, 8, 13, \ldots$

We present two ways of generating the Fibonacci series. First, by appending to a list:

Listing 2.1 Calculating the Fibonacci series in a `list`

```
# eg2-i-fibonacci.py
# Calculates and stores the first n Fibonacci numbers.

n = 100
fib = [1, 1]
for i in range(2, n+1):
    fib.append(fib[i-1] + fib[i-2])
print(fib)
```

Alternatively, we can generate the series without storing more than two numbers at a time as follows:

Listing 2.2 Calculating the Fibonacci series without storing it

```
# eg2-ii-fibonacci.py
# Calculates the first n Fibonacci numbers.

n = 100
```

```
# Keep track of the two most recent Fibonacci numbers.
a, b = 1, 1
print(a, b, end='')
for i in range(2, n+1):
    # The next number (b) is a+b; then a becomes the previous b.
    a, b = b, a+b
    print(' ', b, end='')
```

enumerate

Because range objects can be used to produce a sequence of integers, it is tempting to use them to provide the indexes of lists or tuples when iterating over them in a for loop:

```
>>> mammals = ['kangaroo', 'wombat', 'platypus']
>>> for i in range(len(mammals)):
        print(i, ':', mammals[i])
0 : kangaroo
1 : wombat
2 : platypus
```

This works, of course, but it is more natural to avoid the explicit construction of a range object (and the call to the len built-in) by using enumerate. This method takes an iterable object and produces, for each item in turn, a tuple (count, item), consisting of a counting index and the item itself:

```
>>> mammals = ['kangaroo', 'wombat', 'platypus']
>>> for i, mammal in enumerate(mammals):
        print(i, ':', mammal)
0 : kangaroo
1 : wombat
2 : platypus
```

Note that each (count, item) tuple is unpacked in the for loop into the variables i and mammal. It is also possible to set the starting value of count to something other than 0 (although then it won't be the index of the item in the original list, of course):

```
>>> list(enumerate(mammals, 4))
[(4, 'kangaroo'), (5, 'wombat'), (6, 'platypus')]
```

◇ zip

What if you want to iterate over two (or more) sequences at the same time? This is what the zip built-in function is for: it creates an iterator object in which each item is a tuple of items taken in turn from the sequences passed to it:

```
>>> a = [1, 2, 3, 4]
>>> b = ['a', 'b', 'c', 'd']
>>> zip(a, b)
<builtins.zip at 0x104476998>
>>> for pair in zip(a, b):
...     print(pair)
...
(1, 'a')
(2, 'b')
```

```
(3, 'c')
(4, 'd')
>>> list(zip(a, b))        # convert to list
[(1, 'a'), (2, 'b'), (3, 'c'), (4, 'd')]
```

A nice feature of zip is that it can be used to *unzip* sequences of tuples as well:

```
>>> z = zip(a, b)          # zip
>>> A, B = zip(*z)         # unzip
>>> print(A, B)
(1, 2, 3, 4) ('a', 'b', 'c', 'd')
>>> list(A) == a, list(B) == b
(True, True)
```

zip does not copy the items into a new object, so it is memory-efficient and fast; but this means that you only get to iterate over the zipped items once and you can't index them:

```
>>> z = zip(a, b):
>>> z[0]
TypeError: 'zip' object is not subscriptable

>>> for pair in z:
...     x = 0           # just some dummy operation performed on each iteration
...
>>> for pair in z:
...     print(pair)
...
        # (nothing: we've already exhausted the iterator z)
>>>
```

2.4.4   Exercises

Questions

Q2.4.1 Predict and explain the outcome of the following statements using the variables s = 'hello' and a = [4, 10, 2].

(a)   `print(s, sep='-')`
(b)   `print(*s, sep='-')`
(c)   `print(a)`
(d)   `print(*a, sep='')`
(e)   `list(range(*a))`

Q2.4.2 A list could be used as a simple representation of a polynomial, $P(x)$, with the items as the coefficients of the successive powers of x, and their indexes as the powers themselves. Thus, the polynomial $P(x) = 4 + 5x + 2x^3$ would be represented by the list [4, 5, 0, 2]. Why does the following attempt to differentiate a polynomial fail to produce the correct answer?

```
>>> P = [4, 5, 0, 2]
>>> dPdx = []
>>> for i, c in enumerate(P[1:]):
```

```
...       dPdx.append(i*c)
>>> dPdx
[0, 0, 4]              # wrong!
```

How can this code be fixed?

Q2.4.3 Given an ordered list of test scores, produce a list associating each score with a *rank* (starting with 1 for the highest score). Equal scores should have the same rank. For example, the input list [87, 75, 75, 50, 32, 32] should produce the list of rankings [1,2,2,4,5,5].

Q2.4.4 Use a for loop to calculate π from the first 20 terms of the *Madhava series*:

$$\pi = \sqrt{12}\left(1 - \frac{1}{3\cdot3} + \frac{1}{5\cdot3^2} - \frac{1}{7\cdot3^3} + \cdots\right).$$

Q2.4.5 For what iterable sequences, x, does the expression any(x) and not all(x) evaluate to True?

Q2.4.6 Explain why zip(*z) is the inverse of z = zip(a, b) – that is, while z pairs the items: (a0, b0), (a1, b1), (a2, b2), ..., zip(*z) separates them again: (a0, a1, a2, ...), (b0, b1, b2, ...).

Q2.4.7 Sorting a list of tuples arranges them in order of the first element in each tuple first. If two or more tuples have the same first element, they are ordered by the second element, and so on:

```
>>> sorted([(3, 1), (1, 4), (3, 0), (2, 2), (1, -1)])
[(1, -1), (1, 4), (2, 2), (3, 0), (3, 1)]
```

This suggests a way of using zip to sort one list using the elements of another. Implement this method on the data below to produce an ordered list of the average amount of sunshine in hours in London by month. Output the sunniest month first.

Jan	Feb	Mar	Apr	May	Jun
44.7	65.4	101.7	148.3	170.9	171.4

Jul	Aug	Sep	Oct	Nov	Dec
176.7	186.1	133.9	105.4	59.6	45.8

Problems

P2.4.1 Write a short Python program which, given an array of integers, a, calculates an array of the same length, p, in which p[i] is the product of all the integers in a *except* a[i]. So, for example, if a = [1, 2, 3], then p is [6, 3, 2].

P2.4.2 The *Hamming distance* between two equal-length strings is the number of positions at which the characters are different. Write a Python routine to calculate the Hamming distance between two strings, s1 and s2.

P2.4.3   Using a tuple of strings naming the digits 0–9, create a Python program which outputs the representation of π as read aloud to eight decimal places:

```
three point one four one five nine two six five
```

P2.4.4   Write a program to output a nicely formatted depiction of the first eight rows of Pascal's triangle.

P2.4.5   A DNA sequence encodes each amino acid making up a protein as a three-nucleotide sequence called a *codon*. For example, the sequence fragment AGTCT-TATATCT contains the codons (AGT, CTT, ATA, TCT) if read from the first position (*"frame"*). If read in the second frame it yields the codons (GTC, TTA, TAT) and in the third (TCT, TAT, ATC).

Write some Python code to extract the codons into a list of three-letter strings given a sequence and `frame` as an integer value (0, 1 or 2).

P2.4.6   The factorial function, $n! = 1 \cdot 2 \cdot 3 \cdot \ldots \cdot (n-1)n$ is the product of the first n positive integers and is provided by the `math` module's `factorial` method. The *double factorial* function, $n!!$, is the product of the positive *odd* integers up to and including n (which must itself be odd):

$$n!! = \prod_{i=1}^{(n+1)/2} (2i - 1) = 1 \cdot 3 \cdot 5 \cdot \ldots \cdot (n-2) \cdot n.$$

Write a routine to calculate $n!!$ in Python.

As a bonus exercise, extend the formula to allow for even n as follows:

$$n!! = \prod_{i=1}^{n/2} (2i) = 2 \cdot 4 \cdot 6 \cdot \ldots \cdot (n-2) \cdot n.$$

P2.4.7   *Benford's law* is an observation about the distribution of the frequencies of the first digits of the numbers in many different data sets. It is frequently found that the first digits are not uniformly distributed, but follow the logarithmic distribution

$$P(d) = \log_{10}\left(\frac{d+1}{d}\right).$$

That is, numbers starting with 1 are more common than those starting with 2, and so on, with those starting with 9 the least common. The probabilities follow:

1	0.301
2	0.176
3	0.125
4	0.097
5	0.079
6	0.067
7	0.058
8	0.051
9	0.046

Benford's law is most accurate for data sets which span several orders of magnitude, and can be proved to be exact for some infinite sequences of numbers.

(a) Demonstrate that the first digits of the first 500 Fibonacci numbers (see Example E2.20) follow Benford's law quite closely.

(b) The length of the amino acid sequences of 500 randomly chosen proteins are provided in the file `protein_lengths.py` which can be downloaded from https://scipython.com/ex/bba. This file contains a list, naa, which can be imported at the start of your program with

```
from protein_lengths import naa
```

To what extent does the distribution of protein lengths obey Benford's law?

2.5   Control Flow

Few computer programs are executed in a purely linear fashion, one statement after another as written in the source code. It is more likely that during the program execution, data objects are inspected and blocks of code executed conditionally on the basis of some test carried out on them. Thus, all practical languages have the equivalent of an *if-then-(else)* construction. This section explains the syntax of Python's version of this clause and covers a further kind of loop: the `while` loop.

2.5.1   if ... elif ... else

The `if ... elif ... else` construction allows statements to be executed conditionally, depending on the result of one or more logical tests (which evaluate to the boolean values True or False):

```
if <logical expression 1>:
    <statements 1>
elif <logical expression 2>:
    <statements 2>
...
else:
    <statements>
```

That is, if `<logical expression 1>` evaluates to True, `<statements 1>` are executed; otherwise, if `<logical expression 2>` evaluates to True, `<statements 2>` are executed, and so on; if none of the preceding logical expressions evaluate to True, the statements in the block of code following `else:` are executed. These statement blocks are indented with whitespace, as for the `for` loop. For example,

```
for x in range(10):
    if x <= 3:
        print(x, 'is less than or equal to three')
    elif x > 5:
        print(x, 'is greater than five')
    else:
        print(x, 'must be four or five, then')
```

produces the output:

```
0 is less than or equal to three
1 is less than or equal to three
2 is less than or equal to three
3 is less than or equal to three
4 must be four or five, then
5 must be four or five, then
6 is greater than five
7 is greater than five
8 is greater than five
9 is greater than five
```

It is not necessary to enclose test expressions such as x <= 3 in parentheses, as it is in C, for example, but the colon following the test is mandatory. The test expressions don't, in fact, have to evaluate explicitly to the boolean values True and False: as we have seen, other data types are taken to be equivalent to True unless they are 0 (int) or 0. (float), the empty string, '', empty list, [], the empty tuple, (), and so forth or Python's special type, None (see Section 2.2.4). Consider:

```python
for x in range(10):
    if x % 2:
        print(x, 'is odd!')
    else:
        print(x, 'is even!')
```

This works because x % 2 = 1 for odd integers, which is equivalent to True and x % 2 = 0 for even integers, which is equivalent to False.

There is **no** switch ... case ... finally construction in Python – equivalent control flow can be achieved with if ... elif ... else or with *dictionaries* (see Section 4.2).

Example E2.21   In the Gregorian calendar a year is a *leap year* if it is divisible by 4 with the exceptions that years divisible by 100 are *not* leap years unless they are also divisible by 400. The following Python program determines if year is a leap year.

Listing 2.3 Determining if a year is a leap year

```python
year = 1900

if not year % 400:
    is_leap_year = True
elif not year % 100:
    is_leap_year = False
elif not year % 4:
    is_leap_year = True
else:
    is_leap_year = False

s_ly = 'is a' if is_leap_year else 'is not a'
print('{:4d} {:s} leap year'.format(year, s_ly))
```

Hence the output:

```
1900 is not a leap year
```

2.5.2   `while` Loops

Whereas a `for` loop is established for a fixed number of iterations, statements within the block of a `while` loop execute only and as long as some condition holds:

```
>>> i = 0
>>> while i < 10:
...     i += 1
...     print(i, end='.')
...
>>> print()
1.2.3.4.5.6.7.8.9.10.
```

The counter `i` is initialized to `0`, which is less than 10, so the `while` loop begins. On each iteration, `i` is incremented by one and its value printed. When `i` reaches `10`, on the following iteration `i < 10` is `False`: the loop ends and execution continues after the loop, where `print()` outputs a new line.

Example E2.22   A more interesting example of the use of a `while` loop is given by this implementation of Euclid's algorithm for finding the greatest common divisor of two numbers, gcd(a, b):

```
>>> a, b = 1071, 462
>>> while b:
...     a, b = b, a % b
...
>>> print(a)
21
```

The loop continues until b divides a exactly; on each iteration, b is set to the remainder of `a//b` and then a is set to the old value of b. Recall that the integer 0 evaluates as boolean `False` so `while b:` is equivalent here to `while b != 0:`.

2.5.3   More Control Flow: `break`, `continue`, `pass` and `else`

break

Python provides three further statements for controlling the flow of a program. The `break` command, issued inside a loop, immediately ends that loop and moves execution to the statements following the loop:

```
x = 0
while True:
    x += 1
    if not (x % 15 or x % 25):
        break
print(x, 'is divisible by both 15 and 25')
```

The `while` loop condition here is (literally) always `True` so the only escape from the loop occurs when the `break` statement is reached. This occurs only when the counter x is divisible by both 15 and 25. The output is therefore:

```
75 is divisible by both 15 and 25
```

Similarly, to find the index of the first occurrence of a negative number in a list:

```
alist = [0, 4, 5, -2, 5, 10]
for i, a in enumerate(alist):
    if a < 0:
        break
print(a, 'occurs at index', i)
```

will output:

```
-2 occurs at index 3
```

Note that after escaping from the loop, the variables i and a have the values that they had within the loop at the break statement.

continue

The continue statement acts in a similar way to break but instead of breaking out of the containing loop, it immediately forces the next iteration of the loop without completing the statement block for the current iteration. For example,

```
for i in range(1, 11):
    if i % 2:
        continue
    print(i, 'is even!')
```

prints only the even integers 2, 4, 6, 8, 10: if i is not divisible by 2 (and hence i % 2 is 1, equivalent to True), that loop iteration is canceled and the loop resumed with the next value of i (the print statement is skipped).

pass

The pass command does nothing. It is useful as a "stub" for code that has not yet been written but where a statement is syntactically required by Python's whitespace convention.

```
>>> for i in range(1, 11):
...     if i == 6:
...         pass     # do something special if i is 6
...     if not i % 3:
...         print(i, 'is divisible by 3')
...

3 is divisible by 3
6 is divisible by 3
9 is divisible by 3
```

If the pass statement had been continue the line 6 is divisible by 3 would not have been printed: execution would have returned to the top of the loop and i = 7 instead of continuing to the second if statement.

◇ else

A for or while loop may be followed by an else block of statements, which will be executed only if the loop finished "normally" (that is, *without* the intervention of a break). For for loops, this means these statements will be executed after the loop has reached the end of the sequence it is iterating over; for while loops, they are executed when the while condition becomes False. For example, consider again our program to find the first occurrence of a negative number in a list. This code behaves rather oddly if there aren't any negative numbers in the list:

```
>>> alist = [0, 4, 5, 2, 5, 10]
>>> for i, a in enumerate(alist):
...     if a < 0:
...         break
...
>>> print(a, 'occurs at index', i)
10 occurs at index 5
```

It outputs the index and number of the last item in the list (whether it is negative or not). A way to improve this is to notice when the for loop runs through every item without encountering a negative number (and hence the break) and output a message:

```
>>> alist = [0, 4, 5, 2, 5, 10]
... for i, a in enumerate(alist):
...     if a < 0:
...         print(a, 'occurs at index', i)
...         break
... else:
...     print('no negative numbers in the list')
...

no negative numbers in the list
```

As another example, consider this (not particularly elegant) routine for finding the largest factor of a number a > 2:

```
a = 1013
b = a - 1
while b != 1:
    if not a % b:
        print('the largest factor of', a, 'is', b)
        break
    b -= 1
else:
    print(a, 'is prime!')
```

b is the largest factor not equal to a. The while loop continues as long as b is not equal to 1 (in which case a is prime) and decrements b after testing if b divides a exactly; if it does, b is the highest factor of a, and we break out of the while loop.

Example E2.23 A simple "turtle" virtual robot lives on an infinite two-dimensional plane and its location is always an integer pair of (x, y) coordinates. It can face only in directions parallel to the x and y axes (i.e. "north," "east," "south" or "west") and it understands four commands:

- F: move forward one unit;
- L: turn left (counterclockwise) by 90°;
- R: turn right (clockwise) by 90°;
- S: stop and exit.

The following Python program takes a list of such commands as a string and tracks the turtle"s location. The turtle starts at $(0, 0)$, facing in the direction $(1, 0)$ ("east"). The program ignores (but warns about) invalid commands and reports when the turtle crosses its own path.

Listing 2.4 A virtual turtle robot

```
# eg2-turtle.py
commands = 'FFFFFLFFFLFFFFRRRRFXFFFFFFS'

# Current location, current facing direction.
x, y = 0, 0
dx, dy = 1, 0
# Keep track of the turtle's location in the list of tuples, locs.
locs = [(0, 0)]
```

❶
```
for cmd in commands:
    if cmd == 'S':
        # Stop command.
        break
    if cmd == 'F':
        # Move forward in the current direction.
        x += dx
        y += dy
        if (x, y) in locs:
            print('Path crosses itself at: ({}, {})'.format(x, y))
        locs.append((x, y))
        continue
    if cmd == 'L':
        # Turn to the left (counterclockwise).
        # L => (dx, dy): (1, 0) -> (0, 1) -> (-1, 0) -> (0, -1) -> (1, 0).
        dx, dy = -dy, dx
        continue
    if cmd == 'R':
        # Turn to the right (clockwise).
        # R => (dx, dy): (1, 0) -> (0, -1) -> (-1, 0) -> (0, 1) -> (1, 0).
        dx, dy = dy, -dx
        continue
    # If we're here it's because we don't recognize the command: warn.
    print('Unknown command:', cmd)
```
❷
```
else:
    # We exhausted the commands without encountering an S for STOP.
    print('Instructions ended without a STOP')

# Plot a path of asterisks.
# First find the total range of x and y values encountered.
```
❸
```
x, y = zip(*locs)
xmin, xmax = min(x), max(x)
ymin, ymax = min(y), max(y)
# The grid size needed for the plot is (nx, ny).
nx = xmax - xmin + 1
ny = ymax - ymin + 1
# Reverse the y-axis so that it decreases *down* the screen.
for iy in reversed(range(ny)):
    for ix in range(nx):
        if (ix + xmin, iy + ymin) in locs:
```

```
        print('*', end='')
    else:
        print(' ', end='')
print()
```

❶ We can iterate over the string commands to take its characters one at a time.

❷ Note that the else: clause to the for loop is only executed if we do not break out of it on encountering a STOP command.

❸ We unzip the list of tuples, locs, into separate sequences of the x and y coordinates with zip(*locs).

The output produced from the commands given is:

```
Unknown command: X
Path crosses itself at: (1, 0)
*****
*     *
*     *
******
*
*
*
*
```

2.5.4   Exercises

Questions

Q2.5.1   Write a Python program to normalize a list of numbers, a, such that its values lie between 0 and 1. Thus, for example, the list a = [2, 4, 10, 6, 8, 4] becomes [0.0, 0.25, 1.0, 0.5, 0.75, 0.25].

Hint: use the built-ins min and max, which return the minimum and maximum values in a sequence, respectively; for example, min(a) returns 2 in the earlier mentioned list.

Q2.5.2   Write a while loop to calculate the *arithmetic-geometric mean* (AGM) of two positive real numbers, x and y, defined as the limit of the sequences:

$$a_{n+1} = \tfrac{1}{2}(a_n + b_n)$$
$$b_{n+1} = \sqrt{a_n b_n},$$

starting with $a_0 = x$, $b_0 = y$. Both sequences converge to the same number, denoted agm(x, y). Use your loop to determine *Gauss's constant*, $G = 1/\text{agm}(1, \sqrt{2})$.

Q2.5.3   The game of "Fizzbuzz" involves counting, but replacing numbers divisible by 3 with the word "*Fizz*," those divisible by 5 with "*Buzz*," and those divisible by both 3 and 5 with "*FizzBuzz*." Write a program to play this game, counting up to 100.

Q2.5.4   Straight-chain alkanes are hydrocarbons with the general stoichiometric formula $C_n H_{2n+2}$, in which the carbon atoms form a simple chain: for example, butane, $C_4 H_{10}$, has the structural formula that may be depicted $H_3CCH_2CH_2CH_3$. Write a program to output the structural formula of such an alkane, given its stoichiometry (assume $n > 1$). For example, given stoich = 'C8H18', the output should be

H3C-CH2-CH2-CH2-CH2-CH2-CH2-CH3

Problems

P2.5.1   Modify your solution to Problem P2.4.4 to output the first 50 rows of Pascal's triangle, but instead of the numbers themselves, output an asterisk if the number is odd and a space if it is even.

P2.5.2   The *iterative weak acid* approximation determines the hydrogen ion concentration, $[H^+]$, of an acid solution from the acid dissociation constant, K_a, and the acid concentration, c, by successive application of the formula

$$[H^+]_{n+1} = \sqrt{K_a (c - [H^+]_n)},$$

starting with $[H^+]_0 = 0$. The iterations are continued until $[H^+]$ changes by less than some predetermined, small tolerance value.

Use this method to determine the hydrogen ion concentration, and hence the pH ($= -\log_{10}[H^+]$) of a $c = 0.01$ M solution of acetic acid ($K_a = 1.78 \times 10^{-5}$). Use the tolerance TOL = 1.e-10.

P2.5.3   The *Luhn algorithm* is a simple checksum formula used to validate credit card and bank account numbers. It is designed to prevent common errors in transcribing the number, and detects all single-digit errors and almost all transpositions of two adjacent digits. The algorithm may be written as the following steps:

1.   Reverse the number.
2.   Treating the number as an array of digits, take the even-indexed digits (where the indexes *start at 1*) and double their values. If a doubled digit results in a number greater than 10, add the two digits (e.g. the digit 6 becomes 12 and hence $1 + 2 = 3$).
3.   Sum this modified array.
4.   If the sum of the array modulo 10 is 0 the credit card number is valid.

Write a Python program to take a credit card number as a string of digits (possibly in groups, separated by spaces) and establish if it is valid or not. For example, the string '4799 2739 8713 6272' is a valid credit card number, but any number with a single digit in this string changed is not.

P2.5.4   *Heron's method* for calculating the square root of a number, S, is as follows: starting with an initial guess, x_0, the sequence of numbers $x_{n+1} = \frac{1}{2}(x_n + S/x_n)$ are successively better approximations to $\sqrt{S}$. Implement this algorithm to estimate the square root of $2\,117\,519.73$ to two decimal places and compare with the "exact" answer provided by the math.sqrt method. For the purpose of this exercise, start with an initial guess, $x_0 = 2000$.

P2.5.5   Write a program to determine tomorrow's date given a string representing today's date, today, as either "D/M/Y" or "M/D/Y". Cater for both British and US-style dates when parsing today according to the value of a boolean variable us_date_style. For example, when us_date_style is False and today is '3/4/2014', tomorrow's date should be reported as '4/4/2014'.[25] (*Hint*: use the algorithm for determining if a year is a leap year, which is provided in the example to Section 2.5.1.)

P2.5.6   Write a Python program to determine $f(n)$, the number of trailing zeros in $n!$, using the special case of *de Polignac's formula*:

$$f(n) = \sum_{i=1}^{n} \left\lfloor \frac{n}{5^i} \right\rfloor,$$

where $\lfloor x \rfloor$ denotes the *floor* of x, the largest integer less than or equal to x.

P2.5.7   The *hailstone sequence* starting at an integer $n > 0$ is generated by the repeated application of the three rules:

- if $n = 1$, the sequence ends;
- if n is even, the next number in the sequence is $n/2$;
- if n is odd, the next number in the sequence is $3n + 1$.

(a)   Write a program to calculate the hailstone sequence starting at 27.
(b)   Let the *stopping time* be the number of numbers in a given hailstone sequence. Modify your hailstone program to return the stopping time instead of the numbers themselves. Adapt your program to demonstrate that the hailstone sequences started with $1 \leq n \leq 100$ agree with the *Collatz conjecture* (that all hailstone sequences stop eventually).

P2.5.8   The algorithm known as the *Sieve of Eratosthenes* finds the prime numbers in a list $2, 3, \ldots, n$. It may be summarized as follows, starting at $p = 2$, the first prime number:
Step 1.   Mark all the multiples of p in the list as non-prime (that is, the numbers mp where $m = 2, 3, 4, \ldots$: these numbers are *composite*.
Step 2.   Find the first unmarked number greater than p in the list. If there is no such number, stop.
Step 3.   Let p equal this new number and return to Step 1.
When the algorithm stops, the unmarked numbers are the primes.
    Implement the Sieve of Eratosthenes in a Python program and find all the primes under 10 000.

P2.5.9   *Euler's totient function*, $\phi(n)$, counts the number of positive integers less than or equal to n that are relatively prime to n. (Two numbers, a and b, are relatively prime if the only positive integer that divides both of them is 1; that is, if $\gcd(a, b) = 1$.)
    Write a Python program to compute $\phi(n)$ for $1 \leq n < 100$.
    (*Hint*: you could use Euclid's algorithm for the greatest common divisor given in the example to Section 2.5.2.)

---

[25] In practice, it would be better to use Python's datetime library (described in Section 4.5.3), but avoid it for this exercise.

P2.5.10 The value of π may be approximated by Monte Carlo methods. Consider the region of the xy-plane bounded by $0 \le x \le 1$ and $0 \le y \le 1$. By selecting a large number of random points within this region and counting the proportion of them lying beneath the function $y = \sqrt{1 - x^2}$ describing a quarter-circle, one can estimate $\pi/4$, this being the area bounded by the axes and $y(x)$. Write a program to estimate the value of π by this method.

Hint: use Python's random module. The method `random.random()` generates a (pseudo-)random number between 0. and 1. See Section 4.5.1 for more information.

P2.5.11 Write a program to take a string of text (words, perhaps with punctuation, separated by spaces) and output the same text with the middle letters shuffled randomly. Keep any punctuation at the end of words. For example, the string:

Four score and seven years ago our fathers brought forth on this continent a new nation, conceived in liberty, and dedicated to the proposition that all men are created equal.

might be rendered:

Four sorce and seevn yeras ago our fhtaers bhrogut ftroh on this cnnoientt a new noitan, cvieecond in lbrteiy, and ddicetead to the ptosoiporin that all men are cetaerd euaql.

Hint: `random.shuffle` shuffles a *list* of items in place. See Section 4.5.1.

P2.5.12 The *electron configuration* of an atom is the specification of the distribution of its electrons in atomic orbitals. An atomic orbital is identified by a *principal quantum number*, $n = 1, 2, 3, \ldots$ defining a *shell* comprised of one or more *subshells* defined by the *azimuthal quantum number*, $l = 0, 1, 2, \ldots, n - 1$. The values $l = 0, 1, 2, 3$ are referred to be the letters s, p, d and f respectively. Thus, the first few orbitals are 1s ($n = 1, l = 0$), 2s ($n = 2, l = 0$), 2p ($n = 2, l = 1$), 3s ($n = 3, l = 0$), and each shell has n subshells. A maximum of $2(2l + 1)$ electrons may occupy a given subshell.

According to the *Madelung rule*, the N electrons of an atom fill the orbitals in order of increasing $n + l$ such that whenever two orbitals have the same value of $n + l$, they are filled in order of increasing n. For example, the ground state of titanium ($N = 22$) is predicted (and found) to be $1s^2 2s^2 2p^6 3s^2 3p^6 4s^2 3d^2$.

Write a program to predict the electronic configurations of the elements up to rutherfordium ($N = 104$). The output for titanium should be

```
Ti: 1s2.2s2.2p6.3s2.3p6.4s2.3d2
```

A Python list containing the element symbols in order may be downloaded from https://scipython.com/ex/bbb.

As a bonus exercise, modify your program to output the configurations using the convention that the part of the configuration corresponding to the outermost *closed shell*, a noble gas configuration, is replaced by the noble gas symbol in square brackets; thus,

```
Ti: [Ar].4s2.3d2
```

the configuration of Argon being `1s2.2s2.2p6.3s2.3p6`.

Table 2.13 File modes

mode argument	Open mode
r	Text, read-only (the default)
w	Text, write (an existing file with the same name will be overwritten)
a	Text, append to an existing file
r+	Text, reading and writing
rb	Binary, read-only
wb	Binary, write (an existing file with the same name will be overwritten)
ab	Binary, append to an existing file
rb+	Binary, reading and writing

2.6 File Input/Output

Until now, data have been hard-coded into our Python programs, and output has been to the console (the terminal). Of course, it will frequently be necessary to input data from an external file and to write data to an output file. To achieve this, Python has `file` objects.

2.6.1 Opening and Closing a File

A `file` object is created by opening a file with a given *filename* and *mode*. The filename may be given as an absolute path, or as a path relative to the directory in which the program is being executed. `mode` is a string with one of the values given in Table 2.13. For example, to open a file for text-mode writing:

```
>>> f = open('myfile.txt', 'w')
```

`file` objects are closed with the `close` method: for example, `f.close()`. Python closes any open `file` objects automatically when a program terminates.

2.6.2 Writing to a File

The `write` method of a `file` object writes a *string* to the file and returns the number of characters written:

```
>>> f.write('Hello World!')
12
```

More helpfully, the `print` built-in takes an argument, `file`, to specify where to redirect its output :

```
>>> print(35, 'Cl', 2, sep='', file=f)
```

writes '35Cl2' to the file opened as `file` object f instead of to the console.

Example E2.24 The following program writes the first four powers of the numbers between 1 and 1000 in comma-separated fields to the file `powers.txt`:

```
f = open('powers.txt', 'w')
for i in range(1,1001):
    print(i, i**2, i**3, i**4, sep=', ', file=f)
f.close()
```

The file contents are

```
1, 1, 1, 1
2, 4, 8, 16
3, 9, 27, 81
...
999, 998001, 997002999, 996005996001
1000, 1000000, 1000000000, 1000000000000
```

2.6.3   Reading from a File

To read n bytes from a file, call `f.read(n)`. If n is omitted, the entire file is read in.[26]

`readline()` reads a single line from the file, up to and including the newline character. The next call to `readline()` reads in the next line, and so on. Both `read()` and `readline()` return an empty string when they reach the end of the file.

To read all of the lines into a list of strings in one go, use `f.readlines()`.

`file` objects are iterable, and looping over a (text) `file` returns its lines one at a time:

```
>>> for line in f:
❶ ...       print(line, end='')
...
First line
Second line
...
```

❶ Because `line` retains its newline character when read in, we use `end=''` to prevent print from adding another, which would be output as a blank line.

You probably want to use this method if your file is very large unless you really do want to store every line in memory. See Section 4.3.4 concerning Python's `with` statement for more best practice in file handling.

Example E2.25   To read in the numbers from the file `powers.txt` generated in the previous example, the columns must be converted to lists of integers. To do this, each line must be split into its fields and each field explicitly converted to an int:

```
f = open('powers.txt', 'r')
squares, cubes, fourths = [], [], []
for line in f.readlines():
    fields = line.split(',')
    squares.append(int(fields[1]))
    cubes.append(int(fields[2]))
    fourths.append(int(fields[3]))
f.close()
```

---

[26] To quote the official documentation: "it's your problem if the file is twice as large as your machine's memory."

```
n = 500
print(n, 'cubed is', cubes[n-1])
```

The output is

```
500 cubed is 125000000
```

In practice, it is better to use NumPy (see Chapter 6) to read in data files such as these.

2.6.4　Exercises

Problems

P2.6.1　The coast redwood tree species, *Sequoia sempervirens*, includes some of the oldest and tallest living organisms on Earth. Details concerning individual trees are given in the tab-delimited text file redwood-data.txt, available at https://scipython.com/ex/bbd. (Data courtesy of the Gymnosperm database, www.conifers.org/cu/Sequoia.php)

Write a Python program to read in this data and report the tallest tree and the tree with the greatest diameter.

P2.6.2　Write a program to read in a text file and censor any words in it that are on a list of banned words by replacing their letters with the same number of asterisks. Your program should store the banned words in lower case but censor examples of these words in any case. Assume there is no punctuation.

As a bonus exercise, handle text that contains punctuation. For example, given the list of banned words: ['C', 'Perl', 'Fortran'] the sentence

```
'Some alternative programming languages to Python are C, C++, Perl, Fortran and

Java.'
```

becomes

```
'Some alternative programming languages to Python are *, C++, ****, ******* and

Java.'
```

P2.6.3　The *Earth Similarity Index* (ESI) attempts to quantify the physical similarity between an astronomical body (usually a planet or moon) and Earth. It is defined by

$$\mathrm{ESI}_j = \prod_{i=1}^{n} \left(1 - \left| \frac{x_{i,j} - x_{i,\oplus}}{x_{i,j} + x_{i,\oplus}} \right| \right)^{w_i/n},$$

where the parameters $x_{i,j}$ are described, and their terrestrial values, $x_{i,\oplus}$ and weights, w_i, are given in Table 2.14. The radius, density and escape velocities are taken *relative to* the terrestrial values. The ESI lies between 0 and 1, with the values closer to 1 indicating closer similarity to Earth (which has an ESI of exactly 1: Earth is identical to itself!).

The file ex2-6-g-esi-data.txt available from https://scipython.com/ex/bbc contains the earlier mentioned parameters for a range of astronomical bodies. Use these

Table 2.14 Parameters used in the definition of ESI

i	Parameter	Earth value, $x_{i,\oplus}$	Weight, w_i
1	Radius	1.0	0.57
2	Density	1.0	1.07
3	Escape velocity, v_{esc}	1.0	0.7
4	Surface temperature	288 K	5.58

data to calculate the ESI for each of the bodies. Which has properties "closest" to those of the Earth?

P2.6.4 Write a program to read in a two-dimensional array of strings into a list of lists from a file in which the string elements are separated by one or more spaces. The number of rows, m, and columns, n, may not be known in advance of opening the file. For example, the text file

```
A B C D
E F G H
I J K L
```

should create an object, grid, as

```
[['A', 'B', 'C', 'D'], ['E', 'F', 'G', 'H'], ['I', 'J', 'K', 'L']]
```

Read like this, grid contains a list of the array's *rows*. Once the array has been read in, write loops to output the *columns* of the array:

```
[['A', 'E', 'I'], ['B', 'F', 'J'], ['C', 'G', 'K'], ['D', 'H', 'L']]
```

Harder: also output all its diagonals read in one direction:

```
[['A'], ['B', 'E'], ['C', 'F', 'I'], ['D', 'G', 'J'], ['H', 'K'], ['L']]
```

and the other direction:

```
[['D'], ['C', 'H'], ['B', 'G', 'L'], ['A', 'F', 'K'], ['E', 'J'], ['I']]
```

2.7   Functions

A Python *function* is a set of statements that are grouped together and named so that they can be run more than once in a program. There are two main advantages to using functions. First, they enable code to be reused without having to be replicated in different parts of the program; second, they enable complex tasks to be broken into separate procedures, each implemented by its own function – it is often much easier and more maintainable to code each procedure individually than to code the entire task at once.

2.7.1   Defining and Calling Functions

The def statement defines a function, gives it a name and lists the arguments (if any) that the function expects to receive when called. The function's statements are written in an indented block following this def. If at any point during the execution of this statement block a `return` statement is encountered, the specified values are returned to the caller. For example,

❶
```
>>> def square(x):
...     x_squared = x**2
...     return x_squared
...
>>> number = 2
```
❷
```
>>> number_squared = square(number)
>>> print(number, 'squared is', number_squared)
2 squared is 4
```
❸
```
>>> print('8 squared is', square(8))
8 squared is 64
```

❶ The simple function named `square` takes a single argument, `x`. It calculates `x**2` and returns this value to the caller. Once defined, it can be called any number of times.

❷ In the first example, the return value is assigned to the variable `number_squared`;

❸ In the second example, it is fed straight into the `print` method for output to the console.

To return two or more values from a function, pack them into a `tuple`. For example, the following program defines a function to return both roots of the quadratic equation $ax^2 + bx + c$ (assuming it *has* two real roots):

```
import math

def roots(a, b, c):
    d = b**2 - 4*a*c
    r1 = (-b + math.sqrt(d)) / 2 / a
    r2 = (-b - math.sqrt(d)) / 2 / a
    return r1, r2

print(roots(1., -1., -6.))
```

When run, this program outputs, as expected:

```
(3.0, -2.0)
```

It is not necessary for a function to explicitly return any object: functions that fall off the end of their indented block without encountering a `return` statement return Python's special value, `None`.

Function definitions can appear anywhere in a Python program, but a function cannot be referenced before it is defined. Functions can even be *nested*, but a function defined inside another is not (directly) accessible from outside that function.

Docstrings

A function *docstring* is a string literal that occurs as the first statement of the function definition. It should be written as a triple-quoted string on a single line if the function is simple, or on multiple lines with an initial one-line summary for more detailed descriptions of complex functions. For example,

```
def roots(a, b, c):
    """Return the roots of ax^2 + bx + c."""
    d = b**2 - 4*a*c
    ...
```

The docstring becomes the special $\_\_$doc$\_\_$ attribute of the function:

```
>>> roots.__doc__
'Return the roots of ax^2 + bx + c.'
```

A docstring should provide details about *how to use the function*: which arguments to pass it and which objects it returns,[27] but should not generally include details of the specific *implementation* of algorithms used by the function (these are best explained in *comments*, preceded by #).

Docstrings are also used to provide documentation for classes and modules (see Sections 4.5 and 4.6.2).

Example E2.26 In Python, *functions are "first class" objects*: they can have variable identifiers assigned to them, they can be passed as arguments to other functions, and they can even be returned *from* other functions. A function is given a name when it is defined, but that name can be reassigned to refer to a different object (don't do this unless you mean to!) if desired.

As the following example demonstrates, it is possible for more than one variable name to be assigned to the same function object.

```
>>> def cosec(x):
...     """Return the cosecant of x, cosec(x) = 1/sin(x)."""
...     return 1./math.sin(x)
...
>>> cosec
<function cosec at 0x100375170>
>>> cosec(math.pi/4)
1.4142135623730951
>>> csc = cosec
>>> csc
<function cosec at 0x100375170>
>>> csc(math.pi/4)
1.4142135623730951
```

❶

❶ The assignment csc = cosec associates the identifier (variable name) csc with the same function object as the identifier cosec: this function can then be called with csc() as well as with cosec().

2.7.2   Default and Keyword Arguments

Keyword Arguments

In the previous example, the arguments have been passed to the function in the order in which they are given in the function's definition (these are called *positional* arguments). It is also possible to pass the arguments in an arbitrary order by setting them explicitly as *keyword arguments*:

```
roots(a=1., c=-6., b=-1.)
roots(b=-1., a=1., c=-6.)
```

---

[27] For larger projects, docstrings document an application programming interface (API) for the project.

If you mix nonkeyword (positional) and keyword arguments the former must come first; otherwise Python won't know to which variable the positional argument corresponds:

```
>>> roots(1., c=6., b=-1.)    # OK
(3.0, -2.0)
>>> roots(b=-1., 1., -6.)     # oops: which is a and which is c?
  File "<stdin>", line 1
SyntaxError: non-keyword arg after keyword arg
```

Default Arguments

Sometimes you want to define a function that takes an *optional* argument: if the caller doesn't provide a value for this argument, a default value is used. Default arguments are set in the function definition:

```
>>> def report_length(value, units='m'):
...     return 'The length is {:.2f} {}'.format(value, units)
>>> report_length(33.136, 'ft')
'The length is 33.14 ft'
>>> report_length(10.1)
'The length is 10.10 m'
```

Default arguments are assigned *when the Python interpreter first encounters the function definition*. This can lead to some unexpected results, particularly for mutable arguments. For example,

```
>>> def func(alist = []):
...     alist.append(7)
...     return alist
...
>>> func()
[7]
>>> func()
[7, 7]
>>> func()
[7, 7, 7]
```

The default argument to the function, func, here is an empty list, but it is the specific empty list assigned when the function is defined. Therefore, each time func is called this specific list grows.

Example E2.27   Default argument values are assigned *when the function is defined*. Therefore, if a function is defined with an argument defaulting to some immutable object, subsequently changing that variable *will not change the default*:

```
>>> default_units = 'm'
>>> def report_length(value, units=default_units):
...     return 'The length is {:.2f} {}'.format(value, units)
...
>>> report_length(10.1)
'The length is 10.10 m'
>>> default_units = 'cubits'
>>> report_length(10.1)
'The length is 10.10 m'
```

The default units used by the function `report_length` are unchanged by the reassignment of the variable name `default_units`: the default value is set to the string object referred to by `default_units` when the `def` statement is encountered by the Python compiler (`'m'`) and it cannot be changed subsequently.

This also means that if a default argument is assigned to a *mutable* object, it is always that same object that is used whenever the function is called without providing an alternative: see Question Q2.7.4.

2.7.3 Scope

A function can define and use its own variables. When it does so, those variables are *local* to that function: they are not available outside the function. Conversely, variables assigned outside all function `def`s are *global* and are available everywhere within the program file. For example,

```
>>> def func():
...     a = 5
...     print(a, b)
...
>>> b = 6
>>> func()
5 6
```

The function `func` defines a variable a, but prints out both a and b. Because the variable b isn't defined in the local scope of the function, Python looks in the global scope, where it finds b = 6, so that is what is printed. It doesn't matter that b hasn't been defined when the function is *defined*, but of course it must be before the function is *called*.

What happens if a function defines a variable with the same name as a global variable? In this case, within the function the local scope is searched first when resolving variable names, so it is the object pointed to by the local variable name that is retrieved. For example,

```
>>> def func():
...     a = 5
...     print(a)
...
>>> a = 6
>>> func()
5
>>> print(a)
6
```

Note that the local variable a exists only within the body of the function; it just happens to have the same name as the global variable a. It disappears after the function exits and it doesn't overwrite the global a.

Python's rules for resolving scope can be summarized as "LEGB": first *local* scope, then *enclosing* scope (for nested functions), then *global* scope, and finally *built-ins* – if you happen to give a variable the same name as a built-in function (such as `range` or `len`), then that name resolves to your variable (in local or global scope) and not to the original built-in. It is therefore generally not a good idea to name your variables after built-ins.

◊ The `global` and `nonlocal` Keywords

We have seen that it is possible to access variables defined in scopes other than the local function's. Is it possible to *modify* them ("*rebind*" them to new objects)? Consider the distinction between the behavior of the following functions:

```
>>> def func1():
...     print(x)    # OK, providing x is defined in global or enclosing scope
...
>>> def func2():
...     x += 1       # not OK: can't modify x if it isn't local
...
>>> x = 4
>>> func1()
4
>>> func2()
UnboundLocalError: local variable 'x' referenced before assignment
```

If you really do want to change variables that are defined outside the local scope, you must first declare within the function body that this is your intention with the keywords `global` (for variables in global scope) and `nonlocal` (for variables in enclosing scope, for example, where one function is defined within another). In the previous case:

```
>>> def func2():
...     global x
...     x += 1      # OK now - Python knows we mean x in global scope
...
>>> x = 4
>>> func2()         # no error
>>> x
5
```

The function `func2` really has changed the value of the variable x in global scope.

You should think carefully whether it is really necessary to use this technique (would it be better to pass x as an argument and `return` its updated value from the function?), Especially in longer programs, variable names in one scope that change value (or even type!) within functions lead to confusing code, behavior that is hard to predict and tricky bugs.

Example E2.28   Take a moment to study the following code and predict the result before running it.

Listing 2.5 Python scope rules

```
# eg2-scope.py

def outer_func():
    def inner_func():
        a = 9
        print('inside inner_func, a is {:d} (id={:d})'.format(a, id(a)))
        print('inside inner_func, b is {:d} (id={:d})'.format(b, id(b)))
```

```
    print('inside inner_func, len is {:d} (id={:d})'.format(len,id(len)))

  len = 2
  print('inside outer_func, a is {:d} (id={:d})'.format(a, id(a)))
  print('inside outer_func, b is {:d} (id={:d})'.format(b, id(b)))
  print('inside outer_func, len is {:d} (id={:d})'.format(len,id(len)))
  inner_func()

a, b = 6, 7
outer_func()
print('in global scope, a is {:d} (id={:d})'.format(a, id(a)))
print('in global scope, b is {:d} (id={:d})'.format(b, id(b)))
print('in global scope, len is', len, '(id={:d})'.format(id(len)))
```

This program defines a function, inner_func, nested inside another, outer_func. After these definitions, the execution proceeds as follows:

1. Global variables a = 6 and b = 7 are initialized.
2. outer_func is called:

   a. outer_func defines a local variable, len = 2.
   b. The values of a and b are printed; they don't exist in local scope and there isn't any enclosing scope, so Python searches for and finds them in global scope: their values (6 and 7) are output.
   c. The value of local variable len (2) is printed.
   d. inner_func is called:

      (1) A local variable, a = 9 is defined.
      (2) The value of this local variable is printed.
      (3) The value of b is printed; b doesn't exist in local scope so Python looks for it in enclosing scope, that of outer_func. It isn't found there either, so Python proceeds to look in global scope where it is found: the value b = 7 is printed.
      (4) The value of len is printed: len doesn't exist in local scope, but it is in the enclosing scope since len = 2 is defined in outer_func: its value is output.

3. After outer_func has finished execution, the values of a and b in global scope are printed.
4. The value of len is printed. This is not defined in global scope, so Python searches its own built-in names: len is the built-in function for determining the lengths of sequences. This function is itself an object and it provides a short string description of itself when printed.

```
inside outer_func, a is 6 (id=232)
inside outer_func, b is 7 (id=264)
inside outer_func, len is 2 (id=104)
inside inner_func, a is 9 (id=328)
inside inner_func, b is 7 (id=264)
inside inner_func, len is 2 (id=104)
in global scope, a is 6 (id=232)
```

```
in global scope, b is 7 (id=264)
in global scope, len is <built-in function len> (id=977)
```

Note that in this example outer_func has (perhaps unwisely) redefined (*re-bound*) the name len to the integer object 2. This means that the original len built-in function is not available within this function (and neither is it available within the enclosed function, inner_func).

2.7.4 ◇ Passing Arguments to Functions

A common question from new users of Python who come to it with a knowledge of other computer languages is, are arguments to functions passed "by value" or "by reference?" In other words, does the function make its own copy of the argument, leaving the caller's copy unchanged, or does it receive a "pointer" to the location in memory of the argument, the contents of which the function *can* change? The distinction is important for languages such as C, but does not fit well into the Python *name-object* model. Python function arguments are sometimes (not very helpfully) said to be "references, passed by value". Recall that everything in Python is an object, and the same object may have multiple identifiers (what we have been loosely calling "variables" up until now). When a name is passed to a function, the "value" that is passed is, in fact, the object it points to. Whether the function can change the object or not (from the point of view of the caller) depends on whether the object is mutable or immutable.

A couple of examples should make this clearer. A simple function, func1, taking an integer argument, receives a reference to that integer object, to which it attaches a local name (which may or may not be the same as the global name). The function cannot change the integer object (which is immutable), so any reassignment of the local name simply points to a new object: the global name still points to the original integer object.

```
>>> def func1(a):
...     print('func1: a = {}, id = {}'.format(a, id(a)))
...     a = 7 # reassigns local a to the integer 7
...     print('func1: a = {}, id = {}'.format(a, id(a)))
...
>>> a = 3
>>> print('global: a = {}, id = {}'.format(a, id(a)))

global: a = 3, id = 4297242592

>>> func1(a)
func1: a = 3, id = 4297242592
func1: a = 7, id = 4297242720

>>> print('global: a = {}, id = {}'.format(a, id(a)))

global: a = 3, id = 4297242592
```

func1 therefore prints 3 (inside the function, a is initially the local name for the original integer object); it then prints 7 (this local name now points to a new integer object, with a new id) – see Figure 2.5. After it returns, the global name a still points to the original 3.

Now consider passing a mutable object, such as a `list`, to a function, func2. This time, an assignment to the list changes the original object, and these changes persist after the function call.

```
>>> def func2(b):
...     print('func2: b = {}, id = {}'.format(b, id(b)))
...     b.append(7) # add an item to the list
...     print('func2: b = {}, id = {}'.format(b, id(b)))
...
>>> c = [1, 2, 3]
>>> print('global: c = {}, id = {}'.format(c, id(c)))

global: c = [1, 2, 3], id = 4361122448

>>> func2(c)
func2: b = [1, 2, 3], id = 4361122448
func2: b = [1, 2, 3, 7], id = 4361122448

>>> print('global: c = {}, id = {}'.format(c, id(c)))

global: c = [1, 2, 3, 7], id = 4361122448
```

Note that it doesn't matter what name is given to the list by the function: this name points to the same object, as you can see from its id. The relationship between the variable names and objects is illustrated in Figure 2.6.

So, are Python arguments passed by value or by reference? The best answer is probably that arguments are passed by value, but that value is a reference to an object (which can be mutable or immutable).

Example E2.29 The *Lazy Caterer's Sequence*, $f(n)$, describes the maximum number of pieces a circular pizza can be divided into with an increasing number of cuts, n. Clearly, $f(0) = 1$, $f(1) = 2$ and $f(2) = 4$. For $n = 3$, $f(3) = 7$ (the maximum number of pieces are formed if the cuts do not intersect at a common point). It can be shown that the general recursion formula,

$$f(n) = f(n-1) + n,$$

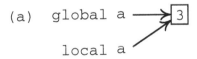

Figure 2.5 Immutable objects. Within func1: (a) before reassigning the local variable a and (b) after reassigning the value of local variable a.

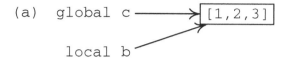

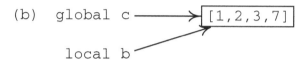

Figure 2.6 Mutable objects. Within func2: (a) before appending to the list pointed to by both global variable c and local variable b and (b) after appending to the list with b.append(7).

applies. Although there is a closed form for this sequence, $f(n) = \frac{1}{2}(n^2 + n + 2)$, we could also define a function to grow a list of consecutive values in the sequence:

```
>>> def f(seq):
...        seq.append(seq[-1] + n)
...
>>> seq = [1]      # f(0) = 1
>>> for n in range(1,16):
...        f(seq)
...
>>> print(seq)
[1, 2, 4, 7, 11, 16, 22, 29, 37, 46, 56, 67, 79, 92, 106, 121]
```

The list seq is mutable and so grows in place each time the function f() is called. The n referred to within this function is the name found in global scope (the for loop counter).

2.7.5   Recursive Functions

A function that can call itself is called a *recursive* function. Recursion is not always necessary but can lead to elegant algorithms in some situations.[28] For example, one way to calculate the factorial of an integer $n \geq 1$ is to define the following recursive function:

```
>>> def factorial(n):
...        if n == 1:
...            return 1
...        return n * factorial(n - 1)
...
>>> factorial(5)
120
```

---

[28] In fact, because of the overhead involved in making a function call, a recursive algorithm can be expected to be slower than a well-designed iterative one.

Here, a call to `factorial(n)` returns n times whatever is returned by the call to `factorial(n - 1)`, which returns $n-1$ times the returned values of `factorial(n - 2)` and so on until `factorial(1)` which is 1 by definition. That is, the algorithm makes use of the fact that $n! = n \cdot (n-1)!$ Care should be taken in implementing such recursive algorithms to ensure that they stop when some condition is met.[29]

Example E2.30 The famous *Tower of Hanoi* problem involves three poles, one of which (pole A) is stacked with n differently sized circular discs in decreasing order of diameter, with the largest at the bottom. The task is to move the stack to the third pole (pole C) by moving one disc at a time in such a way that a larger disc is never placed on a smaller one. It is necessary to use the second pole (pole B) as an intermediate resting place for the discs.

The problem can be solved using the following recursive algorithm. Label the discs D_i with D_1 the smallest disc and D_n the largest.

- Move discs $D_1, D_2, \ldots, D_{n-1}$ from A to B.
- Move disc D_n from A to C.
- Move discs $D_1, D_2, \ldots, D_{n-1}$ from B to C.

The second step is a single move, but the first and last require the movement of a stack of $n-1$ discs from one peg to another – which is exactly what the algorithm itself solves!

In the following code, we identify the discs by the integers $1, 2, 3, \ldots$ stored in one of three lists, A, B and C. The initial state of the system, with all discs on pole A is denoted by, for example, A = [5, 4, 3, 2, 1] where the first indexed item is the "bottom" of the pole and the last indexed item is the "top." The rules of the problem require that these lists must always be *decreasing* sequences.

Listing 2.6 The Tower of Hanoi problem

```
# eg2-hanoi.py

def hanoi(n, P1, P2, P3):
    """ Move n discs from pole P1 to pole P3. """
    if n == 0:
        # No more discs to move in this step.
        return

    global count
    count += 1

    # Move n - 1 discs from P1 to P2.
    hanoi(n - 1, P1, P3, P2)

    if P1:
        # Move disc from P1 to P3.
        P3.append(P1.pop())
        print(A, B, C)
```

---

[29] In practice, an infinite loop is not possible because of the memory overhead involved in each function call, and Python sets a maximum recursion limit.

```
        # Move n - 1 discs from P2 to P3.
        hanoi(n - 1, P2, P1, P3)

    # Initialize the poles: all n discs are on pole A.
    n = 3
    A = list(range(n, 0, -1))
    B, C = [], []

    print(A, B, C)
    count = 0
    hanoi(n, A, B, C)
    print(count)
```

Note that the hanoi function just moves a stack of discs from one pole to another: lists (representing the poles) are passed into it in some order, and it moves the discs from the pole represented by the first list, known locally as P1, to that represented by the third (P3). It does not need to know which list is A, B or C.

2.7.6   Exercises

Questions

Q2.7.1   The following small programs each attempt to output the simple sum:

```
    56
    +44
    -----
    100
    -----
```

Which two programs work as intended? Explain carefully what is wrong with each of the others.

(a)   ```
       def line():
           '-----'

       my_sum = '\n'.join(['   56', '   +44', line(), '   100', line()])
       print(my_sum)
       ```

(b)    ```
       def line():
           return '-----'

       my_sum = '\n'.join(['    56', '    +44', line(), '    100', line()])
       print(my_sum)
       ```

(c)   ```
       def line():
           return '-----'

       my_sum = '\n'.join(['   56', '   +44', line, '   100', line])
       print(my_sum)
       ```

(d)    ```
       def line():
           print('-----')
       ```

```
        print('   56')
        print('  +44')
        print(line)
        print('  100')
        print(line)
```

(e)   def line():
            print('-----')

```
        print('   56')
        print('  +44')
        print(line())
        print('  100')
        print(line())
```

(f)   def line():
            print('-----')

```
        print('   56')
        print('  +44')
        line()
        print('  100')
        line()
```

Q2.7.2 The following code snippet attempts to calculate the balance of a savings account with an annual interest rate of 5% after four years, if it starts with a balance of $100.

```
>>> balance = 100
>>> def add_interest(balance, rate):
...     balance += balance * rate / 100
...
>>> for year in range(4):
...     add_interest(balance, 5)
...     print('Balance after year {}: ${:.2f}'.format(year + 1, balance))
...
Balance after year 1: $100.00
Balance after year 2: $100.00
Balance after year 3: $100.00
Balance after year 4: $100.00
```

Explain why this doesn't work and then provide a working alternative.

Q2.7.3 A *Harshad number* is an integer that is divisible by the sum of its digits (e.g. 21 is divisible by $2 + 1 = 3$ and so is a Harshad number). Correct the following code, which should return True or False if n is a Harshad number, or not, respectively:

```
def digit_sum(n):
    """ Find the sum of the digits of integer n. """

    s_digits = list(str(n))
    dsum = 0
    for s_digit in s_digits:
        dsum += int(s_digit)
```

```
def is_harshad(n):
    return not n % digit_sum(n)
```

When run, the function `is_harshad` raises an error:

```
>>> is_harshad(21)
TypeError: unsupported operand type(s) for %: 'int' and 'NoneType'
```

Q2.7.4   Predict and explain the output of the following code.

```
def grow_list(a, lst=[]):
    lst.append(a)
    return lst

lst1 = grow_list(1)
lst1 = grow_list(2, lst1)

lst2 = grow_list('a')

print(lst1)
print(lst2)
```

Problems

P2.7.1   The word game Scrabble is played on a 15×15 grid of squares referred to by a row index letter (A–O) and a column index number (1–15). Write a function to determine whether a word will fit in the grid, given the position of its first letter as a string (e.g. `'G7'`) a variable indicating whether the word is placed to read *across* or *down* the grid and the word itself.

P2.7.2   Write a program to find the smallest positive integer, n, whose factorial is *not* divisible by the sum of its digits. For example, 6 is not such a number because $6! = 720$ and $7 + 2 + 0 = 9$ divides 720.

P2.7.3   Write two functions which, given two lists of length 3 representing three-dimensional vectors $\mathbf{a}$ and $\mathbf{b}$, calculate the dot product, $\mathbf{a} \cdot \mathbf{b}$ and the vector (cross) product, $\mathbf{a} \times \mathbf{b}$.
  Write two more functions to return the scalar triple product, $\mathbf{a} \cdot (\mathbf{b} \times \mathbf{c})$ and the vector triple product, $\mathbf{a} \times (\mathbf{b} \times \mathbf{c})$.

P2.7.4   A right regular pyramid with height h and a base consisting of a regular n-sided polygon of side length s has a volume $V = \frac{1}{3}Ah$ and total surface area $S = A + \frac{1}{2}nsl$ where A is the base area and l the slant height, which may be calculated from the *apothem* of the base polygon, $a = \frac{1}{2}s \cot \frac{\pi}{n}$ as $A = \frac{1}{2}nsa$ and $l = \sqrt{h^2 + a^2}$.
  Use these formulas to define a function, `pyramid_AV`, returning V and S when passed values for n, s and h.

P2.7.5   The range of a projectile launched at an angle α and speed v on flat terrain is

$$R = \frac{v^2 \sin 2\alpha}{g},$$

where g is the acceleration due to gravity, which may be taken to be 9.81 m s^{-2} for Earth. The maximum height attained by the projectile is given by

$$H = \frac{v^2 \sin^2 \alpha}{2g}.$$

(We neglect air resistance and the curvature and rotation of the Earth.) Write a function to calculate and return the range and maximum height of a projectile, taking α and v as arguments. Test it with the values $v = 10 \text{ m s}^{-1}$ and $\alpha = 30°$.

P2.7.6   Write a function, sinm_cosn, which returns the value of the following definite integral for integers $m, n > 1$.

$$\int_0^{\pi/2} \sin^n \theta \cos^m \theta \, d\theta = \begin{cases} \frac{(m-1)!!(n-1)!!}{(m+n)!!} \frac{\pi}{2} & m, n \text{ both even,} \\ \frac{(m-1)!!(n-1)!!}{(m+n)!!} & \text{otherwise.} \end{cases}$$

Hint: for calculating the double factorial, see Exercise P2.4.6.

P2.7.7   Write a function that determines if a string is a palindrome (that is, reads the same backward as forward) *using recursion*.

P2.7.8   *Tetration* may be thought of as the next operator after exponentiation. Thus, where $x \times n$ can be written as the sum $x + x + x + \ldots + x$ with n terms, and x^n is the multiplication of n factors: $x \cdot x \cdot x \cdot \ldots x$, the expression written $^n x$ is equal to the repeated exponentiation involving n occurrences of x:

$$^n x = x^{x^{\cdot^{\cdot^{\cdot^x}}}}$$

For example, $^4 2 = 2^{2^{2^2}} = 2^{2^4} = 2^{16} = 65536$. Note that the exponential "tower" is evaluated from top to bottom.

Write a recursive function to calculate $^n x$ and test it (for small, positive real values of x and non-negative integers n: tetration generates *very* large numbers)!

How many digits are there in $^3 5$? In $^5 2$?

3 Interlude: Simple Plots and Charts

As Python has grown in popularity, many libraries of packages and modules have become available to extend its functionality in useful ways; Matplotlib is one such library. Matplotlib provides a means of producing graphical plots that can be embedded into applications, displayed on the screen or output as high-quality image files for publication.

Matplotlib has a fully fledged *object-oriented* interface, which is described in more detail in Chapter 7, but for simple plotting in an interactive shell session, its simpler, *procedural* pyplot interface provides a convenient way of visualizing data. This short chapter describes its use alongside some basic NumPy functionality (the NumPy library is described in more detail in Chapter 6).

On a system with Matplotlib and NumPy installed, the recommended imports are:

```
>>> import matplotlib.pyplot as plt
>>> import numpy as np
```

even though this means prefacing method calls with "plt." and "np."[1]

Note: an earlier Python module, pylab, combined the functionality of pyplot and numpy by importing all of their functions into a common namespace to mimic the commercial MATLAB package. Its use is no longer encouraged and we do not describe it here.

3.1 Basic Plotting

3.1.1 Line Plots and Scatter Plots

The simplest (x, y) line plot is achieved by calling plt.plot with two iterable objects of the same length (typically lists of numbers or NumPy arrays). For example,

```
>>> ax = [0., 0.5, 1.0, 1.5, 2.0, 2.5, 3.0]
>>> ay = [0.0, 0.25, 1.0, 2.25, 4.0, 6.25, 9.0]
>>> plt.plot(ax,ay)
>>> plt.show()
```

plt.plot creates a Matplotlib object (here, a Line2D object) and plt.show() displays it on the screen. Figure 3.1 shows the result; by default the line will be in blue.

---

[1] It is better to avoid polluting the global namespace by importing as, e.g. from numpy import *.

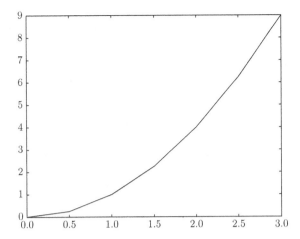

Figure 3.1 A basic (x, y) line plot.

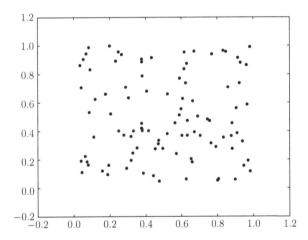

Figure 3.2 A basic scatter plot.

To plot (x, y) points as a scatter plot rather than as a line plot, call `plt.scatter` instead:

```
>>> import random
>>> ax, ay = [], []
>>> for i in range(100):
...     ax.append(random.random())
...     ay.append(random.random())
...
>>> plt.scatter(ax,ay)
>>> plt.show()
```

The resulting plot is shown in Figure 3.2.

The plot can be saved as an image by calling `plt.savefig(`*filename*`)`. The desired image format is deduced from the filename extension. For example,

```
plt.savefig('plot.png')      # save as a PNG image
plt.savefig('plot.pdf')      # save as PDF
plt.savefig('plot.eps')      # save in Encapsulated PostScript format
```

Example E3.1   As an example, let's plot the function $y = \sin^2 x$ for $-2\pi \le x \le 2\pi$. Using only the Python we've covered in the previous chapter, here is one approach:

We calculate and plot 1000 (x, y) points, and store them in the lists `ax` and `ay`. To set up the `ax` list as the abcissa, we can't use `range` directly because that method only produces integer sequences, so first we work out the spacing between each x value as

$$\Delta x = \frac{x_{\max} - x_{\min}}{n - 1}$$

(if our n values are to *include* $x_{\min}$ and $x_{\max}$, there are $n - 1$ intervals of width Δx); the abcissa points are then

$$x_i = x_{\min} + i\Delta x \quad \text{for } i = 0, 1, 2, \ldots, n - 1.$$

The corresponding y-axis points are

$$y_i = \sin^2(x_i).$$

The following program implements this approach, and plots the (x, y) points on a simple line-graph (see Figure 3.3).

Listing 3.1 Plotting $y = \sin^2 x$

```
# eg3-sin2x.py

import math
import matplotlib.pyplot as plt
xmin, xmax = -2. * math.pi, 2. * math.pi
n = 1000
x = [0.] * n
y = [0.] * n
dx = (xmax - xmin)/(n-1)
for i in range(n):
    xpt = xmin + i * dx
    x[i] = xpt
    y[i] = math.sin(xpt)**2

plt.plot(x,y)
plt.show()
```

3.1.2   `linspace` and Vectorization

Plotting the simple function $y = \sin^2 x$ in the previous example involved quite a lot of work, almost all of it to do with setting up the lists `x` and `y`. The NumPy library, described more fully in Chapter 6, can be used to make life much easier.

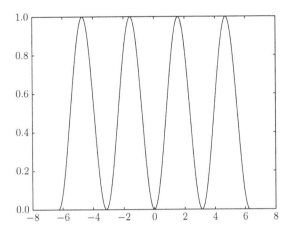

Figure 3.3 A plot of $y = \sin^2 x$.

First, the regularly spaced grid of x-coordinates, x, can be created using `linspace`. This is much like a floating-point version of the `range` built-in: it takes a start value, an end value, and the number of values in the sequence and generates an array of values representing the arithmetic progression between (and *inclusive of*) the two values. For example, `x = np.linspace(-5, 5, 1001)` creates the sequence: $-5.0, -4.99, -4.98, \ldots, 4.99, 5.0$.

Second, the NumPy equivalents of the `math` module's methods can act on iterable objects (such as `lists` or NumPy arrays). Thus, `y = np.sin(x)` creates a sequence of values (actually, a NumPy ndarray), which are $\sin(x_i)$ for each value x_i in the array x:

```
import numpy as np
import matplotlib.pyplot as plt
n = 1000
xmin, xmax = -2*np.pi, 2*np.pi
x = np.linspace(xmin, xmax, n)
y = np.sin(x)**2
plt.plot(x,y)
plt.show()
```

This is called *vectorization* and is described in more detail in Section 6.1.3. Lists and tuples can be turned into array objects supporting vectorization with the array constructor method:

```
>>> w = [1.0, 2.0, 3.0, 4.0]
>>> w = np.array(w)
>>> w * 100      # multiply each element by 100
array([ 100., 200., 300., 400.])
```

To add a second line to the plot, simply call `plt.plot` again:

```
...
x = np.linspace(xmin, xmax, n)
y1 = np.sin(x)**2
y2 = np.cos(x)**2
```

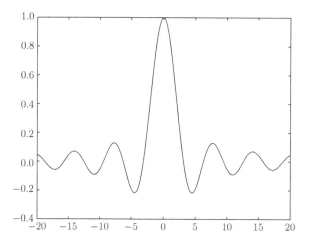

Figure 3.4 A plot of $y = \text{sinc}(x)$.

```
plt.plot(x,y1)
plt.plot(x,y2)
plt.show()
```

Note that after a plot has been displayed with show or saved with savefig, it is no longer available to display a second time – to do this it is necessary to call plt.plot again. This is because of the procedural nature of the pyplot interface: each call to a pyplot method changes the internal *state* of the plot object. The plot object is built up by successive calls to such methods (adding lines, legends and labels, setting the axis limits, etc.), and then the plot object is displayed or saved.

Example E3.2   The sinc function is the function

$$f(x) = \frac{\sin x}{x}.$$

To plot it over $20 \le x \le 20$:

```
>>> x = np.linspace(-20, 20, 1001)
>>> y = np.sin(x)/x

__main__:1: RuntimeWarning: invalid value encountered in true_divide
>>> plt.plot(x,y)
>>> plt.show()
```

Note that even though Python warns of the division by zero at $x = 0$, the function is plotted correctly: the singular point is set to the special value nan (standing for "Not a Number") and is omitted from the plot (Figure 3.4).

```
>>> y[498:503]
array([ 0.99893367,  0.99973335,          nan,  0.99973335,  0.99893367])
```

3.1.3   Exercises

Problems

P3.1.1   Plot the functions

$$f_1(x) = \ln\left(\frac{1}{\cos^2 x}\right) \text{ and}$$

$$f_2(x) = \ln\left(\frac{1}{\sin^2 x}\right).$$

on 1000 points across the range $-20 \leq x \leq 20$. What happens to these functions at $x = n\pi/2$ ($n = 0, \pm 1, \pm 2, \ldots$)? What happens in your plot of them?

P3.1.2   The *Michaelis–Menten* equation models the kinetics of enzymatic reactions as

$$v = \frac{d[P]}{dt} = \frac{V_{max}[S]}{K_m + [S]},$$

where v is the rate of the reaction converting the substrate, S, to product, P, catalyzed by the enzyme. V_{max} is the maximum rate (when all the enzyme is bound to S) and the Michaelis constant, K_m, is the substrate concentration at which the reaction rate is at half its maximum value.

Plot v against [S] for a reaction with $K_m = 0.04$ M and $V_{max} = 0.1$ M s^{-1}. Look ahead to the next section if you want to label the axes.

P3.1.3   The normalized Gaussian function centered at $x = 0$ is

$$g(x) = \frac{1}{\sigma\sqrt{2\pi}} \exp\left(-\frac{x^2}{2\sigma^2}\right).$$

Plot and compare the shapes of these functions for standard deviations $\sigma = 1, 1.5$ and 2.

3.2   Labels, Legends and Customization

3.2.1   Labels and Legends

Plot Legend

Each line on a plot can be given a label by passing a string object to its `label` argument. However, the label won't appear on the plot unless you also call `plt.legend` to add a legend:

```
plt.plot(ax, ay1, label='sin^2(x)')
plt.legend()
plt.show()
```

The location of the legend is, by default, the top right-hand corner of the plot but can be customized by setting the `loc` argument to the `legend` method to any of the string or integer values given in Table 3.1.

Table 3.1 Legend location specifiers

String	Integer
'best'	0
'upper right'	1
'upper left'	2
'lower left'	3
'lower right'	4
'right'	5
'center left'	6
'center right'	7
'lower center'	8
'upper center'	9
'center'	10

The Plot Title Axis Labels

A plot can be given a title above the axes by calling `plt.title` and passing the title as a string. Similarly, the methods `plt.xlabel` and `plt.ylabel` control the labeling of the *x*- and *y*-axes: just pass the label you want as a string to these methods. The optional additional attribute `fontsize` sets the font size in points. For example, the following code produces Figure 3.5.

```
t = np.linspace(0., 0.1, 1000)
Vp_uk, Vp_us = 230 * np.sqrt(2), 120 * np.sqrt(2)
f_uk, f_us = 50, 60
❶ V_uk = Vp_uk * np.sin(2 * np.pi * f_uk * t)
  V_us = Vp_us * np.sin(2 * np.pi * f_us * t)
❷ plt.plot(t*1000, V_uk, label='UK')
  plt.plot(t*1000, V_us, label='US')
  plt.title('A comparison of AC voltages in the UK and US')
  plt.xlabel('Time /ms', fontsize=16.)
  plt.ylabel('Voltage /V', fontsize=16.)
  plt.legend()
  plt.show()
```

❶ We calculate the voltage as a function of time (t, in seconds) in the United Kingdom and in the United States, which have different rms voltages (230 V and 120 V respectively; we have multiplied by $\sqrt{2}$ to get the peak-to-peak voltage) and different frequencies (50 Hz and 60 Hz).

❷ The time is plotted on the *x*-axis in milliseconds (t*1000).

Using LaTeX in pyplot

You can use LaTeX markup in pyplot plots, but this option needs to be enabled in Matplotlib's "rc settings," as follows:

```
plt.rc('text', usetex=True)
```

Then simply pass the LaTeX markup as a string to any label you want displayed in this way. Remember to use raw strings (r'xxx') to prevent Python from escaping any characters followed by LaTeX's backslashes (see Section 2.3.2).

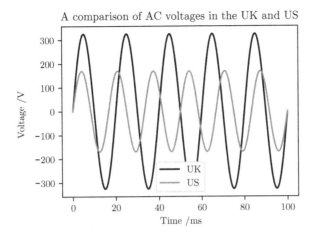

Figure 3.5 A comparison of AC voltages in the United Kingdom and United States.

Example E3.3   To plot the functions $f_n(x) = x^n \sin x$ for $n = 1, 2, 3, 4$:

```
import matplotlib.pyplot as plt
import numpy as np
plt.rc('text', usetex=True)

x = np.linspace(-10,10,1001)
for n in range(1,5):
    y = x**n * np.sin(x)
    y /= max(y)
    plt.plot(x,y, label=r'$x^{}\sin x$'.format(n))
plt.legend(loc='lower center')
plt.show()
```

❶ To make the graphs easier to compare, they have been scaled to a maximum of 1 in the region considered.

The graph produced is given in Figure 3.6.

3.2.2   Customizing Plots

Markers

By default, plot produces a line-graph with no markers at the plotted points. To add a marker on each point of the plotted data, use the marker argument. Several different markers are available and are documented online;[2] some of the more useful ones are listed in Table 3.2.

---

[2] https://matplotlib.org/api/markers_api.html.

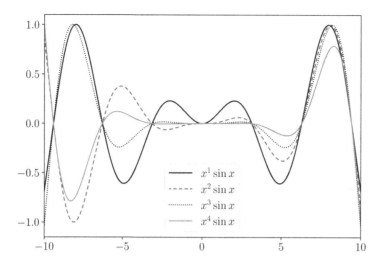

Figure 3.6 $f_n(x) = x^2 \sin x$ for $n = 1, 2, 3, 4$.

Table 3.2 Some Matplotlib marker
styles

Code	Marker	Description
.	·	Point
o	○	Circle
+	+	Plus
x	×	Cross
D	◇	Diamond
v	▽	Downward triangle
^	△	Upward triangle
s	☐	Square
*	★	Star

Colors

The color of a plotted line and/or its markers can be set with the `color` argument. Several formats for specifying the color are supported. First, there are one-letter codes for some common colors, given in Table 3.3. For example, `color='r'` specifies a red line and markers. These colors are somewhat garish and (since Matplotlib 2.0) the default color sequence for a series of lines on the same plot is the more pleasing "Tableau" sequence, whose string identifiers are also given in Table 3.3.

Alternatively, shades of gray can specified as a string representing a `float` in the range 0–1 (`0.` being black and `1.` being white). HTML hex strings giving the red, green and blue (RGB) components of the color in the range `00–ff` can also be passed in the `color` argument (e.g. `color='#ff00ff'` is magenta). Finally, the RGB components can also be passed as a `tuple` of three values in the range 0–1 (e.g. `color=(0.5, 0., 0.)` is a dark red color).

Table 3.3 Matplotlib color code letters

Basic color codes	Tableau colors
b = blue	tab:blue
g = green	tab:orange
r = red	tab:green
c = cyan	tab:red
m = magenta	tab:purple
y = yellow	tab:brown
k = black	tab:pink
w = white	tab:gray
	tab:olive
	tab:cyan

Table 3.4 Matplotlib line styles

Code	Line style
-	Solid
--	Dashed
:	Dotted
-.	Dash-dot

Line Styles and Widths

The default plot line style is a solid line of weight 1.5 pt. To customize this, set the `linestyle` argument (also a string). Some of the possible line style settings are given in Table 3.4.

To draw no line at all, set `linestyle=''` (the empty string). The thickness of a line can be specified in points by passing a `float` to the `linewidth` attribute.

For example,

```
x = np.linspace(0.1, 1., 100)
yi = 1. / x
ye = 10. * np.exp(-2 * x)
plt.plot(x, yi, color='r', linestyle=':', linewidth=4.)
plt.plot(x, ye, color='m', linestyle='--', linewidth=2.)
plt.show()
```

This code produces Figure 3.7.

The following abbreviations for the plot line properties are also valid:

- c for color,
- ls for linestyle,
- lw for linewidth.

For example,

```
plt.plot(x, y, c='g', ls='--', lw=2)        # a thick, green, dashed line
```

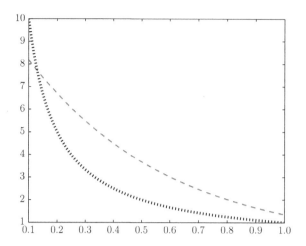

Figure 3.7 Two different line styles on the same plot.

It is also possible to specify the color, line style and marker style in a single string:

```
plt.plot(x, y, 'r:^')        # a red, dotted line with triangle markers
```

Finally, multiple lines can be plotted using a sequence of x, y format arguments:

```
plt.plot(x, y1, 'r--', x, y2, 'k-.')
```

plots a red dashed line for (x, y1) and a black dash-dot line for (x, y2).

Plot Limits

The methods `plt.xlim` and `plt.ylim` set the x- and y-limits of the plot, respectively. They must be called *after* any `plt.plot` statements, before showing or saving the figure. For example, the following code produces a plot of the provided data series between chosen limits (Figure 3.8):

```
t = np.linspace(0, 2, 1000)
f = t * np.exp(t + np.sin(20*t))
plt.plot(t, f)
plt.xlim(1.5,1.8)
plt.ylim(0,30)
plt.show()
```

Example E3.4   *Moore's law* is the observation that the number of transistors on central processing units (CPUs) approximately doubles every 2 years. The following program illustrates this with a comparison between the actual number of transistors on high-end CPUs from between 1972 and 2012, and that predicted by Moore's law, which may be stated mathematically as

$$n_i = n_0 2^{(y_i - y_0)/T_2},$$

where n_0 is the number of transistors in some reference year, y_0, and $T_2 = 2$ is the number of years taken to double this number. Because the data cover 40 years, the

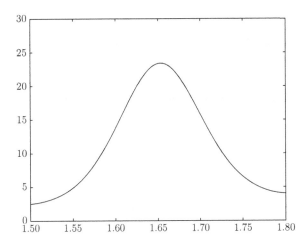

Figure 3.8 A plot produced with explicitly defined data limits.

values of n_i span many orders of magnitude, and it is convenient to apply Moore's law to its logarithm, which shows a linear dependence on y:

$$\log_{10} n_i = \log_{10} n_0 + \frac{y_i - y_0}{T_2} \log_{10} 2.$$

Listing 3.2 An illustration of Moore's law

```python
# eg3-moore.py
import numpy as np
import matplotlib.pyplot as plt

# The data - lists of years:
year = [1972, 1974, 1978, 1982, 1985, 1989, 1993, 1997, 1999, 2000, 2003,
        2004, 2007, 2008, 2012]
# And number of transistors (ntrans) on CPUs in millions:
ntrans = [0.0025, 0.005, 0.029, 0.12, 0.275, 1.18, 3.1, 7.5, 24.0, 42.0,
          220.0, 592.0, 1720.0, 2046.0, 3100.0]
# Turn the ntrans list into a NumPy array and multiply by 1 million.
ntrans = np.array(ntrans) * 1.e6

y0, n0 = year[0], ntrans[0]
# A linear array of years spanning the data's years.
y = np.linspace(y0, year[-1], year[-1] - y0 + 1)
# Time taken in years for the number of transistors to double.
T2 = 2.
moore = np.log10(n0) + (y - y0) / T2 * np.log10(2)

plt.plot(year, np.log10(ntrans), '*', markersize=12, color='r',
            markeredgecolor='r', label='observed')
plt.plot(y, moore, linewidth=2, color='k', linestyle='--', label='predicted')
plt.legend(fontsize=16, loc='upper left')
plt.xlabel('Year')
plt.ylabel('log(ntrans)')
plt.title("Moore's law")
plt.show()
```

In this example, the data are given in two lists of equal length representing the year and representative number of transistors on a CPU in that year. The Moore's law formula above is implemented in logarithmic form, using an array of years spanning the provided data. (Actually, since on a logarithmic scale this will be a straight line, really only two points are needed.)

For the plot, shown in Figure 3.9, the data are plotted as largeish stars and the Moore's law prediction as a dashed black line.

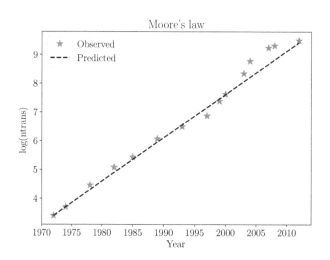

Figure 3.9 Moore's law modeling the exponential growth in transistors on CPUs.

3.2.3   Exercises

Problems

P3.2.1   A molecule, A, reacts to form either B or C with first-order rate constants k_1 and k_2, respectively. That is,

$$\frac{d[A]}{dt} = -(k_1 + k_2)[A],$$

and so

$$[A] = [A]_0 e^{-(k_1+k_2)t},$$

where $[A]_0$ is the initial concentration of A. The product concentrations (starting from 0) increase in the ratio $[B]/[C] = k_1/k_2$ and conservation of matter requires

$[B] + [C] = [A]_0 - [A]$. Therefore,

$$[B] = \frac{k_1}{k_1 + k_2}[A]_0\left(1 - e^{-(k_1+k_2)t}\right)$$

$$[C] = \frac{k_2}{k_1 + k_2}[A]_0\left(1 - e^{-(k_1+k_2)t}\right)$$

For a reaction with $k_1 = 300$ s^{-1} and $k_2 = 100$ s^{-1}, plot the concentrations of A, B and C against time given an initial concentration of reactant $[A]_0 = 2.0$ mol dm^{-3}.

P3.2.2 A *Gaussian integer* is a complex number whose real and imaginary parts are both integers. A *Gaussian prime* is a Gaussian integer $x + iy$ such that either:

- one of x and y is zero and the other is a prime number of the form $4n + 3$ or $-(4n + 3)$ for some integer $n \geq 0$; or
- both x and y are nonzero and $x^2 + y^2$ is prime.

Consider the sequence of Gaussian integers traced out by an imaginary particle, initially at c_0, moving in the complex plane according to the following rule: it takes integer steps in its current direction (± 1 in either the real or imaginary direction), but turns *left* if it encounters a Gaussian prime. Its initial direction is in the positive real direction ($\Delta c = 1 + 0i \Rightarrow \Delta x = 1$, $\Delta y = 0$). The path traced out by the particle is called a *Gaussian prime spiral*.

Write a program to plot the Gaussian prime spiral starting at $c_0 = 5 + 23i$.

P3.2.3 The annual risk of death (given as "1 in N") for men and women in the UK in 2005 for different age ranges is given in the table below. Use `pyplot` to plot these data on a single chart.

Age range	Female	Male
< 1	227	177
1–4	5376	4386
5–14	10417	8333
15–24	4132	1908
25–34	2488	1215
35–44	1106	663
45–54	421	279
55–64	178	112
65–74	65	42
75–84	21	15
> 84	7	6

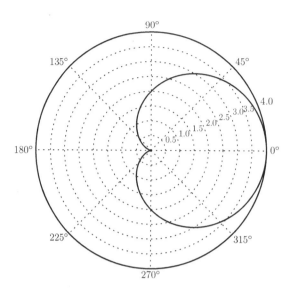

Figure 3.10 The cardioid figure formed with $a = 1$.

3.3   More Advanced Plotting

3.3.1   Polar Plots

`pyplot.plot` produces a plot on Cartesian (x, y) axes. To produce a polar plot using (r, θ) coordinates, use `pyplot.polar`, passing the arguments `theta` (which is usually the independent variable) and `r`.

Example E3.5   A cardioid is the plane figure described in polar coordinates by $r = 2a(1 + \cos\theta)$ for $0 \le \theta \le 2\pi$:

```
theta = np.linspace(0, 2.*np.pi, 1000)
a = 1.
r = 2 * a * (1. + np.cos(theta))
plt.polar(theta, r)
plt.show()
```

The polar graph plotted by this code is illustrated in Figure 3.10.

3.3.2   Histograms

A *histogram* represents the distribution of data as a series of (usually vertical) bars with lengths in proportion to the number of data items falling into predefined ranges (known as *bins*). That is, the range of data values is divided into intervals and the histogram constructed by counting the number of data values in each interval.

The pyplot function `hist` produces a histogram from a sequence of data values. The number of bins can be passed as an optional argument, `bins`; its default value is 10. Also by default the heights of the histogram bars are absolute counts of the data in the

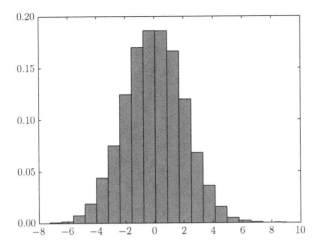

Figure 3.11 A histogram of random, normally distributed data.

corresponding bin; setting the attribute `density=True` normalizes the histogram so that its area (the height times width of each bar summed over the total number of bars) is unity.

For example, take 5000 random values from the normal distribution with mean 0 and standard deviation 2 (see Section 4.5.1):

```
>>> import matplotlib.pyplot as plt
>>> import random
>>> data = []
>>> for i in range(5000):
...     data.append(random.normalvariate(0, 2))
>>> plt.hist(data, bins=20, density=True)
>>> plt.show()
```

The resulting histogram is plotted in Figure 3.11.

3.3.3   Multiple Axes

The command `pyplot.twinx()` starts a new set of axes with the same *x*-axis as the original one, but a new *y*-scale. This is useful for plotting two or more data series, which share an abcissa (*x*-axis) but with *y* values which differ widely in magnitude or which have different units. This is illustrated in the following example.

Example E3.6   As described at https://tylervigen.com/, there is a curious but utterly meaningless correlation over time between the divorce rate in the US state of Maine and the per capita consumption of margarine in that country. The two time series here have different units and meanings and so should be plotted on separate *y*-axes, sharing a common *x*-axis (year).

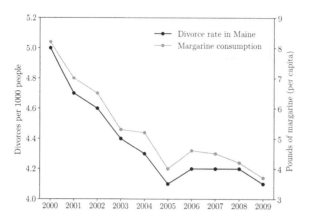

Figure 3.12 The correlation between the divorce rate in Maine and the per capita margarine consumption in the United States.

Listing 3.3 The correlation between margarine consumption in the United States and the divorce rate in Maine

```
# eg3-margarine-divorce.py
import matplotlib.pyplot as plt

years = range(2000, 2010)
divorce_rate = [5.0, 4.7, 4.6, 4.4, 4.3, 4.1, 4.2, 4.2, 4.2, 4.1]
margarine_consumption = [8.2, 7, 6.5, 5.3, 5.2, 4, 4.6, 4.5, 4.2, 3.7]
```
❶
```
line1 = plt.plot(years, divorce_rate, 'b-o',
                     label='Divorce rate in Maine')
plt.ylabel('Divorces per 1000 people')
plt.legend()

plt.twinx()
line2 = plt.plot(years, margarine_consumption, 'r-o',
                     label='Margarine consumption')
plt.ylabel('lb of Margarine (per capita)')

# Jump through some hoops to get labels in the same legend:
```
❷
```
lines = line1 + line2
labels = []
for line in lines:
```
❸
```
    labels.append(line.get_label())

plt.legend(lines, labels)
plt.show()
```

We have a bit of extra work to do in order to place a legend labeled with both lines on the plot: ❶ `pyplot.plot` returns a list of objects representing the lines that are plotted, so we save them as `line1` and `line2`, ❷ concatenate them, and then ❸ loop over them to retrieve their labels. The list of lines and labels can then be passed to `pyplot.legend` directly. The result of this code is the graph plotted in Figure 3.12.

3.3.4 Exercises

Problems

P3.3.1 A spiral may be considered to be the figure described by the motion of a point on an imaginary line as that line pivots around an origin at constant angular velocity. If the point is fixed on the line, then the figure described is a circle.

(a) If the point on the rotating line moves from the origin with constant speed, its position describes an *Archimedean spiral*. In polar coordinates, the equation of this spiral is $r = a + b\theta$. Use pyplot to plot the spiral defined by $a = 0, b = 2$ for $0 \le \theta \le 8\pi$.

(b) If the point moves along the rotating line with a velocity that increases in proportion to its distance from the origin, the result is a *logarithmic spiral*, which may be written as $r = a^\theta$. Plot the logarithmic spiral defined by $a = 0.8$ for $0 \le \theta \le 8\pi$. The logarithmic spiral has the property of *self-similarity*: with each 2π whorl, the spiral grows but maintains its shape.[3] Logarithmic spirals occur frequently in nature, from the arrangements of the chambers of nautilus shells to the shapes of galaxies.

P3.3.2 A simple model for the interaction potential between two atoms as a function of their distance, r, is that of Lennard–Jones:

$$U(r) = \frac{B}{r^{12}} - \frac{A}{r^6},$$

where A and B are positive constants.[4]

For Argon atoms, these constants may be taken to be $A = 1.024 \times 10^{-23}$ J nm^6 and $B = 1.582 \times 10^{-26}$ J nm^{12}.

(a) Plot $U(r)$. On a second y-axis on the same figure, plot the interatomic force

$$F(r) = -\frac{dU}{dr} = \frac{12B}{r^{13}} - \frac{6A}{r^7}.$$

Your plot should show the "interesting" part of these curves, which tend rapidly to very large values at small r.

Hint: life is easier if you divide A and B by Boltzmann's constant, 1.381×10^{-23} J K^{-1} so as to measure $U(r)$ in units of K. What is the depth, ϵ, and location, r_0, of the potential minimum for this system?

(b) For small displacements from the equilibrium interatomic separation (where $F = 0$), the potential may be approximated to the harmonic oscillator function,

$$V(r) = \frac{1}{2}k(r - r_0)^2 + \epsilon,$$

---

[3] The Swiss mathematician Jakob Bernoulli was so taken with this property that he coined the logarithmic spiral *Spira mirabilis*: the "miraculous sprial" and wanted one engraved on his headstone with the phrase "Eadem mutata resurgo" ("Although changed, I shall arise the same"). Unfortunately, an Archimedian spiral was engraved by mistake.

[4] This was popular in the early days of computing because r^{-12} is easy to compute as the square of r^{-6}.

where

$$k = \left| \frac{d^2 U}{dr^2} \right|_{r_0} = \frac{156B}{r_0^{14}} - \frac{42A}{r_0^8}.$$

Plot $U(r)$ and $V(r)$ on the same diagram.

P3.3.3   The seedhead of a sunflower may be modeled as follows. Number the n seeds $s = 1, 2, \ldots, n$ and place each seed a distance $r = \sqrt{s}$ from the origin, rotated $\theta = 2\pi s/\phi$ from the x axis, where ϕ is some constant. The choice nature makes for ϕ is the *golden ratio*, $\phi = (1 + \sqrt{5})/2$, which maximizes the packing efficiency of the seeds as the seedhead grows.

Write a Python program to plot a model sunflower seedhead. (*Hint*: use polar coordinates.)

4   The Core Python Language II

This chapter continues the introduction to the core Python language started in Chapter 2 with a description of Python error handling with exceptions, the data structures known as dictionaries and sets, some convenient and efficient idioms to achieve common tasks, and a survey of some of the modules provided in the Python Standard Library. Finally, a brief introduction to *object-oriented programming* with Python is presented.

4.1   Errors and Exceptions

Python distinguishes between two types of error: *syntax errors* and other *exceptions*. Syntax errors are mistakes in the grammar of the language and are checked for before the program is executed. Exceptions are *runtime* errors: conditions usually caused by attempting an invalid operation on an item of data. The distinction is that syntax errors are always fatal: there is nothing the Python compiler can do for you if your program does not conform to the grammar of the language. Exceptions, however, are conditions that arise during the running of a Python program (such as division by zero) and a mechanism exists for "catching" them and handling the condition gracefully without stopping the program's execution.

4.1.1   Syntax Errors

Syntax errors are caught by the Python compiler and produce a message indicating where the error occurred. For example,

```
>>> for lambda in range(8):
  File "<stdin>", line 1
    for lambda in range(8):
              ^
SyntaxError: invalid syntax
```

Because lambda is a reserved keyword, it cannot be used as a variable name. Its occurrence where a variable name is expected is therefore a syntax error. Similarly,

```
>>> for f in range(8:
  File "<stdin>", line 1
    for f in range(8:
                   ^
SyntaxError: invalid syntax
```

The syntax error here occurs because a single argument to the range built-in must be given as an integer between parentheses: the colon breaks the syntax of calling functions and so Python complains of a syntax error.

Because a line of Python code may be split within an open bracket ("()", "[]", or "{}"), a statement split over several lines can sometimes cause a SyntaxError to be indicated somewhere other than the location of the true bug. For example,

```
>>> a = [1, 2, 3, 4,
... b = 5
  File "<stdin>", line 4
    b = 5
      ^
SyntaxError: invalid syntax
```

Here, the statement b = 5 is syntactically valid: the error arises from failing to close the square bracket of the previous list declaration (the Python shell indicates that a line is a continuation of a previous one with the initial ellipsis ("...")).

There are two special types of SyntaxError that are worth mentioning: an IndentationError occurs when a block of code is improperly indented and TabError is raised when a tabs and spaces are mixed inconsistently to provide indentation.[1]

Example E4.1   A common syntax error experienced by beginner Python programmers is in using the assignment operator "=" instead of the equality operator "==" in a conditional expression:

```
>>> if a = 5:
  File "<stdin>", line 1
    if a = 5:
         ^
SyntaxError: invalid syntax
```

This assignment a = 5 does not return a value (it simply assigns the integer object 5 to the variable name a) and so there is nothing corresponding to True or False that the if statement can use: hence the SyntaxError. This contrasts with the C language in which an assignment returns the value of the variable being assigned (and so the statement a = 5 evaluates to True). This behavior is the source of many hard-to-find bugs and security vulnerabilities and its omission from the Python language is by design.

4.1.2   Exceptions

An exception occurs when a syntactically correct expression is executed and causes a *runtime error*. There are different types of built-in exception, and custom exceptions can be defined by the programmer if required. If an exception is not "caught" using the try ... except clause described later, Python produces a (usually helpful) error message. If the exception occurs within a function (which may have been called, in turn, by another function, and so on), the message returned takes the form of a *stack traceback*:

---

[1] This error can be avoided by using only spaces to indent code.

the history of function calls leading to the error is reported so that its location in the program execution can be determined.

Some built-in exceptions will be familiar from your use of Python so far.

NameError

```
>>> print('4z = ', 4*z)

Traceback (most recent call last):
  File "<stdin>", line 1, in <module>
NameError: name 'z' is not defined
```

A NameError exception occurs when a variable name is used that hasn't been defined: the print statement here is valid, but Python doesn't know what the identifier z refers to.

ZeroDivisionError

```
>>> a, b = 0, 5
>>> b / a

Traceback (most recent call last):
  File "<stdin>", line 1, in <module>
ZeroDivisionError: float division by zero
```

Division by zero is not mathematically defined.

TypeError and ValueError

A TypeError is raised if an object of the wrong type is used in an expression or function. For example,

```
>>> '00' + 7

Traceback (most recent call last):
  File "<stdin>", line 1, in <module>
TypeError: Can't convert 'int' object to str implicitly
```

Python is a (fairly) strongly typed language, and it is not possible to add a string to an integer.[2]

A ValueError, on the other hand, occurs when the object involved has the correct *type* but an invalid *value*:

```
>>> float('hello')

Traceback (most recent call last):
  File "<stdin>", line 1, in <module>
ValueError: could not convert string to float: 'hello'
```

The float built-in does take a string as its argument, so float('hello') is not a TypeError: the exception is raised because the particular string 'hello' does not evaluate to a meaningful floating-point number. More subtly,

---

[2] Unlike in, say, Javascript or PHP, where it seems anything goes.

Table 4.1 Common Python exceptions

Exception	Cause and description
FileNotFoundError	Attempting to open a file or directory that does not exist – this exception is a particular type of OSError.
IndexError	Indexing a sequence (such as a list or string) with a subscript that is out of range.
KeyError	Indexing a dictionary with a key that does not exist in that dictionary (see Section 4.2.2).
NameError	Referencing a local or global variable name that has not been defined.
TypeError	Attempting to use an object of an inappropriate type as an argument to a built-in operation or function.
ValueError	Attempting to use an object of the correct type but with an incompatible value as an argument to a built-in operation or function.
ZeroDivisionError	Attempting to divide by zero (either explicitly (using "/" or "//") or as part of a modulo operation "%").
SystemExit	Raised by the sys.exit function (see Section 4.4.1) – if not handled, this function causes the Python interpreter to exit.

```
>>> int('7.0')

Traceback (most recent call last):
  File "<stdin>", line 1, in <module>
ValueError: invalid literal for int() with base 10: '7.0'
```

A string that looks like a float cannot be directly cast to int: to obtain the result probably intended, use int(float('7.0')).

Table 4.1 provides a list of the more commonly encountered built-in exceptions and their descriptions.

Example E4.2 When an exception is raised but not *handled* (see Section 4.1.3), Python will issue a *traceback report* indicating where in the program flow it occurred. This is particularly useful when an error occurs within nested functions or within imported modules. For example, consider the following short program:[3]

```
# exception-test.py
import math

def func(x):
    def trig(x):
        for f in (math.sin, math.cos, math.tan):
            print('{f}({x}) = {res}'.format(f=f.__name__, x=x, res=f(x)))
    def invtrig(x):
        for f in (math.asin, math.acos, math.atan):
```

---

[3] Note the use of f.__name__ to return a string representation of a function's name in this program; for example, math.sin.__name__ is 'sin'.

❶   `print('{f}({x}) = {res}'.format(f=f.__name__, x=x, res=f(x)))`
      `trig(x)`
❷   `invtrig(x)`

❸ `func(1.2)`

The function `func` passes its argument, `x`, to its two nested functions. The first, `trig`, is unproblematic but the second, `invtrig`, is expected to fail for `x` out of the domain (range of acceptable values) for the inverse trigonometric function, `asin`:

```
sin(1.2) = 0.9320390859672263
cos(1.2) = 0.3623577544766736
tan(1.2) = 2.5721516221263183
Traceback (most recent call last):
  File "exception-test.py", line 14, in <module>
    func(1.2)
  File "exception-test.py", line 12, in func
    invtrig(x)
  File "exception-test.py", line 10, in invtrig
    print('{f}({x}) = {res}'.format(f=f.__name__, x=x, res=f(x)))
ValueError: math domain error
```

Following the traceback backward shows that the `ValueError` exception was raised within `invtrig` (line 10, ❶), which was called from within `func` (line 12, ❷), which was itself called by the `exception-test.py` module (i.e. program) at line 14, ❸.

4.1.3 Handling and Raising Exceptions

Handling Exceptions

Often, a program must manipulate data in a way which might cause an exception to be raised. Assuming such a condition is not to cause the program to exit with an error but to be handled "gracefully" in some sense (an invalid data point ignored, division by a zero value skipped, and so on), there are two approaches to this situation: check the value of the data object before using it, or "handle" any exception that is raised before resuming execution. The Pythonic approach is the latter, summed up in the expression *"It is Easier to Ask Forgiveness than to seek Permission"* (EAFP).

To catch an exception in a block of code, write the code within a `try:` clause and handle any exceptions raised in an `except:` clause. For example,

```
try:
    y = 1 / x
    print('1 /', x, ' = ',y)
except ZeroDivisionError:
    print('1 / 0 is not defined.')
# ... more statements
```

No check is required: we go ahead and calculate `1/x` and handle the error arising from division by zero if necessary. The program execution continues after the `except` block whether the `ZeroDivisionError` exception was raised or not. If a different exception is raised (e.g. a `NameError` because `x` is not defined), then this will not be caught – it is an *unhandled exception* and will trigger an error message.

To handle more than one exception in a single except block, list them in a tuple (which must be within brackets).

```python
try:
    y = 1. / x
    print('1 /', x, ' = ',y)
except (ZeroDivisionError, NameError):
    print('x is zero or undefined!')
# ... more statements
```

To handle each exception separately, use more than one except clause:

```python
try:
    y = 1. / x
    print('1 /', x, ' = ',y)
except ZeroDivisionError:
    print('1 / 0 is not defined.')
except NameError:
    print('x is not defined')
# ... more statements
```

Warning: You may come across the following type of construction:

```python
try:
    [do something]
except:                  # Don't do this!
    pass
```

This will execute the statements in the try block and ignore *any* exceptions raised – in general, it is very unwise to do this as it makes code very hard to maintain and debug (errors, whatever their cause, are silently supressed). Aim to catch specific exceptions and handle them appropriately, allowing any other exceptions to "bubble up" to be handled (or not) by any other except clauses.

The try ... except statement has two more optional clauses (which must follow any except clauses if they are used). Statements in a block following the finally keyword are *always* executed, whether an exception was raised or not. Statements in a block following the else keyword are executed if an exception was *not* raised (see Example E4.5).

◇ **Raising Exceptions**

Usually an exception is raised by the Python interpreter as a result of some behavior (anticipated or not) by the program. But sometimes it is desirable for a program to raise a particular exception if some condition is met. The raise keyword allows a program to force a specific exception and customize the message or other data associated with it. For example,

```python
if n % 2:
    raise ValueError('n must be even!')
# Statements here may proceed, knowing n is even ...
```

A related keyword, assert, evaluates a conditional expression and raises an AssertionError exception if that expression is not equivalent to True. assert statements can be useful to check that some essential condition holds at a specific point in your program's execution and are often helpful in debugging.

```
>>> assert 2 == 2          # [silence]: 2 == 2 is True so nothing happens
>>>
>>> assert 1 == 2          # will raise the AssertionError

Traceback (most recent call last):
  File "<stdin>", line 1, in <module>
AssertionError
```

The syntax assert *expr1, expr2* passes *expr2* (typically an error message) to the AssertionError:

```
>>> assert 1 == 2, 'One does not equal two'

Traceback (most recent call last):
  File "<stdin>", line 1, in <module>
AssertionError: One does not equal two
```

Python is a dynamically typed language and arguments of any type can be legally passed to a function, even if that function is expecting a particular type. It is sometimes necessary to check that an argument object is of a suitable type before using it, and assert could be used to do this.

Example E4.3 The following function returns a string representation of a two-dimensional (2D) or three-dimensional (3D) vector, which must be represented as a list or tuple containing two or three items.

```
>>> def str_vector(v):
...     assert type(v) is list or type(v) is tuple,\
...                 'argument to str_vector must be a list or tuple'
...     assert len(v) in (2, 3),\
...                 'vector must be 2D or 3D in str_vector'
...     unit_vectors = ['i', 'j', 'k']
...     s = []
...     for i, component in enumerate(v):
...         s.append('{}{}'.format(component, unit_vectors[i]))
❶   ...     return '+'.join(s).replace('+-', '-')
```

❶ replace('+-', '-') here converts, for example, '4i+-3j' into '4i-3j'.

Example E4.4 As another example, suppose you have a function that calculates the vector (cross) product of two vectors represented as list objects. This product is only defined for three-dimensional vectors, so calling it with lists of any other length is an error.

```
>>> def cross_product(a, b):
...     assert len(a) == len(b) == 3, 'Vectors a, b must be three-dimensional'
...     return [a[1]*b[2] - a[2]*b[1],
...             a[2]*b[0] - a[0]*b[2],
...             a[0]*b[1] - a[1]*b[0]]
...
>>> cross_product([1, 2, -1], [2, 0, -1, 3])    # Oops!
```

```
Traceback (most recent call last):
  File "<stdin>", line 1, in <module>
  File "<stdin>", line 2, in cross_product
AssertionError: Vectors a, b must be three-dimensional

>>> cross_product([1, 2, -1], [2, 0, -1])
[-2, -1, -4]
```

Example E4.5 The following code gives an example of the use of a try ... except ... else ... finally clause:

```
# try-except-else-finally.py

def process_file(filename):
    try:
        fi = open(filename, 'r')
    except IOError:
        print('Oops: couldn\'t open {} for reading'.format(filename))
        return
    else:
        lines = fi.readlines()
        print('{} has {} lines.'.format(filename, len(lines)))
        fi.close()
    finally:
        print('   Done with file {}'.format(filename))

    print('The first line of {} is:\n{}'.format(filename, lines[0]))
    # further processing of the lines ...
    return

process_file('sonnet0.txt')
process_file('sonnet18.txt')
```

❶ Within the else block, the contents of the file are only read if the file was successfully opened.

❷ Within the finally block, 'Done with file *filename*' is printed whether the file was successfully opened or not.

Assuming that the file sonnet0.txt does not exist but that sonnet18.txt does, running this program prints:

```
Oops: couldn't open sonnet0.txt for reading
   Done with file sonnet0.txt
sonnet18.txt has 14 lines.
   Done with file sonnet18.txt
The first line of sonnet18.txt is:
Shall I compare thee to a summer's day?
```

4.1.4 Exercises

Questions

Q4.1.1 What is the point of `else`? Why not put statements in this block inside the original `try` block?

Q4.1.2 What is the point of the `finally` clause? Why not put any statements you want executed after the `try` block (regardless of whether or not an exception has been raised) after the entire `try` ... `except` clause?

Hint: see what happens if you modify Example E4.5 to put the statements in the `finally` clause after the `try` block.

Problems

P4.1.1 Write a program to read in the data from the file `swallow-speeds.txt` (available at https://scipython.com/ex/bda) and use it to calculate the average air-speed velocity of an (unladen) African swallow. Use exceptions to handle the processing of lines that do not contain valid data points.

P4.1.2 Adapt the function of Example E4.3, which returns a vector in the following form:

```
>>> print(str_vector([-2, 3.5]))
-2i + 3.5j
>>> print(str_vector((4, 0.5, -2)))
4i + 0.5j - 2k
```

to raise an exception if any element in the vector array does not represent a real number.

P4.1.3 Python follows the convention of many computer languages in choosing to define $0^0 = 1$. Write a function, `powr(a, b)`, which behaves the same as the Python expression `a**b` (or, for that matter, `math.pow(a,b)`) but raises a `ValueError` if a and b are both zero.

4.2 Python Objects III: Dictionaries and Sets

A *dictionary* in Python is a type of "associative array" (also known as a "hash" in some languages). A dictionary can contain any objects as its *values*, but unlike sequences such as lists and tuples, in which the items are indexed by an integer starting at 0, each item in a dictionary is indexed by a unique *key*, which may be any *immutable* object.[4] The dictionary therefore exists as a collection of *key-value* pairs; dictionaries themselves are *mutable* objects.

---

[4] Actually, dictionary keys can be any *hashable* object: a hashable object in Python is one with a special method for generating a particular integer from any instance of that object; the idea is that instances (which may be large and complex) that compare as equal should have hash numbers that also compare as equal so they can be rapidly looked up in a *hash table*. This is important for some data structures and for optimizing the speed of algorithms involving their objects.

4.2.1   Defining and Indexing a Dictionary

An dictionary can be defined by giving *key*: *value* pairs between braces:

```
>>> height = {'Burj Khalifa': 828., 'One World Trade Center': 541.3,
              'Mercury City Tower': -1., 'Q1': 323.,
              'Carlton Centre': 223., 'Gran Torre Santiago': 300.,
              'Mercury City Tower': 339.}
>>> height
{'Burj Khalifa': 828.0,
 'One World Trade Center': 541.3,
 'Mercury City Tower': 339.0,
 'Q1': 323.0,
 'Carlton Centre': 223.0,
 'Gran Torre Santiago': 300.0}
```

The command print(height) will return the dictionary in the same format (between braces). If the same key is attached to different values (as 'Mercury City Tower' is here), only the most recent value survives: the keys in a dictionary are unique.

Before Python 3.6, the items in a dictionary were not guaranteed to have any particular order; since this version, the *order of insertion* is preserved. Note that as in the example above, redefining the value attached to a key does not change the key's insertion order: the key 'Mercury City Tower' is the third key to be defined, where it is given the value -1.; it is later reassigned the value 339. but still appears in third position when the dictionary is used.

An individual item can be retrieved by indexing it with its key, either as a literal ('Q1') or with a variable equal to the key:

```
>>> height['One World Trade Center']
541.3
>>> building = 'Carlton Centre'
>>> height[building]
223.0
```

Items in a dictionary can also be *assigned* by indexing it in this way:

```
height['Empire State Building'] = 381.
height['The Shard'] = 306.
```

An alternative way of defining a dictionary is to pass a sequence of (*key*, *value*) pairs to the dict constructor. If the keys are simple strings (of the sort that could be used as variable names), the pairs can also be specified as keyword arguments to this constructor:

```
>>> ordinal = dict([(1, 'First'), (2, 'Second'), (3, 'Third')])
>>> mass = dict(Mercury=3.301e23, Venus=4.867e24, Earth=5.972e24)
>>> ordinal[2]      # NB 2 here is a key, not an index
'Second'
>>> mass['Earth']
5.972e+24
```

A for-loop iteration over a dictionary returns the dictionary *keys* (in order of key insertion):

```
>>> for c in ordinal:
```

```
...      print(c, ordinal[c])
...
1 First
2 Second
3 Third
```

Example E4.6   A simple dictionary of roman numerals:

```
>>> numerals = {'one':'I', 'two':'II', 'three':'III', 'four':'IV', 'five':'V',
                'six':'VI', 'seven':'VII', 'eight':'VIII',
                1: 'I', 2: 'II', 3: 'III', 4:'IV', 5: 'V', 6:'VI', 7:'VII',
                8:'VIII'}
>>> for i in ['three', 'four', 'five', 'six']:
...      print(numerals[i], end=' ')
...
III IV V VI
>>> for i in range(8,0,-1):
...      print(numerals[i], end=' ')
VIII VII VI V IV III II I
```

Note that regardless of the order in which the keys are stored, the dictionary can be indexed in any order. Note also that although the dictionary *keys* must be unique, the dictionary *values* need not be.

4.2.2   Dictionary Methods

get()
Indexing a dictionary with a key that does not exist is an error:

```
>>> mass['Pluto']
Traceback (most recent call last):
  File "<stdin>", line 1, in <module>
KeyError: 'Pluto'
```

However, the useful method get() can be used to retrieve the value, given a key if it exists, or some default value if it does not. If no default is specified, then None is returned. For example,

```
>>> print(mass.get('Pluto'))
None
>>> mass.get('Pluto', -1)
-1
```

keys, values and items
The three methods, keys, values and items, return, respectively, a dictionary's keys, values and key-value pairs (as tuples). In previous versions of Python, each of these were returned in a list, but for most purposes this is wasteful of memory: calling keys, for example, required all of the dictionary's keys to be copied as a list, which in most cases was simply iterated over. That is, storing a whole new copy of the dictionary's keys is not usually necessary. Python 3 solves this by returning an iterable object, which

accesses the dictionary's keys one by one, without copying them to a list. This is faster and saves memory (important for very large dictionaries). For example,

```
>>> planets = mass.keys()
>>> print(planets)
dict_keys(['Mercury', 'Venus', 'Earth'])
>>> for planet in planets:
...      print(planet, mass[planet])
...
Mercury 3.301e+23
Venus 4.867e+24
Earth 5.972e+24
```

A dict_keys object can be iterated over any number of times, but it is not a list and cannot be indexed or assigned:

```
>>> planets = mass.keys()
>>> planets[0]

Traceback (most recent call last):
  File "<stdin>", line 1, in <module>
TypeError: 'dict_keys' object is not subscriptable
```

If you really do want a list of the dictionary's keys, simply pass the dict_keys object to the list constructor (which takes any kind of sequence and makes a list out of it):

```
>>> planet_list = list(mass.keys())
>>> planet_list
['Mercury', 'Venus', 'Earth']
>>> planet_list[0]
'Mercury'
❶ >>> planet_list[1] = 'Jupiter'
>>> planet_list
['Mercury', 'Jupiter', 'Earth']
```

❶ This last assignment only changes the planet_list list; it doesn't alter the original dictionary's keys.

Similar methods exist for retrieving a dictionary's values and items (key-value pairs): the objects returned are dict_values and dict_items.

For example,

```
>>> mass.items()
dict_items([('Mercury', 3.301e+23), ('Venus', 4.867e+24), ('Earth', 5.972e+24)])
>>> mass.values()
dict_values([3.301e+23, 4.867e+24, 5.972e+24])
>>> for planet_data in mass.items():
...      print(planet_data)
...
('Mercury', 3.301e+23)
('Venus', 4.867e+24)
('Earth', 5.972e+24)
```

Example E4.7   A Python dictionary can be used as a simple database. The following code stores some information about some astronomical objects in a dictionary of tuples, keyed by the object name, and manipulates them to produce a list of planet densities.

Listing 4.1 Astronomical data

```
# eg4-astrodict.py
import math

# Mass (in kg) and radius (in km) for some astronomical bodies.
body = {'Sun': (1.988e30, 6.955e5),
        'Mercury': (3.301e23, 2440.),
        'Venus': (4.867e+24, 6052.),
        'Earth': (5.972e24, 6371.),
        'Mars': (6.417e23, 3390.),
        'Jupiter': (1.899e27, 69911.),
        'Saturn': (5.685e26, 58232.),
        'Uranus': (8.682e25, 25362.),
        'Neptune': (1.024e26, 24622.)
        }

planets = list(body.keys())
# The sun isn't a planet!
planets.remove('Sun')

def calc_density(m, r):
    """ Returns the density of a sphere with mass m and radius r. """
    return m / (4/3 * math.pi * r**3)

rho = {}
for planet in planets:
    m, r = body[planet]
    # Calculate the density in g/cm3.
    rho[planet] = calc_density(m*1000, r*1.e5)
```
❶
```
for planet, density in sorted(rho.items()):
    print('The density of {0} is {1:3.2f} g/cm3'.format(planet, density))
```

❶ `sorted(rho.items())` returns a `list` of the rho dictionary's key-value pairs, sorted by key. The keys are strings so in this case the sorting produces a list of the keys in alphabetical order.

The output is

```
The density of Earth is 5.51 g/cm3
The density of Jupiter is 1.33 g/cm3
The density of Mars is 3.93 g/cm3
The density of Mercury is 5.42 g/cm3
The density of Neptune is 1.64 g/cm3
The density of Saturn is 0.69 g/cm3
The density of Uranus is 1.27 g/cm3
The density of Venus is 5.24 g/cm3
```

◇ **Keyword Arguments**

In Section 2.7, we discussed the syntax for passing arguments to functions. In that description, it was assumed that the function would always know what arguments could be passed to it and these were listed in the function definition. For example,

```
def func(a, b, c):
```

Python provides a couple of useful features for handling the case where it is not necessarily known what arguments a function will receive. Including *args (after any "formally defined" arguments) places any additional positional argument into a tuple, args, as illustrated by the following code:

```
>>> def func(a, b, *args):
...     print(args)
...
>>> func(1, 2, 3, 4, 'msg')
(3, 4, 'msg')
```

That is, inside func, in addition to the formal arguments a=1 and b=2, the arguments 3, 4 and 'msg' are available as the items of the tuple args. This tuple can be arbitrarily long. Python's own print built-in function works in this way: it takes an arbitrary number of arguments to output as a string, followed by some optional keyword arguments:

```
def print(*args, sep=' ', end='\n', file=None):
```

It is also possible to collect arbitrary *keyword arguments* (see Section 2.7.2) to a function inside a dictionary by using the **kwargs syntax in the function definition. Python takes any keyword arguments not specified in the function definition and packs them into the dictionary kwargs. For example,

```
>>> def func(a, b, **kwargs):
...     for k in kwargs:
...         print(k, '=', kwargs[k])
...
>>> func(1, b=2, c=3, d=4, s='msg')
d = 4
s = msg
c = 3
```

One can also use *args and **kwargs when *calling* a function, which can be convenient, for example, with functions that take a large number of arguments:

```
>>> def func(a, b, c, x, y, z):
...     print(a, b, c)
...     print(x, y, z)
...
>>> args = [1, 2, 3]
>>> kwargs = {'x': 4, 'y': 5, 'z': 'msg'}
>>> func(*args, **kwargs)
1 2 3
4 5 msg
```

◇ defaultdict

With regular Python dictionaries, an attempt to retrieve a value using a key that does not exist will raise a KeyError exception. There is a useful container, called defaultdict, that subclasses the dict built-in to allow one to specify default_factory, a function which returns the default value to be assigned to the key if it is missing.

Example E4.8   To analyze the word lengths in the first line of the Gettysburg Address with a regular dictionary requires code to catch the KeyError and set a default value:

```
text = 'Four score and seven years ago our fathers brought forth on this
continent, a new nation, conceived in Liberty, and dedicated to the proposition
that all men are created equal'

text = text.replace(',', '').lower()        # remove punctuation

word_lengths = {}
for word in text.split():
    try:
        word_lengths[len(word)] += 1
    except KeyError:
        word_lengths[len(word)] = 1
print(word_lengths)
```

Using `defaultdict` in this case would be more concise and elegant:

```
❶ from collections import defaultdict
❷ word_lengths = defaultdict(int)
for word in text.split():
    word_lengths[len(word)] += 1
print(word_lengths)
```

returns:

```
defaultdict(<class 'int'>, {4: 3, 5: 5, 3: 9, 7: 4, 2: 3, 9: 3, 1: 1, 6: 1, 11: 1})
```

❶ Note that `defaultdict` is not a built-in: it must be imported from the `collections` module.

❷ Here we set the `default_factory` function to int: if a key is missing, it will be inserted into the dictionary and initialized with a call to `int()`, which returns 0.

4.2.3   Sets

A set is an *unordered* collection of *unique* items. As with dictionary keys, elements of a set must be hashable objects. A set is useful for removing duplicates from a sequence and for determining the union, intersection and difference between two collections. Because they are unordered, set objects cannot be indexed or sliced, but they can be iterated over, tested for membership and they support the `len` built-in. A set is created by listing its elements between braces (`{...}`) or by passing an iterable to the `set()` constructor:

```
>>> s = set([1, 1, 4, 3, 2, 2, 3, 4, 1, 3, 'surprise!'])
>>> s
{1, 2, 'surprise!', 3, 4}
>>> len(s)                    # cardinality of the set
5
>>> 2 in s, 6 not in s        # membership, nonmembership
(True, True)
>>> for item in s:
...     print(item)
...
1
2
surprise!
```

3
4

The set method add is used to add elements to the set. To remove elements there are several methods: remove removes a specified element but raises a KeyError exception if the element is not present in the set; discard() does the same but does not raise an error in this case. Both methods take (as a single argument) the element to be removed. pop (with no argument) removes an *arbitrary* element from the set and clear removes *all* elements from the set:

```
>>> s = {2,-2,0}
>>> s.add(1)
>>> s.add(-1)
❶ >>> s.add(1.0)
>>> s
{0, 1, 2, -1, -2}
>>> s.remove(1)
>>> s
{0, 2, -1, -2}
>>> s.discard(3)      # OK - does nothing
>>> s
{0, 2, -1, -2}
>>> s.pop()
0                     # (for example)
>>> s
{2, -1, -2}
>>> s.clear()
set()                 # the empty set
```

❶ This statement will not add a new member to the set, even though the existing 1 is an integer and the item we're adding is a float. The test 1 == 1.0 is True, so 1.0 is considered to be already in the set.

set objects have a wide range of methods corresponding to the properties of mathematical sets; the most useful are illustrated in Table 4.2, which uses the following terms from set theory:

- The *cardinality* of a set, $|A|$, is the number of elements it contains.
- Two sets are *equal* if they both contain the same elements.
- Set A is a *subset* of set B ($A \subseteq B$) if all the elements of A are also elements of B; set B is said to be a *superset* of set A.
- Set A is a *proper subset* of B ($A \subset B$) if it is a subset of B but not equal to B; in this case, set B is said to be a *proper superset* of A.
- The *union* of two sets ($A \cup B$) is the set of all elements from both of them.
- The *intersection* of two sets ($A \cap B$) is the set of all elements they have in common.
- The *difference* of set A and set B ($A \setminus B$) is the set of elements in A that are not in B.
- The *symmetric difference* of two sets, $A \triangle B$, is the set of elements in either but not in both.
- Two sets are said to be *disjoint* if they have no elements in common.

Table 4.2 set methods

Method	Description
isdisjoint(*other*)	Is *set* disjoint with *other*?
issubset(*other*), *set* <= *other*	Is *set* a subset of *other*?
set < *other*	Is *set* a proper subset of *other*?
issuperset(*other*), *set* >= *other*	Is *set* a superset of *other*?
set > *other*	Is *set* a proper superset of *other*?
union(*other*), *set* | *other* | ...	The union of *set* and *other*(s)
intersection(*other*), *set* & *other* & ...	The intersection of *set* and *other*(s)
difference(*other*), *set* - *other* - ...	The difference of *set* and *other*(s)
symmetric_difference(*other*), *set* ^ *other* ^ ...	The symmetric difference of *set* and *other*(s)

There are two forms for most set expressions: the operator-like syntax requires all arguments to be set objects, whereas explicit method calls will convert any iterable argument into a set.

```
>>> A = set((1, 2, 3))
>>> B = set((1, 2, 3, 4))
>>> A <= B
True
>>> A.issubset((1, 2, 3, 4))  # OK: (1, 2, 3, 4) is turned into a set
True
```

Some more examples:

```
>>> C, D = set((3, 4, 5, 6)), set((7, 8, 9))
>>> B | C                # union
{1, 2, 3, 4, 5, 6}
>>> A | C | D            # union of three sets
{1, 2, 3, 4, 5, 6, 7, 8, 9}
>>> A & C                # intersection
{3}
>>> C & D
set()                    # the empty set
>>> C.isdisjoint(D)
True
>>> B - C                # difference
{1, 2}
>>> B ^ C                # symmetric difference
{1, 2, 5, 6}
```

◇ **frozensets**

sets are mutable objects (items can be added to and removed from a set); because of this they are *unhashable* and so cannot be used as dictionary keys or as members of other sets.

```
>>> a = set((1, 2, 3))
>>> b = set(('q', (1, 2), a))

Traceback (most recent call last):
  File "<stdin>", line 1, in <module>
TypeError: unhashable type: 'set'
>>>
```

(In the same way, lists cannot be dictionary keys or set members.) There is, however, a frozenset object which is a kind of immutable (and hashable) set.[5] frozensets are fixed, unordered collections of unique objects and *can* be used as dictionary keys and set members.

```
>>> a = frozenset((1, 2, 3))
>>> b = set(('q', (1, 2), a))       # OK: the frozenset a is hashable
>>> b.add(4)                        # OK: b is a regular set
>>> a.add(4)                        # not OK: frozensets are immutable

Traceback (most recent call last):
  File "<stdin>", line 1, in <module>
AttributeError: 'frozenset' object has no attribute 'add'
```

Example E4.9 A *Mersenne prime*, M_i, is a prime number of the form $M_i = 2^i - 1$. The set of Mersenne primes less than n may be thought of as the intersection of the set of all primes less than n, P_n, with the set, A_n, of integers satisfying $2^i - 1 < n$.

The following program returns a list of the Mersenne primes less than 1 000 000.

Listing 4.2 The Mersenne primes

```
import math

def primes(n):
    """ Return a list of the prime numbers <= n. """

    sieve = [True] * (n // 2)
    for i in range(3, int(math.sqrt(n)) + 1, 2):
        if sieve[i//2]:
            sieve[i*i//2::i] = [False] * ((n - i*i - 1) // (2*i) + 1)
    return [2] + [2*i+1 for i in range(1, n // 2) if sieve[i]]

n = 1000000

❶ P = set(primes(n))


# A list of integers 2^i - 1 <= n.
```

---

[5] In a sense, they are to sets what tuples are to lists.

```
    A = []
❷   for i in range(2, int(math.log(n+1, 2)) + 1):
        A.append(2**i - 1)

    # The set of Mersenne primes as the intersection of P and A.
    M = P.intersection(A)

    # Output as a sorted list of M.
❸   print(sorted(list(M)))
```

The prime numbers are produced in a list by the function `primes`, which implements an optimized version of the Sieve of Eratosthenes algorithm (see Exercise P2.5.8); this is converted into the set, `P` (❶). We can take the intersection of this set with any iterable object using the `intersection` method, so there is no need to explicitly convert our second list of integers, `A`, (❷) into a `set`.

❸ Finally, the set of Mersenne primes we create, `M`, is an unordered collection, so for output purposes we convert it into a sorted `list`.

For $n = 1\,000\,000$, This output is

```
[3, 7, 31, 127, 8191, 131071, 524287]
```

4.2.4   Exercises

Questions

Q4.2.1   Write a one-line Python program to determine if a string is a *pangram* (a string that contains each letter of the alphabet at least once).

Q4.2.2   Write a function, using `set` objects, to remove duplicates from an ordered `list`. For example,

```
>>> remove_dupes([1, 1, 2, 3, 4, 4, 4, 5, 7, 8, 8, 9])
[1, 2, 3, 4, 5, 7, 8, 9]
```

Q4.2.3   Predict and explain the effect of the following statements:

```
>>> set('hellohellohello')
>>> set(['hellohellohello'])
>>> set(('hellohellohello'))
>>> set(('hellohellohello',))
>>> set(('hello', 'hello', 'hello'))
>>> set(('hello', ('hello', 'hello')))
>>> set(('hello', ['hello', 'hello']))
```

Q4.2.4   If `frozenset` objects are *immutable*, how is this possible?

```
>>> a = frozenset((1, 2, 3))
>>> a |= {2, 3, 4, 5}
>>> print(a)
frozenset([1, 2, 3, 4, 5])
```

Table 4.3 Resistor color codes

Color	Abbreviation	Significant figures	Multiplier	Tolerance
Black	bk	0	1	–
Brown	br	1	10	±1%
Red	rd	2	10^2	±2%
Orange	or	3	10^3	–
Yellow	yl	4	10^4	±5%
Green	gr	5	10^5	±0.5%
Blue	bl	6	10^6	±0.25%
Violet	vi	7	10^7	±0.1%
Gray	gy	8	10^8	±0.05%
White	wh	9	10^9	–
Gold	au	–	–	±5%
Silver	ag	–	–	±10%
None	--	–	–	±20%

Q4.2.5   Modify Example E4.8 to use a `defaultdict` to produce a list of words, keyed by their length from the text of the first line of the Gettysburg Address.

Problems

P4.2.1   The values and tolerances of older resistors are identified by four colored bands: the first two indicate the first two significant figures of the resistance in ohms, the third denotes a decimal multiplier (number of zeros) and the fourth indicates the tolerance. The colors and their meanings for each band are listed in Table 4.3.

For example, a resistor with colored bands violet, yellow, red, green has value $74 \times 10^2 = 7400\ \Omega$ and tolerance ±0.5%.

Write a program that defines a function to translate a list of four color abbreviations into a resistance value and a tolerance. For example,

```
In [x]: print(get_resistor_value(['vi', 'yl', 'rd', 'gr']))
Out[x]: (7400, 0.5)
```

P4.2.2   The novel *Moby-Dick* is out of copyright and can be downloaded as a text file from the Project Gutenberg website at www.gutenberg.org/2/7/0/2701/. Write a program to output the 100 words most frequently used in the book by storing a count of each word encountered in a dictionary.

Hint: use Python's string methods to strip out any punctuation. It suffices to replace any instances of the following characters with the empty string: `!?":;,()'.*[]`. When you have a dictionary with words as the keys and the corresponding word counts as the values, create a list of (*count*, *word*) tuples and sort it.

Bonus exercise: compare the frequencies of the top 2000 words in *Moby-Dick* with the prediction of *Zipf's law*:

$$\log f(w) = \log C - a \log r(w),$$

where $f(w)$ is the number of occurrences of word w, $r(w)$ is the corresponding *rank* (1 = most common, 2 = second most common, etc.) and C and a are constants. In the traditional formulation of the law, $C = \log f(w_1)$ and $a = 1$, where w_1 is the most common word, such that $r(w_1) = 1$.

P4.2.3 *Reverse Polish notation* (RPN) (or *postfix* notation) is a notation for mathematical expressions in which each operator follows all of its operands (in contrast to the more familiar *infix* notation, in which the operator appears *between* the operands it acts on). For example, the infix expression 5 + 6 is written in RPN as 5 6 +. The advantage of this approach is that parentheses are not necessary: to evaluate (3 + 7) / 2, it may be written as 3 7 + 2 /. An RPN expression is evaluated left to right with the intermediate values pushed onto a *stack* – a last-in, first-out list of values – and retrieved (popped) from the stack when needed by an operator (see also Example E2.16). Thus, the expression 3 7 + 2 / proceeds with 3 and then 7 pushed to the stack (with 7 on top). The next token is +, so the values are retrieved, added, and the result, 10, pushed onto the (now empty) stack. Next, 2 is pushed to the stack. The final token / pops the two values, 10 and 2 from the stack, and divides them to give the result, 5.

Write a program to evaluate an RPN expression consisting of space-delimited tokens (the operators + - * / ** and numbers).

Hint: parse the expression into a list of string tokens and iterate over it, converting and pushing the numbers to the stack (which may be implemented by appending to a list). Define functions to carry out the operations by retrieving values from the stack with pop. Note that Python does not provide a switch...case syntax, but these function objects can be the values in a dictionary with the operator tokens as the keys.

P4.2.4 Use the dictionary of Morse code symbols in the file morse.py, available from https://scipython.com/ex/bdb, to write a program that can translate a message to and from Morse code, using spaces to delimit individual Morse code "letters" and slashes ("/") to delimit words. For example, 'PYTHON 3' becomes '.-. -.- - .... - -. / ...-'

P4.2.5 The file shark-species.txt, available at https://scipython.com/ex/bdc, contains a list of extant shark species arranged in a hierachy by order, family, genus and species (with the species given as *binomial name : common name*). Read the file into a data structure of nested dictionaries, which can be accessed as follows:

```
>>> sharks['Lamniformes']['Lamnidae']['Carcharodon']['C. carcharias']
Great white shark
```

4.3 Pythonic Idioms: "Syntactic Sugar"

Many computer languages provide syntax to make common tasks easier and clearer to code. Such *syntactic sugar* consists of constructs that could be removed from the language without affecting the language's functionality. We have already seen one example in so-called *augmented assignment*: a += 1 is equivalent to a = a + 1. Another example is negative indexing of sequences: b[-1] is equivalent to and more convenient than b[len(b)-1].

4.3.1   Comparison and Assignment Shortcuts

If more than one variable is to be assigned to the same object, the shortcut

```
x = y = z = -1
```

may be used. Note that if *mutable* objects are assigned this way, the variable names will all refer to the same object, not to distinct copies of it (recall Section 2.4.1).

Similarly, as was shown in Section 2.4.2, multiple assignments to different objects can be achieved in a single line by *tuple unpacking*:

```
a, b, c = x + 1, 'hello', -4.5
```

The tuple on the right-hand side of this expression (parentheses are optional in this case) is unpacked in the assignment to the variable names on the left-hand side. This single line is thus equivalent to the three lines

```
a = x + 1
b = 'hello'
c = -4.5
```

In expressions such as these the right-hand side is evaluated first and then assigned to the left-hand side. As we have already seen, this provides a very useful way of swapping the value of two variables without the need for a temporary variable:

```
a, b = b, a
```

Comparisons may also be chained together in a natural way:

```
if a == b == 3:
    print('a and b both equal 3')
if -1 < x < 1:
    print('x is between -1 and 1')
```

Python supports *conditional assignment*: a variable name can be set to one value or another depending on the outcome of an if ... else expression on the same line as the assignment. For example,

```
y = math.sin(x)/x if x else 1
```

Short examples such as this one, in which the potential division by zero is avoided (recall that 0 evaluates to False) are benign enough, but the idiom should be avoided for anything more complex in favor of a more explicit construct such as

```
try:
    y = math.sin(x)/x
except ZeroDivisionError:
    y = 1
```

4.3.2 List Comprehension

A list comprehension in Python is a construct for creating a list based on another iterable object in a single line of code. For example, given a list of numbers, xlist, a list of the squares of those numbers may be generated as follows:

```
>>> xlist = [1, 2, 3, 4, 5, 6]
>>> x2list = [x**2 for x in xlist]
>>> x2list
[1, 4, 9, 16, 25, 36]
```

This is a faster and syntactically nicer way of creating the same list with a block of code within a for loop:

```
>>> x2list = []
>>> for x in xlist:
...     x2list.append(x**2)
```

List comprehensions can also contain conditional statements:

```
>>> x2list = [x**2 for x in xlist if x % 2]
>>> x2list
[1, 9, 25]
```

Here, x gets fed to the x**2 expression to be entered into the x2list under construction only if x % 2 evaluates to True (i.e. if x is *odd*). This is an example of a *filter* (a single if conditional expression). If you require a more complex *mapping* of values in the original sequence to values in the constructed list, the if .. else expression must appear before the for loop:

```
>>> [x**2 if x % 2 else x**3 for x in xlist]
[1, 8, 9, 64, 25, 216]
```

This comprehension squares the odd integers and cubes the even integers in xlist.

Of course, the sequence used to construct the list does not have to be another list. For example, strings, tuples and range objects are all iterable and can be used in list comprehensions:

```
>>> [x**3 for x in range(1, 10)]
[1, 8, 27, 64, 125, 216, 343, 512, 729]
>>> [w.upper() for w in 'abc xyz']
['A', 'B', 'C', ' ', 'X', 'Y', 'Z']
```

Finally, list comprehensions can be nested. For example, the following code flattens a list of lists:

```
>>> vlist = [[1, 2, 3], [4, 5, 6], [7, 8, 9]]
>>> [c for v in vlist for c in v]
[1, 2, 3, 4, 5, 6, 7, 8, 9]
```

Here, the first loop produces the inner lists, one by one, as v, and each inner list v is iterated over as c to be added to the list being created.

Example E4.10   Consider a 3 × 3 matrix represented by a list of lists:

```
M = [[1, 2, 3],
     [4, 5, 6],
     [7, 8, 9]]
```

Without using list comprehension, the transpose of this matrix could be built up by looping over the rows and columns:

```
MT = [[0, 0, 0],  [0, 0, 0],  [0, 0, 0]]
for ir in range(3):
    for ic in range(3):
        MT[ic][ir] = M[ir][ic]
```

With one list comprehension, the transpose can be constructed as

```
MT = []
for i in range(3):
    MT.append([row[i] for row in M])
```

where rows of the transposed matrix are built from the columns (indexed with i=0, 1, 2) of each row in turn from M. The outer loop here can be expressed as a list comprehension of its own:

```
MT = [[row[i] for row in M] for i in range(3)]
```

Note, however, that NumPy provides a much easier way to manipulate matrices.

4.3.3   lambda Functions

A lambda function in Python is a type of simple *anonymous function*. The executable body of a lambda function must be an *expression* and not a *statement*; that is, it may *not* contain, for example, loop blocks, conditionals or print statements. lambda functions provide limited support for a programming paradigm known as *functional programming*.[6] The simplest application of a lambda function differs little from the way a regular function def would be used:

```
>>> f = lambda x: x**2 - 3*x + 2
>>> print(f(4.))
6.0
```

The argument is passed to x and the result of the function specified in the lambda definition after the colon is passed back to the caller. To pass more than one argument to a lambda function, pass a tuple (without parentheses):

```
>>> f = lambda x,y: x**2 + 2*x*y + y**2
>>> f(2., 3.)
25.0
```

In these examples, not too much is gained by using a lambda function, and the functions defined are not all that anonymous either (because they've been bound to

---

[6] Functional programming is a style of programming in which computation is achieved through the evaluation of mathematical functions with minimal reference to variables defining the *state* of the program.

the variable name f). A more useful application is in creating a list of functions, as in the following example.

Example E4.11 Functions are objects (like everything else in Python) and so can be stored in lists. Without using lambda we would have to define named functions (using def) before constructing the list:

```python
def const(x):
    return 1.
def lin(x):
    return x
def square(x):
    return x**2
def cube(x):
    return x**3
flist = [const, lin, square, cube]
```

Then flist[3](5) returns 125, since flist[3] is the function cube, and is called with the argument 5.

The value of using lambda expressions as anonymous functions is that these functions do not need to be named if they are just to be stored in a list and so can be defined as items "inline" with the list construction:

```python
>>> flist = [lambda x: 1,
...          lambda x: x,
...          lambda x: x**2,
...          lambda x: x**3]
>>> flist[3](5)    # flist[3] is x**3
125
>>> flist[2](4)    # flist[2] is x**2
16
```

Example E4.12 The sorted built-in and sort list method can order lists based on the returned value of a function called on each element prior to making comparisons. This function is passed as the key argument. For example, sorting a list of strings is case-sensitive by default:

```python
>>> sorted('Nobody expects the Spanish Inquisition'.split())
['Inquisition', 'Nobody', 'Spanish', 'expects', 'the']
```

We can make the sorting case-insensitive, however, by passing each word to the str.lower method:

```python
>>> sorted('Nobody expects the Spanish Inquisition'.split(), key=str.lower)
['expects', 'Inquisition', 'Nobody', 'Spanish', 'the']
```

(Of course, key=str.upper would work just as well.) Note that the list elements themselves are not altered: they are being ordered based on a lowercase version of themselves. We do not use parentheses here, as in str.lower(), because we are passing the *function itself* to the key argument, not calling it directly.

It is typical to use lambda expressions to provide simple anonymous functions for this purpose. For example, to sort a list of atoms as (element symbol, atomic number) tuples in order of atomic number (the *second* item in each tuple):

```
>>> halogens = [('At', 85), ('Br', 35), ('Cl', 17), ('F', 9), ('I', 53)]
>>> sorted(halogens, key=lambda e: e[1])
[('F', 9), ('Cl', 17), ('Br', 35), ('I', 53), ('At', 85)]
```

Here, the sorting algorithm calls the function specified by key on each tuple item to decide where it belongs in the sorted list. Our anonymous function simply returns the second element of each tuple, and so sorting is by atomic number.

4.3.4   The with Statement

The with statement creates a block of code that is executed within a certain *context*. A context is defined by a *context manager* that provides a pair of methods describing how to enter and leave the context. User-defined contexts are generally used only in advanced code and can be quite complex, but a common basic example of a built-in context manager involves file input / output. Here, the context is entered by opening the file. Within the context block, the file is read from or written to, and finally the file is closed on exiting the context. The file object is a context manager that is returned by the open() method. It defines an exit method which simply closes the file (if it was opened successfully), so that this does not need to be done explicitly. To open a file within a context, use

```
with open('filename') as f:
    # Process the file in some way, for example:
    lines = f.readlines()
```

The reason for doing this is that you can be sure that the file will be closed after the with block, even if something goes wrong in this block: the context manager handles the code you would otherwise have to write to catch such runtime errors.

4.3.5   Generators

Generators are a powerful feature of the Python language; they allow one to declare a function that behaves like an iterable object. That is, a function that can be used in a for loop and that will yield its values, in turn, on demand. This is often more efficient than calculating and storing all of the values that will be iterated over (particularly if there will be a very large number of them). A generator function looks just like a regular Python function, but instead of exiting with a return value, it contains a yield statement, which returns a value each time it is required to by the iteration.

A very simple example should make this clearer. Let's define a generator, count, to count to n:

```
>>> def count(n):
...     i = 0
...     while i < n:
```

```
...             i += 1
...             yield i
...
>>> for j in count(5):
...     print(j)
...
1
2
3
4
5
```

Note that we can't simply call our generator like a regular function:

```
>>> count(5)
<generator object count at 0x102d8e6e0>
```

The generator count is expecting to be called as part of a loop (here, the for loop) and on each iteration it yields its result and stores its state (the value of i reached) until the loop next calls upon it.

In fact, we have been using generators already because the familiar range built-in function is, in Python 3, a type of generator object.

There is a *generator comprehension* syntax similar to list comprehension (use round brackets instead of square brackets):

```
>>> squares = (x**2 for x in range(5))
>>> for square in squares:
...     print(square)
...
0
1
4
9
16
```

However, once we have "exhausted" our generator comprehension defined in this way, we cannot iterate over it again without redefining it. If we try:

```
>>> for square in squares:
...     print(square)
...
>>>
```

we get nothing as we have already reached the end of the squares generator.

To obtain a list or tuple of a generator's values, simply pass it to list or tuple, as shown in the following example.

Example E4.13   This function defines a generator for the *triangular numbers*, $T_n = \sum_{k=1}^{n} k = 1 + 2 + 3 + \ldots + n$, for $n = 0, 1, 2, \ldots$: that is, $T_n = 0, 1, 3, 6, 10, \ldots$

```
>>> def triangular_numbers(n):
...     i, t = 1, 0
...     while i <= n:
...         yield t
...         t += i
```

```
...              i += 1
...
>>> list(triangular_numbers(15))
[0, 1, 3, 6, 10, 15, 21, 28, 36, 45, 55, 66, 78, 91, 105]
```

Note that the statements after the yield statement are executed each time triangular_num resumes. The call to triangular_numbers(15) returns an iterator that feeds these numbers into list to generate a list of its values.

4.3.6 ◇ map

The built-in function map returns an iterator that applies a given function to every item of a provided sequence, yielding the results as a generator would.[7] For example, one way to sum a list of lists is to map the sum function to it:

```
>>> mylists = [[1, 2, 3], [10, 20, 30], [25, 75, 100]]
>>> list(map(sum, mylists))
[6, 60, 200]
```

(We have to cast explicitly back to a list because map returns a generator-like object.) This statement is equivalent to the list comprehension:

```
>>> [sum(l) for l in mylists]
[6, 60, 200]
```

map is occasionally useful but has the potential to create very obscure code, and list or generator comprehensions are generally to be preferred. The same applies to the filter built-in, which constructs an iterator from the elements of a given sequence for which a provided function returns True. In the following example, the odd integers less than 10 are generated: this function returns x % 2, and this expression evaluates to 0, equivalent to False if x is even:

```
>>> list(filter(lambda x: x%2, range(10)))
[1, 3, 5, 7, 9]
```

Again, the list comprehension is more expressive:

```
>>> [x for x in range(10) if x % 2]
[1, 3, 5, 7, 9]
```

4.3.7 ◇ Assignment Expressions: the *Walrus Operator*

Python 3.8 introduced a new piece of syntax which allows a variable to be assigned within an expression. A conventional Python *expression*, such as 2 + 2 or x == 'a' returns a value (which may be None); Python *statements* are composed of expressions and generally have some effect on the state of the program (e.g. they assign a variable or test a condition). The ability to assign a variable within an expression can lead to more concise code with less repetition. For example, consider the following check that a string is shorter than 10 characters, which produces a meaningful error message:

---

[7] Constructs such as map are frequently used in functional programming.

```
>>> s = 'A string with too many characters'
>>> if len(s) > 10:
...     print(f's has {len(s)} characters. The maximum is 10.')
...
```

```
s has 33 characters. The maximum is 10.
```

The problem with this code is that we evaluate the length of the string twice (once for the check and once for the message). We might assign a variable to avoid this:

```
>>> slen = len(s)
>>> if slen > 10:
...     print(f's has {slen} characters. The maximum is 10.')
...
```

but a more concise way, which saves a line of code, is to use an *assignment expression*. The syntax a := b can be used to assign a to the value of b in the context of an expression (e.g. a conditional expression) rather than a stand-alone statement. That is, it assigns the value *and then returns that value*, in contrast to the usual Python assignment behavior (which doesn't return anything). Hence,

```
>>> if (slen := len(s)) > 10:
...     print(f's has {slen} characters. The maximum is 10.')
...
```

```
s has 33 characters. The maximum is 10.
```

The symbol := supposedly looks like the eyes and tusks of a walrus, and so has become known as the "walrus operator." Note that assignment expressions should generally be enclosed in parentheses.

Example E4.14   A good application of an assignment expression is the reuse of a value that may be expensive to calculate, for example in a list comprehension:

```
filtered_values = [f(x) for x in values if f(x) >= 0]
```

Here, the := operator can be used to assign the returned value of f(x) at the same time as checking if it is positive:

```
filtered_values = [val for x in values if (val := f(x)) >= 0]
```

As a further example, consider the following block of code, which reads in and processes a large file in chunks of 4 kB at a time:

```
CHUNK_SIZE = 4096
chunk = fi.read(CHUNK_SIZE)
while chunk:
    process_chunk(chunk)
    chunk = fi.read(CHUNK_SIZE)
```

This can be written more clearly as

```
while chunk := fi.read(CHUNK_SIZE):
    process_chunk(chunk)
```

(Note that in this case it is not necessary to enclose the assignment expression in parentheses).

Assignment expressions are a controversial addition to the Python language and do not always make code clearer. This book will not use them extensively, since there is always an alternative approach that works on versions of Python 3 prior to 3.8.

4.3.8   Exercises

Questions

Q4.3.1   Rewrite the list of `lambda` functions created in Example E4.11 using a single list comprehension.

Q4.3.2   What does the following code do and how does it work?

```
>>> nmax = 5
>>> x = [1]
>>> for n in range(1,nmax+2):
...     print(x)
...     x = [([0] + x)[i] + (x + [0])[i] for i in range(n+1)]
...
```

Q4.3.3   Consider the lists

```
>>> a = ['A', 'B', 'C', 'D', 'E', 'F', 'G']
>>> b = [4, 2, 6, 1, 5, 0, 3]
```

Predict and explain the output of the following statements:

(a)   `[a[x] for x in b]`
(b)   `[a[x] for x in sorted(b)]`
(c)   `[a[b[x]] for x in b]`
(d)   `[x for (y, x) in sorted(zip(b, a))]`

Q4.3.4   Dictionaries are data structures in which (since Python 3.6) key-value pairs are stored in order of insertion. Write a one-line Python statement returning a list of (*key*, *value*) pairs sorted by the keys themselves. Assume that all keys have the same data type (why is this important?). Repeat the exercise to produce a list ordered by dictionary *values*.

Q4.3.5   In the television series *The Wire*, drug dealers encrypt telephone numbers with a simple substitution cypher based on the standard layout of the phone keypad. Each digit of the number, with the exception of 5 and 0, is replaced with the corresponding digit on the other side of the 5 key ("jump the five"); 5 and 0 are exchanged. Thus, 555-867-5309 becomes 000-243-0751. Devise a one-line statement to encrypt and decrypt numbers encoded in this way.

Q4.3.6   The built-in function `sorted` and sequence method `sort` require that the elements in the sequence be of types that can be compared: they will fail, for example, if a list contains a mixture of strings and numbers. However, it is frequently the case that a list contains numbers and the special value, None (perhaps denoting missing data). Devise a way to sort such a list by passing a lambda function in the argument key; the None values should end up at the end of the sorted list.

Q4.3.7   Use an assignment expression (the walrus operator) (a) in a while loop to determine the smallest Fibonacci number greater than 5000; (b) in a while loop to echo back a lower-case version of the user's input (use the input built-in function) until they enter exit.

Problems

P4.3.1   Use a list comprehension to calculate the *trace* of the matrix M (that is, the sum of its diagonal elements). *Hint*: the sum built-in function takes an iterable object and sums its values.

P4.3.2   The ROT13 substitution cipher encodes a string by replacing each letter with the letter 13 letters after it in the alphabet (cycling around if necessary). For example, a → n and p → c.

(a)   Given a word expressed as a string of lower-case characters only, use a list comprehension to construct the ROT13-encoded version of that string. *Hint*: Python has a built-in function, ord, which converts a character to its Unicode code point (e.g. ord('a') returns 97); another built-in, chr, is the inverse of ord (e.g. chr(122) returns 'z').

(b)   Extend your list comprehension to encode sentences of words (in lower case) separated by spaces into a ROT13 sentence (in which the encoded words are also separated by spaces).

P4.3.3   In *A New Kind of Science*,[8] Stephen Wolfram describes a set of simple one-dimensional cellular automata in which each cell can take one of two values: "on" or "off." A row of cells is initialized in some state (e.g. with a single "on" cell somewhere in the row) and it evolves into a new state according to a rule that determines the subsequent state of a cell ("on" or "off") from its value and that of its two nearest neighbors. There are $2^3 = 8$ different states for these three "parent" cells taken together and so $2^8 = 256$ different automata rules; that is, the state of cell i in the next generation is determined by the states of cells $i - 1$, i and $i + 1$ in the present generation.

These rules are numbered 0–255 according to the binary number indicated by the eight different outcomes each one specifies for the eight possible parent states. For example, rule 30 produces the outcome (off, off, off, on, on, on, on, off) (or 00011110) from the parent states given in the order shown in Figure 4.1. The evolution of the cells can be illustrated by printing the row corresponding to each generation under its parent as shown in this figure.

---

[8] S. Wolfram (2002). *A New Kind of Science*, Wolfram Media.

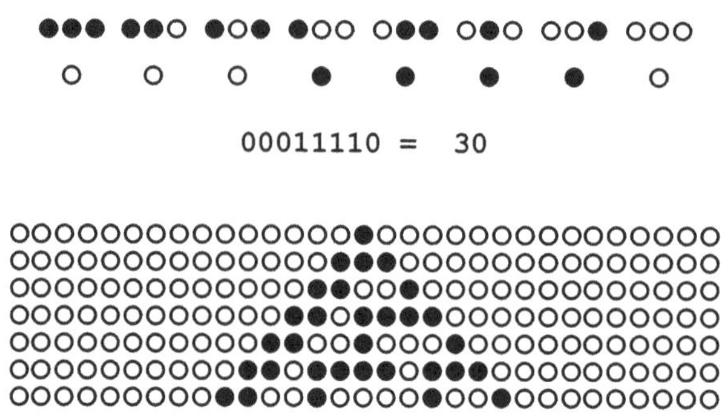

00011110 = 30

Figure 4.1 Rule 30 of Wolfram's one-dimensional two-state cellular automata and the first seven generations.

Write a program to display the first few rows generated by rule 30 on the command line, starting from a single "on" cell in the center of a row 80 cells wide. Use an asterisk to indicate an "on" cell and a space to represent an "off" cell.

P4.3.4 The file iban_lengths.txt, available at https://scipython.com/ex/bdd contains two columns of data: a two-letter country code and the length of that country's International Bank Account Number (IBAN):

```
AL 28
AD 24
...
GB 22
```

The code snippet below parses the file into a dictionary of lengths, keyed by the country code:

```
iban_lengths = {}
with open('iban_lengths.txt') as fi:
    for line in fi.readlines():
        fields = line.split()
        iban_lengths[fields[0]] = int(fields[1])
```

Use a lambda function and list comprehension to achieve the same goal in (a) two lines, (b) one line.

P4.3.5 The *power set* of a set S, $P(S)$, is the set of all subsets of S, including the empty set and S itself. For example,

$$P(\{1, 2, 3\}) = \{\{\}, \{1\}, \{2\}, \{3\}, \{1, 2\}, \{1, 3\}, \{2, 3\}, \{1, 2, 3\}\}.$$

Write a Python program that uses a generator to return the power set of a given set.
Hint: convert your set into an ordered sequence such as a tuple. For each item in this sequence return the power set formed from all subsequent items, inclusive and exclusive of the chosen item. Don't forget to convert the tuples back to sets after you're done.

P4.3.6 The *Brown Corpus* is a collection of 500 samples of (American) English-language text that was compiled in the 1960s for use in the field of computational linguistics. It can be dowloaded from https://nltk.github.com/nltk_data/packages/corpora/brown.zip.

Each sample in the corpus consists of words that have been tagged with their part-of-speech after a forward slash. For example,

```
The/at football/nn opponent/nn on/in homecoming/nn is/bez ,/, of/in
course/nn ,/, selected/vbn with/in the/at view/nn that/cs
```

Here, The has been tagged as an article (/at), football as a noun (/nn) and so on. A full list of the tags is available from the accompanying manual.[9] Write a program that analyzes the Brown Corpus and returns a list of the eight-letter words which feature each possible two-letter combinations *exactly twice*. For example, the two-letter combination *pc* is present in only the words *topcoats* and *upcoming*; *mt* is present only in the words *boomtown* and *undreamt*.

4.4   Operating-System Services

4.4.1   The sys Module

The sys module provides certain system-specific parameters and functions. Many of them are of interest only to fairly advanced users of less-common Python implementations (the details of how floating-point arithmetic is implemented can vary between different systems, for example, but is likely to be the same on all common platforms – see Section 10.1). However, it also provides some that are useful and important: these are described here.

sys.argv

sys.argv holds the command-line arguments passed to a Python program when it is executed. *It is a list of strings.* The first item, sys.argv[0], is the name of the program itself. This allows for a degree of interactivity without having to read from configuration files or requiring direct user input, and means that other programs or shell scripts can call a Python program and pass it particular input values or settings. For example, a simple script to square a given number might be written:

```
# square.py
import sys

n = int(sys.argv[1])
print(n, 'squared is', n**2)
```

(Note that it is necessary to convert the input value into an int, because it is stored in sys.argv as a string.) Running this program from the command line with

```
python square.py 3
```

produces the output

---

[9] This manual is available at http://clu.uni.no/icame/manuals/BROWN/INDEX.HTM though the tags themselves are presented better on the Wikipedia article at https://en.wikipedia.org/wiki/Brown_Corpus.

```
3 squared is 9
```

as expected. But because we did not hard-code the value of n, the same program can be run with

```
python square.py 4
```

to produce the output

```
4 squared is 16
```

sys.exit

Calling sys.exit will cause a program to terminate and exit from Python. This happens "cleanly," so that any commands specified in a try statement's finally clause are executed first and any open files are closed. The optional argument to sys.exit can be any object; if it is an integer, it is passed to the shell which, it is assumed, knows what to do with it.[10] For example, 0 usually denotes "successful" termination of the program and nonzero values indicate some kind of error. Passing no argument or None is equivalent to 0. If any other object is specified as an argument to sys.exit, it is passed to stderr, Python's implementation of the standard error stream. A string, for example, appears as an *error message* on the console (unless redirected elsewhere by the shell).

Example E4.15   A common way to help users with scripts that take command-line arguments is to issue a usage message if they get it wrong, as in the following code example.

Listing 4.3 Issuing a usage message for a script taking command-line arguments

```
# square.py
import sys

try:
    n = int(sys.argv[1])
except (IndexError, ValueError):
    sys.exit('Please enter an integer, <n>, on the command line.\nUsage: '
             'python {:s} <n>'.format(sys.argv[0]))
print(n, 'squared is', n**2)
```

The error message here is reported and the program exits if no command-line argument was specified (and hence indexing sys.argv[1] raises an IndexError) or the command-line argument string does not evaluate to an integer (in which case the int cast will raise a ValueError).

```
$ python square.py hello
Please enter an integer, <n>, on the command line.
Usage: python square.py <n>

$ python square.py 5
5 squared is 25
```

---

[10] At least if it is in the range 0–127; undefined results could be produced for values outside this range.

4.4.2   The os Module

The os module provides various operating-system interfaces in a platform-independent way. Its many functions and parameters are described in full in the official documentation,[11] but some of the more important ones are described in this section.

Process Information

The Python *process* is the particular instance of the Python application that is executing your program (or providing a Python shell for interactive use). The os module provides a number of functions for retrieving information about the context in which the Python process is running. For example, os.uname() returns information about the operating system running Python and the network name of the machine running the process.

One function is of particular use: os.getenv(*key*) returns the value of the environment variable *key* if it exists (or None of it doesn't). Many environment variables are system-specific, but commonly include:

- HOME: the path to the user's home directory;
- PWD: the current working directory;
- USER: the current user's username;
- PATH: the system path environment variable.

For example, on my system:

```
>>> os.getenv('HOME')
'/Users/christian'
```

File-System Commands

It is often useful to be able to navigate the system directory tree and manipulate files and directories from within a Python program. The os module provides the functions listed in Table 4.4 to do just this. There are, of course, inherent dangers: your Python program can do anything that your user can, including renaming and deleting files.

Pathname Manipulations[12]

The os.path module provides a number of useful functions for manipulating pathnames. The version of this library installed with Python will be the one appropriate for the operating system that it runs on (e.g. on a Windows machine, path-name components are separated by the backslash character, "\", whereas on Unix and Linux systems, the (forward) slash character, "/" is used.

Common usage of the os.path module's functions are to find the filename from a path (basename), test to see if a file or directory exists (exists), join strings together to make a path (join), split a filename into a "root" and an "extension" (splitext) and to

---

[11] https://docs.python.org/3/library/os.html.

[12] This section describes the low-level os.path module; since Python 3.4 the Standard Library pathlib module has been available: this offers a higher-level, object-oriented approach to manipulating file-system paths that can be more expressive. See https://docs.python.org/3/library/pathlib.html for details.

Table 4.4 os module: some file-system commands

Function	Description
os.listdir(*path*='.')	List the entries in the directory given by *path* (or the current working directory if this is not specified).
os.remove(*path*)	Delete the file *path* (raises an OSError if *path* is a directory; use os.rmdir instead).
os.rename(*old_name*, *new_name*)	Rename the file or directory *old_name* to *new_name*. If a file with the name *new_name* already exists, *it will be overwritten* (subject to user-permissions).
os.rmdir(*path*)	Delete the directory *path*. If the directory is not empty, an OSError is raised.
os.mkdir(*path*)	Create the directory named *path*.
os.system(*command*)	Execute *command* in a subshell. If the command generates any output, it is redirected to the interpreter standard output stream, stdout.

Table 4.5 os.path module: common pathname manipulations

Function	Description
os.path.basename(*path*)	Return the basename of the pathname *path* giving a relative or absolute path to the file: this usually means the filename.
os.path.dirname(*path*)	Return the directory of the pathname *path*.
os.path.exists(*path*)	Return True if the directory or file *path* exists, and False otherwise.
os.path.getmtime(*path*)	Return the time of last modification of *path*.
os.path.getsize(*path*)	Return the size of *path* in bytes.
os.path.join(*path1*, *path2*, ...)	Return a pathname formed by joining the path components *path1*, *path2*, etc. with the directory separator appropriate to the operating system being used.
os.path.split(*path*)	Split *path* into a directory and a filename, returned as a tuple (equivalent to calling dirname and basename) respectively.
os.path.splitext(*path*)	Split *path* into a "root" and an "extension" (returned as a tuple pair).

find the time of last modification to a file (getmtime). Such common applications are described briefly in Table 4.5.

Some examples referring to a file /home/brian/test.py:

```
>>> os.path.basename('/home/brian/test.py')
'test.py'                    # just the filename

>>> os.path.dirname('/home/brian/test.py')
'/home/brian'                # just the directory

>>> os.path.split('/home/brian/test.py')
('/home/brian', 'test.py')   # directory and filename in a tuple
```

```
>>> os.path.splitext('/home/brian/test.py')
('/home/brian/test', '.py') # file path stem and extension in a tuple

>>> os.path.join(os.getenv('HOME'), 'test.py')
'/home/brian/test.py'       # join directories and/or filename

>>> os.path.exists('/home/brian/test.py')
False                       # file does not exist!
```

Trying to call some of these functions on a path that does not exist will cause a FileNotFoundError exception to be raised (which could be caught within a try ... except clause, of course).

Example E4.16 Suppose you have a directory of data files identified by filenames containing a date in the form data-DD-Mon-YY.txt where DD is the two-digit day number, Mon is the three-letter month abbreviation and YY is the last two digits of the year, for example '02-Feb-10'. The following program converts the filenames into the form data-YYYY-MM-DD.txt so that an alphanumerical ordering of the filenames puts them in chronological order.

Listing 4.4 Renaming data files by date

```
# eg4-osmodule.py
import os
import sys

months = ['jan', 'feb', 'mar', 'apr', 'may', 'jun',
          'jul', 'aug', 'sep', 'oct', 'nov', 'dec']

dir_name = sys.argv[1]
for filename in os.listdir(dir_name):
    # filename is expected to be in the form 'data-DD-MMM-YY.txt'
    d, month, y = int(filename[5:7]), filename[8:11], int(filename[12:14])
❶   m = months.index(month.lower())+1

    newname = 'data-20{:02d}-{:02d}-{:02d}.txt'.format(y, m, d)
    newpath = os.path.join(dir_name, newname)
    oldpath = os.path.join(dir_name, filename)
    print(oldpath, '->', newpath)
    os.rename(oldpath, newpath)
```

❶ We get the month number from the index of corresponding abbreviated month name in the list months, adding 1 because Python list indexes start at 0.

For example, given a directory testdir containing the following files:

```
data-02-Feb-10.txt
data-10-Oct-14.txt
data-22-Jun-04.txt
data-31-Dec-06.txt
```

the command python eg4-osmodule.py testdir produces the output

```
testdir/data-02-Feb-10.txt -> testdir/data-2010-02-02.txt
```

```
testdir/data-10-Oct-14.txt -> testdir/data-2014-10-10.txt
testdir/data-22-Jun-04.txt -> testdir/data-2004-06-22.txt
testdir/data-31-Dec-06.txt -> testdir/data-2006-12-31.txt
```

See also Problem P4.4.4 and the datetime module (Section 4.5.3).

4.4.3   Exercises

Problems

P4.4.1   Modify the hailstone sequence generator of Exercise P2.5.7 to generate the hailstone sequence starting at any positive integer that the user provides on the command line (use sys.argv). Handle the case where the user forgets to provide n or provides an invalid value for n gracefully.

P4.4.2   The *Haversine* formula gives the shortest (great-circle) distance, d, between two points on a sphere of radius, R, from their longitudes (λ_1, λ_2) and latitudes (ϕ_1, ϕ_2):

$$d = 2R \arcsin\left(\sqrt{\text{haversin}(\phi_2 - \phi_1) + \cos\phi_1 \cos\phi_2 \text{haversin}(\lambda_2 - \lambda_1)} \right),$$

where the *haversine* function of an angle is defined by

$$\text{haversin}(\alpha) = \sin^2\left(\frac{\alpha}{2}\right).$$

Write a program to calculate the shortest distance in kilometers between two points on the surface of the Earth (considered as a sphere of radius 6378.1 km) given as two command-line arguments, each of which is a comma-separated pair of latitude, longitude values in degrees. For example, the distance between Paris and Rome is given by executing:

```
python greatcircle.py 48.9,2.4 41.9,12.5
1107 km
```

P4.4.3   Write a Python program to create a directory, test, in the user's home directory and to populate it with 20 Scalable Vector Graphics (SVG) files depicting a small, filled, red circle inside a large, black, unfilled circle. For example,

```
<?xml version="1.0" encoding="utf-8"?>
                <svg xmlns="http://www.w3.org/2000/svg"
                    xmlns:xlink="http://www.w3.org/1999/xlink"
                    width="500" height="500" style="background: #ffffff">
<circle cx="250.0" cy="250.0" r="200" style="stroke: black; stroke-width: 2px;
                                                    fill: none;"/>
<circle cx="430.0" cy="250.0" r="20" style="stroke: red; fill: red;"/>
</svg>
```

Each file should move the red circle around the inside rim of the larger circle so that the 20 files together could form an animation.

One way to achieve this is to use the free ImageMagick software (www.imagemagick.org/). Ensure the SVG files are named fig00.svg, fig01.svg, etc. and issue the following command from your operating system's command line:

```
convert -delay 5 -loop 0 fig*.svg animation.gif
```

to produce an animated GIF image.

P4.4.4   Modify the program of Example E4.16 to catch the following errors and handle them gracefully:

- user does not provide a directory name on the command line (issue a usage message);
- the directory does not exist;
- the name of a file in the directory does not have the correct format;
- the filename is in the correct format but the month abbreviation is not recognized.

Your program should terminate in the first two cases and skip the file in the second two.

4.5   Modules and Packages

As we have seen, Python is quite a modular language and has functionality beyond the core programming essentials (the built-in methods and data structures we have encountered so far), which is made available to a program through the `import` statement. This statement makes reference to *modules* that are ordinary Python files containing definitions and statements. Upon encountering the line

```
import <module>
```

the Python interpreter executes the statements in the file `<module>.py` and enters the module name `<module>` into the current namespace, so that the attributes it defines are available with the "dotted syntax": `<module>.<attribute>`.

Defining your own module is as simple as placing code within a file `<module>.py`, which is somewhere the Python interpreter can find it (for small projects, usually just the same directory as the program doing the importing). Note that because of the syntax of the `import` statement, you should avoid naming your module anything that isn't a valid Python identifier (see Section 2.2.3). For example, the filename `<module>.py` should not contain a hyphen or start with a digit. Do not give your module the same name as any built-in modules (such as `math` or `random`) because these get priority when Python imports.

A Python *package* is simply a structured arrangement of modules within a directory on the file system. Packages are the natural way to organize and distribute larger Python projects. To make a package, the module files are placed in a directory, along with a file named `__init__.py`. This file is run when the package is imported and may perform some initialization and its own imports. It may be an empty file (zero bytes long) if no special initialization is required, but it must exist for the directory to be considered by Python to be a package.

For example, the NumPy package (see Chapter 6) exists as the following directory (some files and directories have been omitted for clarity):

```
numpy/
    __init__.py
    core/
    fft/
        __init__.py
        fftpack.py
        info.py
        ...
    linalg/
        __init__.py
        linalg.py
        info.py
        ...
    polynomial/
        __init__.py
        chebyshev.py
        hermite.py
        legendre.py
        ...
    random/
    version.py
    ...
```

Thus, for example, `polynomial` is a subpackage of the numpy package containing several modules, including `legendre`, which may be imported as

```
import numpy.polynomial.legendre
```

To avoid having to use this full dotted syntax in actually referring to its attributes, it is convenient to use

```
from numpy.polynomial import legendre
```

Table 4.6 lists some of the major, freely available Python modules and packages for general programming applications as well as for numerical and scientific work. Some are installed with the core Python distribution (the *Standard Library*);[13] where indicated others can be downloaded and installed separately. Before implementing your own algorithm, check that it isn't included in an existing Python package.

Whilst other package managers exist,[14] the `pip` application[15] has become the de facto standard. It is usually installed by default with most Python installations and does a pretty good job of managing package versions and dependencies. To install the package *package*, the following syntax is used at the command line:

```
pip install package              # install latest version
pip install package==X.Y.Z       # install version X.Y.Z
pip install 'package>=X.Y.Z'     # install minimum version X.Y.Z
```

To uninstall a package, use:

```
pip uninstall package
```

---

[13] A complete list of the components of the Standard Library is at https://docs.python.org/3/library/index.html.

[14] For example, conda from the Anaconda distribution – see Section 1.3.

[15] See https://pip.pypa.io/en/stable/ for full documentation.

Table 4.6 Python modules and packages. Those marked with an asterisk (*) are not part of the Python Standard Library and must be installed separately, for example with pip.

Module / Package	Description
os, sys	Operating-system services, as described in Section 4.4
math, cmath	Mathematical functions, as introduced in Section 2.2.2
random	Random-number generator (see Section 4.5.1)
collections	Data types for containers that extend the functionality of dictionaries, tuples, etc.
itertools	Tools for efficient iterators that extend the functionality of simple Python loops
glob	Unix-style pathname pattern expansion
datetime	Parsing and manipulating dates and times (see Section 4.5.3)
fractions	Rational-number arithmetic
re	Regular expressions
argparse	Parser for command-line options and arguments
urllib	URL (including web pages) opening, reading and parsing (see Section 4.5.2)
* Django (django)	A popular web application framework
* pyparsing	Lexical parser for simple grammars
pdb	The Python debugger
logging	Python's built-in logging module
xml, lxml	XML parsers
* VPython (visual)	Three-dimensional visualization
unittest	Unit-testing framework for systematically testing and validating individual units of code (see Section 10.3.4)
* NumPy (numpy)	Numerical and scientific computing (described in detail in Chapter 6)
* SciPy (scipy)	Scientific computing algorithms (described in detail in Chapter 8)
* Matplotlib (matplotlib)	Plotting (see Chapters 3 and 7)
* SymPy (sympy)	Symbolic computation (computer algebra)
* pandas	Data manipulation and analysis with table-like data structures
* scikit-learn	Machine learning
* Beautiful Soup 4 (beautifulsoup4)	HTML parser, with handling of malformed documents

4.5.1   The random Module

For simulations, modeling and some numerical algorithms it is often necessary to generate random numbers from some distribution. The topic of random-number generation is a complex and interesting one, but the important aspect for our purposes is that, in common with most other languages, Python implements a *pseudorandom-number generator* (PRNG). This is an algorithm that generates a sequence of numbers that approximates the properties of "truly" random numbers. Such sequences are determined by an originating *seed* state and are always the same following the same seed: in this sense they are deterministic. This can be a good thing (so that a calculation involving random numbers can be reproduced) or a bad thing (e.g. if used for cryptography, where the random sequence must be kept secret). Any PRNG will yield a sequence that eventually repeats, and a good generator will have a long period. The PRNG implemented by Python is the *Mersenne Twister*, a well-respected and much-studied algorithm with a period of $2^{19937} - 1$ (a number with more than 6000 digits in base 10).

Generating Random Numbers

The random-number generator can be seeded with any *hashable* object (e.g. an immutable object such as an integer). When the module is first imported, it is seeded with a representation of the current system time (unless the operating system provides a better source of a random seed). The PRNG can be reseeded at any time with a call to random.seed.

The basic random-number method is random.random. It generates a random number selected from the uniform distribution in the semi-open interval [0, 1) – that is, including 0 but not including 1.

```
>>> import random
>>> random.random()        # PRNG seeded 'randomly'
0.5204514767709216
>>> random.seed(42)        # seed the PRNG with a fixed value
>>> random.random()
0.6394267984578837
>>> random.random()
0.025010755222666936
...
>>> random.seed(42)        # reseed with the same value as before ...
>>> random.random()
0.6394267984578837         # ... and the sequence repeats
>>> random.random()
0.025010755222666936
```

Calling random.seed() with no argument reseeds the PRNG with a "random" value as when the random module is first imported.

To select a random floating-point number, *N*, from a given range, $a \leq N \leq b$, use random.uniform(a, b):

```
>>> random.uniform(-2., 2.)
-0.899882726523523
>>> random.uniform(-2., 2.)
-1.107157047404709
```

The random module has several methods for drawing random numbers from nonuniform distributions – see the documentation[16] – the most important of them are described below.

To return a number from the normal distribution with mean, mu, and standard deviation, sigma, use random.normalvariate(mu, sigma):

```
>>> random.normalvariate(100, 15)
118.82178896586194
>>> random.normalvariate(100, 15)
97.92911405885782
```

---

[16] https://docs.python.org/3/library/random.html.

To select a random *integer*, *N*, in a given range, $a \leq N \leq b$, use the `random.randint` (a, b) method:

```
>>> random.randint(5, 10)
7
>>> random.randint(5, 10)
10
```

Random Sequences

Sometimes you may wish to select an item at random from a sequence such as a `list`. This is what the method `random.choice` does:

```
>>> seq = [10, 5, 2, 'ni', -3.4]
>>> random.choice(seq)
-3.4
>>> random.choice(seq)
'ni'
```

Another method, `random.shuffle`, randomly shuffles (permutes) the items of the sequence *in place*:

```
>>> random.shuffle(seq)
>>> seq
[10, -3.4, 2, 'ni', 5]
```

Note that because the random permutation is made in place, the sequence must be mutable: you can't, for example, shuffle `tuples`.

Finally, to draw a `list` of *k* unique elements from a sequence or set (without replacement) population, there is `random.sample(population, k)`:

```
>>> raffle_numbers = range(1, 100001)
>>> winners = random.sample(raffle_numbers, 5)
>>> winners
[89734, 42505, 7332, 30022, 4208]
```

The resulting list is in selection order (the first-indexed element is the first drawn) so that one could, for example, without bias declare ticket number 89734 to be the jackpot winner and the remaining four tickets second-placed winners.

Example E4.17   The *Monty Hall problem* is a famous conundrum in probability, which takes the form of a hypothetical game show. The contestant is presented with three doors; behind one is a car and behind each of the other two is a goat. The contestant picks a door and then the game show host opens a different door to reveal a goat. The host knows which door conceals the car. The contestant is then invited to switch to the other closed door or stick with their initial choice.

Counterintuitively, the best strategy for winning the car is to switch, as demonstrated by the following simulation.

Listing 4.5 The Monty Hall problem

```
# eg4-montyhall.py
import random
```

```
def run_trial(switch_doors, ndoors=3):
    """

    Run a single trial of the Monty Hall problem, with or without switching
    after the game show host reveals a goat behind one of the unchosen doors.
    (switch_doors is True or False). The car is behind door number 1 and the
    game show host knows that. Returns True for a win, otherwise returns False.

    """

    # Pick a random door out of the ndoors available.
    chosen_door = random.randint(1, ndoors)
    if switch_doors:
        # Reveal a goat.
        revealed_door = 3 if chosen_door==2 else 2
        # Make the switch by choosing any other door than the initially
        # selected one and the one just opened to reveal a goat.
        available_doors = [dnum for dnum in range(1,ndoors+1)
                                 if dnum not in (chosen_door, revealed_door)]
        chosen_door = random.choice(available_doors)

    # You win if you picked door number 1.
    return chosen_door == 1
```
❶

```
def run_trials(ntrials, switch_doors, ndoors=3):
    """

    Run ntrials iterations of the Monty Hall problem with ndoors doors, with
    and without switching (switch_doors = True or False). Returns the number
    of trials which resulted in winning the car by picking door number 1.

    """

    nwins = 0
    for i in range(ntrials):
        if run_trial(switch_doors, ndoors):
            nwins += 1
    return nwins

ndoors, ntrials = 3, 10000
nwins_without_switch = run_trials(ntrials, False, ndoors)
nwins_with_switch = run_trials(ntrials, True, ndoors)

print('Monty Hall Problem with {} doors'.format(ndoors))
print('Proportion of wins without switching: {:.4f}'
        .format(nwins_without_switch/ntrials))
print('Proportion of wins with switching: {:.4f}'
        .format(nwins_with_switch/ntrials))
```

❶ Without loss of generality, we can place the car behind door number 1, leaving the contestant initially to choose any door at random.

To make the code a little more interesting, we have allowed for a variable number of doors in the simulation (but only one car).

```
Monty Hall Problem with 3 doors
Proportion of wins without switching: 0.3334
Proportion of wins with switching: 0.6737
```

4.5.2 ◇   The `urllib` Package

The `urllib` package in Python 3 is a set of modules for opening and retrieving the content referred to by Uniform Resource Locators (URLs), typically web addresses accessed with HTTP(S) (HyperText Transfer Protocol) or FTP (File Transfer Protocol). Here is a very brief introduction to its use.

Opening and Reading URLs

To obtain the content at a URL using HTTP you first need to make an HTTP *request* by creating a `Request` object. For example,

```
import urllib.request
req = urllib.request.Request('https://www.wikipedia.org')
```

The `Request` object allows you to pass data (using GET or POST) and other information about the request (metadata passed through the HTTP headers – see later). For a simple request, however, one can simply open the URL immediately as a file-like object with `urlopen()`:

```
response = urllib.request.urlopen(req)
```

It's a good idea to catch the two main types of exception that can arise from this statement. The first type, `URLError`, results if the server doesn't exist or if there is no network connection; the second type, `HTTPError`, occurs when the server returns an error code (such as *404: Page Not Found*). These exceptions are defined in the `urllib.error` module.

```
from urllib.error import URLError, HTTPError
try:
    response = urllib.request.urlopen(req)
except HTTPError as e:
    print('The server returned error code', e.code)
except URLError as e:
    print('Failed to reach server at {} for the following reason:\n{}'
            .format(url, e.reason))
else:
    # the response came back OK
```

Assuming the `urlopen()` worked, there is often nothing more to do than simply read the content from the response:

```
content = response.read()
```

The content will be returned as a *bytestring*. To decode it into a Python (Unicode) string you need to know how it is encoded. A good resource will include the character set used in the `Content-Type` HTTP header. This can be used as follows:

```
charset = response.headers.get_content_charset()
html = content.decode(charset)
```

where `html` is now a decoded Python Unicode string. If no character set is specified in the headers returned, you may have to guess (e.g. set `charset='utf-8'`).

GET and POST Requests

It is often necessary to pass data along with the URL to retrieve content from a server. For example, when submitting an HTML form from a web page, the values corresponding to the entries you have made are encoded and passed to the server according to either the GET or POST protocols.

The urllib.parse module allows you to encode data from a Python dictionary into a form suitable for submission to a web server. To take an example from the Wikipedia API using a GET request:

```
>>> url = 'https://wikipedia.org/w/api.php'
>>> data = {'page': 'Monty_Python', 'prop': 'text', 'action': 'parse', 'section': 0}
>>> encoded_data = urllib.parse.urlencode(data)
>>> full_url = url + '?' + encoded_data
>>> full_url
'https://wikipedia.org/w/api.php?page=Monty_Python&prop=text&action=parse
    &section=0'
>>> req = urllib.request.Request(full_url)
>>> response = urllib.request.urlopen(req)
>>> html = response.read().decode('utf-8')
```

To make a POST request, instead of appending the encoded data to the string <url>?, pass it to the Request constructor directly:

```
req = urllib.request.Request(url, encoded_data)
```

4.5.3   The datetime Module

Python's datetime module provides classes for manipulating dates and times. There are many subtle issues surrounding the handling of such data (time zones, different calendars, Daylight Saving Time, etc.,) and full documentation is available online;[17] here we provide an overview of only the most common uses.

Dates

A datetime.date object represents a particular day, month and year in an idealized calendar (the current Gregorian calendar is assumed to be in existence for all dates, past and future). To create a date object, pass valid year, month and day numbers explicitly, or call the date.today constructor:

```
>>> from datetime import date
>>> birthday = date(2004, 11, 5)     # OK

>>> notadate = date(2005, 2, 29)     # Oops: 2005 wasn't a leap year!

Traceback (most recent call last):
  File "<stdin>", line 1, in <module>
ValueError: day is out of range for month

>>> today = date.today()
```

---

[17] https://docs.python.org/3/library/datetime.html.

```
>>> today
datetime.date(2014, 12, 6)  # (for example)
```

Dates between 1/1/1 and 31/12/9999 are accepted. Parsing dates to and from strings is also supported (see `strptime` and `strftime`).

Some more useful `date` object methods are used as follows:

```
>>> birthday.isoformat()    # ISO 8601 format: YYYY-MM-DD
'2004-11-05'

>>> birthday.weekday()      # Monday = 0, Tuesday = 1, ..., Sunday = 6
4   # (Friday)

>>> birthday.isoweekday()   # Monday = 1, Tuesday = 2, ..., Sunday = 7
5

>>> birthday.ctime()        # C-standard time output
'Fri Nov  5 00:00:00 2004'
```

date objects can also can be compared (chronologically):

```
>>> birthday < today
True

>>> today == birthday
False
```

Times

A `datetime.time` object represents a (local) time of day to the nearest microsecond. To create a `time` object, pass the number of hours, minutes, seconds and microseconds (in that order; missing values default to zero).

```
>>> from datetime import time
>>> lunchtime = time(hour=13, minute=30)
>>> lunchtime
datetime.time(13, 30)

>>> lunchtime.isoformat()       # ISO 8601 format: HH:MM:SS if no microseconds
'13:30:00'

>>> precise_time = time(4,46,36,501982)
>>> precise_time.isoformat()    # ISO 8601 format: HH:MM:SS.mmmmmm
'04:46:36.501982'

>>> witching_hour = time(24)    # Oops: hour must satisfy 0 <= hour < 24

Traceback (most recent call last):
  File "<stdin>", line 1, in <module>
ValueError: hour must be in 0..23
```

`datetime` Objects

A `datetime.datetime` object contains the information from both the date and time objects: year, month, day, hour, minute, second, microsecond. As well as passing values for these quantities directly to the datetime constructor, the methods today (returning the current date) and now (returning the current date and time) are available:

```
>>> from datetime import datetime    # (a notoriously ugly import)
>>> now = datetime.now()
>>> now
datetime.datetime(2020, 1, 27, 10, 27, 35, 762464)

>>> now.isoformat()
'2020-01-27T10:27:35.762464'

>>> now.ctime()
'Mon Jan 27 10:27:35 2020'
```

Date and Time Formatting

date, time and datetime objects support a method, strftime, to output their values as a string formatted according to a syntax set using the format specifiers listed in Table 4.7.

```
>>> birthday.strftime('%A, %d %B %Y')
'Friday, 05 November 2004'

>>> now.strftime('%I:%M:%S on %d/%m/%y')
'10:27:35 on 27/01/20'
```

The reverse process, parsing a string into a datetime object, is the purpose of the strptime method:

```
>>> launch_time = datetime.strptime('09:32:00 July 16, 1969',
                                    '%H:%M:%S %B %d, %Y')
>>> print(launch_time)
1969-07-16 09:32:00

>>> print(launch_time.strftime('%I:%M %p on %A, %d %b %Y'))
09:32 AM on Wednesday, 16 Jul 1969
```

4.6 ◇ An Introduction to Object-Oriented Programming

4.6.1 Object-Oriented Programming Basics

Structured programming styles may be broadly divided into two categories: *procedural* and *object-oriented*. The programs we have looked at so far in this book have been *procedural* in nature: we have written functions (of the sort that would be called procedures or subroutines in other languages) that are called, passed data, and which return values from their calculations. The functions we have defined do not hold their own data or remember their state in between being called, and we haven't modified them after defining them.

An alternative programming paradigm that has gained popularity through the use of languages such as C++ and Java is *object-oriented programming*. In this context, an *object* represents a concept of some sort: this could be a physical entity, but can also be any abstract collection of components which relate to each other in a semantically

Table 4.7 strftime and strptime format specifiers. Note that many of these are locale-dependent (e.g. on a German-language system, %A will yield Sonntag, Montag, etc.).

Specifier	Description
%a	Abbreviated weekday (Sun, Mon, etc.)
%A	Full weekday (Sunday, Monday, etc.)
%w	Weekday number (0 = Sunday, 1 = Monday, ..., 6 = Saturday)
%d	Zero-padded day of month: 01, 02, 03, ..., 31
%b	Abbreviated month name (Jan, Feb, etc.)
%B	Full month name (January, February, etc.)
%m	Zero-padded month number: 01, 02, ..., 12
%y	Year without century (two-digit, zero-padded): 01, 02, ..., 99
%Y	Year with century (four-digit, zero-padded): 0001, 0002, ... 9999
%H	24-hour clock hour, zero-padded: 00, 01, ..., 23
%I	12-hour clock hour, zero-padded: 00, 01, ..., 12
%p	AM or PM (or locale equivalent)
%M	Minutes (two-digit, zero-padded): 00, 01, ..., 59
%S	Seconds (two-digit, zero-padded): 00, 01, ..., 59
%f	Microseconds (six-digit, zero-padded): 000000, 000001, ..., 999999
%%	The literal % sign

coherent way. An object holds data about itself (*attributes*) and defines functions (*methods*) for manipulating data. That manipulation may cause a change in the object's state (i.e. it may change some of the object's attributes). An object is created (*instantiated*) from a "blueprint" called a *class*, which dictates its behavior by defining its attributes and methods.

In fact, as we have already pointed out, everything in Python is an object. So, for example, a Python string is an instance of the str class. A str object possesses its own data (the sequence of characters making up the string) and provides ("*exposes*") a number of methods for manipulating that data. For example, the capitalize method returns a new string object created from the original string by capitalizing its first letter; the split method returns a list of strings by splitting up the original string:

```
>>> a = 'hello, aloha, goodbye, aloha'
>>> a.capitalize()
'Hello, aloha, goodbye, aloha'
>>> a.split(',')
['hello', ' aloha', ' goodbye', ' aloha']
```

Even indexing a sequence is really to call the method __getitem__:

```
>>> b = [10, 20, 30, 40, 50]
>>> b.__getitem__(4)
50
```

That is, a[4] is equivalent to a.__getitem__(4).[18]

---

[18] The double-underscore syntax usually denotes a name with some special meaning to Python.

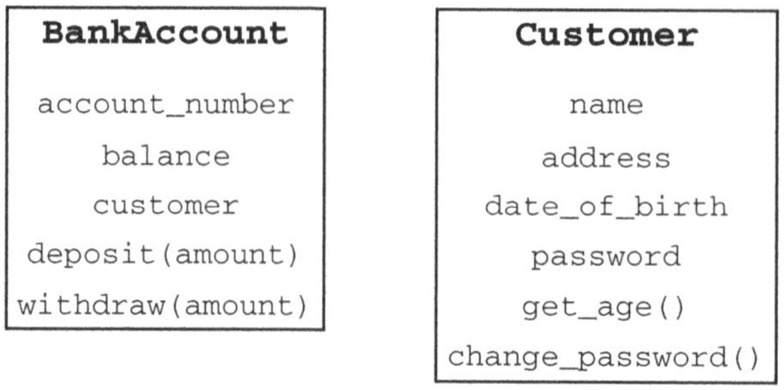

BankAccount
account_number
balance
customer
deposit(amount)
withdraw(amount)

Customer
name
address
date_of_birth
password
get_age()
change_password()

Figure 4.2 Basic classes representing a bank account and a customer.

Part of the popularity of object-oriented programming, at least for larger projects, stems from the way it helps to conceptualize the problem that a program aims to solve. It is often possible to break a problem down into units of data and operations that it is appropriate to carry out on that data. For example, a retail bank deals with people who have bank accounts. A natural object-oriented approach to managing a bank would be to define a BankAccount class, with attributes such as an account number, balance and owner, and a second, Customer, class with attributes such as a name, address and date of birth. The BankAccount class might have methods for allowing (or forbidding) transactions depending on its balance and the Customer class might have methods for calculating the customer's age from their date of birth, for example (see Figure 4.2).

An important aspect of object-oriented programming is *inheritance*. There is often a relationship between objects which takes the form of a hierarchy. Typically, a general type of object is defined by a base class, and then customized classes with more specialized functionality are derived from it. In our bank example, there may be different kinds of bank accounts: savings accounts, current (checking) accounts, etc. Each is derived from a generic base bank account, which might simply define basic attributes such as a balance and an account number. The more specialized bank account classes *inherit* the properties of the base class but may also customize them by overriding (redefining) one or more methods and may also add their own attributes and methods. This helps structure the program and encourages *code reuse* – there is no need to declare an account number separately for both savings and current accounts because both classes inherit one automatically from the base class. If a base class is not to be instantiated itself, but serves only as a template for the derived classes, it is called an *abstract class*.

In Figure 4.3, the relationship between the base class and two derived subclasses is depicted. The base class, BaseAccount, defines some attributes (account_number, balance and customer) and methods (such as deposit and withdraw) common to all types of account, and these are inherited by the subclasses. The subclass SavingsAccount adds an attribute and a method for handling interest payments on the account; the subclass CurrentAccount instead adds two attributes describing the annual account fee and transaction withdrawal limit, and overrides the base withdraw method, perhaps to check that the transaction limit has not been reached before a withdrawal is allowed.

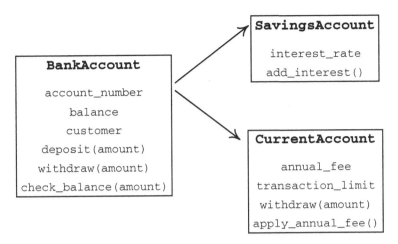

Figure 4.3 Two classes derived from an abstract base class: SavingsAccount and CurrentAccount *inherit* methods and attributes from BankAccount but also customize and extend its functionality.

4.6.2 Defining and Using Classes in Python

A class is defined using the `class` keyword and indenting the body of statements (attributes and methods) in a block following this declaration. It is conventional to give classes names written in *CamelCase*. It is a good idea to follow the `class` statement with a docstring describing what it is that the class does (see Section 2.7.1). Class methods are defined using the familiar `def` keyword, but the first argument to each method should be a variable named `self`[19] – this name is used to refer to the object itself when it wants to call its own methods or refer to attributes, as we shall see.

In our example of a bank account, the base class could be defined as follows:

Listing 4.6 The definition of the abstract base class, BankAccount

```
# bank_account.py

class BankAccount:
    """ An abstract base class representing a bank account."""
    currency = '$'

    def __init__(self, customer, account_number, balance=0):
        """
        Initialize the BankAccount class with a customer, account number
        and opening balance (which defaults to 0.)

        """

        self.customer = customer
        self.account_number = account_number
        self.balance = balance
```

---

[19] Actually, it could be named anything, but `self` is almost universally used.

```
def deposit(self, amount):
    """ Deposit amount into the bank account."""
    if amount > 0:
        self.balance += amount
    else:
        print('Invalid deposit amount:', amount)

def withdraw(self, amount):
    """

    Withdraw amount from the bank account, ensuring there are sufficient
    funds.

    """
    if amount > 0:
        if amount > self.balance:
            print('Insufficient funds')
        else:
            self.balance -= amount
    else:
        print('Invalid withdrawal amount:', amount)
```

To use this simple class, we can save the code defining it as bank_account.py and import it into a new program or the interactive Python shell with

```
from bank_account import BankAccount
```

This new program can now create BankAccount objects and manipulate them by calling the methods described earlier.

Instantiating the Object

An *instance* of a class is created with the syntax *object* = *ClassName(args)*. You may want to require that an object instantiated from a class should initialize itself in some way (perhaps by setting attributes with appropriate values) – such initialization is carried out by the special method _ _ init _ _, which receives any arguments, *args*, specified in this statement.

In our example, an account is opened by creating a BankAccount object, passing the name of the account owner (customer), an account number and, optionally, an opening balance (which defaults to 0 if not provided):

```
my_account = BankAccount('Joe Bloggs', 21457288)
```

We will replace the string customer with a Customer object in Example E4.18.

Methods and Attributes

The class defines two methods: one for depositing a (positive) amount of money and one for withdrawing money (if the amount to be withdrawn is both positive and not greater than the account balance).

The BankAccount class possesses two different kinds of attribute: self.customer, self.account_number and self.balance are *instance variables*: they can take different values for different objects created from the BankAccount class. Conversely, the variable currency is a *class variable*: this variable is defined inside the class but outside any of its methods and is shared by all instances of the class.

Both attributes and methods are accessed using the *object.attr* notation. For example,

```
>>> my_account.account_number    # access an attribute of my_account
21457288
>>> my_account.deposit(64)       # call a method of my_account
>>> my_account.balance
64
```

Let's add a third method, for printing the balance of the account. This must be defined inside the class block:

```
def check_balance(self):
    """ Print a statement of the account balance. """
    print('The balance of account number {:d} is {:s}{:.2f}'
          .format(self.account_number, self.currency, self.balance))
```

Example E4.18 We now define the Customer class described in the class diagram of Figure 4.2: an instance of this class will become the customer attribute of the BankAccount class. Note that it was possible to instantiate a BankAccount object by passing a string literal as customer. This is a consequence of Python's dynamic typing: no check is automatically made that the object passed as an argument to the class constructor is of any particular type.

The following code defines a Customer class and should be saved to a file called customer.py:

```
from datetime import datetime
class Customer:
    """ A class representing a bank customer. """

    def __ init __(self, name, address, date_of_birth):
        self.name = name
        self.address = address
        self.date_of_birth = datetime.strptime(date_of_birth, '%Y-%m-%d')
        self.password = '1234'

    def get_age(self):
        """ Calculates and returns the customer's age. """
        today = datetime.today()
        try:
            birthday = self.date_of_birth.replace(year=today.year)
        except ValueError:
            # birthday is 29 Feb but today's year is not a leap year
            birthday = self.date_of_birth.replace(year=today.year,
                                         day=self.date_of_birth.day - 1)
        if birthday > today:
            return today.year - self.date_of_birth.year - 1
        return today.year - self.date_of_birth.year
```

Then we can pass Customer objects to our BankAccount constructor:

```
>>> from bank_account import BankAccount
>>> from customer import Customer
>>>
>>> customer1 = Customer('Helen Smith', '76 The Warren, Blandings, Sussex',
```

```
                              '1976-02-29')
>>> account1 = BankAccount(customer1, 21457288, 1000)
>>> account1.customer.get_age()
39
>>> print(account1.customer.address)
76 The Warren, Blandings, Sussex
```

4.6.3   Class Inheritance in Python

A subclass may be derived from one or more other base classes with the syntax:

```
class SubClass(BaseClass1, BaseClass2, ...):
```

We will now define the two derived classes (or *subclasses*) illustrated in Figure 4.3 from the base BankAccount class. They can be defined in the same file that defines BankAccount or in a different Python file which imports BankAccount.

```
class SavingsAccount(BankAccount):
    """ A class representing a savings account. """

    def __ init __(self, customer, account_number, interest_rate, balance=0):
        """ Initialize the savings account. """
❶       self.interest_rate = interest_rate
❷       super().__ init __(customer, account_number, balance)

    def add_interest(self):
        """ Add interest to the account at the rate self.interest_rate. """

        self.balance *= (1. + self.interest_rate / 100)
```

❶   The SavingsAccount class adds a new attribute, interest_rate, and a new method, add_interest to its base class, and overrides the __ init __ method to allow interest_rate to be set when a SavingsAccount is instantiated.

❷   Note that the new __ init __ method calls the base class's __ init __ method in order to set the other attributes: the built-in function super allows us to refer to the parent base class.[20] Our new SavingsAccount might be used as follows:

```
>>> my_savings = SavingsAccount('Matthew Walsh', 41522887, 5.5, 1000)
>>> my_savings.check_balance()
The balance of account number 41522887 is $1000
>>> my_savings.add_interest()
>>> my_savings.check_balance()
The balance of account number 41522887 is $1055.00
```

The second subclass, CurrentAccount, has a similar structure:

```
class CurrentAccount(BankAccount):
    """ A class representing a current (checking) account. """
    def __ init __(self, customer, account_number, annual_fee,
                    transaction_limit, balance=0):
```

---

[20] The built-in function super() called in this way creates a "proxy" object that delegates method calls to the parent class (in this case, BankAccount).

```
""" Initialize the current account. """

self.annual_fee = annual_fee
self.transaction_limit = transaction_limit
super().__ init __(customer, account_number, balance)

def withdraw(self, amount):
    """
    Withdraw amount if sufficient funds exist in the account and amount
    is less than the single transaction limit.

    """
    if amount <= 0:
        print('Invalid withdrawal amount:', amount)
        return

    if amount > self.balance:
        print('Insufficient funds')
        return

    if amount > self.transaction_limit:
        print('{0:s}{1:.2f} exceeds the single transaction limit of'
              ' {0:s}{2:.2f}'.format(self.currency, amount,
                                      self.transaction_limit))
        return

    self.balance -= amount

def apply_annual_fee(self):
    """ Deduct the annual fee from the account balance. """

    self.balance = max(0., self.balance - self.annual_fee)
```

Note what happens if we call `withdraw` on a `CurrentAccount` object:

```
>>> my_current = CurrentAccount('Alison Wicks', 78300991, 20., 200.)
>>> my_current.withdraw(220)
Insufficient Funds

>>> my_current.deposit(750)
>>> my_current.check_balance()
The balance of account number 78300991 is $750.00

>>> my_current.withdraw(220)
$220.00 exceeds the transaction limit of $200.00
```

The `withdraw` method called is that of the `CurrentAccount` class, as this method overrides that of the same name in the base class, `BankAccount`.

Example E4.19 A simple model of a polymer in solution treats it as a sequence of randomly oriented segments; that is, one for which there is no correlation between the orientation of one segment and any other (this is the so-called *random-flight* model).

We will define a class, `Polymer`, to describe such a polymer, in which the segment positions are held in a list of `(x,y,z)` tuples. A `Polymer` object will be initialized with the values N and a, the number of segments and the segment length, respectively. The initialization method calls a `make_polymer` method to populate the segment positions list.

The `Polymer` object will also calculate the end-to-end distance, R, and will implement a method `calc_Rg` to calculate and return the polymer's *radius of gyration*, defined as

$$R_{\mathrm{g}} = \sqrt{\frac{1}{N} \sum_{i=1}^{N} (\mathbf{r}_i - \mathbf{r}_{\mathrm{CM}})^2}.$$

Listing 4.7 Polymer class

```python
# polymer.py

import math
import random

class Polymer:
    """ A class representing a random-flight polymer in solution. """

    def __ init __(self, N, a):
        """
        Initialize a Polymer object with N segments, each of length a.

        """

        self.N, self.a = N, a
        # self.xyz holds the segment position vectors as tuples.
        self.xyz = [(None, None, None)] * N
        # End-to-end vector.
        self.R = None
        # Make our polymer by assigning segment positions.
        self.make_polymer()

    def make_polymer(self):
        """
        Calculate the segment positions, center of mass and end-to-end
        distance for a random-flight polymer.

        """

        # Start our polymer off at the origin, (0, 0, 0).
        self.xyz[0] = x, y, z = cx, cy, cz = 0, 0, 0
        for i in range(1, self.N):
            # Pick a random orientation for the next segment.
            theta = math.acos(2 * random.random() - 1)
            phi = random.random() * 2. * math.pi
            # Add on the corresponding displacement vector for this segment.
            x += self.a * math.sin(theta) * math.cos(phi)
            y += self.a * math.sin(theta) * math.sin(phi)
            z += self.a * math.cos(theta)
            # Store it, and update our center of mass sum.
            self.xyz[i] = x, y, z
            cx, cy, cz = cx + x, cy + y, cz + z
        # Calculate the position of the center of mass.
        cx, cy, cz = cx / self.N, cy / self.N, cz / self.N
```

❶

❷

```
    # The end-to-end vector is the position of the last
    # segment, since we started at the origin.
    self.R = x, y, z

    # Finally, re-center our polymer on the center of mass.
    for i in range(self.N):
        self.xyz[i] = (self.xyz[i][0] - cx,
                       self.xyz[i][1] - cy,
                       self.xyz[i][2] - cz)

def calc_Rg(self):
    """

    Calculates and returns the radius of gyration, Rg. The polymer
    segment positions are already given relative to the center of
    mass, so this is just the rms position of the segments.

    """

    self.Rg = 0.
    for x, y, z in self.xyz:
        self.Rg += x**2 + y**2 + z**2
    self.Rg = math.sqrt(self.Rg / self.N)
    return self.Rg
```

❶ One way to pick the location of the next segment is to pick a random point on the surface of the unit sphere and use the corresponding pair of angles in the spherical polar coordinate system, θ and ϕ (where $0 \le \theta < \pi$ and $0 \le \phi < 2\pi$), to set the displacement from the previous segment's position as

$$\Delta x = a \sin \theta \cos \phi$$
$$\Delta y = a \sin \theta \sin \phi$$
$$\Delta z = a \cos \theta$$

❷ We calculate the position of the polymer's center of mass, $\mathbf{r}_{CM}$, and then shift the origin of the polymer's segment coordinates so that they are measured relative to this point (that is, the segment coordinates have their origin at the polymer center of mass).

We can test the `Polymer` class by importing it in the Python shell:

```
>>> from polymer import Polymer
>>> polymer = Polymer(1000, 0.5)    # a polymer with 1000 segments of length 0.5
>>> polymer.R            # end-to-end vector
(5.631332375722011, 9.408046667059947, -1.3047608473668109)
>>> polymer.calc_Rg()    # radius of gyration
5.183761585363432
```

Let's now compare the distribution of the end-to-end distances with the theoretically predicted probability density function:

$$P(R) = 4\pi R^2 \left(\frac{3}{2\pi \langle r^2 \rangle} \right)^{3/2} \exp \left(-\frac{3R^2}{2\langle r^2 \rangle} \right),$$

where the mean square position of the segments is $\langle r^2 \rangle = Na^2$.

Listing 4.8 The distribution of random flight polymers

```
# eg4-c-ii-polymer-a.py
# Compare the observed distribution of end-to-end distances for Np random-
# flight polymers with the predicted probability distribution function.

import matplotlib.pyplot as plt
from polymer import Polymer
pi = plt.pi

# Calculate R for Np polymers.
Np = 3000
# Each polymer consists of N segments of length a.
N, a = 1000, 1.
R = [None] * Np
for i in range(Np):
    polymer = Polymer(N, a)
    Rx, Ry, Rz = polymer.R
    R[i] = plt.sqrt(Rx**2 + Ry**2 + Rz**2)
    # Output a progress indicator every 100 polymers.
    if not (i+1) % 100:
        print(i+1, '/', Np)

# Plot the distribution of Rx as a normalized histogram
# using 50 bins.
plt.hist(R, 50, normed=1)

# Plot the theoretical probability distribution, Pr, as a function of r.
r = plt.linspace(0,200,1000)
msr = N * a**2
Pr = 4.*pi*r**2 * (2 * pi * msr / 3)**-1.5 * plt.exp(-3*r**2 / 2 / msr)
plt.plot(r, Pr, lw=2, c='r')
plt.xlabel('R')
plt.ylabel('P(R)')
plt.show()
```

This program produces a plot that typically looks like Figure 4.4, suggesting agreement with theory.

4.6.4   Classes and Operators

Operators (such as +, * and <=) and built-in functions, such as len and abs act on Python objects by calling special methods these objects define with names beginning and ending with two underscores, __ (so-called "dunder" methods). To implement ("overload") this functionality on custom classes, simply define methods with these names. A complete list of these special methods can be found in the Python language documentation,[21] but Table 4.8 provides a list of the more commonly needed ones. For example, the expression x + y calls x.__add__(y).

Python is a polymorphic language, and there may be circumstances in which x and y have different types. If the object x does not implement the necessary method, then

---

[21] https://docs.python.org/3/reference/datamodel.html.

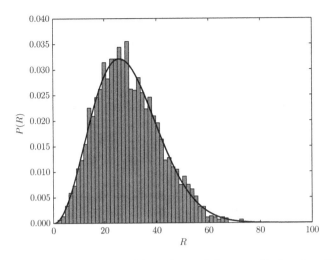

Figure 4.4 Distribution of the end-to-end distances, R, of random flight-polymers with $N = 1000, a = 1$.

Table 4.8 Common Python special methods

Method	Description	Example
__add__	+, addition	x + y
__sub__	-, subtraction	x - y
__mul__	*, multiplication	x * y
__truediv__	/, "true" division	x / y
__floordiv__	//, floor division	x // y
__mod__	%, modulus	x % y
__pow__	**, exponentiation	x ** y
__neg__	negation (unary minus)	-x
__matmul__	@, matrix multiplication	x @ y
__abs__	absolute value	abs(x)
__contains__	membership	y in x
__lt__	less than	y < x
__le__	less than or equal to	y <= x
__eq__	equal to	y == x
__ne__	not equal to[†]	y != x
__gt__	greater than	y > x
__ge__	greater than or equal to	y >= x
__str__	human-readable string representation	str(x)
__repr__	unambiguous string representation	repr(x)

† If not explicitly implemented, __ne__ calls __eq__ and inverts the result.

Python will look for a "reflected" version in the y object. Hence, the expression 'a' * 4 calls 'a'.__mul__(4) on the string object 'a'; the expression 4 * 'a' first tries to call 4.__mul__('a'), and when this fails (int objects do not know how to be multiplied by strs), then tries the reflected version 'a'.__rmul__(4) which returns 'aaaa': str objects know how to be multiplied by ints.

The special methods __str__ and __repr__ deserve special mention. Both return a string representation of the object, but whilst __str__ is expected to return a *human-readable* string, the goal of __repr__ is, as far as possible, to be *unambiguous*. Depending on the class, there may be a natural choice for the return value of __str__ that communicates the essential properties of an instance, whilst the return value of __repr__ should aim to be complete enough that the information it contains could be used for debugging or to create an identical instance. Note that for an object, obj, if __repr__ is defined but __str__ is not, then str(obj) will return obj.__repr__(). A class should always define a __repr__() method, and optionally define a __str__() if an easy to comprehend string is also required.

Example E4.20   Although NumPy (see Chapter 6) offers a faster option, it is still instructive to code a class for vectors in pure Python. The following code defines the Vector2D class and tests it for various operations.

Listing 4.9 A simple class representing a two-dimensional Cartesian vector

```python
import math

class Vector2D:
    """A two-dimensional vector with Cartesian coordinates."""

    def __init__(self, x, y):
        self.x, self.y = x, y

    def __str__(self):
        """Human-readable string representation of the vector."""
        return '{:g}i + {:g}j'.format(self.x, self.y)

    def __repr__(self):
        """Unambiguous string representation of the vector."""
        return repr((self.x, self.y))

    def dot(self, other):
        """The scalar (dot) product of self and other. Both must be vectors."""
        if not isinstance(other, Vector2D):
            raise TypeError('Can only take dot product of two Vector2D objects')
        return self.x * other.x + self.y * other.y
    # Alias the __matmul__ method to dot so we can use a @ b as well as a.dot(b).
    __matmul__ = dot

    def __sub__(self, other):
        """Vector subtraction."""
        return Vector2D(self.x - other.x, self.y - other.y)

    def __add__(self, other):
        """Vector addition."""
        return Vector2D(self.x + other.x, self.y + other.y)

    def __mul__(self, scalar):
        """Multiplication of a vector by a scalar."""
```

❶

❷
```
            if isinstance(scalar, int) or isinstance(scalar, float):
                return Vector2D(self.x*scalar, self.y*scalar)
            raise NotImplementedError('Can only multiply Vector2D by a scalar')

    def __rmul__(self, scalar):
        """Reflected multiplication so vector * scalar also works."""
        return self.__mul__(scalar)

    def __neg__(self):
        """Negation of the vector (invert through origin.)"""
        return Vector2D(-self.x, -self.y)

    def __truediv__(self, scalar):
        """True division of the vector by a scalar."""
        return Vector2D(self.x / scalar, self.y / scalar)

    def __mod__(self, scalar):
        """One way to implement modulus operation: for each component."""
        return Vector2D(self.x % scalar, self.y % scalar)

    def __abs__(self):
        """Absolute value (magnitude) of the vector."""
        return math.sqrt(self.x**2 + self.y**2)

    def distance_to(self, other):
        """The distance between vectors self and other."""
        return abs(self - other)

    def to_polar(self):
        """Return the vector's components in polar coordinates."""
        return self.__abs__(), math.atan2(self.y, self.x)
```

❸
```
if __name__ == '__main__':
    v1 = Vector2D(2, 5/3)
    v2 = Vector2D(3, -1.5)
    print('v1 = ', v1)
    print('repr(v2) = ', repr(v2))
    print('v1 + v2 = ', v1 + v2)
    print('v1 - v2 = ', v1 - v2)
    print('abs(v2 - v1) = ', abs(v2 - v1))
    print('-v2 = ', -v2)
    print('v1 * 3 = ', v1 * 3)
    print('7 * v2 = ', 7 * v1)
    print('v2 / 2.5 = ', v2 / 2.5)
    print('v1 % 1 = ', v1 % 1)
    print('v1.dot(v2) = v1 @ v2 = ', v1 @ v2)
    print('v1.distance_to(v2) = ',v1.distance_to(v2))
    print('v1 as polar vector, (r, theta) =', v1.to_polar())
```

❶ Raise an exception if operands for the dot product are not both vectors.

❷ Only allow multiplication of a vector by a scalar quantity, but support both *av* and *va*.

❸ Code inside this block is only executed if the code is run as the main program, in which case Python will have set the variable `__name__` to the hard-coded string `'__main__'`; if the file is treated as a module and imported (e.g. from `vector2d` import `Vector2D`), this block is ignored.

The output should be:

```
v1 =   2i + 1.66667j
repr(v2) =   (3, -1.5)
v1 + v2 =   5i + 0.166667j
v1 - v2 =   -1i + 3.16667j
abs(v2 - v1) =   3.3208098075285464
-v2 =   -3i + 1.5j
v1 * 3 =   6i + 5j
7 * v2 =   14i + 11.6667j
v2 / 2.5 =   1.2i + -0.6j
v1 % 1 =   0i + 0.666667j
v1.dot(v2) = v1 @ v2 =   3.5
v1.distance_to(v2) =   3.3208098075285464
v1 as polar vector, (r, theta) = (2.6034165586355518, 0.6947382761967033)
```

Example E4.21   The code below uses the above `Vector2D` class to implement a simple molecular dynamics simulation of circular particles with identical masses moving in two dimensions. All particles initially have the same speed; the collisions equilibrate the speeds to the Maxwell–Boltzmann distribution, as demonstrated by the figure produced (Figure 4.5). The website accompanying this book provides further code for an animation of the simulation (https://scipython.com/eg/baa). *Note*: whilst elegant, the object-oriented approach taken here is not the fastest: there is an overhead to instantiating multiple objects, which becomes significant when many particles and collisions need to be considered at each time step. For a faster, NumPy-only approach, see the links from this web page.

Listing 4.10 A simple two-dimensional molecular dynamics simulation

```python
import math
import random
import matplotlib.pyplot as plt
from vector2d import Vector2D

class Particle:
    """A circular particle of unit mass with position and velocity."""

    def __init__(self, x, y, vx, vy, radius=0.01):
        self.pos = Vector2D(x, y)
        self.vel = Vector2D(vx, vy)
```

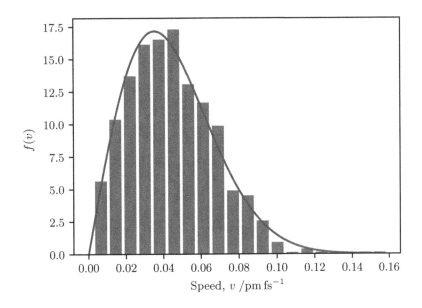

Figure 4.5 Distribution of particle speeds after equilibration to Maxwell–Boltzmann statistics through multiple collisions.

```
        self.radius = radius

    def advance(self, dt):
        """Advance the particle's position according to its velocity."""

        # Use periodic boundary conditions: a Particle that moves across an
        # edge of the domain 0<=x<1, 0<=y<1 magically reappears at the opposite
        # edge.
        self.pos = (self.pos + self.vel * dt) % 1

    def distance_to(self, other):
        """Return the distance from this Particle to other Particle."""
        return self.pos.distance_to(other.pos)

    def get_speed(self):
        """Return the speed of the Particle from its velocity."""
        return abs(self.vel)


class Simulation:
    """A simple simulation of circular particles in motion."""

    def __init__(self, nparticles=100, radius=0.01, v0=0.05):
        self.nparticles = nparticles
        self.radius = radius
        # Randomly initialize the particles' positions and velocity directions.
```

```python
        self.particles = [self.init_particle(v0) for i in range(nparticles)]
        self.t = 0

    def init_particle(self, v0=0.05):
        """Return a new Particle object with random position and velocity.

        The position is chosen uniformly from 0 <= x < 1, 0 <= y < 1;
        The velocity has fixed magnitude, v0, but random direction.

        """

        x, y = random.random(), random.random()
        theta = 2*math.pi * random.random()
        self.v0 = v0
        vx, vy = self.v0 * math.cos(theta), self.v0 * math.sin(theta)
        return Particle(x, y, vx, vy, self.radius)

    def advance(self, dt):
        """Advance the Simulation by dt in time, handling collisions."""

        self.t += dt
        for particle in self.particles:
            particle.advance(dt)

        # Find all distinct pairs of Particles currently undergoing a collision.
        colliding_pair = []
        for i in range(self.nparticles):
            pi = self.particles[i]
            for j in range(i+1, self.nparticles):
                pj = self.particles[j]
                # pi collides with pj if their separation is less than twice
                # their radius.
                if pi.distance_to(pj) < 2 * self.radius:
                    colliding_pair.append((i, j))

        print('ncollisions =', len(colliding_pair))
        # For each pair, the velocities change according to the kinetics of
        # an elastic collision between circles.
        for i,j in colliding_pair:
            p1, p2 = self.particles[i], self.particles[j]
            r1, r2 = p1.pos, p2.pos
            v1, v2 = p1.vel, p2.vel
            dr, dv = r2 - r1, v2 - v1
            dv_dot_dr = dv.dot(dr)
            d = r1.distance_to(r2)**2
            p1.vel = v1 - dv_dot_dr / d * (r1 - r2)
            p2.vel = v2 - dv_dot_dr / d * (r2 - r1)

if __name__ == '__main__':
    import numpy as np

    sim = Simulation(nparticles=1000, radius=0.005, v0=0.05)
    dt = 0.02

    nit = 500
    dnit = nit // 10
```

```
for i in range(nit):
    if not i % dnit:
        print(f'{i}/{nit}')
    sim.advance(dt)

# Plot a histogram of the Particles' speeds.
nbins = sim.nparticles // 50
hist, bins, _ = plt.hist([p.get_speed() for p in sim.particles], nbins,
                         density=True)
v = (bins[1:] + bins[:-1])/2

# The mean kinetic energy per Particle.
KE = sim.v0**2 / 2
# The Maxwell-Boltzmann equilibrium distribution of speeds.
a = 1 / 2 / KE
f = 2*a * v * np.exp(-a*v**2)
plt.plot(v, f)

plt.show()
```

4.6.5   Exercises

Problems

P4.6.1

(a) Modify the base BankAccount class to verify that the account number passed to its __init__ constructor conforms to the Luhn algorithm described in Exercise P2.5.3.

(b) Modify the CurrentAccount class to implement a free overdraft. The limit should be set in the __init__ constructor; withdrawals should be allowed to within the limit.

P4.6.2   Add a method, save_svg to the Polymer class of Example E4.19 to save an image of its polymer as an SVG file. Refer to Exercise P4.4.3 for a template of an SVG file.

P4.6.3   Write a program to create an image of a constellation using the data from the Yale Bright Star Catalog (http://tdc-www.harvard.edu/catalogs/bsc5.html).

Create a class, Star, to represent a star with attributes for its name, magnitude and position in the sky, parsed from the file bsc5.dat which forms part of the catalog. Implement a method for this class which converts the star's position on the celestial sphere as (Right Ascension: α, Declination: δ) to a point in a plane, (x, y), for example using the orthographic projection about a central point (α_0, δ_0):

$$\Delta\alpha = \alpha - \alpha_0$$
$$x = \cos\delta \sin\Delta\alpha$$
$$y = \sin\delta \cos\delta_0 - \cos\delta \cos\Delta\alpha \sin\delta_0$$

Suitably scaled, projected, star positions can be output to an SVG image as `circles` (with a larger radius for brighter stars). For example, the line

```
<circle cx="200" cy="150" r="5" stroke="none" fill="#ffffff"/>
```

represents a white circle of radius 5 pixels, center on the canvas at (200, 150).

Hint: you will need to convert the right ascension from (hr, min, sec) and the declination from (deg, min, sec) to radians. Use the data corresponding to "equinox J2000, epoch 2000.0" in each line of `bsc5.dat`. Let the user select the constellation from the command line using its three-letter abbreviation (e.g. "Ori" for Orion): this is given as part of the star name in the catalog. Don't forget that star magnitudes are *smaller* for *brighter* stars. If you are using the orthographic projection suggested, choose (α_0, δ_0) to be the mean of (α, δ) for stars in the constellation.

P4.6.4 Design and implement a class, `Experiment`, to read in and store a simple series of (x, y) data as NumPy arrays from a text file. Include in your class methods for transforming the data series by some simple function (e.g. $x' = \ln x$, $y' = 1/y$) and to perform a linear least-squares regression on the transformed data (returning the gradient and intercept of the best-fit line, $y'_{\mathrm{fit}} = mx' + c$). NumPy provides methods for performing linear regression (see Section 6.5.3), but for this exercise the following equations can be implemented directly:

$$m = \frac{\overline{xy} - \bar{x}\bar{y}}{\overline{x^2} - \bar{x}^2},$$
$$c = \bar{y} - m\bar{x},$$

where the bar notation, $\bar{\cdot}$, denotes the arithmetic mean of the quantity under it. (*Hint*: use `np.mean(arr)` to return the mean of array `arr`.)

Chloroacetic acid is an important compound in the synthetic production of phamaceuticals, pesticides and fuels. At high concentration under strong alkaline conditions, its hydrolysis may be considered as the following reaction:

$$ClCH_2COO^- + OH^- \rightleftharpoons HOCH_2COO^- + Cl^-.$$

Data giving the concentration of $ClCH_2COO^-$, c (in M), as a function of time, t (in s), are provided for this reaction carried out in excess alkalai at five different temperatures in the data files `caa-T.txt` ($T = 40, 50, 60, 70, 80$ in °C): these may be obtained from https://scipython.com/ex/bde. The reaction is known to be second-order and so obeys the integrated rate law

$$\frac{1}{c} = \frac{1}{c_0} + kt,$$

where k is the effective rate constant and c_0 the initial ($t = 0$) concentration of chloroacetic acid.

Use your `Experiment` class to interpret these data by linear regression of $1/c$ against t, determining $m(\equiv k)$ for each temperature. Then, for each value of k, determine the activation energy of the reaction through a second linear regression of $\ln k$ against $1/T$ in accordance with the Arrhenius law:

$$k = Ae^{-E_a/RT} \quad \Rightarrow \quad \ln k = \ln A - \frac{E_a}{RT},$$

where $R = 8.314 \ \text{J}\,\text{K}^{-1}\,\text{mol}^{-1}$ is the gas constant. Note: the temperature must be in kelvins.

P4.6.5   Create a new class that derives from the `list` object class which re-implements it with one-based indexing instead of zero-based indexing. Overload as many of the special methods listed at https://docs.python.org/3/reference/datamodel.html as necessary and write tests to validate your code.

5 IPython and Jupyter Notebook

The IPython shell and the related interactive, browser-based, Jupyter Notebook are two related, powerful interfaces to the Python language. IPython has several advantages over the native Python shell, including easy interaction with the operating system, introspection and tab completion. Jupyter Notebook (formerly IPython Notebook) is increasingly being adopted by scientists to share their data and the code they write to analyze it in a standardized manner that aids reproducibility and visualization. Its default execution environment ("kernel") is Python, but it can be configured to work with any of several dozen languages.

5.1 IPython

5.1.1 Installing IPython

Comprehensive details on installing IPython are available at the IPython website: see https://ipython.org/install.html, but a summary is provided here.

IPython is included in the Continuum Anaconda Python distribution. To update to the current version within Anaconda, use the conda package manager:

```
conda update conda
conda update ipython
```

If you already have Python installed, there are several alternative options. If you have the pip package manager:

```
pip install ipython
```

It is also possible to manually download the latest IPython version from its GitHub repository at https://github.com/ipython/ipython/releases and compile and install from its top-level source directory with

```
python setup.py install
```

5.1.2 Using the IPython Shell

To start an interactive IPython shell session from the command line, simply type ipython. You should be greeted with a message similar to this one:

```
Python 3.7.3 (default, Mar 27 2019, 16:54:48)
Type 'copyright', 'credits' or 'license' for more information
IPython 7.6.1 -- An enhanced Interactive Python. Type '?' for help.

In [1]:
```

(The precise details of this message will depend on the setup of your system.) The prompt In [1]: is where you type your Python statements and replaces the native Python >>> shell prompt. The counter in square brackets increments with each Python statement or code block. For example,

```
In [1]: 4 + 5
Out[1]: 9
In [2]: print(1)
1
In [3]: for i in range(4):
   ...:     print(i, end='')
   ...:
0123
In [4]:
```

To exit the IPython shell, type quit or exit. Unlike with the native Python shell, no parentheses are required.[1]

Help Commands

As listed in the welcome message, there are various helpful commands to obtain information about using IPython:

- Typing a single "?" outputs an overview of the usage of IPython's main features (page down with the space bar or f; page back up with b; exit the help page with q).
- %quickref provides a brief reference summary of each of the main IPython commands and "magics" (see Section 5.1.3).
- help() or help(*object*) invokes Python's own help system (interactively or for *object* if specified).
- Typing one question mark after an object name provides information about that object: see below.

Possibly the most frequently used help functionality provided by IPython is the *introspection* provided by the object? syntax. For example,

```
In [4]: a = [5, 6]
In [5]: a?
Type:        list
String form: [5, 6]
Length:      2
Docstring:
Built-in mutable sequence.

If no argument is given, the constructor creates a new empty list.
```

---

[1] Some find this alone a good reason to use IPython.

```
The argument must be an iterable if specified.
```

Here, the command a? gives details about the object a: its string representation (which would be produced by, for example, print(a)), its length (equivalent to len(a)) and the docstring associated with the class of which it is an instance: since a is a list, this provides brief details of how to instantiate a list object.[2]

The ? syntax is particularly useful as a reminder of the arguments that a function or method takes. For example,

```
In [6]: import numpy as np
In [7]: np.linspace?

Signature:
np.linspace(
    start,
    stop,
    num=50,
    endpoint=True,
    retstep=False,
    dtype=None,
    axis=0,
)
Docstring:
Return evenly spaced numbers over a specified interval.

Returns `num` evenly spaced samples, calculated over the
interval [`start`, `stop`].

The endpoint of the interval can optionally be excluded.

.. versionchanged:: 1.16.0
    Non-scalar `start` and `stop` are now supported.

Parameters
----------
start : array_like
    The starting value of the sequence.
stop : array_like
    The end value of the sequence, unless `endpoint` is set to False.
    In that case, the sequence consists of all but the last of ``num + 1``
    evenly spaced samples, so that `stop` is excluded.  Note that the step
    size changes when `endpoint` is False.
num : int, optional
    Number of samples to generate. Default is 50. Must be non-negative.
endpoint : bool, optional
    If True, `stop` is the last sample. Otherwise, it is not included.
    Default is True.
retstep : bool, optional
    If True, return (`samples`, `step`), where `step` is the spacing
    between samples.
dtype : dtype, optional
    The type of the output array.  If `dtype` is not given, infer the data
```

---

[2] This is what is meant by introspection: Python is able to inspect its own objects and provide information about them.

```
        type from the other input arguments.

        .. versionadded:: 1.9.0


...
```

For some objects, the syntax object?? returns more advanced information such as the location and details of its source code.

Tab Completion

Just as with many command line shells, IPython supports tab completion: start typing the name of an object or keyword, press the <TAB> key, and it will autocomplete it for you or provide a list of options if more than one possibility exists. For example,

```
In [8]: w<TAB>
        while          %who_ls
        with           %whos
        %who           %%writefile
```

If you resume typing until the word becomes unambiguous (e.g. add the letters hi) and then press <TAB> again, it will be autocompleted to while. The options with percent signs in front of them are "magic functions," described in Section 5.1.3.

History

You may already have used the native Python shell's command history functionality (pressing the up and down arrows through previous statements typed during your current session). IPython stores both the commands you enter and the output they produce in the special variables In and Out (these are, in fact, a list and a dictionary, respectively, and correspond to the prompts at the beginning of each input and output). For example,

```
In [9]: d = {'C': 'Cador', 'G': 'Galahad', 'T': 'Tristan', 'A': 'Arthur'}
In [10]: for a in 'ACGT':
    ....:     print(d[a])
    ....:
Arthur
Cador
Galahad
Tristan
In [11]: d = {'C': 'Cytosine', 'G': 'Guanine', 'T': 'Thymine', 'A': 'Adenine'}
❶ In [12]: In[10]
Out[12]: "for a in 'ACGT':\n    print(d[a])\n    "
❷ In [13]: exec(In[10])
Adenine
Cytosine
Guanine
Thymine
```

❶ Note that In[10] simply holds the string version of the Python statement (here a for loop) that was entered at index 10.

❷ To actually execute the statement (with the *current* dictionary d), we must send it to Python's exec built-in (see also the %rerun magic, Section 5.1.3).

There are a couple of further shortcuts: the alias _i*N* is the same as In[*N*], _*N* is the same as Out[*N*], and the two most recent outputs are returned by the variables _ and __, respectively.

To view the contents of the history, use the %history or %hist magic function. By default only the entered statements are output; it is often more useful to output the line numbers as well, which is achieved using the -n option:

```
In [14]: %history -n
   1: 4 + 5
   2: print(1)
   3:
for i in range(4):
    print(i)
   4: a = [5, 6]
   5: a?
   6: import numpy as np
   7: np.linspace?
   8: d = {'C': 'Cador', 'G': 'Galahad', 'T': 'Tristan', 'A': 'Arthur'}
  10:
for a in 'ACGT':
    print(d[a])
  11: d = {'C': 'Cytosine', 'G': 'Guanine', 'T': 'Thymine', 'A': 'Adenine'}
  12: In[10]
  13: exec(In[10])
  14: %history -n
```

To output a specific line or range of lines, refer to them by number and/or number range when calling %history:

```
In [15]: %history 4
a = [5, 6]

In [16]: %history -n 2-5
   2: print(1)
   3:
for i in range(4):
    print(i)
   4: a = [5, 6]
   5: a?

In [17]: %history -n 1-3 7 12-14
   1: 4 + 5
   2: print(1)
   3:
for i in range(4):
    print(i)
   7: np.linspace?
  12: In[10]
  13: exec(In[10])
  14: %history -n
```

This syntax is also used by several other IPython magic functions (see the following section). The %history function can also take an additional option: -o displays the output as well as the input.

Pressing CTRL-R brings up a prompt, the somewhat cryptic

```
      I-search backward:
```
from which you can search within your command history.[3]

Interacting with the Operating System

IPython makes it easy to execute operating-system commands from within your shell session: any statement preceded by an exclamation mark, !, is sent to the operating-system command line (the "system shell") instead of being executed as a Python statement. For example, you can delete files, list directory contents and even execute other programs and scripts:

```
In [11]: !pwd              # return the current working directory
/Users/christian/research
In [12]: !ls               # list the files in this directory
Meetings        Papers          code            books
databases       temp-file
In [13]: !rm temp-file     # delete temp-file

In [14]: !ls
Meetings        Papers          code            books
databases
```

Note that, for technical reasons,[4] the cd (Unix-like systems) and chdir (Windows) commands must be executed as IPython magic functions:

```
In [15]: %cd /       # change into root directory
In [16]: !ls
Applications    Volumes         usr             Library
bin             net             Network         cores
opt             www             System          dev
private         sbin            Users           home
In [17]: %cd ~/temp   # change directory to temp within user's home directory
In [18]: !ls
output.txt      test.py         readme.txt      utils
zigzag.py
```

If you use Windows and want to include a drive letter (such as C:) in the directory path you should enclose the path in quotes: %cd 'C:\My Documents'.

Help, via !command?, and tab completion, as described above, work within operating-system commands.

You can pass the values of Python variables to operating-system commands by prefixing the variable name with a dollar sign, $:

```
In [19]: python_script = 'zigzag.py'
In [20]: !ls $python_script
zigzag.py
In [21]: text_files = '*.txt'
In [22]: text_file_list = !ls $text_files
In [23]: text_file_list
```
❶

---

[3] This functionality may be familiar to users of the bash shell as (reverse-i-search)`':.

[4] System commands executed via the !command method spawn their own shell, which is discarded immediately afterward; changing a directory occurs only in this spawned shell and is not reflected in the one running IPython.

```
output.txt    readme.txt
In [24]: readme_file = text_file_list[1]
In [25]: !cat $readme_file
This is the file readme.txt
Each line of the file appears as an item
in a list when returned from !cat readme.txt
```

❷ ```
In [26]: readme_lines = !cat $readme_file
```

```
In [27]: readme_lines
Out[28]:
['This is the file readme.txt',
 'Each line of the file appears as an item',
 'in a list when returned from !cat readme.txt']
```

❶ Note that the output of a system command can be assigned to a Python variable, here a list of the .txt files in the current directory.

❷ The cat system command returns the contents of the text file; IPython splits this output on the newline character and assigns the resulting list to readme_lines. See also Section 5.1.3

5.1.3   IPython Magic Functions

IPython provides many "magic" functions (or simply *magics*, those commands prefixed with %) to speed up coding and experimenting within the IPython shell. Some of the more useful ones are described in this section; for more advanced information the reader is referred to the IPython documentation.[5] IPython makes a distinction between *line magics*: those whose arguments are given on a single line, and *cell magics* (prefixed by two percent signs, %%): those which act on a series of Python commands. An example is given in Section 5.1.3 where the %%timeit cell magic is described.

A list of currently available magic functions can be obtained by typing %lsmagic.

The magic function %automagic toggles the "automagic" setting: its default is ON, meaning that typing the name of a magic function without the % will also execute that function, unless you have bound the name as a Python identifier (variable name) to some object. The same principle applies to system commands:

```
In [x]: ls
output.txt         test.py          readme.txt       utils
zigzag.py
In [x]: ls = 0
In [x]: ls            # now ls is an integer; !ls will still work
Out[x]: 0
```

Table 5.1 summarizes some useful IPython magics; the following subsections explain more fully the less straightforward ones.

---

[5] https://ipython.org/documentation.html.
```

Table 5.1 Useful IPython line magics

Magic	Description
%alias	Create an alias to a system command
%alias_magic	Create an alias to an existing IPython magic
%bookmark	Interact with IPython's directory bookmarking system
%cd	Change the current working directory
%dhist	Output a list of visited directories
%edit	Create or edit Python code within a text editor and then execute it
%env	List the system environment variables, such as $HOME
%history	List the input history for this IPython session
%load	Read in code from a provided file and make it available for editing
%macro	Define a named macro from previous input for future reexecution
%paste	Paste input from the clipboard: use this in preference to, for example, CTRL-V, to handle code indenting properly
%recall	Place one or more input lines from the command history at the current input prompt
%rerun	Reexecute previous input from the numbered command history
%reset	Reset the namespace for the current IPython session
%run	Execute a named file as a Python script within the current session
%save	Save a set of input lines or macro (defined with %macro) to a file with a given name
%sx or !!	Shell execute: run a given shell command and store its output
%timeit	Time the execution of a provided Python statement
%who	Output all the currently defined variables
%who_ls	As for %who, but return the variable names as a list of strings
%whos	As for %who, but provides more information about each variable

Aliases and Bookmarks

A system shell command can be given an *alias*: a shortcut for a shell command that can be called as its own magic. For example, on Unix-like systems we could define the following alias to list only the directories on the current path:

```
In [x]: %alias lstdir ls -d */
In [x]: %lstdir
Meetings/        Papers/          code/          books/
databases/
```

Now typing %lstdir has the same effect as !ls -d */. If %automagic is ON this alias can also simply be called with lstdir.

The magic %alias_magic provides a similar functionality for IPython magics. For example, if you want to use %h as an alias to %history, type:

```
In [x]: %alias_magic h history
```

When working on larger projects it is often necessary to switch between different directories. IPython has a simple system for maintaining a list of bookmarks which act as shortcuts to different directories. The syntax for this magic function is

```
%bookmark <name> [directory]
```

If [directory] is omitted, it defaults to the current working directory.

```
In [x]: %bookmark py ~/research/code/python
In [x]: %bookmark www /srv/websites
In [x]: %cd py
/Users/christian/research/code/python
```

It may happen that a directory with the same name as your bookmark is within the current working directory. In that case, this directory takes precedence and you must use %cd -b <name> to refer to the bookmark.

A few more useful commands include:

- %bookmark -l: list all bookmarks;
- %bookmark -d <name>: remove bookmark <name>;
- %bookmark -r: remove all bookmarks.

Timing Code Execution

The IPython magic %timeit <statement> times the execution of the *single-line* statement <statement>. The statement is executed N times in a loop, and each loop is repeated R times. N is a suitable, usually large, number chosen by IPython to yield meaningful results and R is, by default, 3. The average time per loop for the best of the R repetitions is reported. For example, to profile the sorting of a random arrangement of the numbers 1–100:

```
In [x]: import random
In [x]: numbers = list(range(1, 101))
In [x]: random.shuffle(numbers)
In [x]: %timeit sorted(numbers)
100000 loops, best of 3: 13.2 μs per loop
```

Obviously the execution time will depend on the system (processor speed, memory, etc.). The aim of repeating the execution many times is to allow for variations in speed due to other processes running on the system. You can select N and R explicitly by passing values to the options -n and -r respectively:

```
In [x]: %timeit -n 10000 -r 5 sorted(numbers)
10000 loops, best of 5: 11.2 μs per loop
```

The cell magic %%timeit enables one to time a *multiline block* of code. For example, a naive algorithm to find the factors of an integer n can be examined with

```
In [x]: n = 150
In [x]: %%timeit
factors = set()
for i in range(1, n+1):
    if not n % i:
        factors.add(n // i)
 ....:

100000 loops, best of 3: 16.3 μs per loop
```

Recalling and Rerunning Code

To reexecute one or more lines from your IPython history, use %rerun with a line number or range of line numbers:

```
In [1]: import math
In [2]: angles = [0, 30, 60, 90]
In [3]: for angle in angles:
    sine_angle = math.sin(math.radians(angle))
    print('sin({:3d}) = {:8.5f}'.format(angle, sine_angle))
    .....:
sin(  0) =  0.00000
sin( 30) =  0.50000
sin( 45) =  0.70711
sin( 60) =  0.86603
sin( 90) =  1.00000

In [4]: angles = [15, 45, 75]
In [5]: %rerun 3
=== Executing: ===
for angle in angles:
    sine_angle = math.sin(math.radians(angle))
    print('sin({:3d}) = {:8.5f}'.format(angle, sine_angle))

=== Output: ===
sin( 15) =  0.25882
sin( 45) =  0.70711
sin( 75) =  0.96593

In [6]: %rerun 2-3
=== Executing: ===
angles = [0, 30, 45, 60, 90]
for angle in angles:
    sine_angle = math.sin(math.radians(angle))
    print('sin({:3d}) = {:8.5f}'.format(angle, sine_angle))

=== Output: ===
sin(  0) =  0.00000
sin( 30) =  0.50000
sin( 45) =  0.70711
sin( 60) =  0.86603
sin( 90) =  1.00000
```

The similar magic function %recall places the requested lines at the command prompt but does not execute them until you press Enter, allowing you to modify them first if you need to.

If you find yourself reexecuting a series of statements frequently, you can define a named macro to invoke them. Specify line numbers as before:

```
In [7]: %macro sines 3
Macro `sines` created. To execute, type its name (without quotes).
=== Macro contents: ===
for angle in angles:
    sine_angle = math.sin(math.radians(angle))
    print('sin({:3d}) = {:8.5f}'.format(angle, sine_angle))


In [8]: angles = [-45, -30, 0, 30, 45]
In [9]: sines
sin(-45) = -0.70711
```

```
sin(-30) = -0.50000
sin(  0) =  0.00000
sin( 30) =  0.50000
sin( 45) =  0.70711
```

Loading, Executing and Saving Code

To load code from an external file into the current IPython session, use

```
%load <filename>
```

If you want only certain lines from the input file, specify them after the -r option. This magic enters the lines at the command prompt, so they can be edited before being executed.

To load *and execute* code from a file, use

```
%run <filename>
```

Pass any command line options after *filename*; by default IPython treats them the same way that the system shell would. There are a few additional options to %run:

- -i: run the script in the current IPython namespace instead of an empty one (i.e. the program will have access to variables defined in the current IPython session);
- -e: ignore sys.exit() calls and SystemExit exceptions;
- -t: output timing information at the end of execution (pass an integer to the additional option -N to repeat execution that number of times).

For example, to run my_script.py 10 times from within IPython with timing information:

```
In [x]: %run -t -N10 my_script.py
```

To save a range of input lines or a macro to a file, use %save. Line numbers are specified using the same syntax as %history. A .py extension is added if you don't add it yourself, and confirmation is sought before overwriting an existing file. For example,

```
In [x]: %save sines1 1 8 3
The following commands were written to file `sines1.py`:
import math
angles = [-45, -30, 0, 30, 45]
for angle in angles:
    print('sin({:3d}) = {:8.5f}'.format(angle, math.sin(math.radians(angle))))
```

```
In [x]: %save sines2 1-3
The following commands were written to file `sines2.py`:
import math
angles = [0, 30, 60, 90]
for angle in angles:
    print('sin({:3d}) = {:8.5f}'.format(angle, math.sin(math.radians(angle))))
```

Finally, to *append* to a file instead of overwriting it, use the -a option:

```
%save -a <filename> <line numbers>
```

Capturing the Output of a Shell Command

The IPython magic %sx *command*, equivalent to !!*command* executes the shell command *command* and returns the resulting output as a list (split into semantically useful parts on the newline character so there is one item per line). This list can be assigned to a variable to be manipulated later. For example,

```
In [x]: current_working_directory = %sx pwd
In [x]: current_working_directory
['/Users/christian/temp']
In [x]: filenames = %sx ls
In [x]: filenames
Out[x]:
['output.txt',
 'test.py',
 'readme.txt',
 'utils',
 'zigzag.py']
```

Here, filenames is a list of individual filenames.

The returned object is actually an IPython.utils.text.SList string list object. Among the useful additional features provided by SList are a native method for splitting each string into fields delimited by whitespace: fields; for sorting on those fields: sort; and for searching within the string list: grep. For example,

```
In [x]: files = %sx ls -l
In [x]: files
['total 8',
 '-rw-r--r--  1 christian  staff     93  5 Nov 16:30 output.txt',
 '-rw-r--r--  1 christian  staff  23258  5 Nov 16:31 readme.txt',
 '-rw-r--r--  1 christian  staff    218  5 Nov 16:32 test.py',
 'drwxr-xr-x  2 christian  staff     68  5 Nov 16:32 utils',
 '-rw-r--r--  1 christian  staff    365  5 Nov 16:20 zigzag.py']
In [x]: del files[0]    # strip non-file line 'total 8'
In [x]: files.fields()
Out[x]:
[['-rw-r--r--', '1', 'christian', 'staff', '93', '5', 'Nov', '16:30', 'output.txt'],
 ['-rw-r--r--', '1', 'christian', 'staff', '23258', '5', 'Nov', '16:31', 'readme
.txt'],
 ...
 ['-rw-r--r--', '1', 'christian', 'staff', '365', '5', 'Nov', '16:20', 'zigzag.py']]

In [x]: ['{} last modified at {} on {} {}'.format(f[8], f[7], f[5], f[6])
                         for f in files.fields()]
Out[x]:
['output.txt last modified at 16:30 on 5 Nov',
 'readme.txt last modified at 16:31 on 5 Nov',
 'test.py last modified at 16:32 on 5 Nov',
 'utils last modified at 16:32 on 5 Nov',
 'zigzag.py last modified at 16:20 on 5 Nov']
```

The fields method can also take arguments specifying the indexes of the fields to output; if more than one index is given the fields are joined by spaces:

```
In [x]: files.fields(0)    # first field in each line of files
Out[x]: ['-rw-r--r--', '-rw-r--r--', '-rw-r--r--', 'drwxr-xr-x', '-rw-r--r--']
```

```
In [x]: files.fields(-1)    # last field in each line of files
Out[x]: ['output.txt', 'readme.txt', 'test.py', 'utils', 'zigzag.py']

In [x]: files.fields(8, 7, 5, 6)
Out[x]:
['output.txt 16:30 5 Nov',
 'readme.txt 16:31 5 Nov',
 'test.py 16:32 5 Nov',
 'utils 16:32 5 Nov',
 'zigzag.py 16:20 5 Nov']
```

The sort method provided by SList objects can sort by a given field, optionally converting the field from a string to a number, if required (so that, for example, 10 > 9). Note that this method returns a new SList object.

```
In [x]: files.sort(4)      # sort alphanumerically by size (not useful)
Out[x]:
['-rw-r--r--  1 christian  staff    218  5 Nov 16:32 test.py',
 '-rw-r--r--  1 christian  staff  23258  5 Nov 16:31 readme.txt',
 '-rw-r--r--  1 christian  staff    365  5 Nov 16:20 zigzag.py',
 'drwxr-xr-x  2 christian  staff     68  5 Nov 16:32 utils',
 '-rw-r--r--  1 christian  staff     93  5 Nov 16:30 output.txt']

In [x]: files.sort(4, nums=True)       # sort numerically by size (useful)
Out[x]:
['drwxr-xr-x  2 christian  staff     68  5 Nov 16:32 utils',
 '-rw-r--r--  1 christian  staff     93  5 Nov 16:30 output.txt',
 '-rw-r--r--  1 christian  staff    218  5 Nov 16:32 test.py',
 '-rw-r--r--  1 christian  staff    365  5 Nov 16:20 zigzag.py',
 '-rw-r--r--  1 christian  staff  23258  5 Nov 16:31 readme.txt']
```

The grep method returns items from the SList containing a given string;[6] to search for a string in a given field only, use the field argument:

```
In [x]: files.grep('txt')              # search for lines containing 'txt'
Out[x]:
['-rw-r--r--  1 christian  staff     93  5 Nov 16:30 output.txt',
 '-rw-r--r--  1 christian  staff  23258  5 Nov 16:31 readme.txt']

In [x]: files.grep('16:32', field=7)   # search file files created at 16:32
Out[x]:
['-rw-r--r--  1 christian  staff    218  5 Nov 16:32 test.py',
 'drwxr-xr-x  2 christian  staff     68  5 Nov 16:32 utils']
```

Example E5.1   RNA encodes the amino acids of a peptide as a sequence of *codons*, with each codon consisting of three nucleotides chosen from the "alphabet": U (uracil), C (cytosine), A (adenine) and G (guanine).

The Python script, codon_lookup.py, available at https://scipython.com/eg/bab, creates a dictionary, codon_table, mapping codons to amino acids where each amino acid is identified by its one-letter abbreviation (e.g. R = arginine). The stop codons, signaling

---

[6] In fact, its name implies it will match *regular expressions* as well, but we will not expand on this here.

termination of RNA translation, are identified with the single asterisk character, *. The codon AUG signals the start of translation within a nucleotide sequence as well as coding for the amino acid methionine.

This script can be executed within IPython with %run codon_lookup.py (or loaded and then executed with %load codon_lookup.py followed by pressing Enter:

```
In [x]: %run codon_lookup.py
In [x]: codon_table
Out[x]:
{'GCG': 'A',
 'UAA': '*',
 'GGU': 'G',
 'UCU': 'S',
   ...
 'ACA': 'T',
 'ACC': 'T'}
```

Let's define a function to translate an RNA sequence. Type %edit and enter the following code in the editor that appears.

```
def translate_rna(seq):
    start = seq.find('AUG')
    peptide = []
    i = start
    while i < len(seq)-2:
        codon = seq[i:i+3]
        a = codon_table[codon]
        if a == '*':
            break
        i += 3
        peptide.append(a)
    return ''.join(peptide)
```

When you exit the editor it will be executed, defining the function, translate_rna:

```
IPython will make a temporary file named: /var/folders/fj/yv29fhm91v7_6g
7sqsy1z2940000gp/T/ipython_edit_thunq9/ipython_edit_dltv_i.py
Editing... done. Executing edited code...
Out[x]: "def translate_rna(seq):\n    start = seq.find('AUG')\n
peptide = []\
n    i = start\n    while i < len(seq)-2:\n        codon = seq[i:i+3]\n        a
= codon_table[codon]\n        if a == '*':\n            break\n        i += 3\n
        peptide.append(a)\n    return ''.join(peptide)\n"
```

Now feed the function an RNA sequence to translate:

```
In[x]: seq = 'CAGCAGCUCAUACAGCAGGUAAUGUCUGGUCUCGUCCCCGGAUGUCGCUACCCACGAG
ACCCGUAUCCUACUUUCUGGGGAGCCUUUACACGGCGGUCCACGUUUUUCGCUACCGUCGUUUUCCCGGUGC
CAUAGAUGAAUGUU'
In [x]: translate_rna(seq)
Out[x]: 'MSGLVPGCRYPRDPYPTFWGAFTRRSTFFATVVFPVP'
```

To read in a list of RNA sequences (one per line) from a text file, seqs.txt, and translate them, one could use %sx with the system command cat (or, on Windows, the command type):

```
In [x]: seqs = %sx cat seqs.txt
In [x]: for seq in seqs:
   ...:         print(translate_rna(seq))
   ...:
MHMLDENLYDLGMKACHEGTNVLDKWRNMARVCSCDYQFK
MQGSDGQQESYCTLPFEVSGMP
MPVEWRTMQFQRLERASCVKDSTFKNTGSFIKDRKVSGISQDEWAYAMSHQMQPAAHYA
MIVVTMCQ
MGQCMRFAPGMHGMYSSFHPQHKEITPGIDYASMNEVETAETIRPI
```

5.1.4   Exercises

Problems

P5.1.1   Improve on the algorithm to find the number of factors of an integer given in Section 5.1.3 by (a) looping the trial factor, i, up to no greater than the square root of n (why is it not necessary to test values of i greater than this?); and (b) using a generator (see Section 4.3.5). Compare the execution speed of these alternatives using the %timeit IPython magic.

P5.1.2   Using the fastest algorithm from the previous question, devise a short piece of code to determine the *highly composite numbers* less than 100 000 and use the %%timeit cell magic to time its execution. A highly composite number is a positive integer with more factors than any smaller positive integer, for example: 1, 2, 4, 6, 12, 24, 36, 48, . . .

5.2   Jupyter Notebook

Jupyter Notebook provides an interactive environment for Python programming within a web browser.[7] Its main advantage over the more traditional console-based approach of the IPython shell is that Python code can be combined with documentation (including in rendered LaTeX), images and even rich media such as embedded videos. Jupyter Notebooks are increasingly being used by scientists to communicate their research by including the computations carried out on data as well as simply the results of those computations. The format makes it easy for researchers to collaborate on a project and for others to validate their findings by reproducing their calculations on the same data.

5.2.1   Jupyter Notebook Basics

Starting the Jupyter Notebook Server
If you have Jupyter installed, the server that runs the browser-based interface to IPython can be started from the command line with

---

[7] Starting with version 4, the IPython Notebook project was reformulated as Jupyter Notebook, with bindings for other languages as well as for Python.

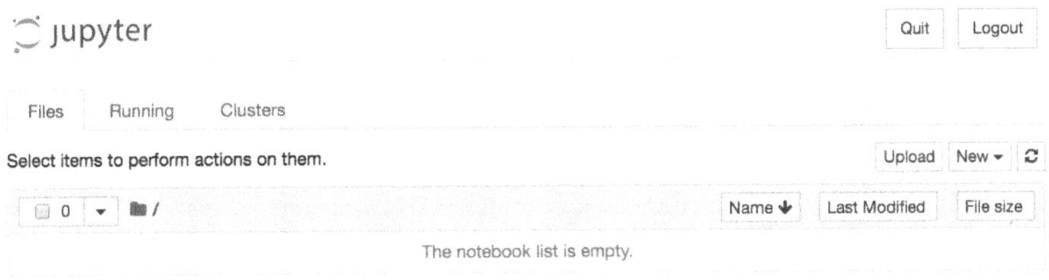

Figure 5.1 The Jupyter Notebook index page.

```
jupyter notebook
```

This will open a web browser window at the URL of the local Jupyter Notebook application. By default this is http://localhost:8888 though it will default to a different port if 8888 is in use.

The Jupyter Notebook index page (Figure 5.1) contains a list of the notebooks currently available in the directory from which the notebook server was started. This is also the default directory to which notebooks will be saved (with the extension .ipynb), so it is a good idea to execute the above command somewhere convenient in your directory hierarchy for the project you are working on.

The index page contains three tabs: *Files* lists all the files, including Jupyter Notebooks and subdirectories within the current working directory; *Running* lists those notebooks that are currently active within your session (even if they are not open in a browser window); *Clusters* provides an interface to IPython's parallel computing engine: we will not cover this topic in this book.

From the index page, one can start a new notebook (by clicking on "New > Notebook: Python 3") or open an existing notebook (by clicking on its name). To import an existing notebook into the index page, either click "Upload" at the top of the page or drag the notebook file into the index listing from elsewhere on your operating system.

To stop the notebook server, press CTRL-C in the terminal window it was started from (and confirm at the prompt).

Editing a Jupyter Notebook

To start a new notebook, click the "New" button and select a notebook kernel (there should at least be one called "Python 3"). This opens a new browser tab containing the interface where you will write your code and connects it to an IPython *kernel*, the process responsible for executing the code and communicating the results back to the browser.

The new notebook document (Figure 5.2) consists of a *title bar*, a *menu bar* and a *tool bar*, under which is an IPython prompt where you will type the code and markup (e.g. explanatory text and documentation) as a series of *cells*.

In the title bar the name of the first notebook you open will probably be "Untitled"; click on it to rename it to something more informative. The menu bar contains options for saving, copying, printing, rearranging and otherwise manipulating the Jupyter Note-

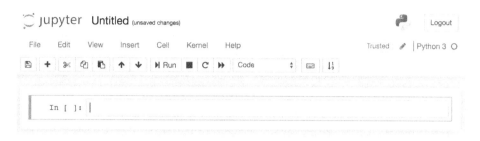

Figure 5.2 Jupyter with a new notebook document.

book document. The tool bar consists of series of icons that act as shortcuts for common operations that can also be achieved through the menu bar.

There are three types of input cells where you can write the content for your notebook:

- Code cells: the default type of cell, this type of cell consists of executable code. As far as this chapter is concerned, the code you write here will be Python, but Jupyter does provide a mechanism of executing code written in other languages such as Julia and R.
- Markdown cells: this type of cell allows for a rich form of documentation for your code. When executed, the input to a markdown cell is converted into HTML, which can include mathematical equations, font effects, lists, tables, embedded images and videos.
- Raw cells: input into this type of cell is not changed by the notebook – its content and formatting is preserved exactly.

Running Cells

Each cell can consist of more than one line of input, and the cell is not interpreted until you "run" (i.e. execute) it. This is achieved either by selecting the appropriate option from the menu bar (under the "Cell" drop-down submenu), by clicking the "Run cell" "play" button on the tool bar, or through the following keyboard shortcuts:

- `Shift-Enter`: Execute the cell, showing any output, and then *move the cursor* onto the cell below. If there is no cell below, a new, empty one will be created.
- `CTRL-Enter`: Execute the cell in place, but *keep the cursor* in the current cell. Useful for quick "disposable" commands to check if a command works or for retrieving a directory listing.
- `Alt-Enter`: Execute the cell, showing any output, and then *insert and move the cursor to a new cell* immediately beneath it.

Two other keyboard shortcuts are useful. When editing a cell the arrow keys navigate the *contents* of the cell (*edit mode*); from this mode, pressing `Esc` enters *command mode* from which the arrow keys navigate *through* the cells. To reenter edit mode on a selected cell, press `Enter`.

The menu bar, under the "Cell" drop-down submenu, provides many ways of running a notebook's cells: usually, you will want to run the current cell individually or run it and all those below it.

Code Cells

You can enter anything into a code cell that you can when writing a Python program in an editor or at the regular IPython shell. Code in a given cell has access to objects defined in other cells (providing they have been run). For example,

```
In [ ]:    n = 10
```

Pressing Shift-Enter or clicking Run Cell executes this statement (defining n but producing no output) and opens a new cell underneath the old one:

```
In [1]:    n = 10
```

```
In [ ]:
```

Entering the following statements at this new prompt:

```
In [ ]:    sum_of_squares = n * (n+1) * (2*n+1) // 6
           print('1**2 + 2**2 + ... + {}**2 = {}'.format(n,
               sum_of_squares))
```

and executing as before produces output and opens a third empty input cell. The whole notebook document then looks like

```
In [1]:    n = 10
```

```
In [2]:    sum_of_squares = n * (n+1) * (2*n+1) // 6
           print('1**2 + 2**2 + ... + {}**2 = {}'.format(n,
               sum_of_squares))
```

```
Out[2]:    1**2 + 2**2 + ... + 10**2 = 385
```

```
In [ ]:
```

You can edit the value of n in input cell 1 and rerun the entire document to update the output. It is worth noting that it is also possible to set a new value for n *after* the calculation in cell 2:

```
In [3]:    n = 15
```

Running cell 3 and then cell 2 then leaves the output to cell 2 as

```
Out[2]:    1**2 + 2**2 + ... + 15**2 = 1240
```

even though the cell above still defines n to be 10. That is, unless you run the entire document from the beginning, the output does not necessarily reflect the output of a script corresponding to the code cells taken in order.

System commands (those prefixed with ! or !!) and IPython magics can all be used within Jupyter Notebook.

Markdown Cells

Markdown cells convert your input text into HTML, applying styles according to a simple syntax illustrated below. The full documentation is at

   https://daringfireball.net/projects/markdown/

Here we explain the most useful features. A complete Jupyter Notebook of these examples can be downloaded from https://scipython.com/book/markdown.

Basic Markdown

- Simple styles can be applied by enclosing text by asterisks or underscores:

In [x]:
```
Surrounding text by two asterisks denotes **bold
style**; using one asterisk denotes *italic
text*, as does _a single underscore_.
```

   Surrounding text by two asterisks denotes **bold style**; using one asterisk denotes *italic text*, as does *a single underscore*.

- Headings at up to six levels (from top-level section titles to paragraph-level text) are denoted with between one and six "#" characters: the text following these characters is rendered at an appropriate font size (in HTML, using the elements <h1> to <h6>).

- Block quotes are indicated by a single angle bracket, >:

In [x]:
```
> "Climb if you will, but remember that courage and
strength are nought without prudence, and that a
momentary negligence may destroy the happiness
of a lifetime. Do nothing in haste; look well to
each step; and from the beginning think what may
be the end." - Edward Whymper
```

   "Climb if you will, but remember that courage and strength are nought without prudence, and that a momentary negligence may destroy the happiness of a lifetime. Do nothing in haste; look well to each step; and from the beginning think what may be the end." – Edward Whymper

- Code *examples* (for illustration rather than execution) are between blank lines and indented by four spaces (or a tab). The following will appear in a monospaced font with the characters as entered:

In [x]:
```
    n = 57
    while n != 1:
        if n % 2:
            n = 3*n + 1
        else:
            n //= 2
```

```
n = 57
while n != 1:
    if n % 2:
        n = 3*n + 1
    else:
```

```
                    n //= 2
```

- Inline code examples are created by surrounding the text with backticks (`):

In [x]:
> Here are some Python keywords: `for`, `while` and
> `lambda`.

Here are some Python keywords: for, while and lambda.

- New paragraphs are started after a blank line.

HTML within Markdown

The markdown used by Jupyter Notebooks encompasses HTML, so valid HTML entities and tags can be used directly (for example, the <em> tag for emphasis), as can CSS styles to produce effects such as underlined text. Even complex HTML such as tables can be marked up directly.

In [x]:
```
The following <em>Punnett table</em> is <span
    style="text-decoration: underline" >marked up</span>
    in HTML.

<table style="text-align: center;">
<tr>
<th style="border-top:none; border-left:none;" rowspan="2"
    colspan="2"></th>
<th colspan="2">Male</th>
</tr>
<tr>
<th>A</th>
<th>a</th>
</tr>
<tr>
<th rowspan="2">Female</th>
<th>a</th>
<td style="background: #aaa;">Aa</td>
<td>aa</td>
</tr>
<tr>
<th>a</th>
<td style="background: #aaa;">Aa</td>
<td>aa</td>
</tr>
</table>
```

The following *Punnett table* is marked up in HTML.

		Male	
		A	**a**
	a	Aa	aa
Female			
	a	Aa	aa

Lists

Itemized (unnumbered) lists are created using any of the markers *, + or -, and nested sublists are simply indented.

In [x]:
```
The inner planets and their satellites:

* Mercury
* Venus
* Earth
    * The Moon
+ Mars
    - Phoebus
    - Deimos
```

The inner planets and their satellites:

- Mercury
- Venus
- Earth

  - The Moon

- Mars

  - Phoebus
  - Deimos

Ordered (that is, numbered) lists are created by preceding items by a number followed by a full stop (period) and a space:

In [x]:
```
1. Symphony No. 1 in C major, Op. 21
2. Symphony No. 2 in D major, Op. 36
3. Symphony No. 3 in E-flat major ("Eroica"), Op. 55
```

1.   Symphony No. 1 in C major, Op. 21
2.   Symphony No. 2 in D major, Op. 36
3.   Symphony No. 3 in E-flat major ("Eroica"), Op. 55

Links

There are three ways of introducing links into markdown text:

- *Inline* links provide a URL in round brackets after the text to be turned into a link in square brackets. For example,

In [x]:
```
Here is a link to the
[IPython website](https://ipython.org/).
```

Here is a link to the IPython website.

- *Reference* links label the text to turn into a link by placing a name (containing letters, numbers or spaces) in square brackets after it. This name is expected to be defined using the syntax [*name*]: *url* elsewhere in the document, as in the following example markdown cell.

In [x]:
```
Some important mathematical sequences are the [prime
    numbers][primes],
[Fibonacci sequence][fib] and the [Catalan
    numbers][catalan_numbers].

...

[primes]: https://oeis.org/A000040
[fib]: https://oeis.org/A000045
[catalan_numbers]: https://oeis.org/A000108]
```

Some important mathematical sequences are the primes, Fibonacci sequence and the Catalan numbers.

- *Automatic* links, for which the clickable text is the same as the URL, are created simply by surrounding the URL by angle brackets:

In [x]:
```
My website is <https://christianhill.co.uk>.
```

My website is https://christianhill.co.uk.

If the link is to a file on your local system, give as the URL the path, relative to the notebook directory, prefixed with `files/`:

In [x]:
```
Here is [a local data file](files/data/data0.txt).
```

Here is a local data file.

Note that links open in a new browser tab when clicked.

Mathematics

Mathematical equations can be written in LaTeX and are rendered using the Javascript library, MathJax. Inline equations are delimited by single dollar signs; "displayed" equations by doubled dollar signs:

In [x]:
```
An inline equation appears within a sentence of text, as
    in the definition of the function $f(x) = \sin(x^2)$;
    displayed equations get their own line(s) between
    lines of text:
$$\int_0^\infty \mathrm{e}^{-x^2}dx =
    \frac{\sqrt{\pi}}{2}.$$
```

An inline equation appears within a sentence of text, as in the definition of the function $f(x) = \sin(x^2)$; displayed equations get their own line(s) between lines of text:

$$\int_0^\infty e^{-x^2} dx = \frac{\sqrt{\pi}}{2}.$$

Images and Video

Links to image files work in exactly the same way as ordinary links (and can be inline or reference links), but are preceded by an exclamation mark, !. The text in square brackets between the exclamation mark and the link acts as *alt text* to the image. For example,

```
In [x]:     ![An interesting plot of the Newton
                fractal](/files/images/newton_fractal.png)
            ![A remote link to a star
                image](https://christianhill.co.uk/static/images/
                star.svg)
```

Video links must use the HTML5 <video> tag, but note that not all browsers support all video formats. For example,

```
In [x]:     <video controls style="width: 500px; margin: 0 auto;
                display: block;" src="files/diffmap-animated.ogv" />
```

The data constituting images, video and other locally linked content are not *embedded* in the notebook document itself: these files must be provided with the notebook when it is distributed.

5.2.2   Converting Notebooks to Other Formats

nbconvert is a tool, installed with Jupyter, to convert notebooks from their native .ipynb format[8] to any of several alternative formats. It is run from the (system) command line as

```
jupyter nbconvert --to <format> <notebook.ipynb>
```

where *notebook.ipynb* is the name of the Jupyter Notebook file to be converted and *format* is the desired output format. The default (if no *format* is given), is to produce a static HTML file, as described below.

Conversion to HTML

The command

```
jupyter nbconvert <notebook.ipynb>
```

converts notebook.ipynb to HTML and produces a file, notebook.html, in the current directory. This file contains all the necessary headers for a stand-alone HTML page,

---

[8] This format is, in fact, just a JSON (JavaScript Object Notation) document.

which will closely resemble the interactive view produced by the Jupyter Notebook server, but as a static document.

If you want just the HTML corresponding to the notebook without the header (`<html>`, `<head>`, `<body>` tags, etc.), suitable for embedding in an existing web page, add the `--template basic` option.

Any supporting files, such as images, are automatically placed in a directory with the same base name as the notebook itself but with the suffix `_files`. For example, `jupyter nbconvert mynotebook.ipynb` generates `mynotebook.html` and the directory `mynotebook_files`.

Conversion to LaTeX

To export the notebook as a LaTeX document, use

```
jupyter nbconvert --to latex <notebook.ipynb>
```

To automatically generate a PDF file by running `pdflatex` on the `notebook.tex` file produced, add the option `--post pdf`.

Conversion to Markdown

```
jupyter nbconvert --to markdown <notebook.ipynb>
```

converts the whole notebook into markdown (see Section 5.2.1): cells that are already in markdown are unaffected and code cells are placed in triple-backtick (``` ``` ```) blocks.

Conversion to Python

The command

```
jupyter nbconvert --to python <notebook.ipynb>
```

converts `notebook.ipynb` into an executable Python script. If any of the notebook's code cells contain IPython magic functions, this script may only be executable from within an IPython session. Markdown and other text cells are converted to comments in the generated Python script code.

5.2.3 JupyterLab

At the time of writing, Project Jupyter is testing a browser-based interactive development environment (IDE) called JupyterLab, which will extend the functionality of Jupyter Notebook and allow real-time collaboration between multiple users, drag-and-drop manipulation of notebook cells, browser-based terminal (console) access, auto-completion, and live preview of markdown. Custom widgets can be installed to allow the loading and exploration of data in different formats within the browser and integration with popular online services such as GitHub, Dropbox and Google Drive. It will be fully backward-compatible with existing Jupyter Notebooks. More information is available from the Project Jupyter website, https://jupyter.org/.

6 NumPy

NumPy has become the de facto standard package for general scientific programming in Python. Its core object is the `ndarray`, a multidimensional array of a single data type, which can be sorted, reshaped, subject to mathematical operations and statistical analysis, written to and read from files, and much more. The NumPy implementations of these mathematical operations and algorithms have two main advantages over the "core" Python objects we have used until now. First, they are implemented as precompiled C code and so approach the speed of execution of a program written in C itself; second, NumPy supports *vectorization*: a single operation can be carried out on an entire array, rather than requiring an explicit loop over the array's elements. For example, compare the multiplication of two one-dimensional lists of n numbers, a and b, in the core python language:

```
c = []
for i in range(n):
    c.append(a[i] * b[i])
```

and using NumPy arrays:[1]

```
c = a * b
```

The elementwise multiplication is handled by optimized, precompiled C and so is very fast (much faster for large n than the core Python alternative). The absence of explicit looping and indexing makes the code cleaner, less error-prone and closer to the standard mathematical notation it reflects.

All of NumPy's functionality is provided by the `numpy` package. To use it, it is strongly advised to import with

```
import numpy as np
```

and then to refer to its attributes with the prefix `np.` (e.g. `np.array`). This is the way we use NumPy in this book.

6.1 Basic Array Methods

The NumPy array class is `ndarray`, which consists of a multidimensional table of elements indexed by a tuple of integers. Unlike Python lists and tuples, the elements cannot

---

[1] The terms "NumPy array" and `ndarray` will be used interchangeably in this book.

be of different types: each element in a NumPy array has the same type, which is specified by an associated *data type* object (dtype). The dtype of an array specifies not only the broad class of element (integer, floating-point number, etc.) but also how it is represented in memory (e.g. how many bits it occupies) – see Section 6.1.2.

The dimensions of a NumPy array are called *axes*; the number of axes an array has is called its *rank*.[2]

6.1.1   Creating an Array

Basic Array Creation

The simplest way to create a small NumPy array is to call the np.array constructor with a list or tuple of values:

```
In [x]: import numpy as np
In [x]: a = np.array( (100, 101, 102, 103) )
In [x]: a
Out[x]: array([100, 101, 102, 103])
In [x]: b = np.array( [[1.,2.], [3.,4.]] )
Out[x]:
array([[ 1.,   2.],
       [ 3.,   4.]])
```

Note that passing a list of lists creates a two-dimensional array (and similarly for higher dimensions).

Indexing a multidimensional NumPy array is a little different from indexing a conventional Python list of lists: instead of b[i][j], refer to the index of the required element as a tuple of integers, b[i,j]:

```
In [x]: b[0,1]           # same as b[(0,1)]
Out[x]: 2.0
In [x]: b[1,1] = 0.      # also for assignment
Out[x]:
array([[ 1.,   2.],
       [ 3.,   0.]])
```

The data type is deduced from the type of the elements in the sequence and "upcast" to the most general type if they are of mixed but compatible types:

```
In [x]: np.array( [-1, 0, 2.])   # mixture of int and float: upcast to float
Out[x]: array([-1.,   0.,   2.])
```

You can also explicitly set the data type using the optional dtype argument (see Section 6.1.2):

```
In [x]: np.array( [0, 4, -4], dtype=complex)
In [x]: array([ 0.+0.j,   4.+0.j,  -4.+0.j])
```

If your array is large or you do not know the element values at the time of creation, there are several methods to declare an array of a particular shape filled with default or arbitrary values. The simplest and fastest, np.empty, takes a tuple of the array's

---

[2] Not to be confused with the concept of *matrix rank* from linear algebra.

shape and creates the array without initializing its elements: the initial element values are undefined (typically, random junk defined from whatever were the contents of the memory that Python allocated for the array).

```
In [x]: np.empty((2,2))
Out[x]:
array([[ -2.31584178e+077,  -1.72723381e-077],
       [  2.15686807e-314,   2.78134366e-309]])
```

There are also helper methods, np.zeros and np.ones, which create an array of the specified shape with elements prefilled with 0 and 1, respectively. np.empty, np.zeros and np.ones also take the optional dtype argument.

```
In [x]: np.zeros((3,2))     # default dtype is 'float'
Out[x]:
array([[ 0.,   0.],
       [ 0.,   0.],
       [ 0.,   0.]])
In [x]: np.ones((3,3), dtype=int)
Out[x]:
array([[1, 1, 1],
       [1, 1, 1],
       [1, 1, 1]])
```

If you already have an array and would like to create another with the same shape, np.empty_like, np.zeros_like and np.ones_like will do that for you:

```
In [x]: a
Out[x]: array([100, 101, 102, 103])
In [x]: np.ones_like(a)
Out[x]: array([1, 1, 1, 1])
In [x]: np.zeros_like(a, dtype=float)
Out[x]: array([ 0.,   0.,   0.,   0.])
```

Note that the array created inherits its dtype from the original array; to set its data type to something else, use the dtype argument.

Initializing an Array from a Sequence

To create an array containing a sequence of numbers there are two methods: np.arange and np.linspace. np.arange is the NumPy equivalent of range, except that it can generate floating-point sequences. It also actually allocates the memory for the elements in an ndarray instead of returning a generator-like object – compare Section 2.4.3.

```
In [x]: np.arange(7)
Out[x]: array([0, 1, 2, 3, 4, 5, 6])
In [x]: np.arange(1.5, 3., 0.5)
Out[x]: array([ 1.5,  2. ,  2.5]))
```

As with range, the array generated in these examples does not include the last elements, 7 and 3. However, arange has a problem: because of the finite precision of floating-point arithmetic it is not always possible to know how many elements will be created. For this reason, and because one often wants the last element of a specifed sequence, the np.linspace function can be a more useful way of creating an sequence.[3] For example, to generate an evenly spaced array of the five numbers between 1 and 20 *inclusive*:

```
In [x]: np.linspace(1, 20, 5)
Out[x]: array([  1.  ,    5.75,   10.5 ,   15.25,   20.  ])
```

np.linspace has a couple of optional boolean arguments. First, setting retstep to True returns the number spacing (step size):

```
In [x]: x, dx = np.linspace(0., 2*np.pi, 100, retstep=True)
In [x]: dx
Out[x]: 0.06346651825433926
```

This saves you from calculating dx = (end-start)/(num-1) separately; in this example, the 100 points between 0 and 2π inclusive are spaced by $2\pi/99 = 0.0634665\ldots$ Finally, setting endpoint to False omits the final point in the sequence, as for np.arange:

```
In [x]: x = np.linspace(0, 5, 5, endpoint=False)
Out[x]: array([ 0.,  1.,  2.,  3.,  4.])
```

Note that the array generated by np.linspace has the dtype of floating-point numbers, even if the sequence generates integers.

Initializing an Array from a Function

To create an array initialized with values calculated using a function, use NumPy's np.fromfunction method, which takes as its arguments a function and a tuple representing the shape of the desired array. The function should itself take the same number of arguments as dimensions in the array: these arguments index each element at which the function returns a value. An example will make this clearer:

```
In [x]: def f(i, j):
   ...:       return 2 * i * j
   ...:
In [x]: np.fromfunction(f,(4,3))
array([[  0.,   0.,   0.],
       [  0.,   2.,   4.],
       [  0.,   4.,   8.],
       [  0.,   6.,  12.]])
```

The function f is called for every index in the specified shape and the values it returns are used to initialize the corresponding elements.[4] A simple expression like this one can be replaced by an anonymous lambda function (see Section 4.3.3) if desired:

```
In [x]: np.fromfunction(lambda i,j: 2*i*j, (4,3))
```

Example E6.1   To create a "comb" of values in an array of length N for which every nth element is one but with zeros everywhere else:

---

[3] We came across linspace in the discussion following Example E3.1.

[4] Note that the indexes are passed as ndarrays and expect the function, f, to use vectorized operations.

```
In [x]: N, n = 101, 5
In [x]: def f(i):
   ...:         return (i % n == 0) * 1
   ...:
In [x]: comb = np.fromfunction(f, (N,), dtype=int)
In [x]: print(comb)
[1 0 0 0 0 1 0 0 0 0 1 0 0 0 0 1 0 0 0 0 1 0 0 0 0 1 0 0 0 0 1 0 0 0 0 1
 0 0 0 0 1 0 0 0 0 1 0 0 0 0 1 0 0 0 0 1 0 0 0 0 1 0 0 0 0 1 0 0 0 0 1 0
 0 0 1 0 0 0 0 1 0 0 0 0 1 0 0 0 0 1 0 0 0 0 1 0 0 0 0 1]
```

ndarray Attributes for Introspection

A NumPy array knows its rank, shape, size, dtype and one or two other properties: these can be determined directly from the attributes described in Table 6.1. For example,

```
In [x]: a = np.array(((1, 0, 1), (0, 1, 0)))
In [x]: a.shape
Out[x]: (2, 3)          # 2 rows, 3 columns
In [x]: a.ndim          # rank (number of dimensions)
Out[x]: 2
In [x]: a.size          # total number of elements
Out[x]: 6
In [x]: a.dtype
Out[x]: dtype('int64')
In [x]: a.data
Out[x]: <memory at 0x102387308>
```

The shape attribute returns the axis dimensions in the same order as the axes are indexed: a two-dimensional array with n rows and m columns has a shape of (n, m).

6.1.2   NumPy's Basic Data Types (dtypes)

So far, the NumPy arrays we have created have contained either integers or floating-point numbers, and we have let Python take care of the details of how these are represented. However, NumPy provides a powerful way of determining these details explicitly using *data type* objects. This is necessary, because in order to interface with the underlying compiled C code the elements of a NumPy array must be stored in a com-

Table 6.1 ndarray attributes

Attribute	Description
shape	The array dimensions: the size of the array along each of its axes, returned as a tuple of integers
ndim	Number of axes (dimensions); note that ndim == len(shape)
size	The total number of elements in the array, equal to the product of the elements of shape
dtype	The array's data type (see Section 6.1.2)
data	The "buffer" in memory containing the actual elements of the array
itemsize	The size in bytes of each element

patible format: that is, each element is represented in a fixed number of bytes that are interpreted in a particular way.

For example, consider an *unsigned* integer stored in 2 bytes (16 bits) of memory (the C-type uint16_t). Such a number can take a value between 0 and $2^{16} - 1 = 65\,535$. No equivalent native Python type exists for this exact representation: Python integers are signed quantities and memory is dynamically assigned for them as required by their size. So NumPy defines a data type object, np.uint16, to describe data stored in this way.

Furthermore, different systems can order the two bytes of this number differently, a distinction known as *endianness*. The *big-endian* convention places the most-significant byte in the smallest memory address; the *little-endian* convention places the least-significant byte in the smallest memory address. In creating your own arrays, NumPy will use the default convention for the hardware your program is running on, but it is essential to set the endianness correctly if reading in a binary file generated by a different computer.

A full list of the numerical data types[5] is given in the NumPy documentation,[6] but the more common ones are listed in Table 6.2. They all exist within the numpy package and so can be referred to as, for example, np.uint16. The data types that get created by default when using the native Python numerical types are those with a trailing underscore: np.float_, np.complex_ and np.bool_.

Apparently higher-precision floating-point number data types such as float96, float128 and longdouble are available but are not to be trusted: their implementation is platform-dependent, and on many systems they do not actually offer any extra precision but simply align array elements on the appropriate byte-boundaries in memory.

To create a NumPy array of values using a particular data type, use the dtype argument of any array constructor function (such as np.array, np.zeros, etc.). This argument takes either a data type object (such as np.uint8) or something that can be converted into one. It is common to specify the dtype using a string consisting of a letter indicating the broad category of data type (integer, unsigned integer, complex number, etc.) optionally followed by a number giving the byte size of the type. For example,

```
In [x]: b = np.zeros((3,3), dtype='u4')
```

creates a 3×3 array of unsigned, 32-bit (4-byte) integers (equivalent to np.uint32). A list of supported data type letters and their meanings is given in Table 6.3.

To specify the endianness, use the prefixes > (big-endian), < (little-endian) or | (endianness not relevant). For example,

```
In [x]: a = np.zeros((3,3), dtype='>f8')
In [x]: b = np.zeros((3,3), dtype='<f')
In [x]: c = np.empty((3,3), dtype='|S4')
```

create arrays of big-endian double-precision numbers, little-endian single-precision numbers and four-character strings, respectively.

---

[5] Strictly speaking, these types are *array scalar types* and not dtypes, but for our use here the distinction is not important.

[6] https://docs.scipy.org/doc/numpy/user/basics.types.html.

Table 6.2 Common NumPy data types

Data Type	Description
int_	The default integer type, corresponding to C's long: *platform-dependent*
int8	Integer in a single byte: -128 to 127
int16	Integer in 2 bytes: $-32\,768$ to $32\,767$
int32	Integer in 4 bytes: $-2\,147\,483\,648$ to $2\,147\,483\,647$
int64	Integer in 8 bytes: -2^{63} to $2^{63} - 1$
uint8	Unsigned integer in a single byte: 0 to 255
uint16	Unsigned integer in 2 bytes: 0 to $65\,535$
uint32	Unsigned integer in 4 bytes: 0 to $4\,294\,967\,295$
uint64	Unsigned integer in 8 bytes: 0 to $2^{64} - 1$
float_	The default floating-point number type, another name for float64
float32	Single-precision, signed float: $\sim 10^{-38}$ to $\sim 10^{38}$ with ~ 7 decimal digits of precision
float64	Double-precision, signed float: $\sim 10^{-308}$ to $\sim 10^{308}$ with ~ 15 decimal digits of precision
complex_	The default complex number type, another name for complex128
complex64	Single-precision complex number (represented by 32-bit floating-point real and imaginary components)
complex128	Double-precision complex number (represented by 64-bit floating-point real and imaginary components)
bool_	The default boolean type represented by a single byte

Table 6.3 Common NumPy data type strings

String	Description
i	Signed integer
u	Unsigned integer
f	Floating-point number[a]
c	Complex floating-point number
b	Boolean value
S, a	String (fixed-length sequence of characters)
U	Unicode

[a] Note that without specifying the byte size, setting dtype='f' creates a *single-precision* floating-point data type, equivalent to np.float32.

In these examples we have passed a *typecode* string to an array constructor's dtype argument, but it is also possible to create a dtype object first and pass that instead:

```
In [x]: dt = np.dtype('f8')
In [x]: dt
dtype('float64')    # double-precision floating-point
In [x]: a = np.array([0., 1., -2.], dtype=dt)
```

dtype objects have a handful of useful introspection methods:

```
In [x]: dt.str       # a string identifying the data type
'<f8'
In [x]: dt.name      # data type name and bit-width
'float64'
In [x]: dt.itemsize # data type size in bytes
8
```

To copy an array to a new array with a different data type, pass the desired dtype or typecode to the astype method:

```
In [x]: a = np.array([1.2345678, 2.5, 3.9])
In [x]: a.astype('float32')      # cast to single-precision float
Out[x]: array([1.2345678, 2.5     , 3.9     ], dtype=float32)

In [x]: a.astype(np.uint8)       # cast to unsigned, 1-byte integer
Out[x]: array([1, 2, 3], dtype=uint8)
```

Strings in NumPy arrays are *bytestrings* of a fixed size: each "character" is represented by a single byte, in contrast to the variable size UTF-8 encoding, commonly used to represent Unicode strings. This is necessary because NumPy arrays have a pre-defined, fixed size in which all the elements occupy the same amount of memory so that they can be indexed efficiently with a constant stride. Unicode strings encoded with UTF-8, however, represent characters as code points with a variable width (see Section 2.3.3). Of course, any string is ultimately stored as a sequence of bytes and Python provides methods for translating between encodings. For example, on a system encoding strings with UTF-8 by default:

```
In [x]: s = 'piñata'        # UTF-8 encoded Unicode string
In [x]: b = s.encode()
In [x]: b
b'pi\xc3\xb1ata'           # bytestring: ñ is stored in two bytes: hex C3B1
In [x]: len(s), len(b)
(6,7)                       # six UTF-8 encoded characters stored in 7 bytes
In [x]: arr = np.empty((2,2), 'S7')
In [x]: arr[:] = b          # store the bytestring b in array arr
In [x]:
array([[b'pi\xc3\xb1ata', b'pi\xc3\xb1ata'],
       [b'pi\xc3\xb1ata', b'pi\xc3\xb1ata']],
      dtype='|S7')
In [x]: arr[0,0]            # returns the bytestring
b'pi\xc3\xb1ata'
In [x]: arr[0,0].decode()   # decode the bytestring back assuming UTF-8
'piñata'
```

6.1.3   Universal Functions (ufuncs)

In addition to the basic arithmetic operations of addition, division and more, NumPy provides many of the familiar mathematical functions that the math module (Section 2.2.2) does, implemented as so-called *universal functions* that act on each element of an array, producing an array in return without the need for an explicit loop. Universal functions are the way NumPy allows for *vectorization*, which promotes clean, efficient and easy-to-maintain code. For example,

```
In [x]: x = np.linspace(1, 5, 5)
In [x]: x**2
Out[x]: array([  1.,   4.,   9.,  16.,  25.])
In [x]: x - 1
Out[x]: array([ 0.,  1.,  2.,  3.,  4.])
In [x]: np.sqrt(x - 1)
Out[x]: array([ 0.,  1.,  1.41421356,  1.73205081,  2.])
In [x]: y = np.exp(-np.linspace(0., 2., 5))
In [x]: np.sin(x - y)
Out[x]: array([ 0.,  0.98431873,  0.48771645, -0.59340065, -0.98842844])
```

Array multiplication occurs *elementwise*: matrix multiplication is implemented by the @ operator[7] or NumPy's dot function:

```
In [x]: a = np.array( ((1, 2), (3, 4)) )
In [x]: b = a
In [x]: a * b         # elementwise multiplication
Out[x]:
array([[ 1,  4],
       [ 9, 16]])
In [x]: a @ b         # matrix multiplication; also a.dot(b) or np.dot(a, b)
Out[x]:
array([[ 7, 10],
       [15, 22]])
```

Comparison and logic operators (~, & and | for *not*, *and* and *or*, respectively) are also vectorized and result in arrays of boolean values:

```
In [x]: a = np.linspace(1, 6, 6)**3
In [x]: print(a)
[   1.    8.   27.   64.  125.  216.]
In [x]: print(a > 100)
[False False False False  True  True]
In [x]: print((a < 10) | (a > 100))
[ True  True False False  True  True]
```

6.1.4 NumPy's Special Values, nan and inf

NumPy defines two special values to represent the outcome of calculations, which are not mathematically defined or not finite. The value np.nan ("Not a Number," NaN) represents the outcome of a calculation that is not a well-defined mathematical operation (e.g. 0/0); np.inf represents infinity.[8] For example,

```
In [x]: a = np.arange(4, dtype='f8')
In [x]: a /= 0        # [0/0 1/0 2/0 3/0]

... RuntimeWarning: invalid value encountered in true_divide ...
... RuntimeWarning: divide by zero encountered in true_divide ...

In [x]: a
Out[x]: array([ nan,  inf,  inf,  inf])
```

---

[7] The @ operator was introduced in Python 35.

[8] These quantities are defined in accordance with the IEEE-754 standard for floating-point numbers.

Do not test nans for equality (np.nan == np.nan is False). Instead, NumPy provides methods np.isnan, np.isinf and np.isfinite:

```
In [x]: np.isnan(a)
Out[x]: array([ True, False, False, False], dtype=bool)
In [x]: np.isinf(a)
Out[x]: array([False,  True,  True,  True], dtype=bool)
In [x]: np.isfinite(a)
Out[x]: array([False, False, False, False], dtype=bool)
```

Note that nan is neither finite nor infinite, and is not equal to itself! (See also Section 10.1.4.)

Example E6.2   A *magic square* is an $N \times N$ grid of numbers in which the entries in each row, column and main diagonal sum to the same number (equal to $N(N^2 + 1)/2$). A method for constructing a magic square for odd N is as follows:

Step 1.   Start in the middle of the top row, and let $n = 1$.

Step 2.   Insert n into the current grid position.

Step 3.   If $n = N^2$ the grid is complete so stop. Otherwise, increment n.

Step 4.   Move diagonally up and right, wrapping to the first column or last row if the move leads outside the grid. If this cell is already filled, move vertically down one space instead.

Step 5.   Return to step 2.

The following program creates and displays a magic square.

Listing 6.1 Creating a magic square

```
# Create an N x N magic square. N must be odd.
import numpy as np

N = 5
magic_square = np.zeros((N, N), dtype=int)

n = 1
i, j = 0, N//2

while n <= N**2:
    magic_square[i, j] = n
    n += 1
    newi, newj = (i - 1) % N, (j + 1)% N
    if magic_square[newi, newj]:
        i += 1
    else:
        i, j = newi, newj

print(magic_square)
```

The 5×5 magic square output by the earlier example is

```
[[17 24  1  8 15]
 [23  5  7 14 16]
 [ 4  6 13 20 22]
 [10 12 19 21  3]
 [11 18 25  2  9]]
```

6.1.5   Changing the Shape of an Array

Whatever the rank of an array, its elements are stored in sequential memory locations that are addressed by a single index (internally the array is one-dimensional, but, knowing the shape of the array, Python is able to resolve a tuple of indexes into a single memory address). NumPy arrays are stored in memory in C-style, *row-major* order, that is, with the elements of the last (rightmost) index stored contiguously. In a two-dimensional array, for example, the element a[0, 0] is followed by a[0, 1]. The array that follows

```
In [x]: a = np.array( ((1, 2), (3, 4)) )
In [x]: print(a)
[[1 2]
 [3 4]]
```

is stored in memory as the sequential elements [1,2,3,4].[9]

flatten and ravel

Suppose you wish to "flatten" a multidimensional array onto a single axis. NumPy provides two methods to do this: flatten and ravel. Both flatten the array into its internal (row-major) ordering, as described earlier. flatten returns an independent *copy* of the elements and is generally slower than ravel, which tries to return a *view* to the flattened array. An array view is a new NumPy array with, in this case, a different shape from the original, but it does not "own" its data elements: it references the elements of another array. Thus, just as with mutable lists (Section 2.4.1), a reassignment of an element of one array affects the other. An example should make this clear:

```
In [x]: a = np.array( [[1, 2, 3], [4, 5, 6], [7, 8, 9]] )
In [x]: b = a.flatten()     # create an independent, flattened copy of a
In [x]: b
Out[x]: array([1, 2, 3, 4, 5, 6, 7, 8, 9])
In [x]: b[3] = 0
In [x]: b
Out[x]: array([1, 2, 3, 0, 5, 6, 7, 8, 9])
In [x]: a          # a is unchanged
Out[x]:
array([[1, 2, 3],
       [4, 5, 6],
       [7, 8, 9]])
```

Assignment to b didn't change a because they are completely independent objects that do not share their data. In contrast, the flattened array created by taking a view on a with ravel refers to the same underlying data:

```
In [x]: c = a.ravel()
In [x]: c
Out[x]:array([1, 2, 3, 4, 5, 6, 7, 8, 9])
In [x]: c[3] = 0
```

---

[9] This contrasts with Fortran's *column-major* ordering, which would store the elements as [1, 3, 2, 4].

```
In [x]: c
Out[x]: array([1, 2, 3, 0, 5, 6, 7, 8, 9])
In [x]: a
Out[x]:
array([[1, 2, 3],
       [0, 5, 6],
       [7, 8, 9]])
```

You should be aware that although the ravel method "does its best" to return a view to the underlying data, various array operations (including *slicing*; see Section 6.1.6) can leave the elements stored in noncontiguous memory locations, in which case ravel has no choice but to make a copy.

resize and reshape

An array may be resized (in place) to a compatible shape[10] with the resize method, which takes the new dimensions as its arguments.

```
In [x]: a = np.linspace(1, 4, 4)
In [x]: print(a)
[1. 2. 3. 4.]

In [x]: a.resize(2, 2)    # reshapes a in place, doesn't return anything
In [x]: print(a)
[[ 1.  2.]
 [ 3.  4.]]
```

The reshape method returns a view on the array with its elements reshaped as required. The original array is not modified, but the objects share the same underlying data.

```
In [x]: a = np.linspace(1, 4, 4)
In [x]: b = a.reshape(2, 2)
In [x]: print(a)
[ 1.  2.  3.  4.]

In [x]: print(b)
[[ 1.  2.]
 [ 3.  4.]]

In [x]: b[0, 0] = -99
In [x]: print(b)
[[-99.  2.
 [  3.   4.]]

In [x]: print(a)
[-99.  2.  3.  4.]
```

Transposing an Array

The method transpose returns a view of an array with the axes transposed. For a two-dimensional array, this is the usual matrix transpose:

---

[10] That is, a shape with the same total number of elements.

```
In [x]: a = np.linspace(1, 6, 6).reshape(3, 2)
In [x]: a
Out[x]:
array([[ 1.,   2.],
       [ 3.,   4.],
       [ 5.,   6.]])
In [x]: a.transpose()          # or simply a.T
Out[x]:
array([[ 1.,   3.,   5.],
       [ 2.,   4.,   6.]])
```

Note that transposing a one-dimensional array returns the array unchanged:

```
In [x]: b = np.array([100, 101, 102, 103])
In [x]: b.transpose()
Out[x]: array([100, 101, 102, 103])
```

See Section 6.1.11 for more on representing vectors with NumPy arrays.

Merging and Splitting Arrays

A clutch of NumPy methods merge and split arrays in different ways. np.vstack, np.hstack and np.dstack stack arrays vertically (in sequential rows), horizontally (in sequential columns) and depthwise (along a third axis). For example,

```
In [x]: a = np.array([0, 0, 0, 0])
In [x]: b = np.array([1, 1, 1, 1])
In [x]: c = np.array([2, 2, 2, 2])
In [x]: np.vstack((a, b, c))
Out[x]:
array([[0, 0, 0, 0],
       [1, 1, 1, 1],
       [2, 2, 2, 2]])
In [x]: np.hstack((a, b, c))
Out[x]:
array([0, 0, 0, 0, 1, 1, 1, 1, 2, 2, 2, 2])
In [x]: np.dstack((a, b, c))
Out[x]:
array([[[0, 1, 2],
        [0, 1, 2],
        [0, 1, 2],
        [0, 1, 2]]])
```

Note that the array created contains an independent *copy* of the data from the original arrays.[11]

The inverse operations, np.vsplit, np.hsplit and np.dsplit, split a single array into multiple arrays by rows, columns or depth. In addition to the array to be split, these methods require an argument indicating how to split the array. If this argument is a *single integer*, the array is split into that number of equal-sized arrays along the appropriate axis. For example,

```
In [x]: a = np.arange(6)
```

---

[11] NumPy has to copy the data because it has to store its data in one contiguous block of memory and the original arrays may be dispersed in different noncontiguous locations.

```
In [x]: a
Out[x]: array([ 0,  1,  2,  3,  4,  5])
In [x]: np.hsplit(a, 3)
Out[x]: [array([ 0,  1]), array([ 2,  3]), array([ 4,  5])]
```

As can be seen, a list of array objects is returned. If the second argument is a sequence of integer indexes, the array is split on those indexes:

```
In [x]: a
Out[x]: array([ 0,  1,  2,  3,  4,  5])
In [x]: np.hsplit(a, (2, 3, 5))
[array([0, 1]), array([2]), array([3, 4]), array([5])]
```

– this is the same as the list `[a[:2], a[2:3], a[3:5], a[5:]]`. Unlike with `np.hstack`, etc., the arrays returned are views on the original data.[12]

Example E6.3 Suppose you have a 3×3 array to which you wish to add a row or column. Adding a row is easy with `np.vstack`:

```
In [x]: a = np.ones((3, 3))
In [x]: np.vstack( (a, np.array((2, 2, 2))) )
Out[x]:
array([[ 1.,  1.,  1.],
       [ 1.,  1.,  1.],
       [ 1.,  1.,  1.],
       [ 2.,  2.,  2.]])
```

Adding a column requires a bit more work, however. You can't use `np.hstack` directly:

```
In [x]: a = np.ones((3, 3))
In [x]: np.hstack( (a, np.array((2, 2, 2))) )

... [Traceback information] ...
ValueError: all the input arrays must have same number of dimensions
```

This is because `np.hstack` cannot concatenate two arrays with different numbers of rows. Schematically:

```
[[ 1.,  1.,  1.],        [2., 2., 2.]
 [ 1.,  1.,  1.],   +                     = ?
 [ 1.,  1.,  1.]]
```

We can't simply transpose our new row, either, because it's a one-dimensional array and its transpose is the same shape as the original. So we need to *reshape* it first:

```
In [x]: a = np.ones((3, 3))
In [x]: b = np.array((2, 2, 2)).reshape(3, 1)
In [x]: b
array([[2],
       [2],
       [2]])
In [x]: np.hstack((a, b))
Out[x]:
```

---

[12] NumPy does this for efficiency reasons – copying large amounts of data is expensive and not necessary to fulfill the function of these splitting methods.

```
array([[ 1.,   1.,   1.,   2.],
       [ 1.,   1.,   1.,   2.],
       [ 1.,   1.,   1.,   2.]])
```

6.1.6   Indexing and Slicing an Array

An array is indexed by a tuple of integers and, as for Python sequences, negative indexes count from the end of the axis. Slicing and striding is supported in the same way as well. Note, however, that slicing a NumPy array returns a *view* on its data, not a *copy* of the data as for Python `lists`. For one-dimensional arrays there is only one index:

```
In [x]: a = np.linspace(1, 6, 6)
In [x]: print(a)
[ 1.   2.   3.   4.   5.   6.]
In [x]: a[1:4:2]      # elements a[1] and a[3] (a stride of 2)
Out[x]: array([ 2.,   4.])
In [x]: a[3::-2]      # elements a[3] and a[1] (a stride of -2)
Out[x]: array([ 4.,   2.]
```

Multidimensional arrays have an index for each axis. If you want to select every item along a particular axis, replace its index with a single colon:

```
In [x]: a = np.linspace(1, 12, 12).reshape(4, 3)
In [x]: a
Out[x]:
array([[  1.,   2.,   3.],
       [  4.,   5.,   6.],
       [  7.,   8.,   9.],
       [ 10.,  11.,  12.]])
In [x]: a[3, 1]
Out[x]: 11.0
In [x]: a[2, :]            # everything in the third row
Out[x]:
array([ 7.,   8.,   9.])
In [x]: a[:, 1]            # everything in the second column
Out[x]: array([  2.,   5.,   8.,  11.])
In [x]: a[1:-1, 1:]       # middle rows, second column onwards
Out[x]:
array([[ 5.,   6.],
       [ 8.,   9.]])
```

These and further examples of NumPy array slicing are illustrated in Figure 6.1.

The special *ellipsis* notation (...) is useful for high-rank arrays: in an index, it represents as many colons as are necessary to represent the remaining axes. For example, for a four-dimensional array, a[3, 1, ...] is equivalent to a[3, 1, :, :] and a[3, ... ,1] is equivalent to a[3, :, :, 1].

The colon and ellipsis syntax also works for assignment:

```
In [x]: a[:, 1] = 0       # set all elements in the second column to zero
In [x]: print(a)
[[  1.   0.   3.]
 [  4.   0.   6.]
 [  7.   0.   9.]
 [ 10.   0.  12.]]
```

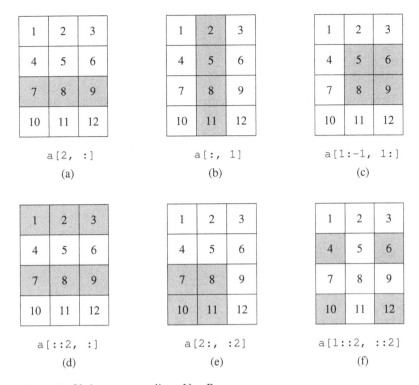

Figure 6.1 Various ways to slice a NumPy array.

Advanced Indexing

NumPy arrays can also be indexed by sequences that aren't simple tuples of integers, including other lists, arrays of integers and tuples of tuples. Such "advanced indexing" creates a new array with its own copy of the data, rather than a view:

```
In [x]: a = np.linspace(0., 0.5, 6)
In [x]: print(a)
[ 0.   0.1  0.2  0.3  0.4  0.5]
In [x]: ia = [1, 4, 5]      # a list of indexes
In [x]: print(a[ia])
[ 0.1  0.4  0.5]
In [x]: ia = np.array( ((1, 2), (3, 4)) )
In [x]: print(a[ia])      # an array to be formed from the specified indexes
[[ 0.1  0.2]
 [ 0.3  0.4]]
```

One can even index a multidimensional array with multidimensional arrays of indexes, picking off individual elements at will to build an array of a specified shape. This can lead to some rather baroque code:

```
In [x]: a = np.linspace(1, 12, 12).reshape(4, 3)
In [x]: print(a)
[[  1.   2.   3.]
 [  4.   5.   6.]
 [  7.   8.   9.]
 [ 10.  11.  12.]]
```

```
In [x]: ia = np.array( ((1, 0), (2, 1)) )
In [x]: ja = np.array( ((0, 1), (1, 2)) )
In [x]: print(a[ia, ja])
[[ 4.   2.]
 [ 8.   6.]]
```

Here we build a 2×2 array (the shape of the index arrays) whose elements are a[1, 0], a[0, 1] on the top row and a[2, 1], a[1, 2] on the bottom row.

Instead of indexing an array with a sequence of integers, it is also possible to use an array of boolean values. The True elements of this indexing array identify elements in the target array to be returned:

```
In [x]: a = np.array([-2, -1, 0, 1, 2])
In [x]: ia = np.array([False, True, False, True, True])
In [x]: print(a[ia])
[-1  1  2]
```

Because comparisons are vectorized across arrays just like mathematical operations, this leads to some useful shortcuts:

```
In [x]: print(a)
[-2 -1  0  1  2]
In [x]: ib = a < 0
In [x]: print(ib)
[ True  True False False False]
In [x]: a[ib] = 0    # set all negative elements to zero
In [x]: print(a)
[0 0 0 1 2]
```

It is not actually necessary to store the intermediate boolean array, ib, and a[a < 0] = 0 does the same job:

```
In [x]: a = np.array([-2, -1, 0, 1, 2])
In [x]: a[a < 0] = 0
In [x]: print(a)
[0 0 0 1 2]
```

The boolean operations *not*, *and* and *or* are implemented on boolean arrays with the operators ~, & and | respectively. For example,

```
In [x]: years = np.array([1900, 1904, 1990, 1993, 2000, 2014, 2016, 2100])
In [x]: leap_year = (years % 400 == 0) | (years % 4 == 0) & ~(years % 100 == 0)
In [x]: print(list(zip(years, leap_year)))
Out[x]: [(1900, False), (1904, True), (1990, False), (1993, False),
         (2000, True), (2014, False), (2016, True), (2100, False)]
```

Adding an Axis

To add an axis (i.e. dimension) to an array, insert np.newaxis in the desired position:

```
In [x]: a = np.linspace(1, 4, 4).reshape(2, 2)
In [x]: print(a)     # a 2 x 2 array (rank = 2)
[[ 1.   2.]
 [ 3.   4.]]
In [x]: a.shape()
(2, 2)
In [x]: b = a[:, np.newaxis, :]
```

```
In [x]: print(b)     # a 2 x 1 x 2 array (rank=3)
[[[ 1.   2.]]

 [[ 3.   4.]]]
In [x]: b.shape
(2, 1, 2)
```

In fact, np.newaxis is the None object, so None can be used directly in its place if desired.

Example E6.4   A *Sudoku* square consists of a 9×9 grid with entries such that each row, column and each of the nine nonoverlapping 3×3 tiles contains the numbers 1–9 once only. The following program verifies that a provided grid is a valid Sudoku square.

Listing 6.2   Verifying the validity of a Sudoku square

```
import numpy as np

def check_sudoku(grid):
    """ Return True if grid is a valid Sudoku square, otherwise False. """
    for i in range(9):
        # j, k index the top left-hand corner of each 3 x 3 tile.
        j, k = (i // 3) * 3, (i % 3) * 3
        if len(set(grid[i,:])) != 9 or len(set(grid[:,i])) != 9\
                or len(set(grid[j:j+3, k:k+3].ravel())) != 9:
            return False
    return True

sudoku = """145327698
            839654127
            672918543
            496185372
            218473956
            753296481
            367542819
            984761235
            521839764"""
# Turn the provided string, sudoku, into an integer array.
grid = np.array([[int(i) for i in line] for line in sudoku.split()])
print(grid)

if check_sudoku(grid):
    print('grid valid')
else:
    print('grid invalid')
```

❶ Here, we use the fact that an array of length nine contains nine unique elements if the *set* formed from these elements has cardinality 9. No check is made that the elements themselves are actually the numbers 1–9.

Meshes

To evaluate a multidimensional function on a grid of points, a *mesh* is useful The function np.meshgrid is passed a series of N one-dimensional arrays representing coordinates along each dimension and returns a set of N-dimensional arrays comprising a mesh of coordinates at which the function can be evaluated. For example, in the two-dimensional case:

```
In [x]: x = np.linspace(0, 5, 6)
In [x]: y = np.linspace(0, 3, 4)
In [x]: X, Y = np.meshgrid(x, y)
In [x]: X
Out[x]:
array([[ 0.,   1.,   2.,   3.,   4.,   5.],
       [ 0.,   1.,   2.,   3.,   4.,   5.],
       [ 0.,   1.,   2.,   3.,   4.,   5.],
       [ 0.,   1.,   2.,   3.,   4.,   5.]])

In [x]: Y
Out[x]:
array([[ 0.,   0.,   0.,   0.,   0.,   0.],
       [ 1.,   1.,   1.,   1.,   1.,   1.],
       [ 2.,   2.,   2.,   2.,   2.,   2.],
       [ 3.,   3.,   3.,   3.,   3.,   3.]])
```

The arrays X and Y can *each* be indexed with indexes i, j: the x array is repeated as rows down X and the y array as columns across Y. A function of two coordinates can therefore be evaluated on the grid as simply f(X, Y).

Setting the optional argument sparse to True will return sparse grid to conserve memory. In the previous example, instead of two arrays, both with shapes (6, 4), arrays with shapes (1, 6) and (4, 1) that can be broadcast against each other (see Section 6.1.7) will be returned:

```
In [X]: X, Y = np.meshgrid(x, y, sparse=True)
In [X]: X
Out[X]: array([[ 0.,   1.,   2.,   3.,   4.,   5.]])
In [X]: Y
Out[X]:
array([[ 0.],
       [ 1.],
       [ 2.],
       [ 3.]])
```

6.1.7 ◊ Broadcasting

We have already seen that simple operations such as addition and multiplication can be carried out elementwise on two arrays of the same shape (vectorization):

```
In [x]: a = np.array([1, 2, 3])
In [x]: b = np.array([0, 10, 100])
In [x]: a * b
Out[x]: array([  0,  20, 300])
```

Broadcasting describes the rules that NumPy uses to carry out such operations when the arrays have *different* shapes. This allows the operation to be carried out using precompiled C loops instead of slower, Python loops, but there are constraints as to which array shapes can be broadcast against each other. The rules are applied on each dimension of the arrays, starting with the last and working backward. Two dimensions compared in this way are said to be *compatible* if they are *equal* or *one of them is 1*.

The simplest example of broadcasting involves the operation between an array and a scalar (which may be considered for this purpose to be a one-dimensional array of length 1). Consider

```
In [x]: a = np.array([[1, 2, 3], [4, 5, 6]])
In [x]: b = 2
In [x]: c = a * b
In [x]: c
Out[x]:
array([[ 2,  4,  6],
       [ 8, 10, 12]])
```

The dimensions of a and b are compatible:

```
a:      2 x 3
b:          1
c:      2 x 3
```

Here, b can be broadcast across the two dimensions of array a by repetition of its value for every element in that array. Similarly, an array of shape (3,) can be broadcast across both rows of a:

```
In [x]: b = np.array([1, 2, 3])
In [x]: c = a * b
In [x]: c
Out[x]:
array([[ 1,  4,  9],
       [ 4, 10, 18]])
```

```
a:      2 x 3
b:          3
c:      2 x 3
```

That is, for each row of a, its entries are multiplied by the corresponding entries of the one-dimensional array b. However, attempting to multiply a by an array whose last dimension is not 1 or 3 is a ValueError here:

```
In [x]: b = np.array([1, 2])
In [x]: a * b
```

```
---------------------------------------------------------------------

...
----> 1 a * b
```

```
ValueError: operands could not be broadcast together with shapes (2,3) (2,)
```

In the example of the sparse mesh created in the previous section, the arrays with shapes (1, 6) and (4, 1) are compatible. For example,

```
In [x]: f = X * Y
Out[x]: f
array([[  0.,   0.,   0.,   0.,   0.,   0.],
       [  0.,   1.,   2.,   3.,   4.,   5.],
       [  0.,   2.,   4.,   6.,   8.,  10.],
       [  0.,   3.,   6.,   9.,  12.,  15.]])
```

The broadcasting process "stretches out" the second axis of Y from 1 to 6 to match that of X and the first axis of X from 1 to 4 to match that of Y:

```
X:       1 x 6
Y:       4 x 1
f:       4 x 6
```

To force a broadcast on an array with insufficient dimensions to meet your require-
ments, you can always add an axis with np.newaxis. For example, one way to take the
outer product of two arrays is by adding a dimension to one of them and broadcasting
the multiplication:

```
In [x]: a = np.array([1, 2, 3])
In [x]: b = np.array([0, 10, 100])
In [x]: c = a[:, np.newaxis] * b
In [x]: c
Out[x]:
array([[  0,  20, 300],
       [  0,  40, 600],
       [  0,  60, 900]])
```

Thus, instead of matching elements in the two arrays with shapes (3,), the extra axis on
a creates an array with shape (3, 1) and this dimension is stretched across the array b:

```
a[:,np.newaxis]:    3 x 1
            b:          3
            c:      3 x 3
```

6.1.8   Maximum and Minimum Values

NumPy arrays have the methods min and max, which return the minimum and maximum
values in the array. By default, a single value for the flattened array is returned; to find
maximum and minimum values along a given axis, use the axis argument:

```
In [x]: a = np.array([[3, 0, -1, 1], [2, -1, -2, 4], [1, 7, 0, 4]])
In [x]: print(a)
[[ 3  0 -1  1]
 [ 2 -1 -2  4]
 [ 1  7  0  4]]
In [x]: a.min()      # "global" minimum
Out[x]: -2
In [x]: a.max()      # "global" maximum
Out[x]: 7
In [x]: print( a.min(axis=0) )
[ 1 -1 -2  1]         # minima in each column
In [x]: print( a.max(axis=1) )
[3 4 7]               # maxima in each row
```

Often, one wants not the maximum (or minimum) value itself but its index in the
array. This is what the methods argmin and argmax do. By default, the index returned
is into the *flattened* array, so the actual value can be retrieved using a view on the array
created by ravel:

```
In [x]: a.argmin()
6
In [x]: a.ravel()[a.argmin()]
-2
```

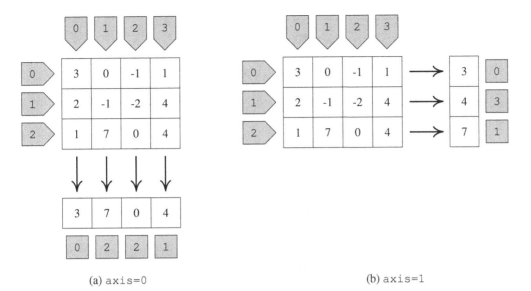

(a) `axis=0`   (b) `axis=1`

Figure 6.2 (a) `a.max(axis=0)` giving the maximum values and `a.argmax(axis=0)` giving the indexes of the maximum values of each column in array a (that is, maintaining the *row* dimension) and (b) The same for `axis=1`: maximum values along each row.

```
In [x]: print(a.argmax(axis=0))
[0 2 2 1]        # row indexes of maxima in each column
In [x]: print(a.argmax(axis=1))
[0 3 1]          # column indexes of maxima in each row
```

Figure 6.2 illustrates the process for `axis=0` and for `axis=1`. Notice that if more than one equal maximum exists in a column, the index of the first is returned.

Example E6.5   Consider the following oscillating functions on the interval $[0, L]$:

$$f_n(x) = x(L - x)\sin\frac{2\pi x}{\lambda_n}; \quad \lambda_n = \frac{2L}{n}, \quad n = 1, 2, 3, \ldots$$

The following code defines a two-dimensional array holding values of these functions for $L = 1$ on a grid of $N = 100$ points (rows) for $n = 1, 2, \ldots, 5$ (columns). The position of the maximum and minimum in each column is calculated with `argmax(axis=0)` and `argmin(axis=0)`. (See Figure 6.3.)

Listing 6.3 argmax and argmin

```python
# eg6-array_maxmin.py
import numpy as np
import matplotlib.pyplot as plt


N = 100
L = 1


def f(i, n):
    x = i * L / N
```

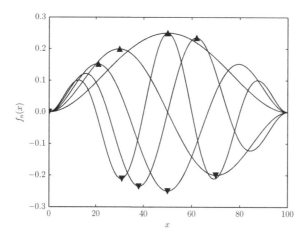

Figure 6.3 Maxima and minima of the functions $f_n(x)$ described in Example E6.5. Note that only the "global" maximum and minimum are returned for each function, and that where more than one point has the same maximum or minimum value, only the first is returned.

```
    lam = 2 * L / (n+1)
    return x * (L-x) * np.sin(2*np.pi*x/lam)

a = np.fromfunction(f, (N+1, 5))
min_i = a.argmin(axis=0)
max_i = a.argmax(axis=0)
plt.plot(a, c='k')
plt.plot(min_i, a[min_i, np.arange(5)], 'v', c='k', markersize=10)
plt.plot(max_i, a[max_i, np.arange(5)], '^', c='k', markersize=10)
plt.xlabel(r'$x$')
plt.ylabel(r'$f_n(x)$')
plt.show()
```

6.1.9   Sorting an Array

NumPy arrays can be sorted in several different ways with the `sort` method, which orders the numbers in an array *in place*. By default, this method sorts multidimensional arrays along their *last* axis. To sort along some other axis, set the `axis` argument. For example,

```
In [x]: a = np.array([5, -1, 2, 4, 0, 4])
In [x]: a.sort()
In [x]: print(a)
[-1  0  2  4  4  5]
In [x]: b = np.array([[0, 3, -2], [7, 1, 3], [4, 0, -1]])
In [x]: print(b)
[[ 0   3  -2]
 [ 7   1   3]
```

```
[ 4  0 -1]]
In [x]: b.sort()            # sort the numbers along each row
In [x]: print(b)
[[-2  0  3]
 [ 1  3  7]
 [-1  0  4]]
```

This is the same as b.sort(axis=1) – "for each row, order the numbers by column." To sort the numbers in each column – "for each column, order the numbers by row," set axis=0:

```
In [x]: b = np.array([[0,  3, -2], [7, 1,  3], [4, 0, -1]])
In [x]: b.sort(axis=0)     # sort the numbers along each column
In [x]: print(b)
[[0  0  -2]
 [4  1  -1]
 [7  3   3]]
```

The sorting algorithm used is the "quicksort" algorithm, which is a good general-purpose choice.[13]

Two other sorting functions are worth mentioning. np.argsort returns the *indexes* that would sort an array rather than the sorted elements themselves:

```
In [x]: a = np.array([3, 0, -1, 1])
In [x]: np.argsort(a)
Out[x]: array([2, 1, 3, 0])
```

Therefore,

```
In [x]: a[np.argsort(a)]
Out[x]: array([-1,  0,  1,  3])
```

The method np.searchsorted takes a *sorted* array, a, and one or more values, v, and returns the indexes in a at which the values should be entered to maintain its order:

```
In [x]: a = np.array([1, 2, 3, 4])
In [x]: np.searchsorted(a, 3.5)
Out[x]: 3
In [x]: np.searchsorted(a, (3.5, 0, 1.1))
Out[x]: array([3, 0, 1])
```

6.1.10 Structured Arrays

Also known as *record arrays*, structured arrays are arrays consisting of rows of values where each value may have its own data type and name. These rows are the "records." This type of array is very much like a table of data with rows (records) consisting of values that fall into columns (fields) and provides a very convenient and natural way to manipulate scientific data that is often obtained or presented in tabular form.

---

[13] Some arrays can be sorted faster with the alternative *mergesort* or *heapsort* algorithms; these can be selected by setting the optional kind argument to the string literal values 'mergesort' and 'heapsort', for example: b.sort(axis=1, kind='heapsort').

Structured arrays are useful for the manipulation of small sets of heterogeneous data, but this functionality is available at a higher level in the pandas library (see Chapter 9), which is often more convenient for large data sets.

Creating a Structured Array

The structure of a record array is defined by its dtype using a more complex syntax than we have used previously. For example,

```
In [x]: a = np.zeros(5, dtype='int8, float32, complex_')
In [x]: print(a)
[(0, 0.0, 0j) (0, 0.0, 0j) (0, 0.0, 0j) (0, 0.0, 0j) (0, 0.0, 0j)]
In [x]: a.dtype
dtype([('f0', '|i1'), ('f1', '<f4'), ('f2', '<c16')])
```

Here, we have created an array of five records, each of which has three fields, defined by constructing a dtype specified by the string 'int8, float32, complex_'.

- The first field is a single-byte, signed integer (int8, which is described by the string '|i1' – clearly the endianness [byte order] is not relevant in a one-byte quantity).
- The second is a single-precision floating-point number, which is stored in memory (on my system) as a little-endian 4-byte sequence, indicated by '<f4'.
- The final field is defined to be a complex number to default precision, which on my system is stored in 16-bytes, little-endian (complex_ is equivalent to complex128 which corresponds to a data type '<c16').

Because we did not explicitly name the fields, they are given the default names 'f0', 'f1' and 'f2'. To name the fields of our structured array explicitly, pass the dtype constructor a list of (*name, dtype descriptor*) tuples: for example,

```
In [x]: dt = np.dtype( [('time', 'f8'), ('signal', 'i4')] )
In [x]: a = np.zeros(10, dtype=dt)
In [x]: a
Out[x]:
array([(0.0, 0), (0.0, 0), ..., (0.0, 0)],
      dtype=[('time', '<f8'), ('signal', '<i4')])
```

A structured array can therefore be visualized as a table of data values with column headings for each field.

Assigning records in a structured array is as expected:

```
In [x]: a[0] = (0., 4)
In [x]: a[1:3] = [(0.5, -3), (1., -5)]
In [x]: a
Out[x]:
array([(0.0, 4), (0.5, -3), (1.0, -5), ..., (0.0, 0)],
      dtype=[('time', '<f8'), ('signal', '<i4')])
```

but the real power of this approach is in the ability to reference a field by its name. For example, to set the 'time' column in our array to a linear sequence:

```
In [x]: a['time'] = np.linspace(0., 4.5, 10)
In [x]: print(a)
```

```
[(0.0, 4) (0.5, -3) (1.0, -5) (1.5, 0) (2.0, 0) (2.5, 0) (3.0, 0) (3.5, 0)
 (4.0, 0) (4.5, 0)]
In [x]: print(a['time'][-1])
4.5
```

Likewise, to obtain a view on a column, refer to it by name:

```
In [x]: print(a['time'])
[ 0.   0.5  1.   1.5  2.   2.5  3.   3.5  4.   4.5]
In [x]: print( a['signal'].min() )
-5
```

More Ways to Create a Structured Array

There are several (arguably, too many) ways to define the `dtype` describing a structured array. So far we have used a string of comma-separated identifiers and a list of tuples. A third way is to use a *dictionary*. The basic usage assigns a list of values to the two keys, `'names'` and `'formats'`, naming the fields and specifying their formats respectively:

```
In [x]: dt = np.dtype({ 'names': ['time', 'signal'],
                        'formats': ['f8', 'i4']
                      })
In [x]: a = np.zeros(10, dtype=dt)
```

defines the same structured array of (`time`, `signal`) records as before. A third key, `'titles'`, can be used to give each field a more detailed description; each title can then be used as an alias to its name in referring to that field in the array.[14]

```
In [x]: dt = np.dtype({'names': ['candidate', 'mark', 'grade'],
                       'formats': ['|S50', 'u1', '|S2'],
                       'titles': ['Candidate Name', 'Percentage Mark', 'Grade: A-F']})
In [x]: a = np.zeros(10, dtype=dt)
In [x]: a[0] = ('John Brown', 64, 'B-')
In [x]: a[1] = ('Jane Smith', 78, 'A')
In [x]: print(a['Candidate Name'])
[b'John Brown' b'Jane Smith' b'' b'' b'' b'' b'' b'' b'' b'']
In [x]: print(a['Percentage Mark'])
[64 78  0  0  0  0  0  0  0  0]
```

Sorting Structured Arrays

Structured arrays can be sorted by giv dimensionsing a specific order to the fields used with the order argument. For example, with the following structured array:

```
In [x]: data = [ ('NiCd', 1.2, 0.14, 2000),
                 ('Lead acid', 2.1, 0.14, 700),
                 ('Lithium ion', 3.6, 0.46, 800) ]
In [x]: dtype = [ ('name', '|S20'),
                  ('voltage', 'f8'),
                  ('specific energy', 'f8'),
                  ('cycle durability', 'i4') ]
In [x]: a = np.array(data, dtype=dtype)
```

---

[14] In fact, `title` can be any Python object and can be used to provide detailed "metadata" concerning the corresponding field.

```
In [x]: a.sort(order='specific energy')
In [x]: print(a)
[(b'Lead acid', 2.1, 0.14, 700) (b'NiCd', 1.2, 0.14, 2000)
 (b'Lithium ion', 3.6, 0.46, 800)]

In [x]: a.sort(order=['specific energy', 'voltage'])
In [x]: print(a)
[(b'NiCd', 1.2, 0.14, 2000) (b'Lead acid', 2.1, 0.14, 700)
 (b'Lithium ion', 3.6, 0.46, 800)]
```

The second sort operation here sorts the records by specific energy, and if this is the same for two or more records, then it sorts by voltage.

6.1.11   Arrays as Vectors

A vector with n components can be defined as a regular one-dimensional array with n elements.

In addition to elementwise operations such as vector addition, subtraction and so on, NumPy array objects implement scalar (dot) product and vector (cross) product methods:

```
In [x]: a = np.array([1, 0, -3])         # vector as a one-dimensional array
In [x]: b = np.array([2, -2, 5])
In [x]: a.dot(b)                          # or a @ b or b.dot(a) or np.dot(a,b)
Out[x]: -13
In [x]: np.cross(a, b)
array([ -6, -11,  -2])
```

You can only take the cross product of an array with two or three elements; the third component is assumed to be zero in the former case. To use dot and cross on two individual vectors, ensure that they are row vectors as described previously and not column vectors represented as an (n, 1) array:

```
In [x]: a = np.array([[1], [0], [-3]])    # a 3 x 1 two-dimensional array
In [x]: b = np.array([[2], [-2], [5]])
In [x]: print(a)
[[ 1]
 [ 0]
 [-3]]
In [x]: np.dot(a,b)     # tries matrix multiplication; won't work
...

ValueError: shapes (3,1) and (3,1) not aligned: 1 (dim 1) != 3 (dim 0)
```

If you do want to take the dot product of two column vectors using np.dot, they need to be turned into row vectors:

```
In [x]: np.dot(a.T[0], b.T[0])     # transpose to row vectors
Out[x]: -13
```

This is a bit tortuous: the index is needed because the transpose of our (n, 1) (two-dimensional) array is a (1, n) array from which we want the first and only row for our vector. Alternatively, we can operate using a flattened view of the column vectors obtained with ravel:

```
In [x]: a.ravel() @ b.ravel()     # the same as a.ravel().dot(b.ravel())
Out[x]: -13
```

To turn a row vector represented by a one-dimensional array of shape (n,) into a column vector of shape (n, 1), add an axis:

```
In [x]: r = np.array([3, 4, 5])
In [x]: c = r[:, np.newaxis]
In [x]: c
array([[3],
       [4],
       [5]])
```

6.1.12   Logic and Comparisons

NumPy provides a set of methods for comparing and performing logical operations on arrays elementwise. The more useful of these are summarized in Table 6.4.

np.all and np.any work the same as Python's built-in functions of the same name[15] (see Section 2.4.3):

```
In [x]: a = np.array([[1, 2, 0, 3], [4, 0, 1, 1]])
In [x]: np.any(a), np.all(a)
Out[x]: (True, False)     # some (but not all) elements are equivalent to True
```

np.isreal and np.iscomplex return boolean arrays:

```
In [x]: b = np.array([1, -1j, 0.5j, 0, 1-2.5j])
In [x]: np.isreal(b)
Out[x]: array([ True, False, False,  True, False], dtype=bool)
In [x]: np.iscomplex(b)
Out[x]: array([False,  True,  True, False,  True], dtype=bool)
```

Because the representation of floating-point numbers is not exact, comparing two float or complex arrays with the == operator is not always reliable and is not recommended Instead, the best we can do is see if two values are "close" to one another within some (typically small) absolute or relative tolerance – NumPy provides the

Table 6.4 ndarray comparison methods

Function	Description
np.all(a)	Determine whether *all* array elements of a evaluate to True.
np.any(a)	Determine whether *any* array element of a evaluates to True.
np.isreal(a)	Determine whether each element of array a is real.
np.iscomplex(a)	Determine whether each element of array a is a complex number.
np.isclose(a, b)	Return a boolean array of the comparison between arrays a and b for equality within some tolerance.
np.allclose(a, b)	Return a True if *all* the elements in the arrays a and b are equal to within some tolerance.

---

[15] Except that they don't work on generator or iterator objects.

function `np.isclose(a, b)` for elementwise comparisons of two arrays: it returns True for elements satisfying

```
abs(a-b) <= (atol + rtol * abs(b))
```

with absolute tolerance, `atol`, and relative tolerance, `rtol`, which are 10^{-8} and 10^{-5}, respectively by default but can be changed by setting the corresponding arguments.[16] An additional argument, `equal_nan`, defaults to `False`, meaning that nan values in corresponding positions in the two arrays are treated as different; to treat such elements as equal, set `equal_nan=True`.

```
In [x]: a = np.array([1.66e-27, 1.38e-23, 6.63e-34, 6.02e23, np.nan])
In [x]: b = np.array([1.66e-27, 1.66e-27, 1.66e-27, 6.00e23, np.nan])
In [x]: np.isclose(a, b)
Out[x]: array([ True,   True,   True, False, False], dtype=bool)
In [x]: np.isclose(a, b, equal_nan=True)
Out[x]: array([ True,   True,   True, False,  True], dtype=bool)
```

Note that small numbers compare as equal even though they may differ by many orders of magnitude – to correct this, set `atol=0` to compare within relative tolerance only:

```
In [x]: np.isclose(a, b, atol=0)
Out[x]: array([ True, False, False, False, False], dtype=bool)
```

Finally, `allclose(a, b)` returns a single value: True only if every element in a is equal to the corresponding element in b (within the tolerance defined by `atol` and `rtol`), and otherwise False.

```
In [x]: x = np.linspace(0, np.pi, 100)
In [x]: np.allclose(np.sin(x)**2, 1 - np.cos(x)**2)
Out[x]: True
```

6.1.13   Exercises

Questions

Q6.1.1 What is the difference between the objects `np.ndarray` and `np.array`?

Q6.1.2 Why doesn't this create a two-dimensional array?

```
>>> np.array((1, 0, 0), (0, 1, 0), (0, 0, 1), dtype=float)
```

What is the correct way?

Q6.1.3 What is the difference, if any, between the following statements:

```
>>> a = np.array([0, 0, 0])
>>> a = np.array([[0, 0, 0]])
```

Q6.1.4 Explain the following behavior:

---

[16] Note that this relation is not symmetric in a and b, so it is possible that `isclose(a, b)` may not equal `isclose(b, a)`.

```
In [x]: a, b = np.zeros((3,)), np.ones((3,))
In [x]: a.dtype = 'int'
In [x]: a
Out[x]: array([0, 0, 0])
In [x]: b.dtype = 'int'
In [x]: b
Out[x]: array([4607182418800017408, 4607182418800017408, 4607182418800017408])
```

What is the correct way to convert an array of one data type to an array of another?

Q6.1.5 A $3 \times 4 \times 4$ array is created with

```
In [x]: a = np.linspace(1, 48, 48).reshape(3, 4, 4)
```

Index or slice this array to obtain the following:

(a)   `20.0`

(b)   `[ 9. 10. 11. 12.]`

(c)   The 4×4 array:

```
[[ 33.  34.  35.  36.]
 [ 37.  38.  39.  40.]
 [ 41.  42.  43.  44.]
 [ 45.  46.  47.  48.]]
```

(d)   The 3×2 array:

```
[[  5.,   6.],
 [ 21.,  22.],
 [ 37.,  38.]]
```

(e)   The 4×2 array:

```
[[ 36.  35.]
 [ 40.  39.]
 [ 44.  43.]
 [ 48.  47.]]
```

(f)   The 3×4 array:

```
[[ 13.   9.   5.   1.]
 [ 29.  25.  21.  17.]
 [ 45.  41.  37.  33.]]
```

(g)   (Harder) Using an array of indexes, the 2×2 array:

```
[[  1.   4.]
 [ 45.  48.]]
```

Q6.1.6 Write an expression, using boolean indexing, which returns only the values from an array that have magnitudes between 0 and 1.

Q6.1.7 Why does the following statement evaluate to `True` even though the two numbers passed to `np.isclose()` differ by more than `atol`?

```
In [x]: np.isclose(-2.00231930436153, -2.0023193043615, atol=1.e-14)
Out[x]: True
```

Q6.1.8 Explain why the following evaluates to True even though the two approximations to π differ by more than 10^{-16}:

```
In [x]: np.isclose(3.1415926535897932, 3.141592653589793, atol=1.e-16, rtol=0)
Out[x]: True
```

whereas this statement works as expected:

```
In [x]: np.isclose(3.14159265358979, 3.1415926535897, atol=1.e-14, rtol=0)
Out[x]: False
```

Q6.1.9 Verify that the magic square created in Example E6.2 satisfies the conditions
that it contains the numbers 1 to N^2 and that its rows, columns and main diagonals sum
to $N(N^2 + 1)/2$.

Q6.1.10 Write a one-line statement that returns True if an array is a monotonically
increasing sequence or False otherwise.

Hint: np.diff returns the *difference* between consecutive elements of a sequence. For
example,

```
In [x]: np.diff([1, 2, 3, 3, 2])
Out[x]: array([ 1,  1,  0, -1])
```

◇ **Q6.1.11** (Harder) The dtype np.uint8 represents an unsigned integer in 8 bits. Its
value may therefore be in the range 0–255. Explain the following behavior:

```
In [x]: x = np.uint8(250)
In [x]: x * 2
Out[x]: 500

In [x]: x = np.array([250,], dtype=np.uint8)
In [x]: x * 2
Out[x]: array([244], dtype=uint8)
```

Problems

P6.1.1 Turn the following data concerning various species of cetacean into a NumPy
structured array and order it by (a) mass and (b) population. Determine in each case the
index at which *Bryde's whale* (population: 100 000, mass: 25 tonnes) should be inserted
to keep the array ordered.

Name	Population	Mass/tonnes
Bowhead whale	9000	60
Blue whale	20 000	120
Fin whale	100 000	70
Humpback whale	80 000	30
Gray whale	26 000	35
Atlantic white-sided dolphin	250 000	0.235
Pacific white-sided dolphin	1 000 000	0.15
Killer whale	100 000	4.5
Narwhal	25 000	1.5
Beluga	100 000	1.5
Sperm whale	2 000 000	50
Baiji	13	0.13
North Atlantic right whale	300	75
North Pacific right whale	200	80
Southern right whale	7000	70

A text file containing these data can be downloaded at https://scipython.com/ex/bfk.

P6.1.2 The *shoelace algorithm* for calculating the area of a simple polygon (that is, one without holes or self-intersections) proceeds as follows: Write down the (x, y) coordinates of the N vertexes in an $N \times 2$ array and then repeat the coordinates of the first vertex as the last row to make an $(N + 1) \times 2$ array. Now (a) multiply each x-coordinate value in the first N rows by the y-coordinate value in the next row down and take the sum, $S_1 = x_1 y_2 + x_2 y_3 + \ldots + x_N y_1$. Then (b) multiply each y-coordinate value in the first N rows by the x-coordinate in the next row down and take the sum, $S_2 = y_1 x_2 + y_2 x_3 + \ldots + y_N x_1$. The area of the polygon is then $\frac{1}{2}|S_1 - S_2|$.

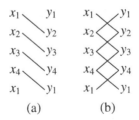

Implement this algorithm as a function that takes a NumPy array of vertexes as its argument and returns the area of the polygon. Do not use Python loops!

P6.1.3 Using NumPy, it is possible to do this exercise without using a single (Python) loop.

The normalized Gaussian function with mean μ and standard deviation σ is

$$g(x) = \frac{1}{\sigma \sqrt{2\pi}} \exp\left(-\frac{(x - \mu)^2}{2\sigma^2}\right).$$

Write a program to calculate and plot the Gaussian functions with $\mu = 0$ and the three values $\sigma = 0.5, 1, 1.5$. Use a grid of 1000 points in the interval $-10 \le x \le 10$.

Verify (by direct summation) that the functions are normalized with area 1.

Finally, calculate the first derivative of these functions on the same grid using the first-order central difference approximation:

$$g'(x) \approx \frac{g(x + h) - g(x - h)}{2h}$$

for some suitably chosen, small h.

6.2   Reading and Writing an Array to a File

Scientific data are frequently read in from a text file, which may contain comments, missing values and blank lines. Columns of values may be either aligned in a fixed-width format or separated by one or more delimiting characters (such as spaces, tabs or commas). Furthermore, there may be a descriptive header and even footnotes to the file, which make it hard to parse directly using Python's string methods.

NumPy provides several functions for reading data from a text file. The simpler `np.loadtxt` handles many common cases; the more sophisticated `np.genfromtxt` allows for better handling of missing values and footers. These are described in the following sections.

6.2.1   `np.save` and `np.load`

There is a platform-independent *binary* format for saving a NumPy array:

```
In [x]: np.save('my-array.npy', a)
```

will save the array a to the binary file `my-array.npy` (the `.npy` extension is appended if it is not provided). The array can then be reloaded using NumPy on any other operating system with

```
In [x]: a = np.load('my-array.npy')
```

(the `.npy` extension must be provided).

6.2.2   `np.loadtxt`

The method prototype for `np.loadtxt` is

```
np.loadtxt(fname, dtype=<class 'float'>, comments='#',
           delimiter=None, converters=None, skiprows=0,
           usecols=None, unpack=False, ndmin=0)
```

The arguments are as follows:

- fname: The only required argument, fname, which can be a filename, an open file or a generator returning the lines of data to be parsed.

- dtype: The data type of the array defaults to float but can be set explicitly by the dtype argument. In particular, this is the place to set up names and types for a structured array (see Section 6.1.10).

- comments: Comments in a file are usually started by some character such as # (as with Python) or %. To tell NumPy to ignore the contents of any line following this character, use the comments argument – by default it is set to #.

- delimiter: The string used to separate columns of data in the file; by default it is None, meaning that any amount of whitespace (spaces, tabs) delimits the data. To read a comma-separated (csv) file, set delimiter=','.

- converters: An optional dictionary mapping the column index to a function converting string values in that column to data (e.g. float).

- skiprows: An integer giving the number of lines at the start of the file to skip over before reading the data (e.g. to pass over header lines). Its default is 0 (no header).

- usecols: A sequence of column indexes determining which columns of the file to return as data; by default it is None, meaning all columns will be parsed and returned.

- unpack: By default, the data table is returned in a single array of rows and columns reflecting the structure of the file read in. Setting unpack=True will transpose this array so that individual columns can be picked off and assigned to different variables.

- ndmin: The minimum number of dimensions the returned array should have. By default, 0 (so a file containing a single number is read in as a scalar); it can also be set to 1 or 2.

For example, to read the first, third and fourth columns from the file data.txt into three separate one-dimensional arrays:

```
col1, col3, col4 = np.loadtxt('data.txt', usecols=(0, 2, 3), unpack=True)
```

Example E6.6 The use of np.loadtxt is best illustrated using an example. Consider the following text file of data relating to a (fictional) population of students. This file can be downloaded as eg6-a-student-data.txt from https://scipython.com/eg/bac.

```
# Student data collected on 17 July 2014.
# Researcher: Dr Wicks, University College Newbury.

# The following data relate to N = 20 students. It
# has been totally made up and so therefore is 100%
# anonymous.
```

Subject	Sex	DOB	Height	Weight	BP	VO2max
(ID)	M/F	dd/mm/yy	m	kg	mmHg	mL.kg-1.min-1
JW-1	M	19/12/95	1.82	92.4	119/76	39.3
JW-2	M	11/1/96	1.77	80.9	114/73	35.5
JW-3	F	2/10/95	1.68	69.7	124/79	29.1
JW-6	M	6/7/95	1.72	75.5	110/60	45.5
# JW-7	F	28/3/96	1.66	72.4	101/68	-
JW-9	F	11/12/95	1.78	82.1	115/75	32.3
JW-10	F	7/4/96	1.60	-	-/-	30.1

```
JW-11    M    22/8/95     1.72    77.2    97/63     48.8
JW-12    M    23/5/96     1.83    88.9    105/70    37.7
JW-14    F    12/1/96     1.56    56.3    108/72    26.0
JW-15    F    1/6/96      1.64    65.0    99/67     35.7
JW-16    M    10/9/95     1.63    73.0    131/84    29.9
JW-17    M    17/2/96     1.67    89.8    101/76    40.2
JW-18    M    31/7/96     1.66    75.1    -/-       -
JW-19    F    30/10/95    1.59    67.3    103/69    33.5
JW-22    F    9/3/96      1.70    -       119/80    30.9
JW-23    M    15/5/95     1.97    89.2    124/82    -
JW-24    F    1/12/95     1.66    63.8    100/78    -
JW-25    F    25/10/95    1.63    64.4    -/-       28.0
JW-26    M    17/4/96     1.69    -       121/82    39.
```

Let's find the average heights of the male and female students. The columns we need are the second and fourth, and there are no missing data in these columns so we can use np.loadtxt. First, construct a record dtype for the two fields, then read the relevant columns after skipping the first nine header lines:

```
In [x]: fname = 'eg6-a-student-data.txt'
In [x]: dtype1 = np.dtype([('gender', '|S1'), ('height', 'f8')])
In [x]: a = np.loadtxt(fname, dtype=dtype1, skiprows=9, usecols=(1,3))
In [x]: a
Out[x]:
array([[(b'M', 1.8200000524520874), (b'M', 1.7699999809265137),
        (b'F', 1.6799999475479126), (b'M', 1.7200000286102295),
        ...
        (b'M', 1.690000057220459)],
       dtype=[('gender', 'S1'), ('height', '<f8')])
```

To find the average heights of the male students, we only want to index the records with the gender field as M, for which we can create a boolean array:

```
In [x]: m = a['gender'] == b'M'
In [x]: m
Out[x]: array([ True,   True,   False,   True, ...,   True], dtype=bool)
```

m has entries that are True or False for each of the 19 valid records (one is commented out) according to whether the student is male or female. So the heights of the male students can be seen to be:

```
In [x]: print(a['height'][m])
[ 1.82000005    1.76999998    1.72000003    1.72000003    1.83000004    1.63
  1.66999996    1.65999997    1.97000003    1.69000006]
```

Therefore, the averages we need are

```
In [x]: m_av = a['height'][m].mean()
In [x]: f_av = a['height'][~m].mean()
In [x]: print('Male average: {:.2f} m, Female average: {:.2f} m'.format(m_av,f_av))
Male average: 1.75 m, Female average: 1.65 m
```

❶ Note that ~m ("not m") is the inverse boolean array of m.

To perform the same analysis on the student weights, we have a bit more work to do because there are some missing values (denoted by "-"). We could use np.genfromtxt (see Section 6.2.3), but let's write a converter method instead. We'll replace the missing values with the nicely unphysical value of −99. The function parse_weight expects a string argument and returns a float:

```
def parse_weight(s):
    try:
        return float(s)
    except ValueError:
        return -99.
```

This is the function we want to pass as a converter for column 4:

```
In [x]: dtype2 = np.dtype([('gender', '|S1'), ('weight', 'f8')])
In [x]: b = np.loadtxt(fname, dtype=dtype2, skiprows=9, usecols=(1, 4),
                       converters={4: parse_weight})
```

Now mask off the invalid data and index the array with a boolean array as before:

```
In [x]: mv = b['weight'] > 0     # elements only True for valid data
In [x]: m_wav = b['weight'][mv & m].mean()       # valid and male
In [x]: f_wav = b['weight'][mv & ~m].mean()      # valid and female
In [x]: print('Male average: {:.2f} kg,
               Female average: {:.2f} kg'.format(m_wav, f_wav))
Male average: 82.44 kg, Female average: 66.94 kg
```

Finally, let's read in the blood pressure data. Here we have a problem, because the systolic and diastolic pressures are not separated by whitespace but by a forward slash (/). One solution is to reformat each line to replace the slash with a space before it is fed to np.loadtxt. Recall that fname can be a generator instead of a filename or open file: we write a suitable generator function, reformat_lines, which takes an open file object and yields its lines to np.loadtxt, one by one, after the replacement. This is going to mess with the column numbering because it has the side effect of splitting up the birth dates into three columns, so in our reformatted lines the blood pressure values are now in the columns indexed at 7 and 8.

Listing 6.4 Reading the blood-pressure column

```
# eg6-a-read-bp.py
import numpy as np

fname = 'eg6-a-student-data.txt'
dtype3 = np.dtype([('gender', '|S1'), ('bps', 'f8'), ('bpd', 'f8')])

def parse_bp(s):
    try:
        return float(s)
    except ValueError:
        return -99.

def reformat_lines(fi):
    for line in fi:
        line = line.replace('/', ' ')
        yield line
with open(fname) as fi:
    gender, bps, bpd = np.loadtxt(reformat_lines(fi), dtype3, skiprows=9,
                   usecols=(1, 7, 8), converters={7: parse_bp, 8: parse_bp},
                   unpack=True)

# Now do something with the data...
```

6.2.3   `np.genfromtxt`

NumPy's `genfromtxt` function is similar to `np.loadtxt` but has a few more options and is able to cope with missing data.

The following arguments to this function are the same as for `np.loadtxt`: fname (the only required argument), `dtype`, `comments`, `converters`, `usecols` and `unpack`.

Headers and Footers

Instead of `np.loadtxt`'s `skiprows`, the `np.genfromtxt` function has two optional arguments, `skip_header` and `skip_footer`, giving the number of lines to skip at the beginning and the end of the file, respectively.

Fixed-Width Fields

The `delimiter` argument works the same as for `np.loadtxt` but can also be provided as a sequence of integers giving the widths of each field to be read in where the data columns do not have delimiters. For example, suppose the following text file, `data.txt`, is to be interpreted as consisting of four columns with widths 2, 1, 9 and 3 characters (spaces are indicated with "␣"):

```
␣12␣␣100.231.03
␣11␣1201.842.04
␣11␣␣␣99.324.02
```

so that the first row is to be split: ' 1', '2', ' 100.231', '.03'. There is no delimiter character, so this isn't possible with `np.loadtxt`, but with `np.genfromtxt`:

```
In [x]: np.genfromtxt(fname='data.txt', delimiter=[2, 1, 9, 3],
                      dtype='i4, i4, f8, f8')
array([(1, 2, 100.231, 0.03), (1, 1, 1201.842, 0.04), (1, 1, 99.324, 0.02)],
      dtype=[('f0', '<i4'), ('f1', '<i4'), ('f2', '<f8'), ('f3', '<f8')])
```

as required.

Missing Data

If a data set is incomplete, `np.loadtxt` will be unable to parse the fields with missing data into valid values for the array and will raise an exception. `np.genfromtxt`, however, sets missing or invalid entries equal to the default values given in Table 6.5.

For example, the comma-separated file here has two ways of indicating missing data: empty fields and entries with "???":

```
10.1,4,-0.1,2
10.2,4,,0
10.3,???,,4
10.4,2,0.,
10.5,-1,???,3
```

Accordingly, `np.genfromtxt` sets the missing fields to its defaults:

```
In [x]: data = np.genfromtxt(fname='data.txt', dtype='f8, i4, f8, i4',
   ...:                      delimiter=',')
```

Table 6.5 Default filling values for missing data used by genfromtxt

Data type	Default value
int	-1
float	np.nan
bool	False
complex	np.nan + 0.j

```
In [x]: print(data)
[(10.1, 4, -0.1, 2) (10.2, 4, nan, 0) (10.3, -1, nan, 4) (10.4, 2, 0.0, -1)
 (10.5, -1, nan, 3)]
```

The missing_values and filling_values arguments allow closer control over which default values to use for which columns. If missing_values is given as a sequence of strings, each string is associated with a column in the data file, in order; if given as a dictionary of string values, the keys denote either column indexes (if they are integers) or column names (if they are strings). The corresponding argument, filling_values, maps these column indexes or names to default values. If filling_values is provided as a single value, this value is used for missing data in all columns.

For example, to replace the invalid values in column 1 (indicated by "???") with 999, the missing or invalid values in column 2 (also indicated by "???") with −99 and the missing values in column 3 with 0:

```
In [x]: data =np.genfromtxt(fname='data.txt', dtype='f8, i4, f8, i4',
   ...:                      delimiter=',', missing_values={1: '???', 2: '???'},
   ...:                      filling_values={1: 999, 2: -99., 3: 0})
   ...:
In [x]: print(data)
[(10.1, 4, -0.1, 2) (10.2, 4, -99.0, 0) (10.3, 999, -99.0, 4)
 (10.4, 2, 0.0, 0) (10.5, -1, -99.0, 3)]
```

Note in particular how the missing entry in the second column has been replaced by 999 instead of the default −1 – this would be particularly important if −1 is a valid value for this column (however, it is now up to the rest of your code to recognize and know what to do with values such as 999.[17]

Column Names

The argument names provides a way of setting names for the columns of data read in. If it is the boolean value True, the names are read from the first valid line after the number of lines skipped over specified by the skip_header argument; if names is a comma-separated string of names or a sequence of strings, those strings will be used as names. By default, names is None and the field names are taken from the dtype, if given.

---

[17] For more advanced handling of missing values, see the genfromtxt documentation for details on the usemask argument and *masked arrays* in general.

Example E6.7   In an experiment to investigate the *Stroop effect*, a group of students were timed reading out 25 randomly ordered color names, first in black ink and then in a color other than the one they name (e.g. the word "red" in blue ink). The results are presented in the text file `stroop.txt`, which can be downloaded from https://scipython.com/eg/baj. Missing data are indicated by the character X.

```
Subject Number, Gender, Time (words in black), Time (words in color)
1,F,18.72,31.11
2,F,21.14,52.47
3,F,19.38,33.92
4,M,22.03,50.57
5,M,21.41,29.63
6,M,15.18,24.86
7,F,14.13,33.63
8,F,19.91,42.39
9,F,X,43.60
10,F,26.56,42.31
11,F,19.73,49.36
12,M,18.47,31.67
13,M,21.38,47.28
14,M,26.05,45.07
15,F,X,X
16,F,15.77,38.36
17,F,15.38,33.07
18,M,17.06,37.94
19,M,19.53,X
20,M,23.29,49.60
21,M,21.30,45.56
22,M,17.12,42.99
23,F,21.85,51.40
24,M,18.15,36.95
25,M,33.21,61.59
```

We can read in this data with `np.genfromtxt` and summarize the results with the code here.

Listing 6.5 Analyzing data from a Stroop effect experiment

```
# eg6-stroop.py
import numpy as np

# Read in the data from stroop.txt, identifying missing values and
# replacing them with NaN.
❶ data = np.genfromtxt('stroop.txt', skip_header=1,
                    dtype=[('student', 'u8'), ('gender', 'S1'),
                          ('black', 'f8'), ('color', 'f8')],
                    delimiter=',',
                    missing_values='X')
nwords = 25


# Remove invalid rows from data set.
❷ filtered_data = data[np.isfinite(data['black']) & np.isfinite(data['color'])]

# Extract rows by gender (M/F) and word color (black/color) and normalize
# to time taken per word.
```

```
fb = filtered_data['black'][filtered_data['gender']==b'F'] / nwords
mb = filtered_data['black'][filtered_data['gender']==b'M'] / nwords
fc = filtered_data['color'][filtered_data['gender']==b'F'] / nwords
mc = filtered_data['color'][filtered_data['gender']==b'M'] / nwords

# Produce statistics: mean and standard deviation by gender and word color.
mu_fb, sig_fb = np.mean(fb), np.std(fb)
mu_fc, sig_fc = np.mean(fc), np.std(fc)
mu_mb, sig_mb = np.mean(mb), np.std(mb)
mu_mc, sig_mc = np.mean(mc), np.std(mc)

print('Mean and (standard deviation) times per word (sec)')
print('gender |   black    |   color     | difference')
print('  F   | {:4.3f} ({:4.3f}) | {:4.3f} ({:4.3f}) |   {:4.3f}'
                  .format(mu_fb, sig_fb, mu_fc, sig_fc, mu_fc - mu_fb))
print('  M   | {:4.3f} ({:4.3f}) | {:4.3f} ({:4.3f}) |   {:4.3f}'
                  .format(mu_mb, sig_mb, mu_mc, sig_mc, mu_mc - mu_mb))
```

❶ In the absence of any provided filling_values, np.genfromtxt will replace the invalid fields with np.nan.

❷ We only want to consider students with times for both parts of the experiment, so create a filtered data set here.

The output shows a significantly slower per-word speed for the false-colored words than for the words in black:

```
Mean and (standard deviation) times per word (sec)
gender |   black    |   color     | difference
  F   | 0.770 (0.137) | 1.632 (0.306) |   0.862
  M   | 0.849 (0.186) | 1.679 (0.394) |   0.830
```

6.2.4   np.savetxt

The np.savetxt function saves a NumPy array as a text file. Its call signature is

```
np.savetxt(fname, X, fmt='%.18e', delimiter=' ',
           newline='\n', header='', footer='', comments='# ')
```

The arguments are as follows:

- fname: The name of the file or an open file handle into which the array data is to be saved.
- X: The array to save.
- fmt: A string defining the C-style format specifier for the array data output (see Section 2.3.7 for details). The default is '%.18e'.
- delimiter: The string delimiting columns in the output file; by default, a single space.
- newline: The string separating lines in the output file; by default, this is the Unix-style '\n'. Windows users may prefer to set newline to the sequence used on their platform: '\r\n'.

- header: A (possibly multiline) string to be written at the start of the output file.
- footer: A (possibly multiline) string to be written at the end of the output file.
- comments: A string that will be added to the header and footer to mark them as comments. The default is '# '. This is useful if the file is to be subsequently read in by np.loadtxt or np.genfromtxt so the number of header and footer lines does not have to be explicitly specified.

Example E6.8   The decay of an ensemble of radioactive nuclei over a period of time can be simulated as follows. Consider the time period to be divided into short, discrete intervals of duration $\Delta t \ll \tau$, where τ is the *lifetime* for the decay (which is related to the half-life, $t_{1/2}$, through $\tau = t_{1/2}/\ln 2$). The probability that a given nucleus will decay in time Δt is $p = \Delta t/\tau$.

At each time-step, the simulation loops over the undecayed nuclei from the previous time-step and draws a random number from the uniform distribution on $[0, 1)$: if this random number is less than p, the nucleus is considered to have decayed.

The code below defines a function to carry out this simulation for a set of $N_0 = 500$ ^{14}C nuclei with half-life $t_{1/2} = 5730$ years. nsims = 10 such simulations are carried out and saved to a comma-separated file, 14C-sim.csv, with a brief, explanatory header.

Listing 6.6 Simulation of the radioative decay of ^{14}C

```python
import random
import numpy as np

def decay_sim(thalf, N0=500, tgrid=None, nhalflives=4):
    """Simulate the radioactive decay of N0 nuclei.

    thalf is the half-life in some units of time.
    If tgrid is provided, it should be a sequence of evenly-spaced time points
    to run the simulation on.
    If tgrid is None, it is calculated from nhalflives, the number of
    half-lives to run the simulation for.

    """

    # Calculate the lifetime from the half-life.
    tau = thalf / np.log(2)

    if tgrid is None:
        # Create a grid of Nt time points up to tmax.
        Nt, tmax = 100, thalf * nhalflives
        tgrid, dt = np.linspace(0, tmax, Nt, retstep=True)
    else:
        # tgrid was provided: deduce Nt and the time step, dt.
        Nt = len(tgrid)
        dt = tgrid[1] - tgrid[0]

    N = np.empty(Nt, dtype=int)
    N[0] = N0
    # The probability that a given nucleus will decay in time dt.
    p = dt / tau
```

```
    for i in range(1, Nt):
        # At each time step, start with the undecayed nuclei from the previous.
        N[i] = N[i-1]
        # Consider each nucleus in turn and decide whether it decays or not.
        for j in range(N[i-1]):
            r = random.random()
            if r < p:
                # This nucleus decays.
                N[i] -= 1
    return tgrid, N


N0 = 500
# Half life of 14C in years.
thalf = 5730

# Use Nt time steps up to tmax years.
Nt, tmax = 100, 20000
tgrid = np.linspace(0, tmax, Nt)

# Repeat the simulation "experiment" nsims times.
nsims = 10
Nsim = np.empty((Nt, nsims))
for i in range(nsims):
    _, Nsim[:, i] = decay_sim(thalf, N0, tgrid)

# Save the time grid, followed by the simulations in columns. We save integer
# values for the data and create a comma-delimited file with a two-line header.
np.savetxt('14C-sim.csv', np.hstack((tgrid[:, None], Nsim)),
    fmt = '%d', delimiter=',',
    header=f'Simulations of the radioactive decay of {N0} 14C nuclei.\n'
           f'Columns are time in years followed by {nsims} decay simulations.'
    )
```

The contents of the output file, `14C-sim.csv`, will resemble:

```
# Simulations of the radioactive decay of 500 14C nuclei.
# Columns are time in years followed by 10 decays.
0,500,500,500,500,500,500,500,500,500,500
202,489,486,487,491,487,486,485,487,490,490
404,479,478,483,479,477,476,480,474,484,482
606,462,467,470,463,464,463,470,454,474,471
...
```

This file can be read in to a NumPy array with:

```
arr = np.loadtxt('14C-sim.csv', delimiter=',')
```

See also Exercise P6.5.7.

6.2.5   Exercises

Problems

P6.2.1   The following text file, which is available to download at https://scipython.com/ex/bfj, gives some data concerning the 8000 m peaks, in alphabetical order.

ex6-2-b-mountain-data.txt This file contains a list of the 14 highest mountains in the world with their names, height, year of first ascent, year of first winter ascent, and location as longitude and latitude in degrees (d), minutes (m) and seconds (s). Note: as of 2019, no winter ascent has been made of K2.

```
-----------------------------------------------------------------
Name           Height  First ascent First winter Location
               m       date         ascent date  (WGS84)
-----------------------------------------------------------------
Annapurna I    8091    3/6/1950     3/2/1987     28d35m46sN 83d49m13sE
Broad Peak     8051    9/6/1957     5/3/2013     35d48m39sN 76d34m06sE
Cho Oyu        8201    19/10/1954   12/2/1985    28d05m39sN 86d39m39sE
Dhaulagiri I   8167    13/5/1960    21/1/1985    27d59m17sN 86d55m31sE
Everest        8848    29/5/1953    17/2/1980    27d59m17sN 86d55m31sE
Gasherbrum I   8080    5/7/1958     9/3/2012     35d43m28sN 76d41m47sE
Gasherbrum II  8034    7/7/1956     2/2/2011     35d45m30sN 76d39m12sE
K2             8611    31/7/1954    -            35d52m57sN 76d30m48sE
Kangchenjunga  8568    25/5/1955    11/1/1986    27d42m09sN 88d08m54sE
Lhotse         8516    18/5/1956    31/12/1988   27d57m42sN 86d56m00sE
Makalu         8485    15/5/1955    9/2/2009     27d53m21sN 87d05m19sE
Manaslu        8163    9/5/1956     12/1/1984    28d33m0sN  84d33m35sE
Nanga Parbat   8126    ,3/7/1953    16/2/2016    35d14m15sN 74d35m21sE
Shishapangma   8027    2/5/1964     14/1/2005    28d21m8sN  85d46m47sE
-----------------------------------------------------------------
```

Use NumPy's `genfromtxt` method to read these data into a suitable structured array to determine the following:

(a)   the lowest 8000 m peak;
(b)   the most northely, easterly, southerly and westerly peaks;
(c)   the most recent first ascent of the peaks;
(d)   the first of the peaks to be climbed in winter.

Also, produce another structured array containing a list of mountains with their height in *feet* and first ascent date, ordered by increasing height.[18]

P6.2.2 The file `busiest_airports.txt`, which can be downloaded from https://scipython.com/ex/bfa , provides details of the 30 busiest airports in the world in 2014. The tab-delimited fields are: three-letter IATA code, airport name, airport location, latitude and longitude (both in degrees).

Write a program to determine the distance between two airports identified by their three-letter IATA code, using the Haversine formula (see, for example, Exercise P4.4.2) and assuming a spherical Earth of radius 6378.1 km.

P6.2.3 The World Bank provides an extensive collection of data sets on a wide range of "indicators," which is searchable at https://data.worldbank.org/. Data sets concerning child immunization rates for BCG (against tuberculosis), Pol3 (Polio) and measles in three Southeast Asian countries between 1960 and 2013 are available at

---

[18] 1 metre = 3.2808399 feet.

https://scipython.com/ex/bfb. Fields are delimited by semicolons and missing values are indicated by '..'.

Use NumPy methods to read in these data and create three plots (one for each vaccine) comparing immunization rates in the three countries.

6.3 Statistical Methods

NumPy provides several methods for performing statistical analysis, either on an entire array or an axis of it.

6.3.1 Ordering Statistics

Maxima and Minima

We have already used np.min and np.max to find the minimum and maximum values of an array (these methods are also available using the names np.amin and np.amax). If the array contains one or more NaN values, the corresponding minimum or maximum value will be np.nan. To ignore NaN values, instead use np.nanmin and np.nanmax:

```
In [x]: a = np.sqrt(np.linspace(-2, 2, 4))
In [x]: print(a)
[       nan        nan 0.          1.          1.41421356]
In [x]: np.min(a), np.max(a)
Out[x]: (nan, nan)
In [x]: np.nanmin(a), np.nanmax(a)
(0.0, 1.4142135623730951)
```

We have also met the functions np.argmin and np.argmax, which return the *index* of the minimum and maximum values in an array; they too have np.nanargmin and np.nanargmax variants:

```
In [x]: np.argmin(a), np.argmax(a)
Out[x]: (0, 0)          # the first nan in the array
In [x]: np.nanargmin(a), np.nanargmax(a)
Out[x]: (2, 4)          # the indexes of 0, 1.41421356
```

The related methods, np.fmin / np.fmax and np.minimum / np.maximum, compare two arrays, *element by element*, and return another array of the same shape. The first pair of methods ignores NaN values and the second pair propagates them into the output array. For example,

```
In [x]: np.fmin([1, -5, 6, 2], [0, np.nan, -1, -1])
array([ 0., -5., -1., -1.])          # NaNs are ignored
In [x]: np.maximum([1, -5, 6, 2], [0, np.nan, -1, -1])
array([ 1.,  nan,  6.,  2.])          # NaNs are propagated
```

Percentiles

The np.percentile method returns a specified percentile, q, of the data along an axis (or along a flattened version of the array if no axis is given). The minimum of an array is the value at q = 0 (0th percentile), the maximum is the value at q = 100 (100th percentile)

and the median is the value at q = 50 (50th percentile). Where no single value in the array corresponds to the requested value of q exactly, a weighted average of the two nearest values is used. For example,

```
In [x]:  a = np.array([[0., 0.6, 1.2], [1.8, 2.4, 3.0]])
In [x]: np.percentile(a, 50)
1.5
In [x]: np.percentile(a, 75)
2.25
In [x]: np.percentile(a, 50, axis=1)
array([ 0.6,  2.4])
In [x]: np.percentile(a, 75, axis=1)
array([ 0.9,  2.7])
```

6.3.2   Averages, Variances and Correlations

Averages

In addition to np.mean, which calculates the arithmetic mean of the values along a specified axis of an array, NumPy provides methods for calculating the weighted average, median, standard deviation and variance. The weighted average is calculated as

$$\bar{x}_w = \frac{\sum_i^N w_i x_i}{\sum_i^N w_i},$$

where the weights, w_i, are supplied as a sequence the same length as the array. For example,

```
In [x]: x = np.array([1., 4., 9., 16.])
In [x]: np.mean(x)
7.5
In [x]: np.median(x)
6.5
In [x]: np.average(x, weights=[0., 3., 1., 0.])
5.25        # i.e. (3.*4. + 1.*9.) / (3. + 1.)
```

If you want the sum of the weights as well as the weighted average, set the returned argument to True. In the following example, we do this and find the weighted averages in each row (axis=1 averages values across *columns* of a two-dimensional array):

```
In [x]: x = np.array( [[1., 8., 27], [-0.5, 1., 0.]] )
In [x]: av, sw = np.average(x, weights=[0., 1., 0.1], axis=1, returned=True)
In [x]: print(av)
[ 9.72727273  0.90909091]
In [x]: print(sw)
[ 1.1  1.1]
```

The averages are therefore $(1 \times 8 + 0.1 \times 27)/1.1 = 9.72727273$ and $(1 \times 1.)/1.1 = 0.90909091$ where 1.1 is the sum of the weights.

Standard Deviations and Variances

The function np.std calculates, by default, the *uncorrected sample standard deviation*:

$$\sigma_N = \sqrt{\frac{1}{N}\sum_i^N (x_i - \bar{x})^2},$$

where x_i are the N observed values in the array and $\bar{x}$ is their mean. To calculate the *corrected sample standard deviation*,

$$\sigma = \sqrt{\frac{1}{N-\delta}\sum_i^N (x_i - \bar{x})^2},$$

pass to the argument ddof the value of δ such that $N - \delta$ is the number of degrees of freedom in the sample. For example, if the sample values are drawn from the population independently with replacement and used to calculate $\bar{x}$ there are $N - 1$ degrees of freedom in the vector of residuals used to calculate σ: $(x_1 - \bar{x}, x_2 - \bar{x}, \dots, x_N - \bar{x})$ and so $\delta = 1$. For example,

```
In [x]: x = np.array([1., 2., 3., 4.])
In [x]: np.std(x)              # or x.std(), uncorrected standard deviation
1.1180339887498949
In [x]: np.std(x, ddof=1)     # corrected standard deviation
1.2909944487358056
```

The function np.nanstd calculates the standard deviation ignoring np.nan values (so that N is the number of non-NaN values in the array). NumPy also has methods for calculating the *variance* of the values in an array: np.var and np.nanvar.

The covariance is returned by the npcov method. In its simplest invocation, it can be passed a single two-dimensional array, X, in which the rows represent variables, x_i, and the columns observations of the value of each variable. np.cov(X) then returns the covariance matrix, C_{ij}, indicating how variable x_i varies with x_j: the element C_{ij} is said to be an estimate of the covariance of variables x_i and x_j:

$$C_{ij} \equiv \mathrm{cov}(x_i, x_j) = \mathrm{E}[(x_i - \mu_i)(x_j - \mu_j)],$$

where μ_i is the mean of the variable x_i and E[ ] denotes the expected value. If there are N observed values for each of the variables, $\mu_i = \frac{1}{N}\sum_k x_{ik}$. The *unbiased* estimate of the covariance is then

$$C_{ij} = \frac{1}{N-1}\sum_k [(x_{ik} - \mu_i)(x_{jk} - \mu_j)].$$

This is the default behavior of np.cov, but if the bias argument is set to 1, then N is used in the denominator here to give the *biased* estimate of the covariance. Finally, the denominator can be set explicitly to $N - \delta$ by passing δ as the argument to the ddof argument of cov.

Example E6.9 As an example, consider the matrix of five observations each of three variables, x_0, x_1 and x_2, whose observed values are held in the three rows of the array X:

```
X = np.array([ [0.1, 0.3, 0.4, 0.8, 0.9],
               [3.2, 2.4, 2.4, 0.1, 5.5],
               [10., 8.2, 4.3, 2.6, 0.9]
             ])
```

The covariance matrix is a 3×3 array of values,

```
In [x]: print( np.cov(X) )
[[  0.115 ,   0.0575,  -1.2325],
 [  0.0575,   3.757 ,  -0.8775],
 [ -1.2325,  -0.8775,  14.525 ]]
```

The diagonal elements, C_{ii}, are the variances in the variables x_i, assuming $N - 1$ degrees of freedom:

```
In [x]: print(np.var(X, axis=1, ddof=1))
[  0.115   3.757  14.525]
```

Although the magnitude of the covariance matrix elements is not always easy to interpret (because it depends on the magnitude of the individual observations, which may be very different for different variables), it is clear that there is a strong anticorrelation between x_0 and x_2 ($C_{02} = -1.2325$: as one increases the other decreases) and no strong correlation between x_0 and x_1 ($C_{01} = 0.0575$: x_0 and x_1 do not trend strongly together).

The *correlation coefficient matrix* is often used in preference to the covariance matrix as it is normalized by dividing C_{ij} by the product of the variables' standard deviations:

$$P_{ij} = \mathrm{corr}(x_i, x_j) = \frac{C_{ij}}{\sigma_i \sigma_j} = \frac{C_{ij}}{\sqrt{C_{ii}C_{jj}}}.$$

This means that the elements P_{ij} have values between -1 and 1 inclusive, and the diagonal elements, $P_{ii} = 1$. In our example, using np.corrcoef gives:

```
In [x]: print( np.corrcoef(X) )
[[ 1.          0.0874779  -0.95363007]
 [ 0.0874779   1.         -0.11878687]
 [-0.95363007 -0.11878687  1.        ]]
```

It is easy to see from this correlation coefficient matrix the strong anticorrelation between x_0 and x_2 ($C_{0,2} = -0.954$) and the lack of correlation between x_1 and the other variables (e.g. $C_{1,0} = 0.087$).

Both the np.cov and np.corrcoef methods can take a second array-like object containing a further set of variables and observations, so they can be called on a pair of one-dimensional arrays without stacking them into a single matrix:

```
In [x]: x = np.array([1., 2., 3., 4., 5.])
In [x]: y = np.array([0.08, 0.31, 0.41, 0.48, 0.62])
In [x]: print( np.corrcoef(x,y) )
[[ 1.          0.97787645]
 [ 0.97787645  1.        ]]
```

That is

```
np.corrcoef(x, y)
```

is a convenient alternative to

```
np.corrcoef(np.vstack((x,y)))
```

Finally, if your observations happen to be in the rows of your matrix, with the variables corresponding to the columns (instead of the other way round) there is no need to transpose the matrix, just pass `rowvar=0` to either `np.cov` or `np.corrcoef` and NumPy will take care of it for you.

Example E6.10 The Cambridge University Digital Technology Group have been recording the weather from the roof of their department building since 1995 and make the data available to download in a single CSV file at www.cl.cam.ac.uk/research/dtg/weather/.

The following program determines the correlation coefficient between pressure and temperature at this site.

Listing 6.7 Calculating the correlation coefficient between air temperature and pressure

```
# eg6-pT.py
import numpy as np
import matplotlib.pyplot as plt

data = np.genfromtxt('weather-raw.csv', delimiter=',', usecols=(1, 4))
# Remove any rows with either missing T or missing p.
data = data[~np.any(np.isnan(data), axis=1)]
# Temperatures are reported after multiplication by a factor of 10 so remove
# this factor.
data[:,0] /= 10

# Get the correlation coefficient.
corr = np.corrcoef(data, rowvar=0)[0, 1]
print('p-T correlation coefficient: {:.4f}'.format(corr))

# Plot the data on a scatter plot: T on x-axis, p on y-axis.
plt.scatter(*data.T, marker='.')
plt.xlabel('$T$ /$\mathrm{^\circ C}$')
plt.ylabel('$p$ /mbar')
plt.show()
```

The output (Figure 6.4) gives a correlation coefficient of 0.0260: as expected, there is little correlation between air temperature and pressure (since the air density also varies).

6.3.3   Histograms

The NumPy function, `np.histogram`, creates a histogram from the values in an array. That is, a set of *bins* is defined with lower and upper limits and each is filled with the number of elements from the array whose value falls within its limits. For example, suppose the following array holds the percentage marks of 10 students in a test:

```
In [x]: marks = np.array([45, 68, 56, 23, 60, 87, 75, 59, 63, 72])
```

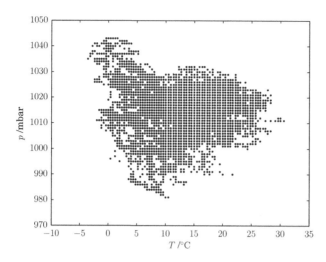

Figure 6.4 There is virtually no correlation between air temperature and air pressure in this data set.

There are several ways to define the histogram bins. If the `bins` argument is a sequence, it defines the boundaries of the sequential bins:

```
In [x]: bins = [20, 40, 60, 80, 100]
```

defines four bins with ranges [20–40%), [40–60%), [60–80%) and [80–100%]. All but the last bin is half open; that is, the first bin includes marks from and including 20% up to but not including 40%. Note that a sequence of $N + 1$ numbers is required to create N bins. The `np.histogram` method returns a tuple consisting of the values of the histogram and the bin edges we defined (both as NumPy arrays).

```
In [x]: hist, bins = np.histogram(marks, bins)
In [x]: hist
Out[x]: array([1, 3, 5, 1])

In [x]: bins
Out[x]: array([ 20, 40, 60, 80, 100])
```

This shows that there is one mark in the 20–40% bin, three in the 40–60% bin and so on.

If you just want a certain number of evenly spaced bins, an integer can be passed as `bins` instead of a sequence:

```
In [x]: np.histogram(marks, bins=5)
Out[x]: (array([1, 1, 3, 3, 2]),
         array([ 23. ,  35.8,  48.6,  61.4,  74.2,  87. ]))
```

By default, the requested number of bins range between the minimum and maximum values of the array (here, 23 and 87); to specify a different minimum and maximum, set the `range` argument tuple:

```
In [x]: np.histogram(marks, bins=5, range=(0, 100))
Out[x]: (array([0, 1, 3, 5, 1]),
         array([  0.,  20.,  40.,  60.,  80., 100.]))
```

The np.histogram method also has an optional boolean argument density: by default it is False, meaning that the histogram array returned contains the *number* of values from the original array in each bin. If density is set to True, the histogram array will contain the *probability density function*, normalized so that the integral over the entire range of the bins is equal to unity:

```
In [x]: hist, bins = np.histogram(marks, bins=5, range=(0,100),
                                     density=True)
In [x]: print(hist)
[ 0.     0.005  0.015  0.025   0.005]
In [x]: bin_width = 100/5
In [x]: print(np.sum(hist) * bin_width)
1.0
```

(By integral here we mean the area under the histogram, which is the sum of each histogram bar height times its corresponding bin width.)

To plot a histogram with pyplot, use pyplot.hist, passing it the same arguments you would to np.histogram:

```
In [x]: import matplotlib.pyplot as plt
In [x]: hist, bins, patches = plt.hist(marks, bins=5, range=(0, 100))
In [x]: hist, bins
Out[x]:
(array([ 0.,   1.,   3.,   5.,   1.]),
 array([   0.,   20.,   40.,   60.,   80.,   100.]))
In [x]: plt.show()
```

❶ In addition to the bin counts (hist) and boundaries (bins), pyplot returns a list of references to the "patches" which appear in the plotted figure (see Section 7.4.4 for more information about this advanced feature).

The resulting histogram is plotted in Figure 6.5. See also Sections 3.3.2 and 7.3.

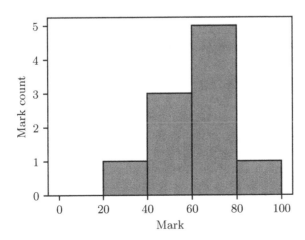

Figure 6.5 An example histogram.

6.3.4   Exercises

Problems

P6.3.1   A certain lottery involves players selecting six numbers without replacement from the range [1, 49]. The jackpot is shared among the players who match all six numbers ("balls") selected in the same way at random in a twice-weekly draw (in any order). If no player matches every drawn number, the jackpot "rolls over" and is added to the following draw's jackpot.

Although the lottery is *fair* in the sense that every combination of drawn numbers is equally likely, it has been observed that many players show a preference in their selection for certain numbers, such as those that represent dates (i.e. more of their numbers are chosen from [1, 31] than would be expected if they chose randomly). Hence, to avoid sharing the jackpot and so maximize one's expected winnings, it would be reasonable to avoid these numbers.

Test this hypothesis by establishing if there is any correlation between the number of balls with values less than 13 (representing a month) and the jackpot winnings per person. Ignore draws immediately following a rollover. The necessary data can be downloaded from https://scipython.com/ex/bfe.

P6.3.2   We have seen how to create a histogram plot from an array with `pyplot.hist`, but suppose you have already created arrays `hist` and `bins` using `np.histogram` and want to plot the resulting histogram from these arrays. You can't use `pyplot.hist` because this function expects to act on the original array of data. Use `pyplot.bar`[19] to plot a `hist` array as a bar chart.

P6.3.3   The heights, in cm, of a sample of 1000 adult men and 1000 adult women from a certain population are collected in the data files `ex6-3-f-male-heights.txt` and `ex6-3-f-female-heights.txt` available at https://scipython.com/ex/bfd. Read in the data and establish the mean and standard deviation for each sex. Create histograms for the two data sets using a suitable binning interval and plot them on the same figure.

Repeat the exercise in imperial units (feet and inches).

6.4   Polynomials

NumPy provides a powerful set of classes for representing polynomials, including methods for evaluation, algebra, root-finding and fitting of several kinds of polynomial basis functions. In this section, the simplest and most familiar basis, the power series, will be described first, before a discussion of a few other classical orthogonal polynomial basis functions.

---

[19] Documentation for this method is at https://matplotlib.org/api/_as_gen/matplotlib.pyplot.bar.html; see also Section 7.3.

6.4.1 Defining and Evaluating a Polynomial

A (finite) polynomial power series has as its basis the powers of x: $1(= x^0), x, x^2, x^3, \cdots, x^N$, with coefficients c_i:

$$P(x) = \sum_{i=0}^{N} c_0 + c_1 x + c_2 x^2 + c_3 x^3 + \ldots + c_N x^N.$$

This section describes the use of the `Polynomial` *convenience class*, which provides a natural interface to the underlying functionality of NumPy's polynomial package.

The polynomial convenience class is `numpy.polynomial.Polynomial`. To import it directly, use

```
In [x]: from numpy.polynomial import Polynomial
```

Alternatively, if the whole NumPy library is already imported as np, then rather than constantly refer to this class as `np.polynomial.Polynomial`, it is convenient to define a variable:

```
In [x]: import numpy as np
In [x]: Polynomial = np.polynomial.Polynomial
```

This is the way we will refer to the `Polynomial` class in this book.

To define a polynomial object, pass the `Polynomial` constructor a sequence of coefficients to increasing powers of x, starting with c_0. For example, to represent the polynomial

$$P(x) = 6 - 5x + x^2,$$

define the object

```
In [x]: p = Polynomial([6, -5, 1])
```

You can inspect the coefficients of a `Polynomial` object with `print` or by referring to its `coef` attribute.

```
In [x]: print(p)
poly([ 6. -5.  1.])
In [x]; p.coef
Out[x]: array([ 6., -5.,  1.])
```

Notice that the integer coefficients used to define the polynomial have been automatically cast to `float`. It is also possible to use `complex` coefficients.

To evaluate a polynomial for a given value of x, "call" it as follows:

```
In [x]: p(4)     # calculate p at a single value of x
2.0
In [x]: x = np.linspace(-5, 5, 11)
In [x]: print(p(x))      # calculate p on a sequence of x values
Out[x]: [ 56.  42.  30.  20.  12.   6.   2.   0.   0.   2.   6.]
```

6.4.2   Polynomial Algebra

The Polynomial convenience class implements the familiar Python operators: +, -, *, //, **, % and divmod on Polynomial objects. These are illustrated in the following examples using the polynomials

$$P(x) = 6 - 5x + x^2,$$

$$Q(x) = 2 - 3x.$$

```
In [x]: p = Polynomial([6, -5, 1])
In [x]: q = Polynomial([2, -3])
In [x]: print(p + q)
poly([ 8. -8.   1.])

In [x]: print(p - q)
poly([ 4. -2.   1.])

In [x]: print(p * q)
poly([ 12. -28.   17.   -3.])

In [x]: print(p // q)
poly([ 1.44444444 -0.33333333])

In [x]: print(p % q)
poly([ 3.11111111])     # i.e. 28/9
```

Division of a polynomial by another polynomial is analogous to integer division (and uses the same // operator): that is, the result is another polynomial (with no reciprocal powers of x), possibly leaving a remainder.

Hence $p = q(-\frac{1}{3}x + \frac{13}{9}) + \frac{28}{9}$ and the // operator returns the quotient polynomial, $-\frac{1}{3}x + \frac{13}{9}$. The remainder (which, in general, will be another polynomial) is returned, as might be expected, by the modulus operator, %. The divmod() built-in returns both quotient and remainder in a tuple:

```
In [x]: quotient, remainder = divmod(p, q)
In [x]: print(quotient)

poly([ 1.44444444 -0.33333333])     # i.e. p(x) // q(x) is 13/9 - x/3

In [x]: print(remainder)
poly([ 3.11111111])
```

Exponentiation is supported through the ** operator; polynomials can only be raised to a non-negative integer power:

```
In [x]: print(q ** 2)
poly([  4. -12.    9.])
```

It isn't always convenient to create a new polynomial object in order to use these operators on one another, so many of the operators described here also work with scalars:

```
In [x]: print(p * 2)     # multiplication by a scalar
poly([ 12. -10.    2.])
```

```
In [x]: print(p / 2)     # division by a scalar
poly([ 3.   -2.5   0.5])
```

and even tuples, lists and arrays of polynomial coefficients. For example, to multiply $P(x)$ by $x^2 - 2x^3$:

```
In [x]: print(p * [0, 0, 1, -2])
poly([  0.    0.    6. -17.   11.   -2.])
```

Finally, one polynomial can be substituted into another. To evaluate $P(Q(x))$, simply use p(q):

```
In [x]: print(p(q))
poly([ 0.   3.   9.])
```

That is, $P(Q(x)) = 3x + 9x^2$.

6.4.3 Root-Finding

The roots of a polynomial are returned by the roots method. Repeated roots are simply repeated in the returned array:

```
In [x]: p.roots()
array([ 2.,   3.])
In [x]: (q * q).roots()
array([ 0.66666667,  0.66666667])
In [x]: Polynomial([5, 4, 1]).roots()
array([-2.-1.j, -2.+1.j])
```

Polynomials can also be created from their roots with Polynomial.fromroots:

```
In [x]: print( Polynomial.fromroots([-4, 2, 1]) )
poly([  8. -10.   1.   1.])
```

That is, $(x + 4)(x - 2)(x - 1) = 8 - 10x + x^2 + x^3$. Note that the way the polynomial is constructed means that the coefficient of the highest power of x will be 1.

Example E6.11   The tanks used in the storage of cryogenic liquids and rocket fuel are often spherical (why?). Suppose a particular spherical tank has a radius R and is filled with a liquid to a height h. It is (fairly) easy to find a formula for the volume of liquid from the height:

$$V = \pi R h^2 - \frac{1}{3}\pi h^3.$$

Suppose that there is a *constant* flow of liquid from the tank at a rate $F = -dV/dt$. How does the height of liquid, h, vary with time? Differentiating the earlier mentioned equation with respect to t leads to

$$(2\pi R h - \pi h^2)\frac{dh}{dt} = -F.$$

If we start with a full tank ($h = 2R$) at time $t = 0$, this ordinary differential equation may be integrated to yield the equation

$$-\frac{1}{3}\pi h^3 + \pi R h^2 + \left(Ft - \frac{4}{3}\pi R^3\right) = 0,$$

a cubic polynomial in h. Because this equation cannot be inverted analytically for h, let's use NumPy's Polynomial class to find $h(t)$, given a tank of radius $R = 1.5$ m from which liquid is being drawn at 200 cm^3 s^{-1}.

The total volume of liquid in the full tank is $V_0 = \frac{4}{3}\pi R^3$. Clearly, the tank is empty when $h = 0$, which occurs at time $T = V_0/F$, since the flow rate is constant. At any particular time, t, we can find h by finding the roots of this equation.

Listing 6.8 Liquid height in a spherical tank

```
# eg6-c-spherical-tank-a.py
import numpy as np
import matplotlib.pyplot as plt
Polynomial  = np.polynomial.Polynomial

# Radius of the spherical tank in m.
R = 1.5
# Flow rate out of the tank, m^3.s-1.
F = 2.e-4
# Total volume of the tank.
V0 = 4/3 * np.pi * R**3
# Total time taken for the tank to empty.
T = V0 / F

# Coefficients of the quadratic and cubic terms
# of p(h), the polynomial to be solved for h.
c2, c3 = np.pi * R, -np.pi / 3

N = 100
# Array of N time points between 0 and T inclusive.
time = np.linspace(0, T, N)
# Create the corresponding array of heights h(t).
h = np.zeros(N)
for i, t in enumerate(time):
    c0 = F*t - V0
    p = Polynomial([c0, 0, c2, c3])
    # Find the three roots to this polynomial.
    roots = p.roots()
    # We want the one root for which 0 <= h <= 2R.
    h[i] = roots[(0 <= roots) & (roots <= 2*R)][0]

plt.plot(time, h, 'o')
plt.xlabel('Time /s')
plt.ylabel('Height in tank /m')
plt.show()
```

❶   We construct an array of time points between $t = 0$ and $t = T$.

❷   For each time point find the roots of the above cubic polynomial. Only one of the roots is physically meaningful, in that $0 \le h \le 2R$ (the height of the level of liquid

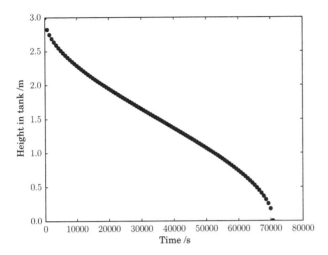

Figure 6.6 The height of liquid as a function of time, $h(t)$, for the spherical tank problem.

cannot be negative or greater than the diameter of the tank), so we extract that root (by boolean indexing) and store it in the array h.

Finally, we plot h as a function of time (Figure 6.6).

6.4.4   Calculus

Polynomials can be differentiated with the `Polynomial.deriv` method. By default, this function returns the first derivative, but the optional argument m can be set to return the mth derivative:

```
In [x]: print(p)
poly([ 6.  -5.   1.]) # 6 - 5x + x^2
In [x]: print(p.deriv())
poly([-5.   2.])
In [x]: print(p.deriv(2))
poly([ 2.])
```

A `Polynomial` object can also be integrated with an optional lower bound, L, and constant of integration, k, treated as shown in the following example:

$$\int_{L}^{x} 2 - 3x\,dx = \left[2x - \tfrac{3}{2}x^2\right]_{L}^{x} = 2x - \tfrac{3}{2}x^2 - 2L + \tfrac{3}{2}L^2,$$

$$\int 2 - 3x\,dx = 2x - \tfrac{3}{2}x^2 + k.$$

By default, L and k are zero, but can be specified by passing the arguments lbnd and k to the `Polynomial.integ` method:

```
In [x]: print(q)
poly([ 2.  -3.])
In [x]: print(q.integ())
poly([ 0.    2.   -1.5])
```

```
In [x]: print(q.integ(lbnd=1))
poly([-0.5  2.  -1.5])
In [x]: print(q.integ(k=2))
poly([ 2.   2.  -1.5])
```

Polynomials can be integrated repeatedly by passing a value to m, giving the number of integrations to perform.[20]

6.4.5 ◇ Classical Orthogonal Polynomials

In addition to the `Polynomial` class representing simple power series such as $a_0 + a_1x + a_2x^2 + \ldots + a_nx^n$, NumPy provides classes to represent a series composed of any of a number of classical orthogonal polynomials. These polynomials and linear combinations of them are widely used in physics, statistics and mathematics. As of NumPy version 1.17, the polynomial convenience classes provided are `Chebyshev`, `Legendre`, `Laguerre`, `Hermite` ("physicists' version") and `HertmiteE` ("probabilists' version"). Many good textbooks exist describing the properties of these polynomial classes; to illustrate their use we will focus here on the Legendre polynomials,[21] denoted $P_n(x)$. These are the solutions to Legendre's differential equation,

$$\frac{d}{dx}\left[(1 - x^2)\frac{d}{dx}P_n(x)\right] + n(n + 1)P_n(x) = 0.$$

The first few Legendre polynomials are

$$P_0(x) = 1,$$
$$P_1(x) = x,$$
$$P_2(x) = \tfrac{1}{2}(3x^2 - 1),$$
$$P_3(x) = \tfrac{1}{2}(5x^3 - 3x),$$
$$P_4(x) = \tfrac{1}{8}(35x^4 - 30x^2 + 3),$$

and are plotted in Figure 6.7.

A useful property of the Legendre polynomials is their orthogonality on the interval $[-1, 1]$:

$$\int_{-1}^{1} P_n(x)P_m(x)\,dx = \frac{2}{2n + 1}\delta_{mn},$$

which is important in their use as a basis for representing suitable functions.[22]

To create a linear combination of Legendre polynomials, pass the coefficients to the `Legendre` constructor, just as for `Polynomial`. For example, to construct the polynomial expansion $5P_1(x) + 2P_2(x)$:

---

[20] Different constants of integration for each can be specified by setting k to an array of values.

[21] The Legendre polynomials are named after the French mathematician Adrien-Marie Legendre (1752–1833); for 200 years until 2005 many publications mistakenly used a portrait of the unrelated French politician Louis Legendre as that of the mathematician.

[22] In particular, in physics, the multipole expansion of electrostatic potentials.

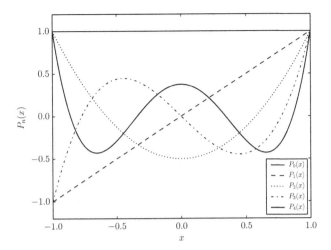

$P_n(x)$

x

Figure 6.7 The first five Legendre polynomials, $P_n(x)$ for $n = 0, 1, 2, 3, 4$.

```
In [x]: Legendre = np.polynomial.Legendre
In [x]: A = Legendre([0, 5, 2])
```

An existing polynomial object can be converted into a Legendre series with the cast method:

```
In [x]: P = Polynomial([0, 1, 1])
In [x]: Q = Legendre.cast(P)
In [x]: print(Q)
leg([ 0.33333333  1.          0.66666667])
```

That is, $x + x^2 = \frac{1}{3}P_0 + P_1 + \frac{2}{3}P_2$.

An instance of a single Legendre polynomial basis function can be created with the basis method:

```
In [x]: L3 = Legendre.basis(3)
```

This creates an object representing $P_3(x)$, and is equivalent to calling Legendre([0, 0, 0, 1]). To obtain a regular power series, we can cast it back to a Polynomial:

```
In [x]: print(Polynomial.cast(L3))
poly([ 0.  -1.5  0.   2.5])
```

In addition to the functions just described for Polynomial, including differentiation and integration of polynomial series, the convenience classes for the classical orthogonal polynomials expose several useful methods.

convert converts between different kinds of polynomials. For example, the linear combination $A(x) = 5P_1(x) + 2P_2(x) = 5x + 2\frac{1}{2}(3x^2 - 1) = -1 + 5x + 3x^2$, as a power series of monomials (a Maclaurin series), is represented by an instance of the Polynomial class as:

```
In [x]: A = Legendre([0, 5, 2])
In [x]: B = A.convert(kind=Polynomial)
In [x]: print(B)
In [x]: poly([-1.   5.   3.])
```

Because the objects A and B represent the same underlying function (just expanded in different basis sets) they evaluate to the same value when given the same x, and have the same roots:

```
In [x]: A(-2) == B(-2)
Out[x]: True
In [x]: print(A.roots(), B.roots(), sep='\n')
[-1.84712709  0.18046042]
[-1.84712709  0.18046042]
```

6.4.6   Fitting Polynomials

A common use of polynomial expansions is in fitting and approximating data series. NumPy's polynomial modules provide methods for the least-squares fitting of functions. The `fit` function of the polynomial convenience classes is described in this section.[23]

The `domain` and `window` Attributes

A typical one-dimensional fitting problem requires the best-fit polynomial to a finite, continuous function over some finite region of the x-axis (the *domain*). However, polynomials themselves can differ from each other wildly and diverge as $x \to \pm\infty$. This makes any attempt to blindly find the least-squares fit on the domain of the function itself potentially risky: the fitted polynomial is frequently subject to numerical instability, overflow, underflow and other types of ill-conditioning (see Section 10.2). As an example, consider the function

$$f(x) = e^{-\sin 40x}$$

in the interval $(100, 100.1)$. There is nothing particularly tricky about this function: it is well-behaved everywhere and $f(x)$ takes very moderate values between e^{-1} and e^1. Yet a straightforward least-squares fit to a fourth-order polynomial on this domain gives

$$-11.881851 + 2379.22228x - 119.741202x^2 - 23828009.7x^3 + 1192894610x^4$$

and clearly the potential for numerical instability and loss of accuracy with even moderate values of x: our approximation to $f(x)$ is built up from difference between very large monomial terms.

Each class of polynomial has a default *window* over which it is optimal to take a linear combination in fitting a function. For example, the Legendre polynomials window is the region $[-1, 1]$ plotted above, on which $P_n(x)$ are orthogonal and everywhere $|P_n(x)| < 1$. The problem is that it is rather unlikely that the function to be fitted falls within the chosen polynomials' window. It is therefore necessary to relate the domain of the function to the window. This is done by shifting and scaling the x-axis: that is, by

---

[23] *Note*: The older `np.poly1d` class representing one-dimensional polynomials is still available (as of NumPy 1.17) for backward-compatibility reasons. It is documented at https://docs.scipy.org/doc/numpy/reference/routines.polynomials.poly1d.html and provides a simpler but less-reliable least-squares fitting method, `np.polyfit`. It is recommended, however, to use the new `Polynomial` class in new code.

mapping points in the function's domain to points in the fitting polynomials' window. The polynomial `fit` function does this automatically, so the fourth-order least-squares fit to the earlier mentioned function yields

```
In [x]: x = np.linspace(100, 100.1, 1001)
In [x]: f = lambda x: np.exp(-np.sin(40*x))
In [x]: p = Polynomial.fit(x, f(x), 4)
In [x]: print(p)
poly([ 1.49422551 -2.54641449  0.63284641  1.84246463 -1.02821956])
```

The domain and window of a polynomial can be inspected as the attributes `domain` and `window` respectively:

```
In [x]: p.domain
array([ 100. ,  100.1])
In [x]: p.window
array([-1.,  1.])
```

It is important to note that the argument x is mapped from the domain to the window whenever a polynomial is evaluated. This means that two polynomials with different domains and/or windows may evaluate to different values *even if they have the same coefficients*. For example, if we create a `Polynomial` object from scratch with the same coefficients as the fitted polynomial p above:

```
In [x]: q = Polynomial([1.49422551, -2.54641449,  0.63284641,
                        1.84246463, -1.02821956])
```

it is created with the default domain and window, which are *both* (-1, 1):

```
In [x]: print(q.domain, q.window)
[-1.  1.] [-1.  1.]
```

and so evaluating q at 100.05, say, maps 100.05 in the domain to 100.05 in the window and gives a very different answer from the evaluation of p at the same point in the domain (which maps to 0. in the window):

```
In [x]: q(100.05), p(100.05)
(-101176442.96772559, 1.4942255113760108)
```

It is easy to show that the mapping function from x in a domain (a, b) to x' in a window (a', b') is

$$x' = m(x) = \chi + \mu x, \quad \text{where } \mu = \frac{b' - a'}{b - a}, \ \chi = b' - b\frac{b' - a'}{b - a}.$$

These are the parameters returned by the polynomial's `mapparms` function:

```
In [x]: chi, mu = p.mapparms()
In [x]: print(chi, mu)
-2001.0, 20.0
```

Therefore,

```
In [x]: print(q(chi + mu*100.05))
1.49422551
```

It is possible to change `domain` and `window` by direct assignment:

```
In [x]: q.domain = np.array((100., 100.1))
In [x]: print(q(100.05))
1.49422551
```

To evaluate a polynomial on a set number of evenly distributed points in its domain, for example, to plot it, use the `Polynomial`'s `linspace` method:

```
In [x]: p.linspace(5)
Out [x]:
(array([ 100.   ,  100.025,  100.05 ,  100.075,  100.1  ]),
 array([ 1.80280222,  2.63107256,  1.49422551,  0.54527422,  0.39490249]))
```

`p.linspace` returns two arrays, with the specified number of samples on the polynomial's domain representing the x points and the values the polynomial takes at those points, $p(x)$.

Polynomial.fit

The `Polynomial` method `fit` returns a least-squares fitted polynomial to data, y, sampled at values x. In its simplest use, `fit` needs only to be passed array-like objects, x and y, and a value for `deg`, the degree of polynomial to fit. It returns the polynomial which minimizes the sum of the squared errors,

$$E = \sum_i |y_i - p(x_i)|^2.$$

For example,

```
In [x]: x = np.linspace(400, 700, 1000)
In [x]: y = 1 / x**4
In [x]: p = Polynomial.fit(x, y, 3)
```

produces the best-fit cubic polynomial to the function x^{-4} on the interval $[400, 700]$.

Weighted least-squares fitting is achieved by setting the argument, w, to a sequence of weighting values that is the same length as x and y. The polynomial returned is that which minimizes the sum of the *weighted* squared errors,

$$E = \sum_i w_i^2 |y_i - p(x_i)|^2.$$

The domain and window of the fitted polynomial may be specified with the arguments domain and window; by default, a minimal domain covering the points x is used.

It is wise to check the *quality* of the fit before using the returned polynomial. Setting the argument `full=True` causes `fit` to return two objects: the fitted polynomial and a list of various statistics about the fit itself:

```
In [x]: deg = 3
In [x]: p, [resid, rank, sing_val, rcond] = Polynomial.fit(x, y, deg, full=True)
In [x]: p
Out[x]:
Polynomial([  1.07041864e-11,  -1.16488662e-11,   1.02545751e-11,
             -5.64068914e-12], [ 400.,  700.], [-1.,  1.])

In [x]: resid
Out[x]: array([  4.57180972e-23])
```

```
In [x]: rank
Out[x]: 4

In [x]: sing_val
Out[x]: array([ 1.3843828 ,  1.32111941,  0.50462215,  0.28893641])

In [x]: rcond
Out[x]: 2.2204460492503131e-13
```

This list can be analyzed to see how well the polynomial function fits the data. `resid` is the sum of the squared residuals,

$$\texttt{resid} = \sum_i |y_i - p(x_i)|^2$$

– a smaller value indicates a better fit. `rank` and `sing_val` are the rank and singular values of the matrix inverted in the least-squares algorithm to find the polynomial coefficients: ill-conditioning of this matrix can lead to poor fits (particularly if the fitted polynomial degree is too high). `rcond` is the cutoff ratio for small singular values within this matrix: values smaller than this value are set to zero in the fit (to protect the fit from spurious artifacts introduced by round-off error) and a `RankWarning` exception is raised. If this happens, the data may be too noisy or not well described by the polynomial of the specified degree. Note that least-squares fitting should always be carried out at double precision and be aware of "over-fitting" the data (attempting to fit a function with too many coefficients, i.e. a polynomial of too high order).

Example E6.12 A straight-line best fit is just a special case of a polynomial least-squares fit (with deg=1). Consider the following data giving the absorbance, A, over a path length of 5 mm of ultraviolet light at 280 nm, by a protein as a function of the concentration, [P]:

[P]/μg mL^{-1}	A
0	2.287
20	3.528
40	4.336
80	6.909
120	8.274
180	12.855
260	16.085
400	24.797
800	49.058
1500	89.400

We expect the absorbance to be linearly related to the protein concentration: $A = m[\mathrm{P}] + A_0$, where A_0 is the absorbance in the absence of protein (e.g. due to the solvent and experimental components).

Listing 6.9 Straight-line fit to absorbance data

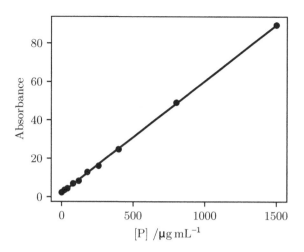

Figure 6.8 Line of least-squares best fit to absorbance data as a function of concentration.

```
# eg6-polyfit.py
import numpy as np
import matplotlib.pyplot as plt
Polynomial = np.polynomial.Polynomial

# The data: conc = [P] and absorbance, A.
conc = np.array([0, 20, 40, 80, 120, 180, 260, 400, 800, 1500])
A = np.array([2.287, 3.528, 4.336, 6.909, 8.274, 12.855, 16.085, 24.797,
              49.058, 89.400])

cmin, cmax = min(conc), max(conc)
pfit, stats = Polynomial.fit(conc, A, 1, full=True, window=(cmin, cmax),
                                             domain=(cmin, cmax))

print('Raw fit results:', pfit, stats, sep='\n')

A0, m = pfit
resid, rank, sing_val, rcond = stats
rms = np.sqrt(resid[0]/len(A))

print('Fit: A = {:.3f}[P] + {:.3f}'.format(m, A0),
      '(rms residual = {:.4f})'.format(rms))

plt.plot(conc, A, 'o', color='k')
plt.plot(conc, pfit(conc), color='k')
plt.xlabel('[P] /$\mathrm{\mu g\cdot mL^{-1}}$')
plt.ylabel('Absorbance')
plt.show()
```

The output shows a good straight-line fit to the data (Figure 6.8):

```
Raw fit results:
poly([ 1.92896129  0.0583057 ])
[array([ 2.47932733]), 2, array([ 1.26633786,  0.62959385]), 2.2204460492503131e-15]
Fit: A = 0.058[P] + 1.929 (rms residual = 0.4979)
```

6.4.7   Exercises

Questions

Q6.4.1   The third derivative of the polynomial function $P(x) = 3x^3 + 2x - 7$ is 18, so why does the following evaluate as `False`?

```
In [x]: Polynomial((-7, 2, 0, 3)).deriv(3) == 18
Out[x]: False
```

Q6.4.2   Find and classify the stationary points of the polynomial

$$f(x) = (x^2 + x - 11)^2 + (x^2 + x - 7)^2.$$

Problems

P6.4.1   The expansion of the spherical ball of fire generated in an explosion may be analyzed to deduce the initial energy, E, released by a nuclear weapon. The British physicist Geoffrey Taylor used dimensional analysis to demonstrate that the radius of this sphere, $R(t)$, should be related to E, the air density, ρ_{air}, and time, t, through

$$R(t) = CE^{\frac{1}{5}}\rho_{air}^{-\frac{1}{5}}t^{\frac{2}{5}},$$

where, using model-shock wave problems, Taylor estimated the dimensionless constant $C \approx 1$. Using the data obtained from declassified timed images of the first New Mexico atomic explosion, Taylor confirmed this law and produced an estimate of the (then unknown) value of E. Use a log–log plot to fit the data in Table 6.6[24] to the model and confirm the time-dependence of R. Taking $\rho_{air} = 1.25 \text{ kg m}^{-3}$, deduce E and express its value in joules and in "kilotons of TNT," where the explosive energy released by 1 ton of TNT (trinitrotoluene) is arbitrarily defined to be 4.184×10^9 J.

P6.4.2   Find the mean and variance of both x and y, the correlation coefficient and the equation of the linear regression line for each of the four data sets given in Table 6.7. Comment on these values in the light of a plot of the data.

Table 6.6 Radius of the ball of fire produced by the "Trinity" nuclear test as a function of time

$t/$ms	$R/$m	$t/$ms	$R/$m	$t/$ms	$R/$m
0.1	11.1	1.36	42.8	4.34	65.6
0.24	19.9	1.50	44.4	4.61	67.3
0.38	25.4	1.65	46.0	15.0	106.5
0.52	28.8	1.79	46.9	25.0	130.0
0.66	31.9	1.93	48.7	34.0	145.0
0.80	34.2	3.26	59.0	53.0	175.0
0.94	36.3	3.53	61.1	62.0	185.0
1.08	38.9	3.80	62.9		
1.22	41.0	4.07	64.3		

Note: these data can be downloaded from
https://scipython.com/ex/bfg.

[24] G. I. Taylor, (1950) *Proc. Roy. Soc. London* **A201**, 159.

Table 6.7 Four sample data sets for analysis of mean, variance and correlation

x_1	y_1	x_2	y_2	x_3	y_3	x_4	y_4
10.0	8.04	10.0	9.14	10.0	7.46	8.0	6.58
8.0	6.95	8.0	8.14	8.0	6.77	8.0	5.76
13.0	7.58	13.0	8.74	13.0	12.74	8.0	7.71
9.0	8.81	9.0	8.77	9.0	7.11	8.0	8.84
11.0	8.33	11.0	9.26	11.0	7.81	8.0	8.47
14.0	9.96	14.0	8.10	14.0	8.84	8.0	7.04
6.0	7.24	6.0	6.13	6.0	6.08	8.0	5.25
4.0	4.26	4.0	3.10	4.0	5.39	19.0	12.50
12.0	10.84	12.0	9.13	12.0	8.15	8.0	5.56
7.0	4.82	7.0	7.26	7.0	6.42	8.0	7.91
5.0	5.68	5.0	4.74	5.0	5.73	8.0	6.89

Note: These data can be downloaded from https://scipython.com/ex/bff.

P6.4.3 The van der Waals equation of state may be written as follows to give the pressure, p, of a gas from its molar volume, V, and temperature, T:

$$p = \frac{RT}{V-b} - \frac{a}{V^2},$$

where a and b are molecule-specific constants and $R = 8.314$ J K^{-1} mol^{-1} is the gas constant. It can readily be rearranged to yield the temperature for a given pressure and volume, but its form giving the molar volume in terms of pressure and temperature is a cubic equation:

$$pV^3 - (pb + RT)V^2 + aV - ab = 0.$$

Of the three roots to this equation, below the *critical point*, (T_c, p_c) all are real: the largest and smallest give the molar volume of the gas phase and liquid phase, respectively; above the critical point, where no liquid phase exists, only one root is real and gives the molar volume of the gas (also known in this region as a *supercritical fluid*). The critical point is given by the condition $(\partial p/\partial V)_T = (\partial^2 p/\partial V^2)_T = 0$ and for a van der Waals gas is given by the formulas

$$T_c = \frac{8a}{27Rb}, \qquad p_c = \frac{a}{27b^2}.$$

For ammonia, the van der Waals constants are $a = 4.225$ L^2 bar mol^{-2} and $b = 0.03707$ L mol^{-1}.

(a)   Find the critical point of ammonia, and then determine the molar volume at room temperature and pressure, (298 K, 1 atm) and at (500 K, 12 MPa).

(b)   An *isotherm* is the set of points (p, V) at a constant temperature satisfying an equation of state. Plot the isotherm (p against V) for ammonia at 350 K using the van der Waals equation of state and compare it with the 350 K isotherm for an ideal gas, which has the equation of state, $p = RT/V$.

P6.4.4 The first stages of the Saturn V rocket that launched the Apollo 11 mission generated an acceleration which increased with time throughout their operation (mostly because of the decrease in mass as it burns its fuel). This acceleration may be modeled (in units of $\mathrm{m\,s^{-2}}$) as a function of time after launch, t in seconds, by the quadratic function:

$$a(t) = 2.198 + (2.842 \times 10^{-2})t + (1.061 \times 10^{-3})t^2.$$

Determine the distance traveled by the rocket at the end of the stage-one center-engine burn, 2 minutes and 15.2 seconds, after launch.

(Harder) Assuming a constant "lapse rate" of $\Gamma = -dT/dz = 6\ \mathrm{K\,km^{-1}}$ from a ground temperature of 302 K, at what time and altitude, z, did the rocket achieve Mach 1? During the relevant phase of the launch, take the average pitch angle to be 12°, and assume that the speed of sound can be calculated as a function of absolute temperature to be

$$c = \sqrt{\frac{\gamma RT}{M}},$$

where the constants $\gamma = 1.4$ and $R = 8.314\ \mathrm{J\,K^{-1}\,mol^{-1}}$, and the mean molar mass of the atmosphere is $M = 0.0288\ \mathrm{kg\,mol^{-1}}$.

6.5   Linear Algebra

6.5.1   Basic Matrix Operations

Matrix operations can be carried out on a regular two-dimensional NumPy array, including scalar multiplication, matrix (dot) product, elementwise multiplication and transpose:

```
In [x]: A = np.array([[0, 0.5], [-1, 2]])
In [x]: A
Out[x]:
array([[ 0. ,  0.5],
       [-1. ,  2. ]])
In [x]: A * 5                  # multiplication by a scalar
Out[x]:
array([[ 0. ,   2.5],
       [ -5. ,  10. ]])
In [x]: B = np.array([[2, -0.5], [3, 1.5]])

In [x]: B
Out[x]:
array([[ 2. , -0.5],
       [ 3. ,  1.5]])

In [x]: A.dot(B)              # or np.dot(A, B): matrix product
Out[x]:
array([[ 1.5 ,  0.75],
       [ 4. ,   3.5 ]])
```

```
In [x]: A * B                # elementwise multiplication
Out[x]:
array([[ 0.  , -0.25],
       [-3.  ,  3.  ]])
In [x]: A.transpose()        # or simply A.T
Out[x]:
array([[ 0. , -1. ],
       [ 0.5,  2. ]])
```

Note that the transpose returns a *view* on the original matrix.

The identity matrix is returned by passing the two dimensions of the matrix to the method np.eye:

```
In [x]: np.eye(3, 3)
Out[x]:
array([[ 1.,  0.,  0.],
       [ 0.,  1.,  0.],
       [ 0.,  0.,  1.]])
```

Matrix Products

NumPy contains further methods for vector and matrix products. For example,

```
In [x]: a = np.array([1, 2, 3])
In [x]: b = np.array([0, 1, 2])
In [x]: np.inner(a, b)           # inner product; here, the same as a.dot(b)
Out[x]: 8

In [x]: np.outer(a, b)           # outer product
Out[x]:
array([[0, 1, 2],
       [0, 2, 4],
       [0, 3, 6]])
```

To raise a matrix to an (integer) power, however, requires a method from the np.linalg module:

```
In [x]: A = np.array([[0, 0.5], [-1, 2]])
In [x]: np.linalg.matrix_power(A, 3)        # the same as A @ A @ A
Out[x]:
array([[-1.  ,  1.75],
       [-3.5 ,  6.  ]])
```

Note that the ** operator performs *elementwise* exponentiation:

```
In [x]: A ** 3                   # the same as A * A * A
Out[x]:
array([[ 0.  ,  0.125],
       [-1.  ,  8.  ]])
```

Example E6.13 One way to create the two-dimensional rotation matrix,

$$\mathbf{R} = \begin{pmatrix} \cos\theta & -\sin\theta \\ \sin\theta & \cos\theta \end{pmatrix},$$

which rotates points in the xy-plane counterclockwise through $\theta = 30°$ about the origin:

```
In [x]: theta = np.radians(30)
In [x]: c, s = np.cos(theta), np.sin(theta)
In [x]: R = np.array([[c, -s], [s, c]])
In [x]: print(R)
[[ 0.8660254 -0.5       ]
 [ 0.5        0.8660254]]
```

The components of the unit vector along the x-axis after this rotation, for example, are given by the dot product:

```
In [x]: v = np.array([1, 0])
In [x]: R @ v
Out[x]: array([0.8660254, 0.5      ])
```

Other Matrix Properties

The norm of a matrix or vector is returned by the function `np.linalg.norm`. It is possible to calculate several different norms (see the documentation), but the ones used by default are the Frobenius norm for two-dimensional arrays:

$$\|A\| = \left(\sum_{i,j} |a_{ij}|^2 \right)^{1/2}$$

and the Euclidean norm for one-dimensional arrays:

$$\|a\| = \left(\sum_i |z_i|^2 \right)^{1/2} = \sqrt{|z_0|^2 + |z_1|^2 + \cdots + |z_{n-1}|^2}.$$

Thus,

```
In [x]: np.linalg.norm(A)
Out[x]: 2.2912878474779199
```

```
In [x]: c = np.array([1, 2j, 1 - 1j])
In [x]: np.linalg.norm(c)
Out[x]: 2.6457513110645907     # sqrt(1 + 4 + 2)
```

The function `np.linalg.det` returns the determinant of a matrix, and the regular NumPy function `np.trace` returns its trace (the sum of its diagonal elements):

```
In [x]: np.linalg.det(A)
Out[x]: 0.5
```

```
In [x]: np.trace(A)
Out[x]: 2.0
```

The *rank* of a matrix is obtained using np.linalg.matrix_rank:

```
In [x]: np.linalg.matrix_rank(A)        # matrix A has full rank
Out[x]: 2
In [x]: D = np.array([[1,1],[2,2]])      # a rank-deficient matrix

In [x]: np.linalg.matrix_rank(D)
Out[x]: 1
```

To find the inverse of a square matrix, use np.linalg.inv. A LinAlgError exception is raised if the matrix inversion fails:

```
In [x]: np.linalg.inv(A)
Out[x]:
array([[ 4., -1.],
       [ 2.,  0.]])

In [x]: np.linalg.inv(D)
...
LinAlgError: Singular matrix
```

Example E6.14   The currents flowing in the closed regions labeled I_1, I_2 and I_3 of the circuit given here may be analyzed by *mesh analysis*.

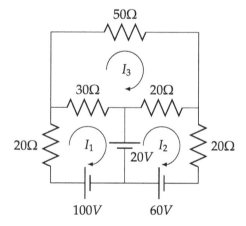

For each closed loop, we can apply Kirchoff's voltage law ($\sum_k V_k = 0$) in conjunction with Ohm's law ($V = IR$), to give three simultaneous equations:

$$50I_1 - 30I_3 = 80,$$

$$40I_2 - 20I_3 = 80,$$

$$-30I_1 - 20I_2 + 100I_3 = 0.$$

These can be expressed in matrix form as **RI = V**:

$$\begin{pmatrix} 50 & 0 & -30 \\ 0 & 40 & -20 \\ -30 & -20 & 100 \end{pmatrix} \begin{pmatrix} I_1 \\ I_2 \\ I_3 \end{pmatrix} = \begin{pmatrix} 80 \\ 80 \\ 0 \end{pmatrix},$$

We could use the numerically stable `np.linalg.solve` method (Section 6.5.3) to find the loop currents, $\mathbf{I}$, here, but in this well-behaved system,[25] let's find them through left multiplication by the matrix inverse, $\mathbf{R}^{-1}$:

$$\mathbf{R}^{-1}\mathbf{R}\mathbf{I} = \mathbf{I} = \mathbf{R}^{-1}\mathbf{V}.$$

Using NumPy's array methods:

```
In [x]: R = np.array([[50, 0, -30], [0, 40, -20], [-30, -20, 100]])
In [x]: V = np.array([80, 80, 0])
In [x]: I = np.linalg.inv(R) @ V
In [x]: print(I)
[[ 2.33333333]
 [ 2.61111111]
 [ 1.22222222]]
```

Thus, $I_1 = 2.33$ A, $I_2 = 2.61$ A, $I_3 = 1.22$ A.

Example E6.15   The matrix $\mathbf{B}$, defined here, may be manipulated as follows:

$$\mathbf{B} = \begin{pmatrix} 1 & 3-j \\ 3j & -1+j \end{pmatrix}, \quad \mathbf{B}^{\mathrm{T}} = \begin{pmatrix} 1 & 3j \\ 3-j & -1+j \end{pmatrix}$$

$$\mathbf{B}^{\dagger} = \begin{pmatrix} 1 & -3j \\ 3+j & -1-j \end{pmatrix}, \quad \mathbf{B}^{-1} = \begin{pmatrix} -\frac{1}{20} - \frac{3}{20}j & \frac{1}{20} - \frac{7}{20}j \\ \frac{3}{10} + \frac{3}{20}j & -\frac{1}{20} + \frac{1}{10}j \end{pmatrix}.$$

```
In [x]: B = np.array([[1, 3 - 1j], [3j, -1 + 1j]])
In [x]: print(B)
[[ 1.+0.j  3.-1.j]
 [ 0.+3.j -1.+1.j]]

In [x]: print(B.T)                    # matrix transpose
[[ 1.+0.j  0.+3.j]
 [ 3.-1.j -1.+1.j]]

In [x]: print(B.conj().T)             # Hermitian conjugate
[[ 1.-0.j  0.-3.j]
 [ 3.+1.j -1.-1.j]]

In [x]: print(np.linalg.inv(B))       # matrix inverse
[[-0.05-0.15j  0.05-0.35j]
 [ 0.30+0.15j -0.05+0.1j ]]
```

A few other common matrix operations are also available from within the NumPy package, including the trace, determinant, eigenvalues and (right) eigenvectors:

```
In [x]: print(np.trace(B))
1j
In [x]: print(np.linalg.det(B))
```

---

[25] In general, matrix inversion may be an ill-conditioned problem, but this particular matrix is easy to invert accurately. See Section 10.2.2 for more on conditioning.

```
(-4-8j)
In [x]: eigenvalues, eigenvectors = np.linalg.eig(B)
In [x]: print(eigenvalues, eigenvectors, sep='\n\n')
[ 2.50851535+2.09456868j  -2.50851535-1.09456868j]

[[ 0.77468569+0.j          -0.52924821+0.38116633j]
 [ 0.18832434+0.60365224j   0.75802940+0.j          ]]
```

6.5.2   Eigenvalues and Eigenvectors

To calculate the eigenvalues and (right) eigenvectors of a general square array with shape (n, n), use np.linalg.eig, which returns the eigenvalues, w, as an array of shape (n,) and the normalized eigenvectors, v, as a complex array of shape (n, n). The eigenvalues are not returned in any particular order, but the eigenvalue w[i] corresponds to the eigenvector v[:, i]. Note that the eigenvectors are arranged in columns. If the eigenvalue calculation does not converge for some reason, a LinAlgError is raised.

```
In [x]: vals, vecs = np.linalg.eig(A)
In [x]: vals
Out[x]: array([ 0.29289322,  1.70710678])
```

❶
```
In [x]: np.isclose(np.sum(vals), A.trace())
Out[x]: True
```

```
In [x]: vecs
Out[x]:
array([[-0.86285621, -0.28108464],
       [-0.50544947, -0.95968298]])
```

❶ Verify that the sum of the eigenvalues is equal to the matrix trace.

If the matrix is Hermitian or real-symmetric, the function np.linalg.eigh may be used instead. This method takes an additional argument, UPLO, which can be 'L' or 'U' according to whether the lower or upper triangular part of the matrix is used. The default is 'L'.

Two additional methods, np.linalg.eigvals and np.linalg.eigvalsh, return *only the eigenvalues* (and not the eigenvectors) of a general and Hermitian matrix, respectively.

Since NumPy version 1.8, these and most other linalg methods follow the usual broadcasting rules so that several matrices can be operated on at once: each matrix is assumed to be stored in the last two dimensions. For example, we may work with an array with shape (3, 2, 2), representing the three 2×2 Pauli matrices:

$$\sigma_x = \begin{pmatrix} 0 & 1 \\ 1 & 0 \end{pmatrix} \quad \sigma_y = \begin{pmatrix} 0 & -i \\ i & 0 \end{pmatrix} \quad \sigma_z = \begin{pmatrix} 1 & 0 \\ 0 & -1 \end{pmatrix}.$$

```
In [x]: pauli_matrices = np.array((
            ((0, 1), (1, 0)),        # sigma_x
            ((0, -1j), (1j, 0)),     # sigma_y
            ((1, 0), (0, -1))        # sigma_z
```

```
                            ))
In [x]: np.linalg.eigh(pauli_matrices)
Out[x]:
(array([[-1.,   1.],
        [-1.,   1.],
        [-1.,   1.]]),
 array([[[-0.70710678+0.j       ,  0.70710678+0.j       ],
         [ 0.70710678+0.j       ,  0.70710678+0.j       ]],

        [[-0.70710678-0.j       , -0.70710678+0.j       ],
         [ 0.00000000+0.70710678j,  0.00000000-0.70710678j]],

        [[ 0.00000000+0.j       ,  1.00000000+0.j       ],
         [ 1.00000000+0.j       ,  0.00000000+0.j       ]]]))
```

Example E6.16   A linear transformation in two dimensions can be visualized through its effect on the unit square defined by the two orthonormal basis vectors, $\hat{\imath}$ and $\hat{\jmath}$. In general, it can be represented by a 2×2 matrix, **T**, which acts on a vector v to map it from a vector space spanned by one basis onto a different vector space spanned by another basis: $v' = \mathbf{T}v$. *Eigenvectors* under such a tranformation may be scaled but do not change orientation, as illustrated by the following code for the transformation matrix:

$$\mathbf{T} = \begin{pmatrix} \frac{3}{2} & \frac{1}{2} \\ \frac{1}{2} & \frac{3}{2} \end{pmatrix}.$$

The effect of the transformation on a set of points in the Cartesian plane is also visualized in the output plot (Figure 6.9).

Listing 6.10 Linear transformations in two dimensions

```
import numpy as np
import matplotlib.pyplot as plt

# Set up a Cartesian grid of points.
XMIN, XMAX, YMIN, YMAX = -3, 3, -3, 3
N = 16
xgrid = np.linspace(XMIN, XMAX, N)
ygrid = np.linspace(YMIN, YMAX, N)
grid = np.array(np.meshgrid(xgrid, ygrid)).reshape(2, N**2)

# Our untransformed unit basis vectors, i and j:
basis = np.array([[1,0], [0,1]])

def plot_quadrilateral(basis, color='k'):
    """Plot the quadrilateral defined by the two basis vectors."""
    ix, iy = basis[0]
    jx, jy = basis[1]
    plt.plot([0, ix, ix+jx, jx, 0], [0, iy, iy+jy, jy, 0], color)

def plot_vector(v, color='k', lw=1):
    """Plot vector v as a line with a specified color and linewidth."""
    plt.plot([0, v[0]], [0, v[1]], c=color, lw=lw)
```

```
def plot_points(grid, color='k'):
    """Plot the grid points in a specified color."""
    plt.scatter(*grid, c=color, s=2, alpha=0.5)

def apply_transformation(basis, T):
    """Return the transformed basis after applying transformation T."""
    return (T @ basis.T).T

# The untransformed grid and unit square.
plot_points(grid)
plot_quadrilateral(basis)

# Apply the transformation matrix, S, to the scene.
S = np.array(((1.5, 0.5),(0.5, 1.5)))
tbasis = apply_transformation(basis, S)
plot_quadrilateral(tbasis, 'r')
tgrid = S @ grid
plot_points(tgrid, 'r')

# Find the eigenvalues and eigenvectors of S...
vals, vecs = np.linalg.eig(S)
print(vals, vecs)
if all(np.isreal(vals)):
    # ... if they're all real, indicate them on the diagram.
    v1, v2 = vals
    e1, e2 = vecs.T
    plot_vector(v1*e1, 'r', 3)
    plot_vector(v2*e2, 'r', 3)
    plot_vector(e1, 'k')
    plot_vector(e2, 'k')

# Ensure the plot has 1:1 aspect (i.e. squares look square) and set the limits.
plt.axis('square')
plt.xlim(XMIN, XMAX)
plt.ylim(YMIN, YMAX)

plt.show()
```

❶ We need to reshape the meshgrid of $N \times N$ points into an array of $2 \times N^2$ coordinates...

❷ ... which can be transformed in a single line of code by the vectorized operation, S @ grid.

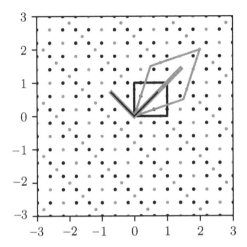

Figure 6.9 The effect of a linear transformation given by a matrix, **T**, on the usual Cartesian basis: the unit square (black) is stretched along one diagonal and squeezed along the other; the scaled eigenvectors are also indicated.

6.5.3   Solving Equations

Linear Scalar Equations

NumPy provides an efficient and numerically stable method for solving systems of linear scalar equations. The set of equations

$$m_{11}x_1 + m_{12}x_2 + \ldots + m_{1n}x_1 = b_1$$
$$m_{21}x_1 + m_{22}x_2 + \ldots + m_{2n}x_2 = b_2$$
$$\ldots$$
$$m_{n1}x_1 + m_{n2}x_2 + \ldots + m_{nn}x_n = b_n$$

can be expressed as the matrix equation $\mathbf{Mx} = \mathbf{b}$:

$$\begin{pmatrix} m_{11} & m_{12} & \cdots & m_{1n} \\ m_{21} & m_{22} & \cdots & m_{2n} \\ \vdots & & \ddots & \vdots \\ m_{n1} & m_{n2} & \cdots & m_{nn} \end{pmatrix} \begin{pmatrix} x_1 \\ x_2 \\ \vdots \\ x_n \end{pmatrix} = \begin{pmatrix} b_1 \\ b_2 \\ \vdots \\ b_n \end{pmatrix}.$$

The solution of this system of equations (the vector **x**) is returned by the np.linalg. solve method. For example, the three simultaneous equations

$$3x - 2y = 8,$$
$$-2x + y - 3z = -20,$$
$$4x + 6y + z = 7$$

can be represented as the matrix equation **Mx** = **b**:

$$\begin{pmatrix} 3 & -2 & 0 \\ -2 & 1 & -3 \\ 4 & 6 & 1 \end{pmatrix} \begin{pmatrix} x \\ y \\ z \end{pmatrix} = \begin{pmatrix} 8 \\ -20 \\ 7 \end{pmatrix}$$

and solved by passing arrays corresponding to matrix **M** and vector **b** to np.linalg.solve:

```
In [x]: M = np.array([[3, -2, 0], [-2, 1, -3], [4, 6, 1]])
In [x]: b = np.array([8, -20, 7])
In [x]: np.linalg.solve(M, b)
Out[x]: array([ 2., -1.,  5.])
```

That is, $x = 2, y = -1, z = 5$.

If no unique solution exists (for nonsquare or singular matrix, **M**), a LinAlgError is raised.

Linear Least-Squares Solutions ("Best Fit")

Where a set of equations, **Mx** = **b**, does not have a unique solution, a least-squares solution that minimizes the L^2 norm, $\|b - Mx\|^2$ (sum of squared residuals), may be sought using the np.linalg.lstsq method. This is the type of problem described as *overdetermined* (more data points than the two unknown quantities, m and c). Passed M and b, np.linalg.lstsq returns the solution array x, the sum of squared residuals, the rank of M and the singular values of M.

The rank of the matrix M is determined by the number of its singular values: by default, a singular value is considered to be equal to zero if it is less than the optional parameter, rcond, times the largest singular value of M. If rcond is explicitly set to None, it will be taken to be the machine precision times the largest dimension of M. Before version 1.14 of NumPy, the default value of rcond was taken to be machine precision itself (set rcond = -1 to use this default). At the time of writing, if no value is provided for rcond, a FutureWarning is given.

A typical use of this method is to find the "line of best-fit," $y = mx + c$, through some data thought to be linearly related, as in the following example.

Example E6.17   The Beer–Lambert law relates the concentration, c, of a substance in a solution sample to the intensity of light transmitted through the sample, I_t, across a given path length, l, at a given wavelength, λ:

$$I_t = I_0 e^{-\alpha c l},$$

where I_0 is the incident light intensity and α is the absorption coefficient at λ.

Given a series of measurements of the fraction of light transmitted, I_t/I_0, α may be determined through a least-squares fit to the straight line:

$$y = \ln \frac{I_t}{I_0} = -\alpha c l.$$

Although this line passes through the origin ($y = 0$ for $c = 0$), we will fit the more general linear relationship:

$$y = mc + k,$$

where $m = -\alpha l$, and verify that k is close to zero.

Given a sample with path length $l = 0.8$ cm, the following data were measured for I_t/I_0 at five different concentrations:

c /M	I_t/I_0
0.4	0.886
0.6	0.833
0.8	0.784
1.0	0.738
1.2	0.694

The matrix form of the least-squares equation to be solved is

$$\begin{pmatrix} c_1 & 1 \\ c_2 & 1 \\ c_3 & 1 \\ c_4 & 1 \\ c_5 & 1 \end{pmatrix} \begin{pmatrix} m \\ k \end{pmatrix} = \begin{pmatrix} T_1 \\ T_2 \\ T_3 \\ T_4 \\ T_5 \end{pmatrix},$$

where $T = \ln(I_t/I_0)$. The code here determines m and hence α using np.linalg.lstsq:

Listing 6.11 Linear least-squares fitting of the Beer–Lambert law

```
# eg6-beer-lambert-lstsq.py
import numpy as np
import matplotlib.pyplot as plt
# Path length, cm.
path = 0.8
# The data: concentrations (M) and It/I0.
c = np.array([0.4, 0.6, 0.8, 1.0, 1.2])
It_over_I0 = np.array([0.891, 0.841, 0.783, 0.744, 0.692])

n = len(c)
A = np.vstack((c, np.ones(n))).T
T = np.log(It_over_I0)

x, resid, _, _ = np.linalg.lstsq(A, T, rcond=None)
m, k = x
alpha = - m / path
print('alpha = {:.3f} M-1.cm-1'.format(alpha))
print('k =', k)
print('rms residual = ', np.sqrt(resid[0]))
```

❶

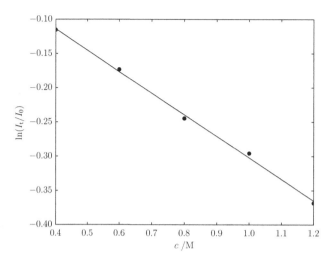

Figure 6.10 Line of least-squares best fit to absorbance data as a function of concentration.

```
plt.plot(c, T, 'o')
plt.plot(c, m*c + k)
plt.xlabel('$c\;/\mathrm{M}$')
plt.ylabel('$\ln(I_\mathrm{t}/I_0)$')
plt.show()
```

❶ Here, _ is the dummy variable name conventionally given to an object we do not need to store or use.

The output produces a best-fit value of $\alpha = 0.393$ $M^{-1}\,cm^{-1}$ and a value of k compatible with experimental error:

```
alpha = 0.393 M-1.cm-1
k = 0.0118109033334
rms residual =  0.0096843591966
```

Figure 6.10 shows the data and fitted line.

6.5.4   Exercises

Questions

Q6.5.1 Demonstrate that the three Pauli matrices given in Section 6.5.2 are *unitary*. That is, that $\sigma_p^\dagger \sigma_p = I_2$ for $p = x, y, z$, where I_2 is the 2×2 identity matrix and $\dagger$ denotes the Hermitian conjugate (conjugate transpose).

Q6.5.2 The ticker timer, much used in school physics experiments, is a device that marks dots on a strip of paper tape at evenly spaced intervals of time as the tape moves through it at some (possibly variable) speed. The following data relate to the positions (in cm) of marks on a tape pulled through a ticker timer by a falling weight. The marks are made every 1/10 s.

```
x = [1.3, 6.0, 20.2, 43.9, 77.0, 119.6, 171.7, 233.2, 304.2, 384.7,
     474.7, 574.1, 683.0, 801.3, 929.2, 1066.4, 1213.2, 1369.4, 1535.1,
     1710.3, 1894.9]
```

Fit these data to the function $x = x_0 + v_0 t + \frac{1}{2}gt^2$ and determine an approximate value for the acceleration due to gravity, g.

Problems

P6.5.1   In physics, the *Planck units* of measurement are those defined such that the five universal physical constants, c (the speed of light), G (the gravitational constant), $\hbar$ (the reduced Planck constant), $(4\pi\epsilon_0)^{-1}$ (the Coulomb constant) and k_B (the Boltzmann constant) are set to unity. The dimensions of these quantities in terms of length (L), mass (M), time (T), charge (Q) and thermodynamic temperature (Θ) are given in Table 6.8, along with their values in SI units.

This suggests the following matrix relationship between the constants and their dimensions:

$$
\begin{array}{cccccc}
 & L & M & T & Q & \Theta \\
c & 1 & 0 & -1 & 0 & 0 \\
G & 3 & -1 & -2 & 0 & 0 \\
\hbar & 2 & 1 & -1 & 0 & 0 \\
(4\pi\epsilon_0)^{-1} & 3 & 1 & -2 & -2 & 0 \\
k_B & 2 & 1 & -2 & 0 & -1
\end{array}
$$

Using the inverse of this matrix, determine the SI values of length, mass, time, charge and temperature in the base Planck units; that is, the combination of these physical constants yielding the dimensions L, M, T, Q and Θ. For example, the *Planck length* is found to be $l_P = \sqrt{\hbar G/c^3} = 1.616199 \times 10^{-35}$ m.

Table 6.8 Some physical constants and their dimensions

c	Speed of light	$2.99792458 \times 10^8 \ \mathrm{m\,s^{-1}}$	$L\,T^{-1}$
G	Gravitational constant	$6.67384 \times 10^{-11} \ \mathrm{m^3\,kg^{-1}\,s^{-2}}$	$L^3\,M^{-1}\,T^{-2}$
$\hbar$	Reduced Planck constant	$1.054571726 \times 10^{-34} \ \mathrm{J\,s}$	$L^2\,M\,T^{-1}$
$(4\pi\epsilon_0)^{-1}$	Coulomb constant	$8.9875517873681764 \times 10^9 \ \mathrm{N\,m^2\,C^{-2}}$	$L^3\,M\,T^{-2}\,Q^{-2}$
k_B	Boltzmann constant	$1.3806488 \times 10^{-23} \ \mathrm{J\,K^{-1}}$	$L^2\,M\,T^{-2}\,\Theta^{-1}$

P6.5.2 The (symmetric) matrix representing the inertia tensor of a collection of masses, m_i, with positions (x_i, y_i, z_i) relative to their center of mass is

$$\mathbf{I} = \begin{pmatrix} I_{xx} & I_{xy} & I_{xz} \\ I_{xy} & I_{yy} & I_{yz} \\ I_{xz} & I_{yz} & I_{zz} \end{pmatrix},$$

where

$$I_{xx} = \sum_i m_i(y_i^2 + z_i^2), \qquad I_{yy} = \sum_i m_i(x_i^2 + z_i^2), \qquad I_{zz} = \sum_i m_i(x_i^2 + y_i^2),$$

$$I_{xy} = -\sum_i m_i x_i y_i, \qquad I_{yz} = -\sum_i m_i y_i z_i, \qquad I_{xz} = -\sum_i m_i x_i z_i.$$

There exists a transformation of the coordinate frame such that this matrix is diagonal: the axes of this transformed frame are called the *principal* axes and the diagonal inertia matrix elements, $I_a \leq I_b \leq I_c$, are the *principal moments of inertia*.

Write a program to calculate the principal moments of inertia of a molecule, given the position and masses of its atoms *relative to some arbitrary origin*. Your program should first relocate the atom coordinates relative to its center of mass and then determine the principal moments of inertia as the eigenvalues of the matrix $\mathbf{I}$.

A molecule may be classified as follows according to the relative values of I_a, I_b and I_c:

- $I_a = I_b = I_c$: spherical top;
- $I_a = I_b < I_c$: oblate symmetric top;
- $I_a < I_b = I_c$: prolate symmetric top;
- $I_a < I_b < I_c$: asymmetric top.

Determine the principal moments of inertia and classify the molecules NH_3, CH_4, CH_3Cl and O_3 given the data available at https://scipython.com/ex/bfh. Also determine the *rotational constants*, A, B and C, related to the moments of inertia through $Q = h/(8\pi^2 c I_q)$ ($Q = A, B, C; q = a, b, c$) and usually expressed in cm^{-1}.

◇   **P6.5.3** The NumPy method `numpy.linalg.svd` returns the *singular value decomposition* (SVD) of a matrix, $\mathbf{M}$, as the arrays, $\mathbf{U}$, $\mathbf{\Sigma}$ and $\mathbf{V}$, satisfying the factorization $\mathbf{M} = \mathbf{U\Sigma V}^\dagger$, where $\dagger$ denotes the Hermitian conjugate (the conjugate transpose).

The SVD and the eigendecomposition are related in that the left-singular row vectors, $\mathbf{U}$ are the eigenvectors of $\mathbf{MM}^*$ and the right-singular column vectors, $\mathbf{V}$, are the eigenvectors of $\mathbf{M}^*\mathbf{M}$. Furthermore, the diagonal entries of $\mathbf{\Sigma}$ are the square roots of the nonzero eigenvalues of both $\mathbf{MM}^*$ and $\mathbf{M}^*\mathbf{M}$.

Show that this is the case for the special case of $\mathbf{M}$, a 3×3 matrix with random real entries, by comparing the output of `numpy.linalg.svd` with that of `numpy.linalg.eig`.

Hint: the singular values of $\mathbf{M}$ are sorted in descending order, but the eigenvalues returned by `numpy.linalg.eig` are in no particular order. Both methods produce normalized eigenvectors, but may differ by sign (ignore the possibility that any of the eigenvalues could have an eigenspace with dimension greater than 1).

P6.5.4 Let the column matrix

$$\mathbf{F_n} = \begin{pmatrix} p_n \\ q_n \end{pmatrix}$$

describe the number of non-negative integers less than 10^n ($n \geq 0$) that do (p_n) and do not (q_n) contain the digit 5. Hence, for $n = 1$, $p_1 = 1$ and $q_1 = 9$. Devise a matrix-based recursion relation for finding $\mathbf{F_{n+1}}$ from $\mathbf{F_n}$.

How many numbers less than 10^{10} contain the digit 5?

For each $n \leq 10$, find p_n and verify that $p_n = 10^n - 9^n$.

P6.5.5 The matrix

$$\mathbf{F} = \begin{pmatrix} 1 & 1 \\ 1 & 0 \end{pmatrix}$$

can be used to produce the Fibonacci sequence by repeated multiplication: the element F_{11}^n of the matrix $\mathbf{F}^n$ is the $(n+1)$th Fibonacci number (for $n = 0, 1, 2 \ldots$). Use a NumPy representation of $\mathbf{F}$ to calculate the first 10 Fibonacci numbers.

One can show that

$$\mathbf{F}^n = \mathbf{CD}^n\mathbf{C}^{-1}, \quad \text{where} \quad \mathbf{D} = \mathbf{C}^{-1}\mathbf{FC}$$

is the diagonal matrix related to $\mathbf{F}$ through the similarity transformation associated with matrix $\mathbf{C}$. Use this relationship to find the 1100th Fibonacci number.

P6.5.6 The *implicit* formula for a conic section may be written as the second-degree polynomial,

$$Q = Ax^2 + Bxy + Cy^2 + Dx + Ey + F = 0,$$

or in matrix form using the homogeneous coordinate vector,

$$\mathbf{x} = \begin{pmatrix} x \\ y \\ 1 \end{pmatrix},$$

as $\mathbf{x}^T\mathbf{Qx} = 0$, where

$$\mathbf{Q} = \begin{pmatrix} A & B/2 & D/2 \\ B/2 & C & E/2 \\ D/2 & E/2 & F \end{pmatrix}.$$

Conic sections may be classified according to the following properties of $\mathbf{Q}$, where the submatrix $\mathbf{Q_{33}}$ is

$$\mathbf{Q_{33}} = \begin{pmatrix} A & B/2 \\ B/2 & C \end{pmatrix}.$$

- If $\det\mathbf{Q} = 0$, the conic is *degenerate* in one of the following forms:

    - if $\det\mathbf{Q_{33}} < 0$, the equation represents two intersecting lines;
    - if $\det\mathbf{Q_{33}} = 0$, the equation represents two parallel lines;

- if $\det\mathbf{Q}_{33} > 0$, the equation represents a single point.

- if $\det\mathbf{Q} \neq 0$:

  - if $\det\mathbf{Q}_{33} < 0$, the conic is a *hyperbola*;
  - if $\det\mathbf{Q}_{33} = 0$, the conic is a *parabola*;
  - if $\det\mathbf{Q}_{33} > 0$, the conic is an *ellipse*:

    - if $A = C$ and $B = 0$, the ellipse is a *circle*.

Write a program to classify the conic section represented by the six coefficients A, B, C, D, E and F.

Some test-cases (coefficients not given are zero):

- hyperbola: $B = 1, F = -9$;
- parabola: $A = \frac{1}{2}, D = 2, E = -\frac{1}{2}$;
- circle: $A = \frac{1}{2}, C = \frac{1}{2}, D = -2, E = -3, F = 2$;
- ellipse: $A = 9, C = 4, F = -36$;
- two parallel lines: $A = 1, F = -1$;
- a single point: $A = 1, C = 1$.

P6.5.7 Example E6.8 produced a comma-separated text file containing 10 simulations of the radioactive decay of an ensemble of 500 ^{14}C nuclei. For each simulation column, the number of undecayed nuclei as a function of time, $N(t)$, is given; the grid of time points (in years) is in the first column.

Average the simulation data, which is available at https://scipython.com/ex/bac , and use NumPy's np.linalg.lstsq function to perform a linear least-squares fit. Retrieve the half-life of ^{14}C, $t_{1/2} = \tau \ln 2$, where:

$$N(t) = N(0)e^{-t/\tau} \quad \Rightarrow \quad \ln[N(t)] = \ln[N(0)] - \frac{t}{\tau}.$$

6.6   Random Sampling

NumPy's random module provides methods for obtaining random numbers from any of several distributions as well as convenient ways to choose random entries from an array and to randomly shuffle the contents of an array.

As with the Standard Library's random module (Section 4.5.1), np.random uses a Mersenne Twister *pseudorandom*-number generator (PRNG) The way it seeds itself is operating-system dependent, but it can be reseeded with any hashable object (e.g. an immutable object such as an integer) by calling np.random.seed. For example, using the randint method described here:

```
In [x]: np.random.seed(42)
In [x]: np.random.randint(1, 10, 10)    # 10 random integers in [1, 10)
array([7, 4, 8, 5, 7, 3, 7, 8, 5, 4])
In [x]: np.random.randint(1, 10, 10)
array([8, 8, 3, 6, 5, 2, 8, 6, 2, 5])
```

```
In [x]: np.random.randint(1, 10, 10)
array([1, 6, 9, 1, 3, 7, 4, 9, 3, 5])
In [x]: np.random.seed(42)              # reseed the PRNG       .
In [x]: np.random.randint(1,10, 10)
array([7, 4, 8, 5, 7, 3, 7, 8, 5, 4])   # same as before
```

6.6.1   Uniformly Distributed Random Numbers

Random Floating-Point Numbers

The basic random method, random_sample,[26] takes the shape of an array as its argument and creates an array of the corresponding shape filled with numbers sampled randomly from the uniform distribution over $[0, 1)$; that is, the interval between 0 and 1 inclusive of 0 but exclusive of 1:

```
In [x]: np.random.random_sample((3,2))
array([[ 0.92338355,  0.2978852 ],
       [ 0.75175429,  0.88110707],
       [ 0.16759816,  0.32203783]])
```

(called without an argument, it returns a single random number). If you want numbers sampled from the uniform distribution over $[a, b)$, you need to do a bit of work:

```
In [x]: a, b = 10, 20
In [x]: a + (b - a) * np.random.random_sample((3, 2))
array([[ 18.07084068,  12.11591797],
       [ 14.08171741,  19.34857282],
       [ 13.06759203,  11.07003867]])
```

In a uniform distribution, every number has the same probability of being sampled, as can be seen from a histogram of a large number of samples (Figure 6.11):

```
In [x]: plt.hist(np.random.random_sample(10000), bins=100)
In [x]: plt.show()
```

The np.random.rand method is similar, but is passed the dimensions of the desired array as separate arguments. For example,

```
In [x]: np.random.rand(2, 3)
Out[x]:
array([[ 0.61075227,  0.37459455,  0.95670676],
       [ 0.25276732,  0.1601836 ,  0.3746576 ]])
```

Random Integers

Sampling random integers is supported through a couple of methods. The np.random. randint method takes up to three arguments: low, high and size.

* If both low and high are supplied, then the random number(s) are sampled from the discrete half-open interval [low, high).[27]

---

[26] np.random.random_sample is also available under the aliases np.random.random, np.random.ranf and np.random.sample.

[27] Note that this is different from the behavior of the Standard Library's random.randint(a, b) method (see Section 4.5.1), which picks numbers uniformly from the *closed* interval, [a, b].

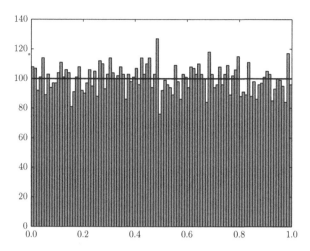

Figure 6.11 Histogram of 10 000 random samples from the uniform distribution on $[0, 1)$ provided by `np.random.random_sample()`.

- If `low` is supplied but `high` is not, then the sampled interval is `[0, low)`.
- `size` is the shape of the array of random integers desired. If it is omitted, as with `np.random.rand`, a single random integer is returned.

```
In [x]: np.random.randint(4)                # random integer from [0, 4)
2
In [x]: np.random.randint(4, size=10)       # 10 random integers from [0, 4)
array([3, 2, 2, 2, 0, 2, 2, 1, 3, 1])
In [x]: np.random.randint(4, size=(3, 5))   # array of random integers from [0, 4)
array([[0, 1, 1, 2, 2],
       [2, 0, 3, 3, 0],
       [0, 1, 0, 1, 1]])
In [x]: np.random.randint(1, 4, (3, 5))     # array of random integers from [1, 4)
array([[1, 1, 1, 3, 2],
       [1, 1, 2, 1, 3],
       [1, 3, 1, 3, 1]])
```

`np.random.randint` can be useful for selecting random elements (with replacement) from an array by picking random indexes:

```
In [x]: a = np.array([6, 6, 6, 7, 7, 7, 7, 7, 7])
In [x]: a[np.random.randint(len(a), size=5)]
array([7, 7, 7, 6, 7])
```

The other method for sampling random integers, `np.random.random_integers`, has the same syntax but returns integers sampled from the uniform distribution over the *closed* interval `[low, high]` (if `high` is supplied) or `[0, low]` (if it is not).

Example E6.18   These random-integer methods can be used for sampling from a set of evenly spaced real numbers, though it requires a bit of extra work: to pick a number from n evenly spaced real numbers between a and b (inclusive), use

```
In [x]: a + (b - a) * (np.random.random_integers(n) - 1) / (n - 1.)
```

For example, to sample from $[\frac{1}{2}, \frac{3}{2}, \frac{5}{2}, \frac{7}{2}]$,

```
In [x]: a, b, n = 0.5, 3.5, 4
In [x]: a + (b - a) * (np.random.random_integers(n, size=10) - 1) / (n - 1.)
array([ 1.5,  0.5,  1.5,  1.5,  3.5,  2.5,  3.5,  3.5,  3.5,  3.5])
```

Example E6.19   In a famous experiment, a group of volunteers are asked to toss a fair coin 100 times and note down the results of each toss (heads, H, or tails, T). It is generally easy to spot the participants who fake the results by writing down what they think is a random sequence of Hs and Ts instead of actually tossing the coin because they tend not to include as many "streaks" of repeated results as would be expected by chance.

If they had access to a Python interpreter, here's how they could produce a more plausibly random set of results:

```
In [x]: res = ['H', 'T']
In [37]: tosses = ''.join([res[i] for i in np.random.randint(2, size=100)])
In [38]: tosses
Out[38]: 'TTHHTHHTTHHHTHTTHHHTHHTHTTHHTHHTTTTHHHHHHHHTTTHTTHHHHHHHHTHHHTHHHH
THTTTHTTHHHHTHTTTTHTTTHTHTHHTHHHHHHH'
```

This virtual experiment features a run of eight heads in a row, and two runs of seven heads in a row:

```
  TAILS     | i |    HEADS
------------------------------
            | 8 | *
            | 7 | **
            | 6 |
            | 5 |
         ** | 4 | **
        *** | 3 | ***
    ******* | 2 | ******
  ********* | 1 | ********
```

6.6.2   Random Numbers from Non-Uniform Distributions

The full range of random distributions supported by NumPy is described in the official documentation.[28] This section describes in detail only the *normal*, *binomial* and *Poisson* distributions.

The Normal Distribution

The normal probability distribution is described by the Gaussian function,

$$P(x) = \frac{1}{\sigma \sqrt{2\pi}} \exp\left(-\frac{(x - \mu)^2}{2\sigma^2}\right),$$

---

[28] https://docs.scipy.org/doc/numpy-1.14.0/reference/routines.random.html.

where μ is the mean and σ the standard deviation. The NumPy function, np.random.normal, selects random samples from the normal distribution. The mean and standard deviation are specified by loc and scale respectively, which default to 0 and 1. The shape of the returned array is specified with the size attribute.

```
In [x]: np.random.normal()
-0.34599057326978105
In [x]: np.random.normal(scale=5., size=3)
array([  4.38196707,  -5.17358738,  11.93523167])
In [x]: np.random.normal(100., 8., size=(4, 2))
array([[ 107.730434  ,  101.06221195],
       [ 100.75627505,   88.79995561],
       [  88.82658615,   94.89630767],
       [ 105.91254312,   98.21190741]])
```

It is also possible to draw numbers from the standard normal distribution (that with $\mu = 0$ and $\sigma = 1$) with the np.random.randn method. Like random.rand, this takes the dimensions of an array as its arguments:

```
In [x]: np.random.randn(2, 2)
array([[-1.25092263,  2.6291925 ],
       [ 0.34158642,  0.40339403]])
```

Although np.random.randn does not provide a way to set the mean and standard deviation explicitly, the standard distribution can be rescaled easily enough:

```
In [x]: mu, sigma = 100., 8.
In [x]: mu + sigma * np.random.randn(4, 2)
array([[ 104.92454826,   98.84646729],
       [ 109.43568726,   92.9568489 ],
       [  90.21632016,   96.25271625],
       [ 102.65745451,   89.94890264]])
```

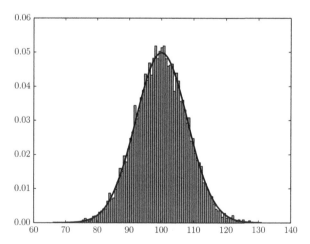

Figure 6.12 Histogram of 10 000 random samples from the normal distribution provided by np.random.normal.

Example E6.20   The normal distribution may be plotted from sampled data as a histogram (Figure 6.12):

```
In [x]: mu, sigma = 100., 8.
In [x]: samples = np.random.normal(loc=mu, scale=sigma, size=10000)
In [x]: counts, bins, patches = plt.hist(samples, bins=100, density=True)
In [x]: plt.plot(bins, 1/(sigma * np.sqrt(2 * np.pi)) *
...                 np.exp( -(bins - mu)**2 / (2 * sigma**2) ), lw=2)
In [x]: plt.show()
```

The Binomial Distribution

The binomial probability distribution describes the number of particular outcomes in a sequence of n *Bernoulli trials* – that is, n independent experiments, each of which can yield exactly two possible outcomes (e.g. *yes/no, success/failure, heads/tails*). If the probability of a single particular outcome (say, *success*) is p, the probability that such a sequence of trials yields exactly k such outcomes is

$$\binom{n}{k}p^k(1 - p)^{n-k}, \quad \text{where} \quad \binom{n}{k} = \frac{n!}{k!(n-k)!}.$$

For example, when a fair coin is tossed, the probability of it coming up heads each time is $\frac{1}{2}$. The probability of getting exactly three heads out of four tosses is therefore $4(\frac{1}{2})(\frac{1}{2})^3 = \frac{1}{4}$, where the factor of $\binom{4}{3} = 4$ accounts for the four possible equivalent outcomes: THHH, HTHH, HHTH, HHHT.

To sample from the binomial distribution described by parameters n and p, use np.random.binomial(n, p). Again, the shape of an array of samples can be specified with the third argument, size:

```
In [x]: np.random.binomial(4, 0.5)
2
In [x]: np.random.binomial(4, 0.5, (4, 4))
array([[1, 2, 2, 4],
       [2, 1, 3, 2],
       [2, 3, 1, 1],
       [2, 4, 2, 3]])
```

Example E6.21   There are two stable isotopes of carbon, ^{12}C and ^{13}C (the radioactive ^{14}C nucleus is present in nature in only trace amounts of the order of parts per trillion). Taking the abundance of ^{13}C to be $x = 0.0107$ (i.e. about 1%), we will calculate the relative amounts of buckminsterfullerene, C_{60}, with exactly zero, one, two, three and four ^{13}C atoms. (This is important in nuclear magnetic resonance studies of fullerenes, for example, because only the ^{13}C nucleus is magnetic and so detectable by NMR.)

The number of ^{13}C atoms in a population of carbon atoms sampled at random from a population with natural isotopic abundance follows a binomial distribution: the probability that, out of n atoms, m will be ^{13}C (and therefore $n - m$ will be ^{12}C) is

$$p_m(n) = \binom{n}{m}x^m(1 - x)^{n-m}.$$

We can, of course, calculate $p_m(60)$ exactly from this formula for $0 \leq m \leq 4$, but we can also simulate the sampling with the `np.random.binomial` method.

Listing 6.12 Modeling the distribution of ^{13}C atoms in C_{60}

```
# eg6-e-c13-a.py
import numpy as np


n, x = 60, 0.0107
mmax = 4
m = np.arange(mmax + 1)


# Estimate the abundances by random sampling from the binomial distribution.
ntrials = 10000
pbin = np.empty(mmax + 1)
for r in m:
    pbin[r] = np.sum(np.random.binomial(n, x, ntrials) == r)/ntrials


# Calculate and store the binomial coefficients nCm.
nCm = np.empty(mmax + 1)
nCm[0] = 1
for r in m[1:]:
    nCm[r] = nCm[r - 1] * (n - r + 1) / r
# The "exact" answer from binomial distribution.
p = nCm * x**m * (1-x)**(n-m)

print('Abundances of C60 as (13C)[m](12C)[60-m]')
print('m   "Exact"    Estimated')
print('-'*24)
for r in m:
    print('{:1d}    {:6.4f}      {:6.4f}'.format(r, p[r], pbin[r]))
```

❶ For each value of r in the array m, we sample a large number of times (ntrial) from the binomial distribution described by $n = 60$ and probability, $x = 0.0107$. The comparison of these sample values with a given value of r yields a boolean array which can be summed (remembering that `True` evaluates to 1 and `False` evaluates to 0); division by `ntrials` then gives an estimate of the probability of exactly r atoms being of type ^{13}C and the remainder of type ^{12}C.

The explicit loop over m could be removed by creating an array of shape (ntrials, mmax + 1) containing all the samples, and summing over the first axis of this array in the comparison with the m array:

```
samples = np.random.binomial(n, x, (ntrials, mmax + 1))
pbin = np.sum(samples == m, axis=0) / ntrials
```

The abundances of $^{13}C_m{}^{12}C_{60-m}$ produced by our program are given as the following output.

```
Abundances of C60 as (13C)[m](12C)[60-m]
m   "Exact"    Estimated
------------------------
0    0.5244      0.5199
1    0.3403      0.3348
2    0.1086      0.1093
3    0.0227      0.0231
```

4   0.0035   0.0031

That is, almost 48% of C_{60} molecules contain at least one magnetic nucleus.

The Poisson Distribution

The Poisson distribution describes the probability of a particular number of independent events occurring in a given interval of time if these events occur at a known average rate. It is also used for occurrences in specified intervals over other domains such as distance or volume. The Poisson probability distribution of the number of events, k, is

$$P(k) = \frac{\lambda^k e^{-\lambda}}{k!},$$

where the parameter λ is the expected (average) number of events occurring within the considered interval.[29] The NumPy implementation np.random.poisson takes λ as its first argument (which defaults to 1) and, as before, the shape of the desired array of samples can be specified with a second argument, size. For example, if I receive an average of 2.5 emails an hour, a sample of the number of emails I receive each hour over the next 8 hours could be obtained as:

```
In [x]: np.random.poisson(2.5, 8)
array([4, 1, 3, 0, 4, 1, 3, 2])
```

Example E6.22   The endonuclease enzyme *Eco*RI is used as a restriction enzyme which cuts DNA at the nucleic acid sequence GAATTC. Suppose a given DNA molecule contains 12 000 base pairs and a 50% G + C content. The Poisson distribution can be used to predict the probability that *Eco*RI will fail to cleave this molecule as follows:

The recognition site, GAATTC, consists of six nucleotide base pairs; the probability that any given six-base sequence corresponds to GAATTC is $1/4^6 = 1/4096$ and so the expected number of cleavage sites for *Eco*RI in this DNA molecule is $\lambda = 12\,000/4096 = 2.93$. From the Poisson distribution, we expect the probability that the endonuclease will fail to cleave this molecule is therefore

$$P(0) = \frac{\lambda^0 e^{-\lambda}}{0!} = 0.053,$$

or about 5.3%. To simulate the possibilities stochastically:

```
In [x]: lam = 12000 / 4**6
In [x]: N = 100000
In [x]: np.sum(np.random.poisson(lam, N) == 0) / N
Out[x]: 0.053699999999999998
```

---

[29] The Poisson distribution is the limit of the binomial distribution as $n \rightarrow \infty$ and $p \rightarrow 0$ such that $\lambda = np$ tends to some finite constant value.

6.6.3   Random Selections, Shuffling and Permutations

It is often the case that given an array of values, you wish to pick one or more at random (with or without replacement). This is the purpose of the np.random.choice method. Given a single argument, a one-dimensional sequence, it returns a random element drawn from the sequence:

```
In [x]: np.random.choice([ 1,   5,   2,  -5,   5,   2,   0])
2
In [x]: np.random.choice(np.arange(10))
7
```

A second argument, size, controls the shape of the array of random samples returned, as before. By default, the elements of the sequence are drawn randomly with a uniform distribution and *with* replacement; to draw the sample *without* replacement, set replace=False.

```
In [x]: a = np.array([1, 2, 0, -1, 1])
In [x]: np.random.choice(a, 6)   # six random selections from a
array([ 1, -1,   2,   1, -1,   1])
In [x]: np.random.choice(a, (2, 2), replace=False)
array([[ 2, -1],
       [ 1,   0]])
In [x]: np.random.choice(a, (3, 2), replace=False)
... <some traceback information> ...
ValueError: Cannot take a larger sample than population when 'replace=False'
```

This last example shows that, as you might expect, it is not possible to draw a larger number of elements than there are in the original population if you are sampling without replacement.

To specify the probability of each element being selected, pass a sequence of the same length as the population to be sampled as the argument p. The probabilities should sum to 1.

```
In [x]: a = np.array([1, 2, 0, -1, 1])
In [x]: np.random.choice(a, 5, p=[0.1, 0.1, 0., 0.7, 0.1])
Out[x]: array([-1, -1, -1, -1,   1])
In [x]: np.random.choice(a, 2, False, p=[0.1, 0.1, 0., 0.8, 0.])
Out[x]: array([-1,   2])            # sample without replacement
```

There are two methods for permuting the contents of an array: np.random.shuffle randomly rearranges the order of the elements *in place* whereas np.random.permutation makes a copy of the array first, leaving the original unchanged:

```
In [x]: a = np.arange(6)
In [x]: np.random.permutation(a)
array([4, 2, 5, 1, 3, 0])
In [x]: a
array([0, 1, 2, 3, 4, 5])
In [x]: np.random.shuffle(a)
In [x]: a
array([5, 4, 1, 3, 0, 2])
```

These methods only act on the first dimension of the array:

```
In [x]: a = np.arange(6).reshape(3, 2)
In [x]: a
array([[0, 1],
       [2, 3],
       [4, 5]])
In [x]: a.random.permutation(a)  # permutes the rows, but not the columns
array([[2, 3],
       [4, 5],
       [0, 1]])
```

6.6.4   Exercises

Questions

Q6.6.1   Explain the difference between

```
In [x]: a = np.array([6, 6, 6, 7, 7, 7, 7, 7, 7])
In [x]: a[np.random.randint(len(a), size=5)]
array([7, 7, 7, 6, 7])       # (for example)
```

and

```
In [x]: np.random.randint(6, 8, 5)
array([6, 6, 7, 7, 7])       # (for example)
```

Q6.6.2   In Example E6.18 we used random.random_integers to sample from the uniform distribution on the floating-point numbers $[\frac{1}{2}, \frac{3}{2}, \frac{5}{2}, \frac{7}{2}]$. How can you do the same using the np.random.randint instead?

Q6.6.3   The American lottery, Mega Millions, at the time of writing, involves the selection of five numbers out of 70 *and* one from 25. The jackpot is shared among the players who match all of their numbers in a corresponding random draw. What is the probability of winning the jackpot? Write a single line of Python code using NumPy to pick a set of random numbers for a player.

Q6.6.4   Suppose an n-page book is known to contain m misprints. If the misprints are independent of one another, the probability of a misprint occurring on a particular page is $p = 1/n$ and their distribution may be considered to be binomial. Write a short program to conduct a number of trial virtual "printings" of a book with $n = 500, m = 400$ and determine the probability, Pr, that a single given page will contain two or more misprints.

Compare with the result predicted by the Poisson distribution with rate parameter $\lambda = m/n$, $\text{Pr} = 1 - e^{-\lambda}\left(\frac{\lambda^0}{0!} + \frac{\lambda}{1!}\right)$.

Problems

P6.6.1   Simulate an experiment carried out ntrials times in which, for each experiment, n coins are tossed and the total number of heads each time is recorded.

Plot the results of the simulation on a suitable histogram and compare with the expected binomial distribution of heads.

P6.6.2 A classic problem, first posed by Georges-Louis Leclerc, Comte de Buffon, can be stated as follows:

Given a plane ruled with parallel lines a distance d apart, what is the probability that a needle of length $l \le d$ dropped at random onto the plane will cross a line?

The problem can be solved analytically, yielding the answer $2l/\pi d$; show that this solution is given approximately for the case $l = d$ using a random simulation (Monte Carlo) method, that is, by simulating the experiment with a large number of random orientations of the needle.

A related problem involves dropping a circular coin of radius a onto a floor consisting of square tiles, each of side d. Show that the probability of a coin crossing a tile edge is $1 - (d - 2a)^2/d^2$ and confirm it with a Monte Carlo simulation.

P6.6.3 Some bacteria, such as *Escherichia coli*, possess helical flagella which enable them to move toward attractants such as nutrients, a process known as chemotaxis. When the flagella rotate counterclockwise, the bactrium is propelled forward; when they rotate clockwise, it tumbles randomly, changing its orientation. A combination of such movements enables the bacterium to perform a *biased random walk*: if the bacterium senses it is moving up a concentration gradient toward an attractant it will rotate its flagella counterclockwise more often than clockwise so as to continue moving in that direction; conversely, if it is moving away it is more likely to rotate its flagella clockwise so as to tumble, with the aim of randomly changing its orientation to one that points it toward the attractant.

The chemotaxis of *E. coli* may be modeled (very) simplistically by considering a bacterium to move in a two-dimensional "world" populated by an attractant with a constant concentration gradient away from some location. At each of a series of time steps, a model bacterium detects whether it is moving up or down this gradient and either continues moving or tumbles according to some pair of probabilities.

Write a Python program to implement this simple model of chemotaxis for a world consisting of the unit square with an attractant at its center. Plot the locations of 10 model bacteria that start off evenly spaced around the unit circle centered on the attractant location.

P6.6.4 One way to simulate the meanders in a river is as the average of a large number of a random walks.[30] Using a coordinate system (x, y), start at point $A = (0, 0)$ and aim to finish at $B = (b, 0)$. Starting from an initial heading of ϕ_0 from the AB direction, at each step change this angle by a random amount drawn from a normal distribution with mean $\mu = 0$ and standard deviation σ, and proceed by unit distance in this direction. Discard any walks which do not, after n steps, finish within one unit of B (this will be the majority!).

---

[30] B. Hayes, *American Scientist* **94**, 490 (2006); H. von Schelling, *General Electric Report* No. 64GL92.

Write a program to find the average path meeting the above constraints for $b = 10$, using $\phi_0 = 110°$, $\sigma = 17°$, $n = 40$ and 10^6 random-walk trials. Plot the accepted walks and their average, which should resemble a meander.

6.7   Discrete Fourier Transforms

6.7.1   One-dimensional Fast Fourier Transforms

numpy.fft is NumPy's Fast Fourier Transform (FFT) library for calculating the discrete Fourier transform (DFT) using the ubiquitous Cooley and Tukey algorithm.[31] The definition for the DFT of a function defined on n points, $f_m, m = 1, 2, \ldots, n - 1$ used by NumPy is

$$F_k = \sum_{m=0}^{n-1} f_m \exp\left(-\frac{2\pi i m k}{n}\right), \quad k = 0, 1, 2, \ldots, n - 1 \tag{6.1}$$

NumPy's basic DFT method, for real and complex functions, is np.fft.fft. If the input signal function, f, is considered to be in the time domain, the output Fourier Transform, F, is in the frequency domain and is returned by the fft(f) function call in a standard order: F[:n/2] are the positive-frequency terms in increasing order, F[n/2+1:] contains the negative-frequency terms in decreasing order, and F[n/2] is the (positive and negative) Nyquist frequency.[32] np.abs(F), np.abs(F)**2 and np.angle(F) are the *amplitude spectrum*, *power spectrum* and *phase spectrum*, respectively.

The frequency bins corresponding to the values of F are given by np.fft.fftfreq n, d) where d is the sample spacing. For even n, this is equivalent to

$$0, \frac{1}{dn}, \frac{2}{dn}, \ldots, \frac{n/2 - 1}{dn}, -\frac{n/2}{dn}, -\frac{n/2 - 1}{dn}, \ldots, -2, -1$$

To shift the spectrum so that the zero-frequency component is at the center, call np.fft.fftshift. To undo that shift, call np.fft.ifftshift.

For example, consider the following waveform in the time domain with some synthetic Gaussian noise added:

$$f(t) = 2 \sin(20\pi t) + \sin(100\pi t).$$

```
In [x]: A1, A2 = 2, 1
In [x]: freq1, freq2 = 10, 50
In [x]: fsamp = 500
In [x]: t = np.arange(0, 1, 1/fsamp)
In [x]: n = len(t)
In [x]: f = A1*np.sin(2*np.pi*freq1*t) + A2*np.sin(2*np.pi*freq2*t)
In [x]: f += 0.2 * np.random.randn(n)
In [x]: plt.plot(t, f)
In [x]: plt.xlabel('Time /s')
```

---

[31] J. W. Cooley and J. W. Tukey, *Math. Comput.* **19**, 297–301 (1965).
[32] Here, n is assumed to be even.

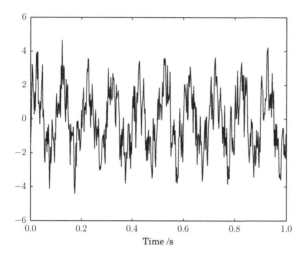

Figure 6.13 The noisy waveform referred to in the text.

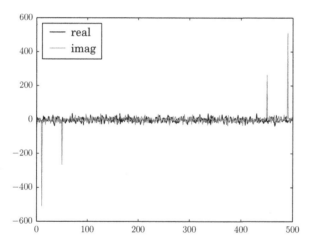

Figure 6.14 The Fourier transform of a noisy waveform with two frequency components, as returned by `np.fft.fft`.

```
In [x]: plt.show()
```

The plot of this waveform is depicted in Figure 6.13.

The Fourier transform of this function is complex; its real and imaginary components are plotted in Figure 6.14.

```
In [x]: F = np.fft.fft(f)
In [x]: plt.plot(F.real, 'k', label='real')
In [x]: plt.plot(F.imag, 'gray', label='imag')
In [x]: plt.legend(loc=2)
In [x]: plt.show()
```

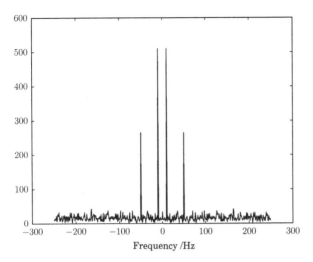

Figure 6.15 The Fourier Transform of a noisy waveform with two frequency components plotted against frequency.

Now look at the shifted amplitude spectrum with the zero-frequency component at the center:[33]

```
In [x]: freq = np.fft.fftfreq(n, 1/fsamp)
In [x]: F_shifted = np.fft.fftshift(F)
In [x]: freq_shifted = np.fft.fftshift(freq)
In [x]: plt.plot(freq_shifted, np.abs(F_shifted))
In [x]: plt.xlabel('Frequency /Hz')
In [x]: plt.show()
```

This plot is given in Figure 6.15.

Now, because our input function is real, its Fourier transform is Hermitian: the negative frequency components are the complex conjugates of the positive frequency components so they don't contain any further information. Therefore, we only need to deal with the first half of the F array. Plotted against its (positive) frequencies as an amplitude spectrum (Figure 6.16):

❶
```
In [x]: spec = 2/n * np.abs(F[:n//2])
In [x]: plt.plot(freq[:n//2], spec, 'k')
In [x]: plt.xlabel('Frequency /Hz')
In [x]: plt.show()
```

❶ Note that because of the way this DFT has been defined, a normalization factor of $\frac{2}{n}$ is required to faithfully regenerate the original amplitudes of each component.

The amplitudes of the 10 Hz and 50 Hz signals are easily resolved in this spectrum.

---

[33] The shifting here is for illustration: note that it isn't really necessary to shift both `freq` and `F` arrays simply to plot one against the other.

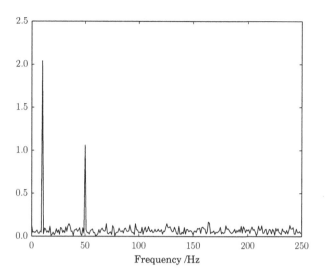

Figure 6.16 The positive-frequency components of the Fourier transform of a noisy waveform, normalized to show their intensities.

The inverse Fourier Transform defined through

$$f_m = \frac{1}{n} \sum_{k=0}^{n-1} F_k \exp\left(\frac{2\pi i m k}{n}\right) \quad m = 0, 1, 2, \ldots, n-1$$

is returned by the method `np.fft.ifft`.

If, as mentioned earlier, the input function array is real and only the non-negative frequency components are needed, the `np.fft` methods `rfft`, `irfft`, `rfftfreq` can be used.

6.7.2   Two-Dimensional Fast Fourier Transforms

Discrete Fourier transforms and their inverses in two and higher dimensions are possible using the `np.fft` methods `fft2`, `ifft2`, `fftn` and `ifftn`. The two-dimensional DFT is defined as

$$F_{jk} = \sum_{p=0}^{m-1} \sum_{q=0}^{n-1} f_{pq} \exp\left[-2\pi i \left(\frac{pj}{m} + \frac{qk}{n}\right)\right],$$

$$j = 0, 1, 2, \ldots, m-1; \ k = 0, 1, 2, \ldots, n-1.$$

and higher dimensions follow similarly.

Example E6.23   The two-dimensional DFT is widely used in image processing.[34] For example, multiplying the DFT of an image by a two-dimensional Gaussian function is a common way to blur an image by decreasing the magnitude of its high-frequency components.

The following code produces an image of randomly arranged squares and then blurs it with a Gaussian filter.

Listing 6.13 Blurring an image with a Gaussian filter

```python
# eg6-fft2-blur.py
import numpy as np
import matplotlib.pyplot as plt

# Image size, square side length, number of squares.
ncols, nrows = 120, 120
sq_size, nsq = 10, 20

# The image array (0=background, 1=square) and boolean array of allowed places
# to add a square so that it doesn't touch another or the image sides.
image = np.zeros((nrows, ncols))
sq_locs = np.zeros((nrows, ncols), dtype=bool)
sq_locs[1:-sq_size-1:,1:-sq_size-1] = True


def place_square():
    """ Place a square at random on the image and update sq_locs. """
    # Valid_locs is an array of the indexes of True entries in sq_locs.
    valid_locs = np.transpose(np.nonzero(sq_locs))
    # Pick one such entry at random, and add the square so its top left
    # corner is there; then update sq_locs.
    i, j = valid_locs[np.random.randint(len(valid_locs))]
    image[i:i+sq_size, j:j+sq_size] = 1
    imin, jmin = max(0, i-sq_size-1), max(0, j-sq_size-1)
    sq_locs[imin:i+sq_size+1, jmin:j+sq_size+1] = False

# Add the required number of squares to the image.
for i in range(nsq):
    place_square()
plt.imshow(image)
plt.show()

# Take the two-dimensional DFT and center the frequencies.
ftimage = np.fft.fft2(image)
ftimage = np.fft.fftshift(ftimage)
plt.imshow(np.abs(ftimage))
plt.show()


# Build and apply a Gaussian filter.
sigmax, sigmay = 10, 10
```

---

[34] Note that there is an entire SciPy subpackage, `scipy.ndimage`, not described in this book, devoted to image processing. This example serves simply to illustrate the syntax and format of NumPy's two-dimensional FFT implementation.

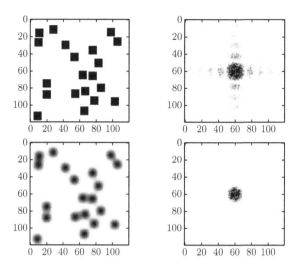

Figure 6.17 Blurring an image with a Gaussian filter applied to its two-dimensional Fourier transform.

```
cy, cx = nrows/2, ncols/2
x = np.linspace(0, nrows, nrows)
y = np.linspace(0, ncols, ncols)
X, Y = np.meshgrid(x, y)
gmask = np.exp(-(((X-cx)/sigmax)**2 + ((Y-cy)/sigmay)**2))

ftimagep = ftimage * gmask
plt.imshow(np.abs(ftimagep))
plt.show()

# Finally, take the inverse transform and show the blurred image.
imagep = np.fft.ifft2(ftimagep)
plt.imshow(np.abs(imagep))
plt.show()
```

The results are shown in Figure 6.17.

6.7.3   Exercises

Questions

Q6.7.1   Compare the speed of execution of NumPy's `np.fft.fft` algorithm and that of the direct implementation of Equation 6.1.

Hint: treat the direct equation as a matrix multiplication (dot product) of an array of n function values (random ones will do) with the $n \times n$ array with entries $\exp(-2\pi imk/n)$ $(m, k = 0, 1, \ldots, n-1)$. Use IPython's `%timeit` magic function.

Problems

P6.7.1   Consider a signal in the time domain defined by the function

$$f(t) = \cos(2\pi vt)e^{-t/\tau},$$

with frequency $v = 250$ Hz decaying exponentially with a lifetime $\tau = 0.2$ s. Plot the function, sampled at 1000 Hz, and its discrete Fourier transform against frequency. Examine, by means of a suitable plot, the effect of *apodization* on the DFT by truncating the time sequence after (a) 0.5 s, (b) 0.2 s.

P6.7.2   A square wave of period T may be defined through the following function

$$f_{sq}(t) = \begin{cases} 1 & t < T/2 \\ -1 & t \geq T/2 \end{cases},$$

with $f(t) = f(t + nT)$ for $n = \pm 1, \pm 2, \ldots$

Plot the square wave with $T = 1$ (and hence cycle frequency, $v = 1$) for $0 \leq t < 2$ taking a grid of 2048 time points over this interval. Calculate and plot its discrete Fourier transform.

The Fourier *expansion* of this function is the infinite series

$$f_{sq}(t) = \frac{4}{\pi} \sum_{k=1}^{\infty} \frac{1}{2k-1} \sin[2\pi(2k-1)vt].$$

Compare the square wave function with this Fourier expansion truncated at 3, 9 and 18 terms. Also compare their (suitably normalized) Fourier transforms: the missing frequencies in each truncated series should appear as zeros in its Fourier transform, whereas the present terms will have intensities $4/[\pi(2k-1)]$.

P6.7.3   The scipy library provides a routine for reading in .wav files as NumPy arrays:

```
In [x]: from scipy.io import wavfile
In [x]: sample_rate, wav = wavfile.read(<filename>)
```

For a stereo file, the array wav has shape (n, 2) where n is the number of samples.

Use the routines of np.fft to identify the chords present in the sound file chords.wav, which may be downloaded from https://scipython.com/ex/bfi. Which major chord do they comprise?

The frequencies of musical notes on an equal-tempered scale for which $A_4 = 440$ Hz are provided as a dictionary in the file notes.py.

7 Matplotlib

Matplotlib is probably the most popular Python package for plotting data. It can be used through the procedural interface `pyplot` in very quick scripts to produce simple visualizations of data (see Chapter 3) but, as described in this chapter, with care it can also produce high-quality figures for journal articles, books and other publications. Although there is some limited functionality for producing three-dimensional plots (see Section 7.6), it is primarily a two-dimensional plotting library.

7.1   Line Plots and Scatter Plots

Matplotlib is a large package organized in a hierarchy: at the highest level is the `matplotlib.pyplot` module. This provides a "state-machine environment" with a similar interface to MATLAB and allows the user to add plot elements (data points, lines, annotations, etc.) through simple function calls. This is the interface introduced in Chapter 3.

At a lower level, which allows more advanced and customizable use, Matplotlib has an object-oriented interface that allows one to create a *figure* object to which one or more *axes* objects are attached. Most plotting, annotation and customization then occurs through these axes objects. This is this approach adopted in this chapter.

To use Matplotlib in this way, we use the following recommended imports:

```
import matplotlib.pyplot as plt
import numpy as np
```

7.1.1   Plotting on a Single Axes Object

The top-level object, containing all the elements of a plot is called `Figure`. To create a figure object, call `plt.figure`. No arguments are necessary, but optional customization can be specified by setting the values described in Table 7.1. For example,

```
In [x]: # a default figure, with title "Figure 1"
In [x]: fig = plt.figure()

In [x]: # a small (4.5" x 2") figure with red background
In [x]: fig = plt.figure('Population density', figsize=(4.5, 2.),
   ....:                   facecolor='red')
```

Table 7.1 Arguments to `plt.figure`

Argument	Description
num	An identifier for the figure – if none is provided, an integer, starting at 1, is used and incremented with each figure created; alternatively, using a string will set the window title to that string when the figure is displayed with `plt.show()`
figsize	A tuple of figure (`width, height`), unfortunately in inches
dpi	Figure resolution in dots per inch
facecolor	Figure background color
edgecolor	Figure border color

To actually plot data, we need to create an `Axes` object – a region of the figure containing the axes, tick-marks, labels, plot lines and markers, and so on. The simplest figure, consisting of a single `Axes` object, is created and returned with

```
In [x]: ax = fig.add_subplot()
```

The `Axes` object, ax, is the one on which we can actually plot the data with `ax.plot`. The essential features of this `plot` method were described in Chapter 3. Here, however, note that the `plot` method actually returns a `list` of objects representing the plotted lines. In its simplest usage, only a single line is plotted, and so this list consists of one `Line2D` object that we may assign to a variable if desired. As a full example, consider the following comparison of the catenary $y = \cosh(x)$ and its parabolic approximation, $y = 1 + x^2/2$.

```
import matplotlib.pyplot as plt
import numpy as np

fig = plt.figure()
ax = fig.add_subplot()

x = np.linspace(-2, 2, 1000)
```
❶
```
line_cosh, = ax.plot(x, np.cosh(x))
line_quad, = ax.plot(x, 1 + x**2 / 2)

plt.show()
```

❶ Note the syntax `line_cosh, = ...` to assign the returned line object to the variable `line_cosh` rather than the `list` containing that object.

The two plotted lines are shown in Figure 7.1.

If no arguments need to be passed to `fig.add_subplot()`, the fig and ax objects can be created in a single line with the convenience function `plt.subplots()`:[1]

```
fig, ax = plt.subplots()
```

---

[1] This function can also be passed arguments nrows and ncols for a grid of multiple subplots, and subplot_kw, a dictionary of keyword arguments which will be passed to the add_subplot call for each subplot; additional keyword arguments are passed to the plt.figure() call.

Table 7.2 Matplotlib line
styles

	(No line)
-	Solid
--	Dashed
:	Dotted
-.	Dash-dot

7.1.2   Plot Limits

By default, Matplotlib plots all of the data passed to plot and sets the axis limits accordingly. To set the axis limits to something else, use the ax.set_xlim and ax.set_ylim methods. Either both limits can be set or an individual limit can be set with the arguments left, right (or xmin, xmax) and bottom, top (or ymin, ymax). Unspecified limits are left unchanged. For example,

```
x = np.linspace(-3, 3, 1000)
y = x**3 + 2 * x**2 - x - 1
fig = plt.figure()
ax = fig.add_subplot()
ax.plot(x, y)

ax.set_xlim(-1, 2)      # x-limits are -1 to 2
ax.set_ylim(bottom=0)   # ymin=0: plot will be "clipped" at the bottom
```

If bottom is greater than top or right less than left, the corresponding axis will be reversed; that is, values on this axis will *decrease* from left to right (or from bottom to top) (see Exercise P7.4.5).

   If you wish to invert the axis direction without changing the limit values, the method calls ax.invert_xaxis() and ax.invert_yaxis() will do that for you.

7.1.3   Line Styles, Markers and Colors

As we have seen previously, the plot style can be specified by passing extra arguments to the plot() method. The default line style is a solid, 1.5 pt weight line in a color determined by the order in which it is added to the plot.

   An alternative line style can be selected from the predefined options with the linestyle (or simply ls) argument. Possible string values to pass to this argument (including the empty string for plotting no line) are shown in Table 7.2.

   Further customization is possible by setting the dashes argument to a sequence of values describing the repeated dash pattern in points. For example, dashes=[2, 4, 8, 4, 2, 4]] represents a pattern of dot (2 pts), space (4 pts), dash (8 pts), space (4 pts), dot (2 pts), space (4 pts) to be repeated as the line style. Equivalently, one can call a plotted line's set_dashes method, as in the following code snippet:

```
x = np.linspace(-np.pi, np.pi, 1000)
line, = plt.plot(x, np.sin(x))
line.set_dashes([2, 4, 8, 4, 2, 4])      # dot-dash-dot
```

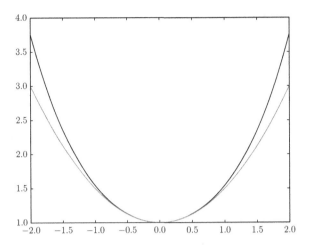

Figure 7.1 A simple plot of two lines on a single **Axes** object.

Table 7.3 Matplotlib color code
letters

Basic color codes	Tableau colors
b = blue	`tab:blue`
g = green	`tab:orange`
r = red	`tab:green`
c = cyan	`tab:red`
m = magenta	`tab:purple`
y = yellow	`tab:brown`
k = black	`tab:pink`
w = white	`tab:gray`
	`tab:olive`
	`tab:cyan`

The line weight is customized by setting the `lineweight` (or simply `lw`) argument to a number of points.

Line colors are specified with the `color` (or simply `c`) argument used in one of several ways:

- *string*: by letter or name, one of the values given in Table 7.3; single-letter colors are rather bright and the (default) "tableau" colors are more pleasing;
- *string*: by HTML six-digit hex-string preceded by '`#`', for example '`#ffff00`' is yellow;
- *string*: a string representation of a `float` between 0. and 1. (for example '`0.4`') gives a gray-scale between black (0.) and white(1.);
- *tuple of floats* between 0. and 1.: RGB components, for example (`0.5, 0., 0.`) is a dark red color.

Table 7.4 Some Matplotlib marker
styles (single-character string codes)

Code	Marker	Description
.	·	Point
o	○	Circle
+	+	Plus
x	×	Cross
D	◇	Diamond
v	▽	Downward triangle
^	△	Upward triangle
s	☐	Square
*	★	Star

Table 7.5 Matplotlib marker properties

Argument	Abbreviation	Description
markersize	ms	Marker size, in points
markevery		Set to a positive integer, N, to print a marker every N points; the default, None, prints a marker for every point
markerfacecolor	mfc	Fill color of the marker
markeredgecolor	mec	Edge color of the marker
markeredgewidth	mew	Edge width of the marker, in points

By default, the Line2D object created by calling plot on an Axes object does not include *markers*: symbols printed at each point on the plot. To add them, specify one of the single-character marker codes given in Table 7.4 using the marker argument

```
ax.plot(x, y, marker='v')   # downward pointing triangles
```

Other marker properties can be set with the arguments listed in Table 7.5.

Matplotlib markers can be further customized; see the documentation for details.[2]

7.1.4   Scatter Plots

A typical two-dimensional scatter plot depicts the data as points on a Cartesian axes system. Sometimes there is no meaningful or helpful ordering to the data and so no need to join data points by lines. The pyplot.scatter function creates a scatter plot. In addition to one-dimensional sequences of *x*- and *y*-data, as for pyplot.plot, the data point marker colors and sizes can be set individually by passing a sequence of appropriate values of the same length as the data to the arguments s and c, respectively. The marker sizes are in points[2] (*points squared*) so that their *area* is proportional to the

---

[2] https://matplotlib.org/api/markers_api.html.

values passed to s. Manipulating the size of the markers is a common way of indicating a third dimension to the data, as in the following example.

Example E7.1 To explore the correlation between birth rate, life expectancy and per capita income, we may use a scatter plot. The marker sizes are set in proportion to the countries' per-capita gross domestic product (GDP) but have to be scaled a little so they don't get too large (see Figure 7.2).

Listing 7.1 Scatter plot of demographic data for eight countries

```python
# eg7-scatter.py

import numpy as np
import matplotlib.pyplot as plt

countries = ['Brazil', 'Madagascar', 'S. Korea', 'United States',
             'Ethiopia', 'Pakistan', 'China', 'Belize']
# Birth rate per 1000 population.
birth_rate = [16.4, 33.5, 9.5, 14.2, 38.6, 30.2, 13.5, 23.0]
# Life expectancy at birth, years.
life_expectancy = [73.7, 64.3, 81.3, 78.8, 63.0, 66.4, 75.2, 73.7]
# Per person income fixed to US Dollars in 2000.
GDP = np.array([4800, 240, 16700, 37700, 230, 670, 2640, 3490])

fig, ax = plt.subplots()

# Some arbitrary colors:
colors = range(len(countries))
ax.scatter(birth_rate, life_expectancy, c=colors, s=GDP/20)
ax.set_xlim(5, 45)
ax.set_ylim(60, 85)
ax.set_xlabel('Birth rate per 1000 population')
ax.set_ylabel('Life expectancy at birth (years)')

plt.show()
```

7.2 Plot Customization and Refinement

7.2.1 Gridlines

Gridlines are vertical (for the x-axis) and horizontal (for the y-axis) lines running across the plot to aid with locating the numerical values of data points. By default, no gridlines are drawn, but they may be turned on by calling the grid method on an Axes object (to add both horizontal and vertical gridlines) or the xaxis or yaxis objects of a given Axes (to select the gridlines to use). For example,

```python
ax.yaxis.grid(True)  # turn on horizontal gridlines
```

or

```python
ax.grid(True)        # turn on all gridlines
```

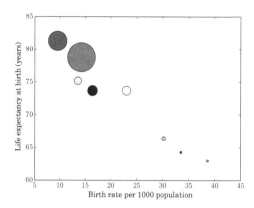

Figure 7.2 A scatter plot with variable marker sizes indicating each country's GDP.

The line properties of the gridlines are set with the `linestyle`, `linewidth`, `color`, etc. arguments as for plot lines.

Two sorts of gridlines correspond to the major and minor tick marks (see below): these can be selected with the `which` argument, which takes on of the values `'major'`, `'minor'` or `'both'`. The default (if not specified) is `which='major'`.

```
ax.xaxis.grid(True, which='minor', c='b')    # minor x-axis gridlines in blue
```

7.2.2   Log Scales

By default, Matplotlib plots data on a linear scale. To set a logarithmic scale, call one or both of the following on your `Axes` object:

```
ax.set_xscale('log')
ax.set_yscale('log')
```

Base-10 logarithms are used by default, but the (integer) base can be set with the optional arguments `basex` or `basey`. Nonpositive values in the data will be masked as invalid by default. If you want negative values to be handled "symmetrically" with positive ones, such that $\log(-|x|) = -\log(|x|)$, then use `'symlog'` instead of `'log'`. See also Question Q7.4.1.

7.2.3   Adding Titles, Labels and Legends

Axis labels may be added to the subplot `Axes` object with `ax.set_xlabel` and `ax.set_ylabel`.

Plot-line legend labels are defined by adding the `label` attribute to the `plt.plot` function call. However, the legend itself will not appear unless `legend` is called on the plot `Axes` object (e.g. with `ax.legend()`). The appearance of the legend itself can be customized extensively, but the most common additional argument you may wish to pass to `legend()` is `loc`, defining the location of the legend on the plot (see Table 3.1).

There are two types of title you may want to give your figure: `fig.suptitle` adds a centered title to the entire figure, which may contain more than one subplot; `ax.title` adds a title to a single subplot.[3]

Example E7.2   The data read in from the file `eg7-marriage-ages.txt`, which can be downloaded from https://scipython.com/eg/bag, giving the median age at first marriage in the United States for 13 decades since 1890, are plotted by the program below. Gridlines are turned on for both axes with `ax.grid()`, and custom markers are used for the data points themselves (see Figure 7.3).

Listing 7.2 The median age at first marriage in the US over time

```python
# eg7-marriage-ages.py
import numpy as np
import matplotlib.pyplot as plt

year, age_m, age_f = np.loadtxt('eg7-marriage-ages.txt', unpack=True, skiprows=3)
fig, ax = plt.subplots()

# Plot ages with male or female symbols as markers.
ax.plot(year, age_m, marker='$\u2642$', markersize=14, c='blue', lw=2,
        mfc='blue', mec='blue')
ax.plot(year, age_f, marker='$\u2640$', markersize=14, c='magenta', lw=2,
        mfc='magenta', mec='magenta')
ax.grid()

ax.set_xlabel('Year')
ax.set_ylabel('Age')
ax.set_title('Median age at first marriage in the USA, 1890-2010')

plt.show()
```

Example E7.3   The historical populations of five US cities are given in the files `boston.tsv`, `houston.tsv`, `detroit.tsv`, `san_jose.tsv`, `phoenix.tsv` as tab-separated columns of (year, population). They can be downloaded from https://scipython.com/eg/baf.

The following program plots these data on one set of axes with a different line style for each.

Listing 7.3 The populations of five US cities over time

```python
# eg7-populations.py
import matplotlib.pyplot as plt
import numpy as np

fig, ax = plt.subplots()
```

---

[3] See the documentation at https://matplotlib.org/api/legend_api.html for more details.

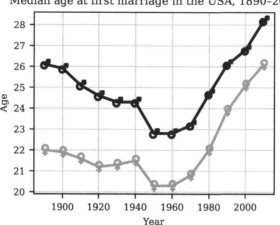

Figure 7.3 Median age at first marriage in the USA, 1890–2010.

```
cities = ['Boston', 'Houston', 'Detroit', 'San Jose', 'Phoenix']
# Line styles: solid, dashes, dots, dash-dots, and dot-dot-dash.
linestyles = [{'ls': '-'}, {'ls': '--'}, {'ls': ':'}, {'ls': '-.'},
              {'dashes': [2, 4, 2, 4, 8, 4]}]

for i, city in enumerate(cities):
    filename = '{}.tsv'.format(city.lower()).replace(' ', '_')
    yr, pop = np.loadtxt(filename, unpack=True)
    line, = ax.plot(yr, pop/1.e6, label=city, c='k', **linestyles[i])
ax.legend(loc='upper left')
ax.set_xlim(1800, 2020)
ax.set_xlabel('Year')
ax.set_ylabel('Population (millions)')
plt.show()
```

❶ Note how the city name is used to deduce the corresponding filename.
The plot produced is shown in Figure 7.4.

7.2.4 Font Properties

The text elements of a plot (titles, legend, axis labels, etc.) can be customized with the arguments given in Table 7.6. For example,

```
ax.title('Plot Title', fontsize=18, fontname='Times New Roman', color='blue')
```

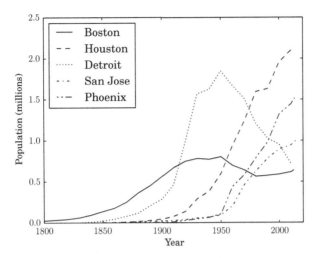

Figure 7.4 Population trends for five US cities.

Table 7.6 Font property arguments for text elements of a plot

Argument	Description
fontsize	The size of the font in points (e.g. 12, 16)
fontname	The font name (e.g. 'Courier', 'Arial')
family	The font family (e.g. 'sans-serif', 'cursive', 'monospace')
fontweight	The font weight (e.g. 'normal', 'bold')
fontstyle	The font style (e.g. 'normal', 'italic')
color	Any Matplotlib color specifier (e.g. 'r', '#ff00ff')

To use the same font properties for all text elements, it is easiest to set Matplotlib's rc settings using a dictionary of values. This involves a separate import first:[4]

```
from matplotlib import rc
font_properties = {'family' : 'monospace',
                   'weight' : 'bold',
                   'size'   : 22}
```
❶ ```
rc('font', **font_properties)
# All text will now be rendered in 22-point, bold monospace in plots.
```

❶ Recall that the syntax **kwargs passes the (key, value) pairs of dictionary kwargs and passes them to a function as keyword arguments (see Section 4.2.2).

---

[4] It is also possible to edit Matplotlib's configuration file, matplotlibrc, to set many kinds of plot preferences and styles: see https://matplotlib.org/users/customizing.html.
```

7.2.5   Tick Marks

Matplotlib does its best to label representative values (*tick marks*) on each axis appropriately, but there are some occasions when you want to customize them, for example, to make the tick marks more or less frequent, or to label them differently.

Most commonly, one simply wants to set the tick mark values to a given sequence of values: this is accomplished by calling `ax.set_xticks` and `ax.set_yticks` on the `Axes` object of the plot. For example,

```
ax.set_xticks([0, 1, 3.5, 6.5, 15])
```

Note that the ticks do not have to be evenly spaced.

To replace the actual numbered labels, pass a sequence of strings of a suitable length to `ax.set_xticklabels` and `ax.set_yticklabels`, as in the following example.[5]

Example E7.4   The following program plots the exponential decay described by $y = Ne^{-t/\tau}$, labeled by lifetimes ($n\tau$ for $n = 0, 1, \ldots$), such that after each lifetime the value of y falls by a factor of e. The plot is given as Figure 7.5.

Listing 7.4 Exponential decay illustrated in terms of lifetimes

```python
# eg7-ticks-exp-decay.py
import numpy as np
import matplotlib.pyplot as plt

# Initial value of y at t=0, lifetime (s).
N, tau = 10000, 28
# Maximum time to consider (s).
tmax = 100
# A suitable grid of time points, and the exponential decay itself.
t = np.linspace(0, tmax, 1000)
y = N * np.exp(-t/tau)

fig = plt.figure()
ax = fig.add_subplot()
ax.plot(t, y)

# The number of lifetimes that fall within the plotted time interval.
ntau = tmax // tau + 1
# xticks at 0, tau, 2*tau, ..., ntau*tau; yticks at the corresponding y-values.
xticks = [i*tau for i in range(ntau)]
yticks = [N * np.exp(-i) for i in range(ntau)]
ax.set_xticks(xticks)
ax.set_yticks(yticks)

# xtick labels: 0, tau, 2tau, ...
❶ xtick_labels = [r'$0$', r'$\tau$'] + [r'${}\tau$'.format(k) for k in range(2, ntau)]
ax.set_xticklabels(xtick_labels)
# Corresponding ytick labels: N, N/e, N/2e, ...
```

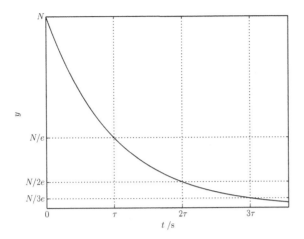

Figure 7.5 An exponential decay with customized tick labels.

❷
```
ytick_labels = [r'$N$', r'$N/e$'] + [r'$N/{}e$'.format(k) for k in range(2, ntau)]
ax.set_yticklabels(ytick_labels)

ax.set_xlabel(r'$t\;/\mathrm{s}$')
ax.set_ylabel(r'$y$')
ax.grid()
plt.show()
```

❶ The x-axis tick labels are $0, \tau, 2\tau, \ldots$
❷ The y-axis tick labels are $N, N/e, N/2e, \ldots$

Note that the length of the sequence of tick labels must correspond to that of the list of tick values required.

To remove the tick labels altogether set them to the empty list, for example

```
ax.set_yticklabels([])
```

This retains the tick marks themselves. If you want neither tick marks nor tick labels on the axis use:

```
ax.set_yticks([])
```

There are two kinds of ticks: major ticks and minor ticks. Only major ticks are turned on by default; the smaller and more frequent minor ticks can most easily be enabled with

```
ax.minorticks_on()
```

More advanced customization of tick marks and their labels, including showing minor tick marks for one axis only, can be achieved using the `ax.tick_params` convenience function, which takes the arguments described in Table 7.7.

Finally, `ax.xaxis` and `ax.yaxis` have a method, `set_ticks_position`, which takes a single argument used to determine where the ticks appear: for `ax.xaxis`, 'top',

Table 7.7 Common arguments to `ax.tick_params`

Argument	Description
`axis`	Which axis to customize: `'x'`, `'y'` or `'both'`; default is `'both'`.
`which`	Which tick mark set to customize: `'major'`, `'minor'` or `'both'`; default is `'major'`
`direction`	Tick mark direction: `'in'`, `'out'` or `'inout'`; default is `'in'`
`length`	Length of the tick marks in points
`width`	Width of the tick marks in points
`pad`	Distance between the tick mark and its label in points
`labelsize`	Tick label size in points
`color`	Tick mark color (a Matplotlib specifier)
`labelcolor`	Tick mark label color (a Matplotlib specifier)

`'bottom'`, `'both'` (the default) or `'none'`; for `ax.yaxis`, `'left'`, `'right'`, `'both'` (the default) or `'none'`.

Example E7.5   The following program creates a plot with both major and minor tick marks, customized to be thicker and wider than the default, where the major tick marks point into and out of the plot area.

Listing 7.5 Customized tick marks

```
# eg7-tick-customization.py

import numpy as np
import matplotlib.pyplot as plt

# A selection of functions on rn abcissa points for 0 <= x < 1.
rn = 100
rx = np.linspace(0, 1, rn, endpoint=False)

def tophat(rx):
    """ Top hat function: y = 1 for x < 0.5, y = 0 for x >= 0.5 """
    ry = np.ones(rn)
    ry[rx>=0.5] = 0
    return ry

# A dictionary of functions to choose from.
ry = {'half-sawtooth': lambda rx: rx.copy(),
      'top-hat': tophat,
      'sawtooth': lambda rx: 2 * np.abs(rx-0.5)}

# Repeat the chosen function nrep times.
nrep = 4
x = np.linspace(0, nrep, nrep*rn, endpoint=False)
❶ y = np.tile(ry['top-hat'](rx), nrep)

fig, ax = plt.subplots()
ax.plot(x, y, 'k', lw=2)
```

```
# Add a bit of padding around the plotted line to aid visualization.
ax.set_ylim(-0.1, 1.1)
ax.set_xlim(x[0]-0.5, x[-1]+0.5)
# Customize the tick marks and turn the grid on.
ax.minorticks_on()
ax.tick_params(which='major', length=10, width=2, direction='inout')
ax.tick_params(which='minor', length=5, width=2, direction='in')
ax.grid(which='both')
plt.show()
```

❶ This np.tile method constructs an array by repeating a given array nrep times. To plot a different periodic function, choose 'half-sawtooth' or 'sawtooth' here.

The resulting plot is shown in Figure 7.6.

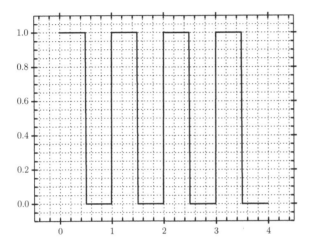

Figure 7.6 A periodic function plotted on a graph with gridlines and customized tick marks.

7.2.6   Error Bars

To produce a plotted line with error bars, use the method ax.errorbar instead of ax.plot. In addition to the usual arguments of the plot function, errorbar allows the specification of errors in the *x*- and *y*-coordinates by passing the following types of value to the arguments xerr and yerr:

- None: no error bars for this coordinate;
- a scalar (e.g. xerr=0.2): all values are associated with symmetric error bars at plus and minus this value (i.e. ±0.2);
- an array-type object of length n or shape (n, 1) (e.g. yerr=[0.1, 0.15, 0.1]): the symmetric error bars are plotted at plus and minus the values in this sequence for each of the n data points (i.e. ±0.1, ±0.15, ±0.1);
- an array-type object of shape (2, n) (i.e. two rows for each of n data points): error bars, which may be asymmetric, are plotted using minus-values from the first row and plus-values from the second.

Table 7.8 Common arguments to `ax.errorbar`

Argument	Description
x, y	The data to plot
yerr, xerr	Errors on the x and y data coordinates, as described in the text
fmt	The plot format symbol (marker for the data point); set to None or the empty string, '', to display only the error bars
ecolor	A Matplotlib color specifier for the error bars; the default, None, uses the same color as the connecting line between data markers
elinewidth	The width of the error bar lines in points; use None to use the same line width as the plotted data
capsize	The length of the error bar caps, in points; by default, None: no caps to the error bars
errorevery	A positive integer giving the subsampling for the error bars; for example, errorevery=10 draws error bars on every 10th data point only

The appearance of the error bars may be customized using the arguments summarized in Table 7.8. For example,

```
# Some data:
x = np.array([ 0.3,   0.5,   0.7,   0.9])
y = np.array([ 1. ,   2. ,   2.5,   3.9])
# Constant, symmetric errors of +/- 0.05 on x-data.
xerr = 0.05
# Asymmetric, variable errors on y-data.
yerr = np.array([[ 0.1 ,   0.25,   0.5 ,   0.4 ],
                 [ 0.1 ,   0.15,   0.2 ,   0.  ]])
ax.errorbar(x, y, yerr, xerr, fmt='o', ls='')
```

Example E7.6 Before fledging, some species of birds lose weight relative to the surface area of their wings to maximize their aerodynamic efficiency. The file `fledging-data.csv`, available at https://scipython.com/eg/bad gives wing-loading values (body mass per wing area) as averages for two broods of swifts in the two weeks prior to fledging, with their uncertainties.[6]

In the program below, we perform a weighted fit to the data and plot it, with error bars.

Listing 7.6 Wing-loading variation in swifts prior to fledging

```
# eg7-fledging.py
import numpy as np
import matplotlib.pyplot as plt

# Read in the data: day before fledging, wing loading and error for two broods.
dt = np.dtype([('day', 'i2'), ('wl1', 'f8'), ('wl1-err', 'f8'),
               ('wl2', 'f8'), ('wl2-err', 'f8')])
data = np.loadtxt('fledging-data.csv', dtype=dt, delimiter=',')
```

---

[6] J. Wright et al., *Proc. R. Soc. B* **273**, 1895 (2006).

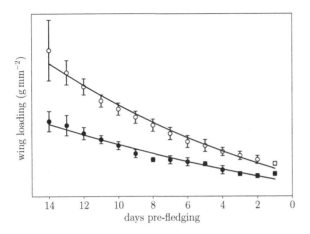

Figure 7.7 Fitted time series for wing-loading values in two cohorts of swift nestlings.

```
# Weighted fit of exponential decay to the data. This is a linear least-squares
# problem because y = Aexp(-Bx) => ln y = ln A - Bx = mx + c.
p1_fit = np.poly1d(np.polyfit(data['day'], np.log(data['wl1']), 1,
                   w=np.log(data['wl1'])**-2))
p2_fit = np.poly1d(np.polyfit(data['day'], np.log(data['wl2']), 1,
                   w=np.log(data['wl2'])**-2))
wl1fit = np.exp(p1_fit(data['day']))
wl2fit = np.exp(p2_fit(data['day']))

# Plot the data points with their uncertainties and the fits.
fig, ax = plt.subplots()

# wl1 data: white circles, black borders, with error bars.
ax.errorbar(data['day'], data['wl1'], yerr=data['wl1-err'], ls='', marker='o',
            color='k', mfc='w', mec='k', capsize=3)
ax.plot(data['day'], wl1fit, 'k', lw=1.5)

# wl2 data: black filled circles, with error bars.
ax.errorbar(data['day'], data['wl2'], yerr=data['wl2-err'], ls='', marker='o',
            color='k', mfc='k', mec='k'), capsize=3)
ax.plot(data['day'], wl2fit, 'k', lw=1.5)

ax.set_xlim(15, 0)
ax.set_ylim(0.003, 0.012)
ax.set_xlabel('days pre-fledging')
ax.set_ylabel('wing loading ($\mathrm{g\,mm^{-2}}$)')
plt.show()
```

❶ The data points are weighted in the fit by $1/\sigma^2$ where σ is the estimated one-standard deviation error of the measurement.

Figure 7.7 shows the results of the fit. The broods, initially with different average wing-loading values, are seen to converge prior to fledging.

7.2.7   Multiple Subplots

To create a figure with more than one subplot (that is, Axes), call add_subplot on your Figure object, setting its argument to indicate where the subplot should be placed. Each call returns an Axes object. Single figures with more than 10 subplots are uncommon, so the usual argument is a three-digit number where each digit indicates the number of rows, number of columns and subplot number. The subplot number increases along the columns in each row and then down the rows. For example, a figure consisting of three rows of two columns of subplots can be constructed by adding Axes objects:

```
In [x]: fig = plt.figure()
In [x]: ax1 = fig.add_subplot(321)   # top left subplot
In [x]: ax2 = fig.add_subplot(322)   # top right subplot
In [x]: ax3 = fig.add_subplot(323)   # middle left subplot
...
In [x]: ax6 = fig.add_subplot(326)   # bottom right subplot
```

Alternatively, to create a figure and add all its subplots to it at the same time, call plt.subplots, which takes arguments nrows and ncols (in addition to those listed in Table 7.1) and returns a Figure and an array of Axes objects, which can be indexed for each individual axis:

```
In [x]: fig, axes = plt.subplots(nrows=3, ncols=2)
In [x]: axes.shape
Out[x]: (3, 2)

In [x]: ax1 = axes[0, 0]      # top left subplot
In [x]: ax2 = axes[2, 1]      # bottom right subplot
```

In fact, a useful idiom to create a plot with a single Axes object is to call subplots() with no arguments:

```
In [x]: fig, ax = plt.subplots()
In [x]: ax.plot(x, y)         # no need to index the single Axes object created
```

Figures with subplots run the risk of their labels, titles and ticks overlapping each other – if this happens, call the method tight_layout on the Figure object and Matplotlib will do its best to arrange them so that there is sufficient space between them.

Example E7.7   Consider a metal bar of cross-sectional area, A, initially at a uniform temperature, θ_0, which is heated instantaneously at the exact center by the addition of an amount of energy, H. The subsequent temperature of the bar (relative to θ_0) as a function of time, t, and position, x, is governed by the one-dimensional diffusion equation:

$$\theta(x, t) = \frac{H}{c_p A} \frac{1}{\sqrt{Dt}} \frac{1}{\sqrt{4\pi}} \exp\left(-\frac{x^2}{4Dt}\right),$$

where c_p and D are the metal's specific heat capacity per unit volume and thermal diffusivity (which we assume are constant with temperature). The following code plots $\theta(x, t)$ for three specific times and compares the plots between two metals, with different thermal diffusivities but similar heat capacities, copper and iron.

Listing 7.7 The one-dimensional diffusion equation applied to the temperature of two different metal bars

```python
# eg7-diffusion1d.py
import numpy as np
import matplotlib.pyplot as plt

# Cross-sectional area of bar in m3, heat added at x = 0 in J.
A, H = 1.e-4, 1.e3
# Temperature in K at t = 0.
theta0 = 300

# Metal element symbol, specific heat capacities per unit volume (J.m-3.K-1),
# Thermal diffusivities (m2.s-1) for Cu and Fe.
metals = np.array([('Cu', 3.45e7, 1.11e-4), ('Fe', 3.50e7, 2.3e-5)],
                   dtype=[('symbol', '|S2'), ('cp', 'f8'), ('D', 'f8')])

# The metal bar extends from -xlim to xlim (m).
xlim, nx = 0.05, 1000
x = np.linspace(-xlim, xlim, nx)

# Calculate the temperature distribution at these three times.
times = (1e-2, 0.1, 1)
# Create our subplots: three rows of times, one column for each metal.
fig, axes = plt.subplots(nrows=3, ncols=2, figsize=(7, 8))
for j, t in enumerate(times):
    for i, metal in enumerate(metals):
        symbol, cp, D = metal
        ax = axes[j, i]
        # The solution to the diffusion equation.
        theta = theta0 + H/cp/A/np.sqrt(D*t * 4*np.pi) * np.exp(-x**2/4/D/t)
        # Plot, converting distances to cm and add some labeling.
        ax.plot(x*100, theta, 'k')
        ax.set_title('{}, $t={}$ s'.format(symbol.decode('utf8'), t))
        ax.set_xlim(-4, 4)
        ax.set_xlabel('$x\;/\mathrm{cm}$')
        ax.set_ylabel('$\Theta\;/\mathrm{K}$')

# Set up the y-axis so that each metal has the same scale at the same t.
for j in (0, 1, 2):
    ymax = max(axes[j, 0].get_ylim()[1], axes[j, 1].get_ylim()[1])
    print(axes[j, 0].get_ylim(), axes[j, 1].get_ylim())
    for i in (0, 1):
        ax = axes[j, i]
        ax.set_ylim(theta0, ymax)
        # Ensure there are only three y-tick marks.
        ax.set_yticks([theta0, (ymax + theta0)/2, ymax])
# We don't want the subplots to bash into each other: tight_layout() fixes this.
fig.tight_layout()
plt.show()
```

Because copper is a better conductor, the temperature increase is seen to spread more rapidly for this metal (see Figure 7.8).

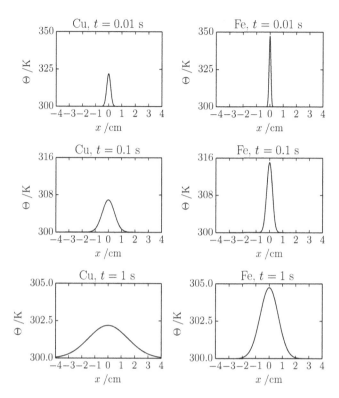

Figure 7.8 Numerical solutions of the one-dimensional diffusion equation for the temperatures of two metal bars.

To further customize the subplot spacing, call `fig.subplots_adjust()`. This method takes any of the keywords `left`, `bottom`, `right`, `top`, `wspace` and `hspace`, which can be set to fractional values of the figure's height and width as appropriate, to determine the positions of the subplots' left side (default 0.125), right side (0.9), bottom (0.1), top (0.9), vertical spacing (0.2) and horizontal spacing (0.2). A practical use of this function is to create "ganged" subplots that share a common axis, as in the following example.

Example E7.8   This code generates a figure of 10 subplots depicting the graph of $\sin(n\pi x)$ for $n = 0, 1, \ldots, 9$. The subplot spacing is configured so that they "run into" each other vertically (see Figure 7.9).

Listing 7.8 Ten subplots with zero vertical spacing

```python
import numpy as np
import matplotlib.pyplot as plt

nrows = 10
fig, axes = plt.subplots(nrows, 1)
# Zero vertical space between subplots.
fig.subplots_adjust(hspace=0)

x = np.linspace(0, 1, 1000)
```

```
for i in range(nrows):
    # n = nrows for the top subplot, n = 0 for the bottom subplot.
    n = nrows - i
    axes[i].plot(x, np.sin(n * np.pi * x), 'k', lw=2)
    # We only want ticks on the bottom of each subplot.
    axes[i].xaxis.set_ticks_position('bottom')
    if i < nrows-1:
        # Set ticks at the nodes (zeros) of our sine functions.
        axes[i].set_xticks(np.arange(0, 1, 1/n))
        # We only want labels on the bottom subplot x-axis.
        axes[i].set_xticklabels('')
    axes[i].set_yticklabels('')
plt.show()
```

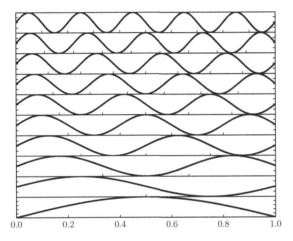

Figure 7.9 Ten subplots of $\sin(n\pi x)$ for $n = 0, 1, \ldots 9$ adjusted to remove vertical space between them.

7.2.8   Saving Figures

Saving Figures for Printing

As mentioned in Section 7.1.1, the size of a Matplotlib figure can be controlled (in inches) with the `figsize` argument to `plt.figure()`. For high-quality line-figures (as opposed to images, heatmaps, and so on), saving to a vector file format will produce the best printed appearance which can be rescaled without a loss of quality. Encapsulated PostScript (EPS) or PDF are good, general-purpose choices:

```
# A figure, 6" wide and 5" high.
fig = plt.figure(figsize=(6, 4))
ax = fig.add_subplot()

# Draw the figure ...

plt.savefig('my-figure.eps')
```

For figures which cannot be effectively vectorized (for example, images), a raster format is necessary. In this case, to control the resolution of the printed version there is an additional argument, dpi (*dots per inch*), which must be specified on both `plt.figure()` and `plt.savefig()`. Suitable formats include JPG and PNG. For example, a publication-quality figure for a journal article might be created with

```
DPI = 300   # a minimum resolution of 300 dpi is generally sufficient
fig = plt.figure(figsize=(3.5, 3), dpi=DPI)
ax = fig.add_subplot()

# Draw the figure ...

plt.savefig('my-figure.png', dpi=DPI)
```

Saving Figures for Online Use

The first thing to note is that different monitors have different resolutions (pixel densities), so there is no way to fix the physical dimensions of the image as it appears on a user's screen. However, it is possible to produce an image file of a figure with specified dimensions as the number of pixels (which is often all that is required for web pages). There is a hoop to jump through: since `figsize` expects its dimensions in inches, we pick a reasonable DPI and calculate these dimensions from the desired width and height in pixels:

```
DPI = 100   # this is a reasonable choice
WIDTH, HEIGHT = 800, 800
fig = plt.figure(figsize=(WIDTH / DPI, HEIGHT / DPI), dpi=DPI)

# Draw the figure ...

plt.savefig('my-figure.png', dpi=DPI)   # important: respecify the DPI here
```

7.3   Bar Charts, Pie Charts and Polar Plots

7.3.1   Bar Charts and Histograms

The basic `pyplot` function for plotting a bar chart is `ax.bar`, which makes a plot of rectangular bars defined by their *left* edges and height. For example,

```
ax.bar([0, 1, 2], [40, 80, 20])
```

The width of the rectangles is, by default, 0.8 but can be set with the (third) argument, `width`. If you want the bars vertically centered, either set the argument `align` to `'center'` or calculate where their left edges should be:

```
w = 0.5
x, y = np.array([0, 1, 2]), np.array([40, 80, 20])
ax.bar(x, y, w, align='center') # easiest way of centering the bars
ax.bar(x - w/2, y, w)           # or calculate the left edges
```

Additional arguments, including the provision of error bars, are given in Table 7.9.

Table 7.9 Common arguments to `ax.bar` and `barh`

Argument	Description
left	A sequence of *x*-coordinates of the left edges of the bars (but see `align`)
width	Width of the bars; if a scalar, all bars have the same width – can be array-like for variable widths
bottom	The *y*-coordinates of the bottom of the bars
height	A sequence of heights for the bars
color	Colors of the bar faces (scalar or array-like)
edgecolor	Colors of the bar edges (scalar or array-like)
linewidth	Line widths of the bar edges, in points (scalar or array-like)
xerr, yerr	Error bar limits, as for `errorbar` (scalar or array-like)
error_kw	A dictionary of keyword arguments corresponding to customization of the appearance of the errorbars (see Table 7.8)
align	The default, `'edge'`, aligns the bars by their left edges (for vertical bars) or bottom edges (for horizontal bars); `'center'` centers the bars on this axis instead
log	Set to `True` to use a logarithmic axis scale
orientation	`'vertical'` (the default) or `'horizontal'`
hatch	Set the hatching pattern for the bars: one of `'/'`, `'\'`, `'|'`, `'-'`, `'+'`, `'x'`, `'o'`, `'O'`, `'.'`, `'*'`; repeat the character for a denser pattern

By default, `ax.bar` produces a vertical bar chart. Horizontal bar charts are catered for either by setting `orientation='horizontal'` or by using the analogous `ax.barh` method.

Example E7.9 The following program produces a bar chart of letter frequencies in the English language, estimated by analysis of the text of *Moby-Dick*.[7] The vertical bars are centered and labeled by letter (Figure 7.10).

Listing 7.9 Letter frequencies in the text of *Moby-Dick*.

```python
# eg7-charfreq.py
import numpy as np
import matplotlib.pyplot as plt

text_file = 'moby-dick.txt'

letters = 'ABCDEFGHIJKLMNOPQRSTUVWXYZ'
# Initialize the dictionary of letter counts: {'A': 0, 'B': 0, ...}.
lcount = dict([(letter, 0) for letter in letters])

# Read in the text and count the letter occurrences.
for letter in open(text_file).read():
    try:
        lcount[letter.upper()] += 1
    except KeyError:
        # Ignore characters that are not letters.
```

---

[7] See, for example, www.gutenberg.org/ebooks/2701 for a free text file of this novel.

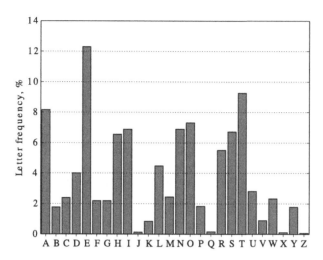

Figure 7.10 Letter frequencies in the novel *Moby-Dick*.

```
        pass
# The total number of letters.
norm = sum(lcount.values())

fig, ax = plt.subplots()
# The bar chart, with letters along the horizontal axis and the calculated
# letter frequencies as percentages as the bar height.
x = range(26)
ax.bar(x, [lcount[letter]/norm * 100 for letter in letters], width=0.8,
       color='g', alpha=0.5, align='center')
ax.set_xticks(x)
ax.set_xticklabels(letters)
ax.tick_params(axis='x', direction='out')
ax.set_xlim(-0.5, 25.5)
ax.yaxis.grid(True)
ax.set_ylabel('Letter frequency, %')
plt.show()
```

For monochrome plots, it is sometimes preferable to distinguish bars by patterns. The hatch argument can be used to do this, using any of several predefined patterns (see Table 7.9) as illustrated in the example below.

Example E7.10   The file germany-energy-sources.txt, which can be downloaded from https://scipython.com/eg/bae contains data on the renewable sources of electricity produced in Germany from 1990 to 2018:

```
Renewable electricity generation in Germany in GWh (million kWh)
Year      Hydro     Wind     Biomass   Photovoltaics
2018      17974     109951    50851        45784
2017      20150     105693    50917        39401
2016      20546      79924    50928        38098
```

. . .

The program below plots these data as a *stacked bar chart*, using Matplotlib's hatch patterns to distinguish between the different sources (Figure 7.11).

Listing 7.10 Visualizing renewable electricity generation in Germany

```
# eg7-germany-alt-energy.py
import numpy as np
import matplotlib.pyplot as plt

data = np.loadtxt('germany-energy-sources.txt', skiprows=2, dtype='f8')
years = data[:, 0]
n = len(years)

# GWh to TWh.
data[:, 1:] /= 1000

fig, ax = plt.subplots()
sources = ('Hydroelectric', 'Wind', 'Biomass', 'Photovoltaics')
hatch = ['oo', '', 'xxxx', '//']
bottom = np.zeros(n)
bars = [None]*n
for i, source in enumerate(sources):
    bars[i] = ax.bar(years, bottom=bottom, height=data[:, i+1], color='w',
                     hatch=hatch[i], align='center', edgecolor='k')
    bottom += data[:, i+1]

ax.set_xticks(years[::2])    # for clarity, label every other year
plt.xticks(rotation=90)
ax.set_xlim(1989, 2019)
ax.set_ylabel('Renewable Electricity (TWh)')
ax.set_title('Renewable Electricity Generation in Germany, 1990-2018')
plt.legend(bars, sources, loc='best')
plt.show()
```

❶ To include a legend, each bar chart object[8] must be stored in a list, bars, which ❷ is passed to the ax.legend method with a corresponding sequence of labels, sources.

7.3.2   Pie Charts

It is straightforward to draw a pie chart in Matplotlib by passing an array of values to ax.pie. The values will be normalized by their sum if this sum is greater than 1, or otherwise treated directly as fractions. Labels, percentages, "exploded" segments and other effects are handled as described in Table 7.10 and illustrated in the following example.

---

[8] Actually a Container of *artists*.

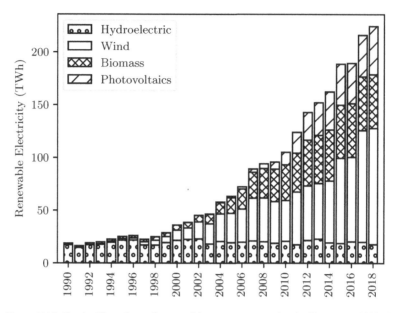

Figure 7.11 Stacked bar chart of renewable energy generation in Germany, 1990–2018.

Table 7.10 Common arguments to `ax.pie`

Argument	Description
`colors`	A sequence of Matplotlib color specifiers for coloring the segments
`labels`	A sequence of strings for labeling the segments
`explode`	A sequence of values specifying the fraction of the pie chart radius to offset each wedge by (0 for no explode effect)
`shadow`	`True` or `False`: specifies whether to draw an attractive shadow under the pie
`startangle`	Rotate the "start" of the pie chart by this number of degrees counter-clockwise from the horizontal axis
`autopct`	A format string to label the segments by their percentage fractional value, or a function for generating such a string from the data
`pctdistance`	The radial position of the `autopct` text, relative to the pie radius; the default is 0.6 (i.e. within the pie, which can be awkward for narrow segments)
`labeldistance`	The radial position of the `label` text, relative to the pie radius; the default is 1.1 (just outside the pie)
`radius`	The radius of the pie (the default is 1); this is useful when creating overlapping pie charts with different radii

Example E7.11 The following program depicts the emissions of greenhouse gases by mass of "carbon equivalent" (data from the 2007 IPCC report).[9]

Listing 7.11 Pie chart of greenhouse gas emissions

```python
# eg7-pie.py
import numpy as np
import matplotlib.pyplot as plt

# Annual greenhouse gas emissions, billion tons carbon equivalent (GtCe).
gas_emissions = np.array([(r'$\mathrm{CO_2}$-d', 2.2),
                          (r'$\mathrm{CO_2}$-f', 8.0),
                          ('Nitrous\nOxide', 1.0),
                          ('Methane', 2.3),
                          ('Halocarbons', 0.1)],
                          dtype=[('source', 'U17'), ('emission', 'f4')])

# Five colors beige.
colors = ['#C7B299', '#A67C52', '#C69C6E', '#754C24', '#534741']

explode = [0, 0, 0.1, 0, 0]                                          ❶

fig, ax = plt.subplots()
ax.axis('equal')        # So our pie looks round!
ax.pie(gas_emissions['emission'], colors=colors, shadow=True, startangle=90,
       explode=explode, labels=gas_emissions['source'], autopct='%.1f%%',   ❷
       pctdistance=1.15, labeldistance=1.3)

plt.show()
```

❶ The segment corresponding to nitrous oxide has been "exploded" by 10%.

❷ The percentage values are formatted to one decimal place (`autopct='%.1f%%'`). The resulting pie chart is shown in Figure 7.12.

Pie charts have fallen out of favor in recent years ("the Comic Sans of data visualization") and they should certainly be avoided in comparing a large number of categories or very similar values. Mercifully, Matplotlib does not support three-dimensional pie charts.

7.3.3 Polar Plots

A polar plot, one of a radius value, r, given as a function of angle, θ, is created either using `pyplot.polar` as in Section 3.3.1, or by specifying the default *projection* in adding a subplot to a figure:

```python
fig = plt.figure()
ax = fig.add_subplot(projection='polar')
```

---

[9] IPCC, *Climate Change 2007: Synthesis Report. Contribution of Working Groups I, II and III to the Fourth Assessment Report of the Intergovernmental Panel on Climate Change* [Core Writing Team, Pachauri, R. K and Reisinger, A. (eds.)]. Geneva, Switzerland (2007).

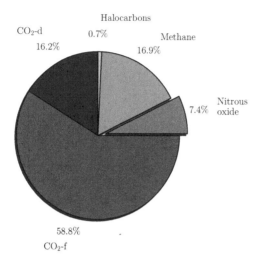

Figure 7.12 Greenhouse gas emissions by percentage for five different sources. CO_2-d denotes CO_2 emissions from deforestation; CO_2-f denotes CO_2 emissions from fossil fuel burning.

The following example illustrates both approaches.

Example E7.12   An antenna array can be used to direct radio waves in a particular direction by adjusting their number, geometrical arrangement, and relative amplitudes and phases.[10] Consider an array of n isotropic antennas at positions, d_i, evenly spaced by d along the x-axis from the origin:

$$d_0 = 0, d_1 = d\hat{x}, \dots, d_{n-1} = (n-1)d\hat{x}.$$

If a single antenna produces a radiation vector, $F(k)$, where $k = k\hat{r} = (2\pi/\lambda)\hat{r}$, the total radiation vector due to all n antennas is

$$F_{\text{tot}}(k) = \sum_{j=0}^{n-1} w_j e^{jik \cdot d_j} F(k) = A(k)F(k),$$

where w_j is the *feed coefficient* of the jth antenna, representing its amplitude and phase, and $A(k)$ is known as the *array factor*. We can choose $w_0 = 1$ to specify the feed coefficients relative to the antenna at the origin. If we further choose to look only at the azimuthal (ϕ) contribution to the radiation in the xy plane, setting the polar angle $\theta = \pi/2$, we have:

$$A(\phi) = \sum_{j=0}^{n-1} w_j e^{jikd \cos\phi}.$$

---

[10] S. J. Orfanidis, Electromagnetic Waves and Antennas, Rutgers University, http://eceweb1.rutgers.edu/~orfanidi/ewa/.

The relative radiation power pattern ("gain") is the square of this quantity. For two identical antennas,

$$g(\phi) = |A(\phi)|^2 = |w_0 + w_1 e^{ikd \cos \phi}|^2.$$

In the code below, the related quantity, the *directive gain*, $10 \log_{10}(g/g_{max})$, is plotted in Figure 7.13 as a function of ϕ for the two-antenna case on a polar plot for $d = \lambda$ and $w_0 = 1, w_1 = -i$.

Listing 7.12 Plotting the directive gain of a two-antenna system

```
import numpy as np
import matplotlib.pyplot as plt


def gain(d, w):
    """Return the power as a function of azimuthal angle, phi."""
    phi = np.linspace(0, 2*np.pi, 1000)
    psi = 2*np.pi * d / lam * np.cos(phi)
    A = w[0] + w[1]*np.exp(1j*psi)
    g = np.abs(A)**2
    return phi, g


def get_directive_gain(g, minDdBi=-20):
    """Return the "directive gain" of the antenna array producing gain g."""
    DdBi = 10 * np.log10(g / np.max(g))
    return np.clip(DdBi, minDdBi, None)          ❶


# Wavelength, antenna spacing, feed coefficients.
lam = 1
d = lam
w = np.array([1, -1j])
# Calculate gain and directive gain; plot on a polar chart.
phi, g = gain(d, w)
DdBi = get_directive_gain(g)

plt.polar(phi, DdBi)
ax = plt.gca()                                   ❷
ax.set_rticks([-20, -15, -10, -5])               ❸
ax.set_rlabel_position(45)                       ❹
plt.show()
```

❶ To better show the interesting region of the plot, where the power is highest, we "clip" the values less than minDdBi to that value.

❷ To customize the plot we need the Axes object in the current plot context; this is returned by plt.gca() ("get current axes").

❸ set_rticks sets the position of the radial tick marks.

❹ set_rlabel_position defines the angular position of the radial ticks.

NumPy's broadcasting methods (see Section 6.1.7) provide a natural way to extend this code to an arbitrary number of antennas; in the following example the figure method add_subplot is called with the argument projection='polar' and returns a corresponding Axes object, ax.

Listing 7.13 Plotting the directive gain of a three-antenna system

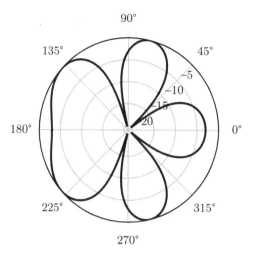

Figure 7.13 The directive gain of a two-antenna array with $d = \lambda$, $w_0 = 1$, $w_1 = -i$.

```python
import numpy as np
import matplotlib.pyplot as plt

def gain(d, w):
    """Return the power as a function of azimuthal angle, phi."""
    phi = np.linspace(0, 2*np.pi, 1000)
    psi = 2*np.pi * d / lam * np.cos(phi)
    j = np.arange(len(w))
    A = np.sum(w[j] * np.exp(j * 1j * psi[:, None]), axis=1)
    g = np.abs(A)**2
    return phi, g

def get_directive_gain(g, minDdBi=-20):
    """Return the "directive gain" of the antenna array producing gain g."""
    DdBi = 10 * np.log10(g / np.max(g))
    return np.clip(DdBi, minDdBi, None)

# Wavelength, antenna spacing, feed coefficients.
lam = 1
d = lam / 2
w = np.array([1, -1, 1])
# Calculate gain and directive gain; plot on a polar chart.
phi, g = gain(d, w)
DdBi = get_directive_gain(g)

fig = plt.figure()
ax = fig.add_subplot(projection='polar')
ax.plot(phi, DdBi)
ax.set_rticks([-20, -15, -10, -5])
ax.set_rlabel_position(45)
plt.show()
```

❶ annotates `A = np.sum(w[j] * np.exp(j * 1j * psi[:, None]), axis=1)`

❷ annotates `ax.plot(phi, DdBi)`

❶ The sum here is over the terms in the array factor expression: adding an axis to the psi array calculates this sum for each of the angular positions, ϕ.

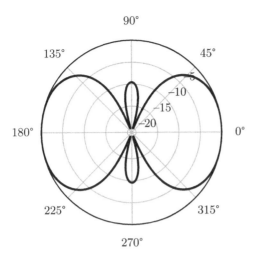

Figure 7.14 The directive gain of a many-antenna array with $d = \lambda/2$, $w_0 = 1$, $w_1 = -1$, $w_2 = 1$.

❷ Note that with the plot projection already defined, we need `ax.plot` here, not `ax.polar`.

The resulting plot is shown in Figure 7.14.

7.4 Annotating Plots

Matplotlib provides several ways to add different kinds of annotation to your plots. The most important methods for adding text, arrows, lines and shapes are described below.

7.4.1 Adding Text

The method `ax.text(x, y, s)` is a basic method used to add a text string s at position `(x, y)` (in *data* coordinates) to the axes. The font properties can be determined by additionally passing a dictionary of (keyword, value) pairs to `fontdict` (see Table 7.6). Individual keyword arguments (such as `fontsize=20`) can also be used to customize the font in this way.

If the text annotation refers to a feature of the data, you will usually want the default behavior, placing it using data coordinates so that it maintains the same relative position to the data even if the plot limits are changed. If, instead, you want to place the text in *axis* coordinates, such that (0, 0) is the lower left of the axes and (1, 1) is the upper right, set the keyword argument `transform=ax.transAxes` where ax is the `Axes` object the coordinates refer to.

7.4.2   Arrows and Text

The `ax.annotate` method is similar to `ax.text` (although with an annoyingly different syntax) but draws an arrow from the text to a specified point in the plot. The important arguments to `ax.annotate` are:

- s, the string to output as a text label;
- xy, a tuple (x, y) giving the coordinates of the position to annotate (i.e. where the arrow points *to*);
- xytext, a tuple (x, y) giving the coordinates of the text label (i.e. where the arrow points *from*);
- xycoords, an optional string determining the type of coordinates referred to by the argument xy: several options are available,[11] but the most commonly used ones are:

   - 'data': data coordinates, the default,
   - 'figure fraction': fractional coordinates of the *figure size* – (0, 0) is lower left, (1, 1) is upper right,
   - 'axes fraction': fractional coordinates of the *axes* – (0, 0) is lower left, (1, 1) is upper right;

- textcoords: as for xycoords, an optional string determining the type of coordinates referred to by xytext; an additional value is permitted for this string: 'offset points' specifies that the tuple xytext is an offset *in points* from the xy position;
- arrowprops: if present, determines the properties and style of the arrow drawn between xytext and xy (see below).

Additional keyword arguments are interpreted as properties of the Text object produced as the label (e.g. fontsize and color). An important pair is verticalalignment (or va) and horizontalalignment (or ha), which determine how the label is aligned relative to its xytext position. Valid values are 'center', 'right', 'left', 'top', 'bottom' and 'baseline', as appropriate.

In its simplest usage, `ax.annotate` just adds a text label to the plot (without an arrow). For example,

```
ax.annotate('My Label', xy=(0.5, 0.8), fontsize=16, xycoords='axes fraction',
            ha='center')
```

which adds 'My Label' at the center, near the top of the axes in 16-point text. Note that if there is no arrow or line, xytext is not necessary and the label is placed directly at xy.

The argument arrowprops is a dictionary determining the style of line or arrow joining the label at xytext to the specified xy point. There is a somewhat bewildering array of possible items to put in this dictionary, but the important ones are illustrated by the following example.

---

[11] See the documentation at https://matplotlib.org/api/text_api.html.

Example E7.13   The following program produces a plot with eight arrows with different styles (Figure 7.15).

Listing 7.14 Annotations with arrows in Matplotlib

```python
# eg7-arrows.py
import numpy as np
import matplotlib.pyplot as plt

fig, ax = plt.subplots()
x = np.linspace(0, 1)
ax.plot(x, x, 'o')

ax.annotate('default line', xy=(0.15, 0.1), xytext=(0.6, 0.1),
            arrowprops={'arrowstyle': '-'}, va='center')
ax.annotate('dashed line', xy=(0.25, 0.2), xytext=(0.6, 0.2),
            arrowprops={'arrowstyle': '-', 'ls': 'dashed'}, va='center')
ax.annotate('default arrow', xy=(0.35, 0.3), xytext=(0.6, 0.3),
            arrowprops={'arrowstyle': '->'}, va='center')
ax.annotate('thick blue arrow', xy=(0.45, 0.4), xytext=(0.6, 0.4),
            arrowprops={'arrowstyle': '->', 'lw': 4, 'color': 'blue'},
            va='center')
ax.annotate('double-headed arrow', xy=(0.45, 0.5), xytext=(0.01, 0.5),
            arrowprops={'arrowstyle': '<->'}, va='center')
ax.annotate('arrow with closed head', xy=(0.55, 0.6), xytext=(0.1, 0.6),
            arrowprops={'arrowstyle': '-|>'}, va='center')
ax.annotate('a really thick red arrow\nwith not much space', xy=(0.65, 0.7),
            xytext=(0.1, 0.7), va='center', multialignment='right',
            arrowprops={'arrowstyle': '-|>', 'lw': 8, 'ec': 'r'})
ax.annotate('a really thick red arrow\nwith space between\nthe tail and the'
            'label', xy=(0.85, 0.9), xytext=(0.1, 0.9), va='center',
            multialignment='right',
            arrowprops={'arrowstyle': '-|>', 'lw': 8, 'ec': 'r', 'shrinkA': 10})

plt.show()
```

Example E7.14   Another example of an annotated plot, this time of the share price of BP plc (LSE: BP) with a couple of notable events added to it. The necessary data for this example can be downloaded from Yahoo! Finance.[12]

Listing 7.15 Plotting a share price time series on an annotated chart

```python
import datetime
import numpy as np
import matplotlib.pyplot as plt
from matplotlib.dates import strpdate2num
from datetime import datetime

def date_to_int(s):
    epoch = datetime(year=1970, month=1, day=1)
```
❶

---

[12] https://uk.finance.yahoo.com/q/hp?s=BP.L.

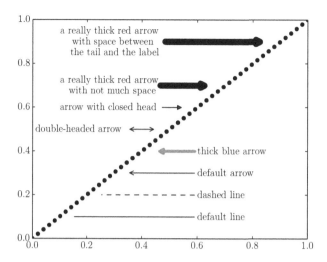

Figure 7.15 An example of different arrow styles.

```
        date = datetime.strptime(s, '%Y-%m-%d')
        return (date - epoch).days

def bindate_to_int(bs):
    return date_to_int(bs.decode('ascii'))

dt = np.dtype([('daynum','i8'), ('close', 'f8')])
share_price = np.loadtxt('bp-share-prices.csv', skiprows=1, delimiter=',',
                         usecols=(0, 4), converters={0: bindate_to_int},
                         dtype=dt)
fig, ax = plt.subplots()
ax.plot(share_price['daynum'], share_price['close'], c='g')
ax.fill_between(share_price['daynum'], 0, share_price['close'], facecolor='g',
                alpha=0.5)

daymin, daymax = share_price['daynum'].min(), share_price['daynum'].max()
ax.set_xlim(daymin, daymax)

price_max = share_price['close'].max()

def get_xy(date):
    """ Return the (x, y) coordinates of the share price on a given date. """
    x = date_to_int(date)
    return share_price[np.where(share_price['daynum']==x)][0]

# A horizontal arrow and label.
x, y = get_xy('1999-10-01')
ax.annotate('Share split', (x, y), xytext = (x+1000, y), va='center',
            arrowprops=dict(facecolor='black', shrink=0.05))
# A vertical arrow and label.
x, y = get_xy('2010-04-20')
ax.annotate('Deepwater Horizon\noil spill', (x, y), xytext = (x, price_max*0.9),
            arrowprops=dict(facecolor='black', shrink=0.05), ha='center')

years = range(1989, 2015, 2)
```

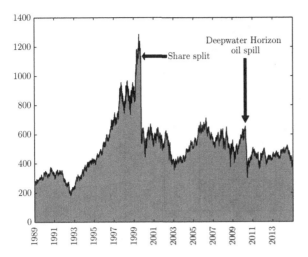

Figure 7.16 BP plc's share price on an annotated chart.

```
   ax.set_xticks([date_to_int('{:4d}-01-01'.format(year)) for year in years])
❸ ax.set_xticklabels(years, rotation=90)

   plt.show()
```

❶ We need to do some work to read in the date column: first decode the bytestring read in from the file to ASCII (`bindate_to_int`), then use `datetime` (see Section 4.5.3) to convert it into an integer number of days since some reference date (epoch): here we choose the Unix epoch, 1 January 1970 (`date_to_int`).

❷ `ax.fill_between` fills the region below the plotted line with a single color.

❸ We rotate the year labels so there's enough room for them (reading bottom to top). Figure 7.16 shows the plotted chart.

7.4.3   Lines and Span Rectangles

Adding an arbitrary straight line to a Matplotlib plot can be achieved by simply plotting the data corresponding to its start and end points with `ax.plot`; for example,

```
ax.plot([x1, x2], [y1, y2], color='k', lw=2)
```

draws a line between (`x1`, `y1`) and (`x2`, `y2`). Of course, this approach would be tedious for drawing a large number of disconnected lines, so for horizontal and vertical lines there are a pair of convenient methods, `ax.hlines` and `ax.vlines`. `ax.hlines` takes mandatory arguments, `y`, `xmin`, `xmax`, and draws horizontal lines with y-coordinates at each of the values given by the sequence `y` (if `y` is passed as a scalar, a single line is drawn). `xmin` and `xmax` specify the start and end of each line; they can be scalars (in which case all the lines will have the same start and end x-coordinates) or a sequence (with one value for each of the y-coordinates specified by `y`). `ax.vlines` draws vertical lines; its mandatory arguments, `x`, `ymin` and `ymax`, are entirely analogous.

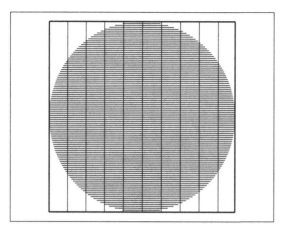

Figure 7.17 A figure generated from vertical and horizontal lines.

Example E7.15 The code below illustrates some different uses of `ax.vlines` and `ax.hlines` (see Figure 7.17).

Listing 7.16 Some different ways to use `ax.vlines` and `ax.hlines`

```
# eg7-circle-lines.py
import numpy as np
import matplotlib.pyplot as plt

fig, ax = plt.subplots()
ax.axis('equal')

# A circle made of horizontal lines.
y = np.linspace(-1, 1, 100)
xmax = np.sqrt(1 - y**2)
ax.hlines(y, -xmax, xmax, color='g')

# Draw a box of thicker lines around the circle.
ax.vlines(-1, -1, 1, lw=2, color='r')
ax.vlines(1, -1, 1, lw=2, color='r')
ax.hlines(-1, -1, 1, lw=2, color='r')
ax.hlines(1, -1, 1, lw=2, color='r')
# Some evenly spaced vertical lines.
ax.vlines(y[::10], -1, 1, color='b')

# Remove tick marks and labels.
ax.xaxis.set_visible(False)
ax.yaxis.set_visible(False)
# A bit of padding around the outside of the box.
ax.set_xlim(-1.1, 1.1)
ax.set_ylim(-1.1, 1.1)

plt.show()
```

Table 7.11 Keyword arguments for styling patches

Argument	Description
alpha	Set the alpha transparency (0–1)
color	Set both the `facecolor` and the `edgecolor` of the patch
edgecolor, ec	Set the edge (border) color
facecolor, fc	Set the patch face color
fill	Indicate whether to fill the patch or not (`True` or `False`)
hatch	Set the hatching pattern for the patch: one of `'/'`, `'\'`, `'|'`, `'-'`, `'+'`, `'x'`, `'o'`, `'O'`, `'.'`, `'*'`; repeat the character for a denser pattern
linestyle, ls	Set the patch line style: `'solid'`, `'dashed'`, `'dashdot'`, `'dotted'`
linewidth, lw	Set the patch line width, in points

On static plots such as figures for printing, `ax.hlines` and `ax.vlines` work well, but note that the line limits don't change upon changing the axes' limits in an interactive plot. There are two further methods, `ax.axhline` and `ax.axvline`, which simply plot a horizontal or vertical line across the axis, whatever its current limits. `ax.axhline` takes arguments y, xmin, xmax, but these must be scalar values (so multiple lines require repeated calls) and xmin, xmax are given in *fractional* coordinates such that 0 is the far left of the plot and 1 the far right. Again, the `ax.axvline` arguments, x, ymin, ymax, are analogous. Some examples:

```
ax.axhline(100, 0, 1)    # horizontal line across whole of x-axis at y = 100
ax.axhline(100)          # same thing: xmin and xmax default to 0 and 1
# A thick, blue, dashed vertical line at x = 5, around the center of the y-axis.
ax.axvline(5, 0.4, 0.6, c='b', lw=4, ls='--')
```

The methods `ax.axhspan` and `ax.axvspan` are similar but produce a horizontal or vertical *spanning rectangle* across the axis. `ax.axhspan` is passed arguments ymin, ymax (in *data* coordinates), and xmin, xmax (in *fractional axes* units). `ax.axvspan` takes the arguments xmin, xmax, ymin and ymax in the same way. Extra keywords can be used to style the spanning rectangle (which is a type of `Patch` object; see Table 7.11).

Example E7.16 The program below annotates a simple wave plot to indicate the different regions of the electromagnetic spectrum, using `text`, `axvline`, `axhline` and `axvspan` (see Figure 7.18).

Listing 7.17 A representation of the electromagnetic spectrum, 250–1000 nm

```
# eg7-annotate.py
import numpy as np
import matplotlib.pyplot as plt

# Wavelength range, nm.
lmin, lmax = 250, 1000
x = np.linspace(lmin, lmax, 1000)
# A wave with a smoothly increasing wavelength.
wv = (np.sin(10 * np.pi * x / (lmax+lmin-x)))[::-1]
```

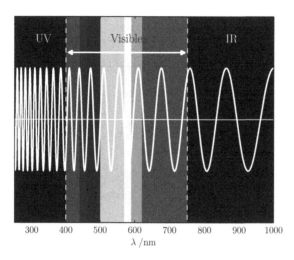

Figure 7.18 A representation of the electromagnetic spectrum.

```python
fig = plt.figure()
ax = fig.add_subplot(facecolor='k')
ax.plot(x, wv, c='w', lw=2)
ax.set_xlim(250, 1000)
ax.set_ylim(-2, 2)

# Label and delimit the different regions of the electromagnetic spectrum.
ax.text(310, 1.5, 'UV', color='w', fontdict={'fontsize': 20})
ax.text(530, 1.5, 'Visible', color='k', fontdict={'fontsize': 20})
ax.annotate('', (400, 1.3), (750, 1.3), arrowprops={'arrowstyle': '<|-|>',
                                              'color': 'w', 'lw': 2})
ax.text(860, 1.5, 'IR', color='w', fontdict={'fontsize': 20})
ax.axvline(400, -2, 2, c='w', ls='--')
ax.axvline(750, -2, 2, c='w', ls='--')
# Horizontal "axis" across the center of the wave.
ax.axhline(c='w')
# Remove the y-axis ticks and labels; label the x-axis.
ax.yaxis.set_visible(False)
ax.set_xlabel(r'$\lambda\;/\mathrm{nm}$')

# Finally, add some colorful rectangles representing a rainbow in the
# visible region of the spectrum.
# Dictionary mapping of wavelength regions (nm) to approximate RGB values.
rainbow_rgb = { (400, 440): '#8b00ff', (440, 460): '#4b0082',
                (460, 500): '#0000ff', (500, 570): '#00ff00',
                (570, 590): '#ffff00', (590, 620): '#ff7f00',
                (620, 750): '#ff0000'}
for wv_range, rgb in rainbow_rgb.items():
    ax.axvspan(*wv_range, color=rgb, ec='none', alpha=1)
plt.show()
```

7.4.4 ◊   Circles, Polygons and Other Patches

Almost everything that gets rendered in a Matplotlib figure is a subclass of the abstract base class, `Artist`. This includes lines (through `Line2D`) and text (through `Text`).[13] An important collection of rendered objects is further derived from the `Artist` subclass `Patch`: a two-dimensional shape. The wedges of a pie chart (Section 7.3) and the arrows of an annotation (Section 7.4) are examples we have met before.

To add a shape to an `Axes` object, create a patch using one of the classes described in full in the Matplotlib documentation[14] and call `ax.add_patch(patch)`. To set the color, line widths, transparency, etc. of the patch, pass one or more of the keywords listed in Table 7.11 when creating the patch.

This usage for a few types of `Patch` objects is described below.

Circles and Ellipses

A `Circle` centered at `xy = (x, y)` (in data coordinates) and with radius `r` is created with:

```
from matplotlib.patches import Circle
circle = Circle(xy, r, **kwargs)
```

It is added to the Axes with `ax.add_patch`:

```
ax.add_patch(circle)
```

The supported keyword arguments indicated by `**kwargs` are the usual patch styling ones, summarized in Table 7.11.

`Ellipse` patches are similar but take arguments `width` and `height` (the total length of the horizontal and vertical axes of the ellipse before rotation) and `angle` (the angle of counterclockwise rotation of the ellipse *in degrees*).

```
from matplotlib.patches import Ellipse
ellipse = Ellipse(xy, width, height, angle, **kwargs)
```

Example E7.17   The following code reads in the heights and masses of 260 women and 247 men from the data set published by Heinz *et al.*[15] and available for download at https://scipython.com/eg/bai. It plots the (height, mass) pairs for each individual on a scatter plot and, for each sex, draws a 3σ covariance ellipse around the mean point. The dimensions of this ellipse are given by the (scaled) eigenvalues of the covariance matrix and it is rotated such that its semi-major axis lies along the largest eigenvector.

Listing 7.18 An analysis of the height–mass relationship in 507 healthy individuals

```
# eg7-body-mass-height.py
```

---

[13] In fact, there are two kinds of `Artist`: *primitives* and *containers*. Primitives are the graphical objects (such as `Line2D` themselves) and containers are the elements of a figure onto which they are rendered (for example `Axes`).

[14] https://matplotlib.org/api/artist_api.html.

[15] G. Heinz *et al.*, *Journal of Statistical Education* 11(2), (2003). This article is available at https://doi.org/10.1080/10691898.2003.11910711.

```
import numpy as np
import matplotlib.pyplot as plt
from matplotlib.patches import Ellipse

FEMALE, MALE = 0, 1
dt = np.dtype([('mass', 'f8'), ('height', 'f8'), ('gender', 'i2')])
data = np.loadtxt('body.dat.txt', usecols=(22, 23, 24), dtype=dt)

fig, ax = plt.subplots()

def get_cov_ellipse(cov, center, nstd, **kwargs):
    """

    Return a matplotlib Ellipse patch representing the covariance matrix
    cov centered at center and scaled by the factor nstd.

    """

    # Find and sort eigenvalues and eigenvectors into descending order.
    eigvals, eigvecs = np.linalg.eigh(cov)
    order = eigvals.argsort()[::-1]
    eigvals, eigvecs = eigvals[order], eigvecs[:, order]

    # The counterclockwise angle to rotate our ellipse by.
    vx, vy = eigvecs[:, 0][0], eigvecs[:, 0][1]
    theta = np.arctan2(vy, vx)

    # Width and height of ellipse to draw.
    width, height = 2 * nstd * np.sqrt(eigvals)
    return Ellipse(xy=center, width=width, height=height,
                   angle=np.degrees(theta), **kwargs)

labels, colors =['Female', 'Male'], ['magenta', 'blue']
for gender in (FEMALE, MALE):
    sdata = data[data['gender']==gender]
    height_mean = np.mean(sdata['height'])
    mass_mean = np.mean(sdata['mass'])
    cov = np.cov(sdata['mass'], sdata['height'])
    ax.scatter(sdata['height'], sdata['mass'], color=colors[gender],
               label=labels[gender])
    e = get_cov_ellipse(cov, (height_mean, mass_mean), 3,
                        fc=colors[gender], alpha=0.4)
    ax.add_patch(e)

ax.set_xlim(140, 210)
ax.set_ylim(30, 120)
ax.set_xlabel('Height /cm')
ax.set_ylabel('Mass /kg')
ax.legend(loc='upper left', scatterpoints=1)
plt.show()
```

❶ The function np.arctan2 returns the "two-argument arctangent": np.arctan2 (*y*, *x*) is the angle in radians between the positive *x*-axis and the point (x, y).

Figure 7.19 shows the resulting plot.

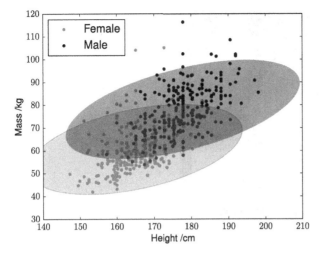

Figure 7.19 Scatter plots for each gender of mass and height for a total of 507 students, with their covariance ellipses annotated.

Rectangles

`Rectangle` patches are created in a similar way to `Ellipses`:

```
from matplotlib.patches import Rectangle
rectangle = Rectangle(xy, width, height, angle, **kwargs)
```

Here, however, the tuple `xy=(x, y)` gives the coordinates of the *lower left* corner of the rectangle. A square is simply a rectangle with the same `width` and `height`, of course.

Polygons

A `Polygon` patch is created by passing an array of shape `(N, 2)`, in which each row represents the (x, y) coordinates of a vertex. If the additional argument, `closed`, is `True` (the default), the polygon will be closed so that the start and end points are the same. This is illustrated in the following example.

Example E7.18 This code produces an image (Figure 7.20) of some colorful shapes.

Listing 7.19 Some colorful shapes

```
import numpy as np
import matplotlib.pyplot as plt
from matplotlib.patches import Polygon, Circle, Rectangle

red, blue, yellow, green = '#ff0000', '#0000ff', '#ffff00', '#00ff00'
square = Rectangle((0.7, 0.1), 0.25, 0.25, facecolor=red)
circle = Circle((0.8, 0.8), 0.15, facecolor=blue)
triangle = Polygon(((0.05, 0.1), (0.396, 0.1), (0.223, 0.38)), fc=yellow)
rhombus = Polygon(((0.5, 0.2), (0.7, 0.525), (0.5, 0.85), (0.3, 0.525)), fc=green)
```

Figure 7.20 Some colorful shapes using Matplotlib patches.

```
fig = plt.figure(facecolor='k')
ax = fig.add_subplot(aspect='equal')
for shape in (square, circle, triangle, rhombus):
    ax.add_patch(shape)
ax.axis('off')

plt.show()
```

Questions

Q7.4.1   Compare plots of $y = x^3$ for $-10 \leq x \leq 10$ using a logarithmic scale on the x-axis, y-axis and both axes. What is the difference between using `ax.set_xscale('log')` and `ax.set_xscale('symlog')`?

Q7.4.2   Adapt Example E7.9 to produce a *horizontal* bar chart, with the bars in order of decreasing letter frequency (i.e. with the most common letter, E, at the bottom).

Problems

P7.4.1   *The Economist*'s Big Mac Index is a lighthearted measure of purchasing power parity between two currencies. Its premise is that the difference between the price of a McDonald's Big Mac hamburger in one currency, converted into US dollars (USD) at the prevailing exchange rate, and its price in the United States is a measure of the extent to which that currency is over- or under-valued (relative to the dollar).

The files at https://scipython.com/ex/bga provide the historical Big Mac prices and exchange rates for four currencies. For each currency, calculate the percentage over- or under-valuation of each currency as

$$\frac{\text{(local price converted to USD} - \text{US price)}}{\text{(US price)}} \times 100$$

and plot it as a function of time.

P7.4.2 Plot, as a histogram, the data in the table below concerning the number of cases of West Nile virus disease in the United States between 1999 and 2008. The two types of disease, neuroinvasive and non-neuroinvasive, should be plotted as separate bars on the same chart for each year.

Year	Neuroinvasive cases	Non-neuroinvasive cases
1999	59	3
2000	19	2
2001	64	2
2002	2946	1210
2003	2866	6996
2004	1148	1391
2005	1309	1691
2006	1495	2774
2007	1227	117
2008	689	667

P7.4.3 A bubble chart is a type of scatter plot that can depict three dimensions of data through the position (*x*- and *y*-coordinates) and size of the marker. The plt.scatter method can produce bubble charts by passing the marker size to its s attribute (in points-squared such that the area of the marker is proportional to the magnitude of the third dimension – see Example E7.1).

The files gdp.tsv, bmi_men.tsv and population_total.tsv, available at https://scipython.com/ex/bgc, contain the following data from 2007 for each country: the GDP per person per capita in international dollars fixed at 2005 prices, the body mass index (BMI) of men (in $kg\,m^{-2}$) and the total population. Generate a bubble chart of BMI against GDP, in which the population is depicted by the size of the bubble markers. Beware: some data are missing for some countries.

Bonus exercise: color the bubbles by continent using the list provided in the file continents.tsv.

P7.4.4 The US National Oceanic and Atmospheric Administration (NOAA) makes a data set of atmospheric carbon dioxide (CO_2) concentrations since 1958 freely available to the public at ftp://aftp.cmdl.noaa.gov/products/trends/co2/co2_mm_mlo.txt. Using these data, plot the "interpolated" and "trend" CO_2 concentration against time on the same graph.

P7.4.5 Write a program to plot the Planck function, $B(\lambda)$, for the spectral radiance of a Black body at temperature, T, as a function of wavelength, λ, for the Sun ($T = 5778$ K):

$$B(\lambda) = \frac{2hc^2}{\lambda^5} \frac{1}{\exp\left(hc/\lambda k_B T\right) - 1}.$$

Use a NumPy array to store values of $B(\lambda)$ from 100 to 5000 nm, but set the wavelength range to *decrease* from 4000 nm to 0. The necessary physical constants may be taken

Figure 7.21 An image produced using Matplotlib `Circle` patches.

to have the values $h = 6.626 \times 10^{-34}$ J s, $c = 2.998 \times 10^8$ m s^{-1} and $k_B = 1.381 \times 10^{-23}$ J K^{-1}.

P7.4.6 Reproduce Figure 7.21 using `Circle` patches.

7.5   Contour Plots and Heatmaps

Until now, we have looked only at plotting one-dimensional data (that is, functions of one coordinate only). Matplotlib also supports several ways to plot data that is a function of two dimensions.

7.5.1   Contour Plots

The `pyplot` method `contour` makes a contour plot of a provided two-dimensional array. In its simplest invocation, `contour(Z)`, no further arguments are required: the (x, y) values are indexes into the two-dimensional array Z and contour intervals are selected automatically. To explicitly include (x, y) coordinates, pass them as `contour(X, Y, Z)`. The arrays X and Y must have the same shape as Z (for example, as produced by `np.meshgrid`: see Section 6.1.6) or be one-dimensional such that X has the same length as the number of *columns* in Z, and Y has the same length as the number of *rows* in Z.

The contour levels can be controlled by a further argument: either a scalar, N, giving the total number of contour levels, or a sequence, V, explicitly listing the values of Z at which to draw contours.

The contours are colored according to Matplotlib's default *colormap*. In this process, the data are normalized linearly onto the interval [0, 1], which is then mapped onto a list of colors that are used to style the contours at the corresponding values. The module `matplotlib.cm` provides several colormap schemes:[16] some of the more practical ones are `cm.viridis` (since Matplotlib 2.0, the default), `cm.hot`, `cm.bone`, `cm.winter`, `cm.jet`, `cm.Greys` and `cm.hsv`. If you want to use a colormap with its colors reversed, tack a `_r` on the end of its name (e.g `cm.hot_r`).

As an alternative, `contour` supports the `colors` argument, which takes either a single Matplotlib color specifier or a sequence of such specifiers. For single-color contour plots, contours corresponding to negative values are plotted in dashed lines. The widths of the contour lines can be styled individually or all together with the argument `linewidths`.

Example E7.19   The following code produces a plot of the electrostatic potential of an electric dipole $\mathbf{p} = (qd, 0, 0)$ in the (x, y) plane for $q = 1.602 \times 10^{-19}$ C, $d = 1$ pm, using the point dipole approximation (see Figure 7.22).

Listing 7.20 The electrostatic potential of a point dipole

```
# eg7-elec-dipole-pot.py
import numpy as np
import matplotlib.pyplot as plt

# Dipole charge (C), Permittivity of free space (F.m-1).
q, eps0 = 1.602e-19, 8.854e-12
# Dipole +q, -q distance (m) and a convenient combination of parameters.
d = 1.e-12
k = 1/4/np.pi/eps0 * q * d

# Cartesian axis system with origin at the dipole (m).
X = np.linspace(-5e-11, 5e-11, 1000)
Y = X.copy()
X, Y = np.meshgrid(X, Y)

# Dipole electrostatic potential (V), using point dipole approximation.
Phi = k * X / np.hypot(X, Y)**3

fig, ax = plt.subplots()
# Draw contours at values of Phi given by levels.
levels = np.array([10**pw for pw in np.linspace(0, 5, 20)])
levels = sorted(list(-levels) + list(levels))
# Monochrome plot of potential.
ax.contour(X, Y, Phi, levels=levels, colors='k', linewidths=2)
plt.show()
```

To add labels to the contours, store the `ContourSet` object returned by the call to `ax.contour` and pass it to `ax.clabel` (perhaps with some additional parameters dictating the font properties). A further method, `ax.contourf`, which takes the same arguments

---

[16] See the page https://matplotlib.org/tutorials/colors/colormaps.html for a complete list.

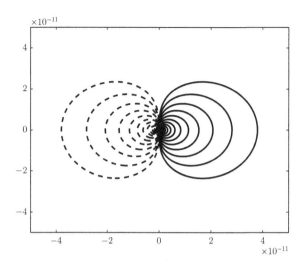

Figure 7.22 A contour plot of the electrostatic potential of a point dipole.

as contour, draws *filled* contours. contour and ax.contourf can be used together, as in the following example.

Example E7.20 This program produces a filled contour plot of a function, labels the contours and provides some custom styling for their colors (see Figure 7.23).

Listing 7.21 An example of filled and styled contours

```python
# eg7-2dgau.py
import numpy as np
import matplotlib.pyplot as plt
import matplotlib.cm as cm

X = np.linspace(0, 1, 100)
Y = X.copy()
X, Y = np.meshgrid(X, Y)
alpha = np.radians(25)
cX, cY = 0.5, 0.5
sigX, sigY = 0.2, 0.3
rX = np.cos(alpha) * (X-cX) - np.sin(alpha) * (Y-cY) + cX
rY = np.sin(alpha) * (X-cX) + np.cos(alpha) * (Y-cY) + cY

Z = (rX-cX)*np.exp(-((rX-cX)/sigX)**2) * np.exp(-((rY-cY)/sigY)**2)
fig = plt.figure()
ax = fig.add_subplot()

# Reversed Greys colormap for filled contours.
cpf = ax.contourf(X, Y, Z, 20, cmap=cm.Greys_r)
# Set the colors of the contours and labels so they're white where the
# contour fill is dark (Z < 0) and black where it's light (Z >= 0).
colors = ['w' if level<0 else 'k' for level in cpf.levels]
cp = ax.contour(X, Y, Z, 20, colors=colors)
ax.clabel(cp, fontsize=12, colors=colors)
```

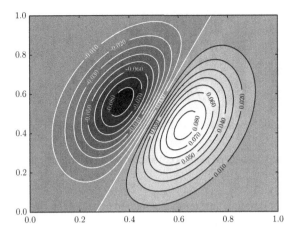

Figure 7.23 A two-dimensional plot with labeled contours.

```
plt.show()
```

7.5.2   Heatmaps

Another way to depict two-dimensional data is as a *heatmap*: an image in which the color of each pixel is determined by the corresponding value in the array of data. The use of Matplotlib's functions `ax.imshow`, `ax.pcolor` and `ax.pcolormesh` is described in this section.

`ax.imshow`

The `Axes` method `ax.imshow` displays an image on the axes. In its basic usage, it takes a two-dimensional array and maps its values to the pixels on an image according to some interpolation scheme and normalization. If the array data are taken from an image read in with the Matplotlib method `image.imread`, this is usually all that is required:

```
In [x]: import matplotlib.pyplot as plt
In [x]: import matplotlib.image as mpimg
In [x]: im = mpimg.imread('image.jpg')
In [x]: plt.imshow(im)
In [x]: plt.show()
```

(In this case, `im` is a three-dimensional array of shape (`n`, `m`, 3) in which the "depth" coordinate corresponds to the red, green and blue components of each pixel in the n-by-m image.)

imshow is frequently used to visualize matrices or other two-dimensional arrays of data. If the image produced for the figure has a different size to the array dimensions, some kind of interpolation scheme is employed: for example, to visualize a 10×10 matrix as a 100×100 pixel image, a lot of intermediate points need to be approximated. The default interpolation scheme is `'nearest'`, which is the most faithful to the underlying data but can look "blocky." There are many alternative interpolation schemes (see

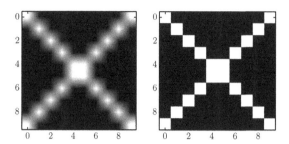

Figure 7.24 A small matrix visualized using `ax.imshow` with two different interpolation schemes.

the documentation[17] for details): an example using `interpolation='bilinear'` is given below: these can produce somewhat blurry-looking images for small arrays, however. Note that `imshow` takes a `cmap` argument that assigns a colormap in the same way as it does for `ax.contourf`.

Example E7.21   The following code compares two interpolation schemes: `'bilinear'` and `'nearest'` (the default), which should look "blocky" (i.e. more faithful to the data): see Figure 7.24.

Listing 7.22 A comparison of interpolation schemes for a small array visualized with `imshow()`

```
# eg7-matrix-show.py
import numpy as np
import matplotlib.pyplot as plt
import matplotlib.cm as cm

# Make an array with ones in the shape of an 'X'.
a = np.eye(10, 10)
a += a[::-1,:]

fig = plt.figure()
ax1 = fig.add_subplot(121)
# Bilinear interpolation - this will look blurry.
ax1.imshow(a, interpolation='bilinear', cmap=cm.Greys_r)

ax2 = fig.add_subplot(122)
# 'nearest' interpolation - faithful but blocky.
ax2.imshow(a, cmap=cm.Greys_r)

plt.show()
```

Recent versions of Matplotlib include an `Axes` method, `matshow`, which can be used in place of `imshow` and sets desirable defaults for visualizing matrices.

---

[17] https://matplotlib.org/api/_as_gen/matplotlib.axes.Axes.imshow.html

Example E7.22   The *Barnsley fern* is a fractal that resembles the black spleenwort species of fern. It is constructed by plotting a sequence of points in the (x, y) plane, starting at $(0, 0)$, generated by the following affine transformations, f_1, f_2, f_3 and f_4, where each transformation is applied to the previous point and chosen at random with probabilities $p_1 = 0.01$, $p_2 = 0.85$, $p_3 = 0.07$ and $p_4 = 0.07$:

$$f_1(x, y) = \begin{pmatrix} 0 & 0 \\ 0 & 0.16 \end{pmatrix} \begin{pmatrix} x \\ y \end{pmatrix},$$

$$f_2(x, y) = \begin{pmatrix} 0.85 & 0.04 \\ -0.04 & 0.85 \end{pmatrix} \begin{pmatrix} x \\ y \end{pmatrix} + \begin{pmatrix} 0 \\ 1.6 \end{pmatrix},$$

$$f_3(x, y) = \begin{pmatrix} 0.2 & -0.26 \\ 0.23 & 0.22 \end{pmatrix} \begin{pmatrix} x \\ y \end{pmatrix} + \begin{pmatrix} 0 \\ 1.6 \end{pmatrix},$$

$$f_4(x, y) = \begin{pmatrix} -0.15 & 0.28 \\ 0.26 & 0.24 \end{pmatrix} \begin{pmatrix} x \\ y \end{pmatrix} + \begin{pmatrix} 0 \\ 0.44 \end{pmatrix}.$$

This algorithm is implemented in the program below and the result is depicted in Figure 7.25.

Listing 7.23 Barnsley's fern

```
# eg7-fern.py
import numpy as np
import matplotlib.pyplot as plt
import matplotlib.cm as cm

f1 = lambda x, y: (0., 0.16*y)
f2 = lambda x, y: (0.85*x + 0.04*y, -0.04*x + 0.85*y + 1.6)
f3 = lambda x, y: (0.2*x - 0.26*y, 0.23*x + 0.22*y + 1.6)
f4 = lambda x, y: (-0.15*x + 0.28*y, 0.26*x + 0.24*y + 0.44)
fs = [f1, f2, f3, f4]

npts = 50000
# Canvas size (pixels).
width, height = 300, 300
aimg = np.zeros((width, height))

x, y = 0, 0
for i in range(npts):
    # Pick a random transformation and apply it.
    f = np.random.choice(fs, p=[0.01, 0.85, 0.07, 0.07])
    x, y = f(x, y)
    # Map (x, y) to pixel coordinates.
    # NB we "know" that -2.2 < x < 2.7 and 0 <= y < 10.

    ix, iy = int(width / 2 + x * width / 10), int(y * height / 12)
    # Set this point of the array to 1 to mark a point in the fern.
    aimg[iy, ix] = 1

plt.imshow(aimg[::-1,:], cmap=cm.Greens)
plt.show()
```

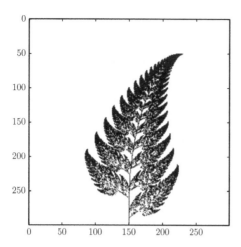

Figure 7.25 The Barnsley fern fractal.

ax.pcolor and ax.pcolormesh

There are a couple of other similar Matplotlib methods that you will come across: ax.pcolor and ax.pcolormesh. These are very similar. The precise differences are beyond the scope of this book, but pcolormesh is very much faster than pcolor and is the recommended alternative to imshow for this reason. The most noticeable difference is that imshow follows the convention used in the image-processing community that places the origin in the *top left* corner; the pcolor methods associate the origin with the bottom left corner.

Colorbars

It is often useful to have a legend indicating how the colors of the plot relate to the values of the array used to derive it. This is added with the fig.colorbar method. In its most simple usage, simply call fig.colorbar(*mappable*) where *mappable* is the Image, ContourSet or other suitable object to which the colorbar applies and a new Axes object holding the colorbar will be created (and room made in the figure to accommodate it). This object can be further customized and labeled, as shown in the following examples.

Example E7.23   The following code reads in a data file of maximum daily tempera-tures in Boston for 2019 and plots them on a heatmap, with a labeled colorbar legend (see Figure 7.26). The data file may be downloaded from https://scipython.com/eg/bah .

Listing 7.24 Heatmap of Boston's temperatures in 2019

```
# eg7-heatmap.py

import numpy as np
import matplotlib.pyplot as plt

# Read in the relevant data from our input file.
dt = np.dtype([('month', np.int), ('day', np.int), ('T', np.float)])
data = np.genfromtxt('boston2019.dat', dtype=dt, usecols=(1, 2, 3),
```

```
                                delimiter=(4, 2, 2, 6))

# In our heatmap, nan will mean "no such date", e.g. 31 June.
heatmap = np.empty((12, 31))
heatmap[:] = np.nan

for month, day, T in data:
    # NumPy arrays are zero-indexed; days and months are not!
    heatmap[month-1, day-1] = T

# Plot the heatmap, customize and label the ticks.
fig = plt.figure()
ax = fig.add_subplot()
im = ax.imshow(heatmap, interpolation='nearest')
ax.set_yticks(range(12))
ax.set_yticklabels(['Jan', 'Feb', 'Mar', 'Apr', 'May', 'Jun',
                    'Jul', 'Aug', 'Sep', 'Oct', 'Nov', 'Dec'])
days = np.array(range(0, 31, 2))
ax.set_xticks(days)
ax.set_xticklabels(['{:d}'.format(day+1) for day in days])
ax.set_xlabel('Day of month')
ax.set_title('Maximum daily temperatures in Boston, 2019')

# Add a colorbar along the bottom and label it.
cbar = fig.colorbar(ax=ax, mappable=im, orientation='horizontal')
cbar.set_label('Temperature, $^\circ\mathrm{C}$')

plt.show()
```

❶ The "mappable" object passed to `fig.colorbar` is the `AxesImage` object returned by `ax.imshow`.

Example E7.24 The two-dimensional diffusion equation is

$$\frac{\partial U}{\partial t} = D\left(\frac{\partial^2 U}{\partial x^2} + \frac{\partial^2 U}{\partial y^2}\right),$$

where D is the diffusion coefficient. A simple numerical solution on the domain of the unit square $0 \leq x < 1, 0 \leq y < 1$ approximates $U(x, y; t)$ by the discrete function $u_{i,j}^{(n)}$, where $x = i\Delta x$, $y = j\Delta y$ and $t = n\Delta t$. Applying finite difference approximations yields

$$\frac{u_{i,j}^{(n+1)} - u_{i,j}^{(n)}}{\Delta t} = D\left[\frac{u_{i+1,j}^{(n)} - 2u_{i,j}^{(n)} + u_{i-1,j}^{(n)}}{(\Delta x)^2} + \frac{u_{i,j+1}^{(n)} - 2u_{i,j}^{(n)} + u_{i,j-1}^{(n)}}{(\Delta y)^2}\right],$$

and hence the state of the system at time step $n + 1$, $u_{i,j}^{(n+1)}$ may be calculated from its state at time step n, $u_{i,j}^{(n)}$ through the equation

$$u_{i,j}^{(n+1)} = u_{i,j}^{(n)} + D\Delta t\left[\frac{u_{i+1,j}^{(n)} - 2u_{i,j}^{(n)} + u_{i-1,j}^{(n)}}{(\Delta x)^2} + \frac{u_{i,j+1}^{(n)} - 2u_{i,j}^{(n)} + u_{i,j-1}^{(n)}}{(\Delta y)^2}\right].$$

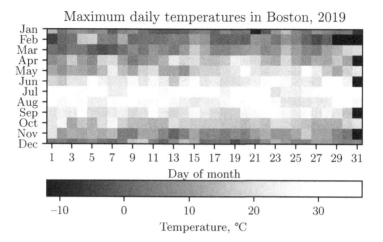

Figure 7.26 A heatmap of maximum daily temperatures in Boston during 2019.

Consider the diffusion equation applied to a metal plate initially at temperature T_{cold}, apart from a disc of a specified size, which is at temperature T_{hot}. We suppose that the edges of the plate are held fixed at T_{cool}. The following code applies the above formula to follow the evolution of the temperature of the plate. It can be shown that the maximum time step, Δt, that we can allow without the process becoming unstable is

$$\Delta t = \frac{1}{2D} \frac{(\Delta x \Delta y)^2}{(\Delta x)^2 + (\Delta y)^2}.$$

In the code below, each call to do_timestep updates the numpy array u from the results of the previous time step, u0. The simplest approach to applying the partial difference equation is to use a Python loop:

```
for i in range(1, nx-1):
    for j in range(1, ny-1):
        uxx = (u0[i+1,j] - 2*u0[i,j] + u0[i-1,j]) / dx2
        uyy = (u0[i,j+1] - 2*u0[i,j] + u0[i,j-1]) / dy2
        u[i,j] = u0[i,j] + dt * D * (uxx + uyy)
```

However, this runs extremely slowly and using vectorization will farm out these explicit loops to the much faster precompiled C code underlying NumPy's array implementation.

The state of the system is plotted as an image at four different stages of its evolution (see Figure 7.27).

Listing 7.25 The two-dimensional diffusion equation applied to the temperature of a steel plate

```
# eg7-diffusion2d.py
import numpy as np
import matplotlib.pyplot as plt

# plate size, mm.
w = h = 10.
# intervals in x-, y- directions, mm.
dx = dy = 0.1
```

```python
# Thermal diffusivity of steel, mm2.s-1.
D = 4.

Tcool, Thot = 300, 700

nx, ny = int(w/dx), int(h/dy)

dx2, dy2 = dx*dx, dy*dy
dt = dx2 * dy2 / (2 * D * (dx2 + dy2))

u0 = Tcool * np.ones((nx, ny))
u = u0.copy()

# Initial conditions - ring of inner radius r, width dr centered at (cx, cy) (mm)
r, cx, cy = 2, 5, 5
r2 = r**2
for i in range(nx):
    for j in range(ny):
        p2 = (i*dx-cx)**2 + (j*dy-cy)**2
        if p2 < r2:
            u0[i, j] = Thot

def do_timestep(u0, u):
    # Propagate with forward-difference in time, central-difference in space.
    u[1:-1, 1:-1] = u0[1:-1, 1:-1] + D * dt * (
        (u0[2:, 1:-1] - 2*u0[1:-1, 1:-1] + u0[:-2, 1:-1])/dx2
        + (u0[1:-1, 2:] - 2*u0[1:-1, 1:-1] + u0[1:-1, :-2])/dy2 )

    u0 = u.copy()
    return u0, u

# Number of time steps.
nsteps = 101
# Output 4 figures at these time steps.
mfig = [0, 10, 50, 100]
fignum = 0
fig = plt.figure()
for m in range(nsteps):
    u0, u = do_timestep(u0, u)
    if m in mfig:
        fignum += 1
        print(m, fignum)
        ax = fig.add_subplot(220 + fignum)
        im = ax.imshow(u.copy(), cmap=plt.get_cmap('hot'), vmin=Tcool, vmax=Thot)
        ax.set_axis_off()
        ax.set_title('{:.1f} ms'.format(m*dt*1000))
fig.subplots_adjust(right=0.85)
```
❶ `cbar_ax = fig.add_axes([0.9, 0.15, 0.03, 0.7])`
```python
cbar_ax.set_xlabel('$T$ / K', labelpad=20)
fig.colorbar(im, cax=cbar_ax)
plt.show()
```

❶ To set a common colorbar for the four plots we define its own Axes, `cbar_ax` and make room for it with `fig.subplots_adjust`. The plots all use the same color range, defined by `vmin` and `vmax`, so it doesn't matter which one we pass in the first argument to `fig.colorbar`.

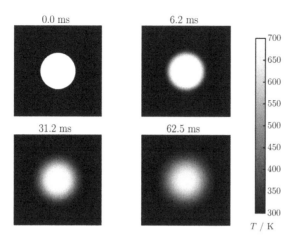

0.0 ms   6.2 ms

31.2 ms   62.5 ms

T / K

Figure 7.27 A representation of the temperature of a circular disc at four times after its instantaneous heating.

7.5.3   Exercises

Questions

Q7.5.1   Generate an image plot of the sinc function in the Cartesian plane, $\mathrm{sinc}(r) = \sin r/r$, where $r = \sqrt{x^2 + y^2}$.

Q7.5.2   The data provided in the comma-separated file `birthday-data.csv`, available at https://scipython.com/ex/bgd gives the number of births recorded by the US Centers for Disease Control and Prevention's National Center for Health Statistics for each day of the year as a total from years 1969–1988. The columns are month number (1 = January, 12 = December), day number and number of live births.

Use NumPy to estimate, for each day of the year, the probability of a particular individual's birthday being on that day. Plot the probabilities as a heatmap like that of Example E7.23 and investigate any features of interest.

Hint: the data need "cleaning" to a small extent – inspect the data file first to establish the presence of any incorrect entries.

Problems

P7.5.1   The so-called *chaos game* is an algorithm for generating a fractal. First define the n vertexes of a regular polygon and an initial point, (x_0, y_0), selected at random within the polygon. Then generate a sequence of points, starting with (x_0, y_0), where each point is a fraction r of the distance between the previous one and a polygon vertex chosen at random. For example, the algorithm applied with parameters $n = 3, r = 0.5$ generates a Sierpinski triangle.

Write a program to draw fractals using the chaos game algorithm.

P7.5.2 Extend the code in Example E7.17 to include contours of body mass index, defined by BMI = $(\text{mass/kg})/(\text{height/m})^2$. Plot these contours to delimit the supposed categories of "under-weight" (<18.5), "over-weight" (>25) and "obese" (>30). Manually place the contour labels so that they are out of the way of the scatter plotted data points and format them to one decimal place.

P7.5.3 The two-dimensional *advection equation* may be written

$$\frac{\partial U}{\partial t} = -v_x \frac{\partial U}{\partial x} - v_y \frac{\partial U}{\partial y},$$

where $\mathbf{v} = (v_x, v_y)$ is the vector velocity field, giving the velocity components v_x and v_y, which may vary as a function of position, (x, y). In a similar way to the approach taken in Example E7.24, this equation may be discretized and solved numerically. With forward-differences in time and central-differences in space, we have

$$u_{i,j}^{(n+1)} = u_{i,j}^{(n)} - \Delta t \left[v_{x;i,j} \frac{u_{i+1,j}^{(n)} - u_{i-1,j}^{(n)}}{2\Delta x} + v_{y;i,j} \frac{u_{i,j+1}^{(n)} - u_{i,j-1}^{(n)}}{2\Delta y} \right].$$

Implement this approximate numerical solution on the domain $0 \le x < 10, 0 \le y < 10$ discretized with $\Delta x = \Delta y = 0.1$ with the initial condition

$$u_0(x, y) = \exp\left(-\frac{(x - c_x)^2 + (y - c_y)^2}{\alpha^2} \right),$$

where $(c_x, c_y) = (5, 5)$ and $\alpha = 2$. Take the velocity field to be a circulation at constant speed 0.1 about an origin at $(7, 5)$.

P7.5.4 The *Julia set* associated with the complex function $f(z) = z^2 + c$ may be depicted using the following algorithm.

For each point, z_0, in the complex plane such that $-1.5 \le \text{Re}[z_0] \le 1.5$ and $-1.5 \le \text{Im}[z_0] \le 1.5$, iterate according to $z_{n+1} = z_n^2 + c$. Color the pixel in an image corresponding to this region of the complex plane according to the number of iterations required for $|z|$ to exceed some critical value, $|z|_{max}$ (or black if this does not happen before a certain maximum number of iterations n_{max}).

Write a program to plot the Julia set for $c = -0.1 + 0.65j$, using $|z|_{max} = 10$ and $n_{max} = 500$.

P7.5.5 The mean altitudes of the 10 km $\times$ 10 km *hectad* squares used by the UK's Ordnance Survey in mapping Great Britain are given in the NumPy array file gb-alt.npy, available at https://scipython.com/ex/bgb.

Plot a map of the island using this data with ax.imshow and plot further maps assuming a mean sea-level rise of (a) 25 m, (b) 50 m, (c) 200 m. In each case, deduce the percentage of land area remaining, relative to its present value.

7.6   Three-Dimensional Plots

Matplotlib is primarily a two-dimensional plotting library, but it does support three-dimensional (3D) plotting functionality that is good enough for many purposes. The easiest way to set up a three-dimensional plot is to import `Axes3D` from the `mpl_toolkits.mplot3d` module and to set the subplot's `projection` argument to `'3d'`:

```
import matplotlib.pyplot as plt
from mpl_toolkits.mplot3d import Axes3D

fig = plt.figure()
ax = fig.add_subplot(projection='3d')
```

The corresponding `Axes` object can then depict data in three dimensions as a line plot, scatter plot, wireframe plot or surface plot.[18]

7.6.1   Wireframe Plots and Surface Plots

The simplest kind of surface plot is a wireframe plot that draws lines in three-dimensional perspective joining the provided two-dimensional array of points, `Z`, on a grid of data values provided as two-dimensional arrays, `X` and `Y` (as for `imshow` and `contour`). By default, wires are drawn for every point in the array: if this is too many, set the arguments `rstride` and `cstride` to specify the array row step size and column step size.

The `ax.plot_surface` method is similar but produces a surface plot of filled patches. The patch colors can be set to a single color with the `color` argument or styled to a specifed colormap with the `cmap` argument. `rstride` and `cstride` default to 10 for the `ax.plot_surface` method. Both methods are illustrated in the following example.

Example E7.25   Some of the different options for producing surface plots are illustrated by the code below, which produces Figure 7.28.

Listing 7.26 Four three-dimensional plots of a simple two-dimensional Gaussian function

---

```
# eg7-3d-surface-plots.py
import numpy as np
import matplotlib.pyplot as plt
from mpl_toolkits.mplot3d import Axes3D
import matplotlib.cm as cm

L, n = 2, 400
x = np.linspace(-L, L, n)
y = x.copy()
X, Y = np.meshgrid(x, y)
Z = np.exp(-(X**2 + Y**2))

fig, ax = plt.subplots(nrows=2, ncols=2, subplot_kw={'projection': '3d'})
```

---

[18] It is even possible to produce three-dimensional contour plots and bar charts, though these are of doubtful use in practice.

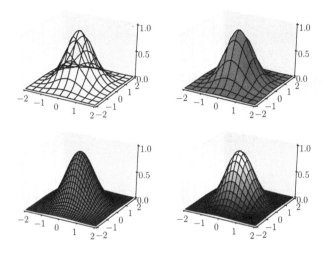

Figure 7.28 Four different three-dimensional surface plots of the same function.

```
ax[0, 0].plot_wireframe(X, Y, Z, rstride=40, cstride=40)
ax[0, 1].plot_surface(X, Y, Z, rstride=40, cstride=40, color='m')
ax[1, 0].plot_surface(X, Y, Z, rstride=12, cstride=12, color='m')
ax[1, 1].plot_surface(X, Y, Z, rstride=20, cstride=20, cmap=cm.hot)
for axes in ax.flatten():
    axes.set_xticks([-2, -1, 0, 1, 2])
    axes.set_yticks([-2, -1, 0, 1, 2])
    axes.set_zticks([0, 0.5, 1])
fig.tight_layout()
plt.show()
```

In an interactive plot, the viewing direction can be changed by clicking and dragging on the plot. To fix a particular viewing direction for a static plot image, pass the required elevation and azimuthal angles (in degrees, in that order) to `ax.view_init`, as in the following example.

Example E7.26 The parametric description of a torus with major radius c and minor radius a is

$$x = (c + a\cos\theta)\cos\phi$$
$$y = (c + a\cos\theta)\sin\phi$$
$$z = a\sin\theta$$

for θ and ϕ each between 0 and 2π. The code below outputs two views of a torus rendered as a surface plot (Figure 7.29).

Listing 7.27 A three-dimensional surface plot of a torus

```
# eg7-torus-surface.py
```

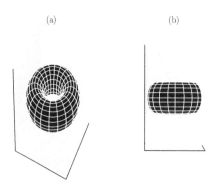

Figure 7.29 Two views of the same torus: (a) $\theta = 36°, \phi = 26°$, (b) $\theta = 0°, \phi = 0°$.

```
import numpy as np
import matplotlib.pyplot as plt
from mpl_toolkits.mplot3d import Axes3D

n = 100

theta = np.linspace(0, 2.*np.pi, n)
phi = np.linspace(0, 2.*np.pi, n)
❶ theta, phi = np.meshgrid(theta, phi)
c, a = 2, 1
x = (c + a*np.cos(theta)) * np.cos(phi)
y = (c + a*np.cos(theta)) * np.sin(phi)
z = a * np.sin(theta)

fig = plt.figure()
ax1 = fig.add_subplot(121, projection='3d')
ax1.set_zlim(-3, 3)
❷ ax1.plot_surface(x, y, z, rstride=5, cstride=5, color='k', edgecolors='w')
❸ ax1.view_init(36, 26)
ax2 = fig.add_subplot(122, projection='3d')
ax2.set_zlim(-3, 3)
ax2.plot_surface(x, y, z, rstride=5, cstride=5, color='k', edgecolors='w')
ax2.view_init(0, 0)
ax2.set_xticks([])
plt.show()
```

❶   We need θ and ϕ to range over the interval $(0, 2\pi)$ independently, so we use a meshgrid.

❷   Note that we can use keywords such as `edgecolors` to style the polygon patches created by `ax.plot_surface`.

❸   Elevation angle above the xy-plane of $36°$, azimuthal angle in the xy-plane of $26°$.

7.6.2   Line Plots and Scatter Plots

Line plots and scatter plots work in three dimensions in a way similar to that in which they work in two dimensions: the basic method call is `ax.plot(x, y, z)` and

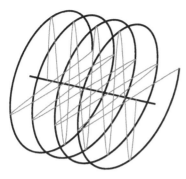

Figure 7.30 A depiction of circularly polarized light as a helix on a three-dimensional plot.

`ax.scatter(x, y, z)`, where x, y and z are equal-length, one-dimensional arrays. Only limited annotation of such plots is possible without using advanced methods, however.

Example E7.27 Below is a simple example of a three-dimensional plot of a helix, which could represent circularly polarized light, for example. See Figure 7.30.

Listing 7.28 A depiction of a helix on a three-dimensional plot

```
# eg7-circular-polarization.py
import numpy as np
import matplotlib.pyplot as plt
from mpl_toolkits.mplot3d import Axes3D

n = 1000
fig = plt.figure()
ax = fig.add_subplot(projection='3d')

# Plot a helix along the x-axis.
theta_max = 8 * np.pi
theta = np.linspace(0, theta_max, n)
x = theta
z =  np.sin(theta)
y =  np.cos(theta)
ax.plot(x, y, z, 'b', lw=2)

# A line through the center of the helix.
ax.plot((-theta_max*0.2, theta_max * 1.2), (0, 0), (0, 0), color='k', lw=2)
# sin/cos components of the helix (e.g. electric and magnetic field
# components of a circularly polarized electromagnetic wave..
ax.plot(x, y, 0, color='r', lw=1, alpha=0.5)
ax.plot(x, [0]*n, z, color='m', lw=1, alpha=0.5)

# Remove axis planes, ticks and labels.
ax.set_axis_off()
plt.show()
```

7.7   Animation

This section provides a brief introduction to the use of `FuncAnimation` to produce animated plots and charts from a Python script or within a Jupyter notebook. Matplotlib's animation functionality is provided by the `animation` module, which must be explicitly imported before it can be used:

```
import matplotlib.animation as animation
```

7.7.1   Animating Plotted Data

A Simple Animated Line

The `FuncAnimation` class makes an animation by repeatedly calling a provided function, `func`, which updates the plotted objects on a Matplotlib `Figure` object, `fig`. Additional arguments are described in Table 7.12

The `Figure` and its Axes should be set up before calling `FuncAnimation`, and any references to plotted objects need to be retained so that their data can be manipulated by the animation function. For example, the (x, y) data plotted in a `Line2D` object can be (re)set using its `set_data`. This is illustrated in the code below, which animates a decaying sine curve.

Example E7.28   The following code animates a decaying sine curve that could, for example, represent the decaying chime of a struck tuning fork at a fixed frequency:

$$M(t) = \sin(2\pi f t)e^{-\alpha t}$$

Listing 7.29 An animation of a decaying sine curve

```
import numpy as np
import matplotlib.pyplot as plt
import matplotlib.animation as animation

# Time step for the animation (s), max time to animate for (s).
dt, tmax = 0.01, 5
# Signal frequency (s-1), decay constant (s-1).
f, alpha = 2.5, 1
# These lists will hold the data to plot.
t, M = [], []

# Draw an empty plot, but preset the plot x- and y-limits.
fig, ax = plt.subplots()
line, = ax.plot([], [])                                    ❶
ax.set_xlim(0, tmax)
ax.set_ylim(-1, 1)
ax.set_xlabel('t /s')
ax.set_ylabel('M (arb. units)')

def animate(i):
    """Draw the frame i of the animation."""
```

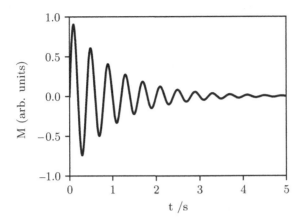

Figure 7.31 The final frame of an animation of a decaying sine curve.

❷
```
    global t, M
    # Append this time point and its data and set the plotted line data.
    _t = i*dt
    t.append(_t)
    M.append(np.sin(2*np.pi*f*_t) * np.exp(-alpha*_t))
    line.set_data(t, M)

# Interval between frames in ms, total number of frames to use.
interval, nframes = 1000 * dt, int(tmax / dt)
# Animate once (set repeat=False so the animation doesn't loop).
ani = animation.FuncAnimation(fig, animate, frames=nframes, repeat=False,
                              interval=interval)
plt.show()
```
❸ appears at the line `interval, nframes = 1000 * dt, int(tmax / dt)`

❶ Recall that the `ax.plot` method returns a tuple of `Line2D` objects, even if there is only one plotted line. We need to retain a reference to it so we can set its data in the animation function, `animate`.

❷ By declaring the `t` and `M` lists to be `global` objects we can modify them from inside the `animate` function.

❸ By setting the time interval between frames to be the same (in milliseconds) as the time step, the animation is made to appear "in real time."

The final frame of the animation is depicted in Figure 7.31.

Blitting

In the previous example, the entire line object had to be redrawn for each frame. Where there are a lot of data or a complex figure, this can slow down the animation. In this case, it can help to use *blitting*, a technique from computer graphics that enables the animation loop to only redraw parts of the figure that have changed between frames instead of having to redraw all the data.

Table 7.12 Arguments to `FuncAnimation`

Argument	Description
`fig`	The Matplotlib `Figure` object to animate
`func`	A function called to set each frame of the animation by manipulating objects on the `Figure`
`frames`	The source of the object passed to `func` for each frame; if `None` (the default), an increasing integer index is passed; this can also be an iterable or generator function
`init_func`	The function called to produce an empty frame of the animation; must be provided if `blit=True`
`fargs`	Any additional arguments to be passed to `func`
`interval`	The time delay between frames in milliseconds (by default, 200)
`repeat`	Boolean flag determining whether the animation should loop (repeat) or not (by default, `True`)
`blit`	Boolean flag determining whether to use blitting to optimize the animation (by default, `False`); see text

There is some extra code necessary to use blitting: we must define a method to pass to `FuncAnimation`'s `init_func` argument that creates a blank frame but returns a sequence of the artist objects that will be redrawn for each frame. The function `func`, which is called for each frame to be rendered, must also return a sequence of the altered artists.

Example E7.29   This code repeats the animation of the previous example, but using the blitting technique and passing arguments explicitly to the animation function instead of declaring them to be `globals` within it.

Listing 7.30 An animation of a decaying sine curve, using `blit=True`

```
import numpy as np
import matplotlib.pyplot as plt
import matplotlib.animation as animation

# Time step for the animation (s), max time to animate for (s).
dt, tmax = 0.01, 5
# Signal frequency (s-1), decay constant (s-1).
f, alpha = 2.5, 1
# These lists will hold the data to plot.
t, M = [], []

# Draw an empty plot, but preset the plot x- and y-limits.
fig, ax = plt.subplots()
line, = ax.plot([], [])
ax.set_xlim(0, tmax)
ax.set_ylim(-1, 1)
ax.set_xlabel('t /s')
ax.set_ylabel('M (arb. units)')

def init():
    return line,
```

```
def animate(i, t, M):
    """Draw the frame i of the animation."""

    # Append this time point and its data and set the plotted line data.
    _t = i*dt
    t.append(_t)
    M.append(np.sin(2*np.pi*f*_t) * np.exp(-alpha*_t))
    line.set_data(t, M)
    return line,

# Interval between frames in ms, total number of frames to use.
interval, nframes = 1000 * dt, int(tmax / dt)
# Animate once (set repeat=False so the animation doesn't loop).
ani = animation.FuncAnimation(fig, animate, frames=nframes, init_func=init,
                    fargs=(t, M), repeat=False, interval=interval, blit=True)
plt.show()
```

❶ Any objects assigned to the `fargs` argument of `FuncAnimation` will be handed on to the animation function.

7.7.2   Animating Other Matplotlib Objects

To animate other Matplotlib objects such as patches and annotation labels, a reference must be kept and manipulated for each frame. Just as `Line2D` objects have a `set_data` method, these other classes have "setter" methods (for example, `set_center`, `set_radius` for `Circle` patches), as demonstrated in the following example.

Example E7.30   The following program animates a bouncing ball, starting from a position $(0, y_0)$ with velocity $(v_{x0}, 0)$. The ball's position, trajectory history and height label all change with each frame.

Here, the `frames` argument to `FuncAnimation` is set to a generator function, `get_pos`, which returns the next position of the ball each time it iterates. This position is handed on to the `animate` function instead of the integer index of the frame.

Listing 7.31 An animation of a bouncing ball

```
import matplotlib.pyplot as plt
import matplotlib.animation as animation

# Acceleration due to gravity, m.s-2.
g = 9.81
# The maximum x-range of ball's trajectory to plot.
XMAX = 5
# The coefficient of restitution for bounces (-v_up/v_down).
cor = 0.65
# The time step for the animation.
dt = 0.005

# Initial position and velocity vectors.
x0, y0 = 0, 4
vx0, vy0 = 1, 0
```

```
def get_pos(t=0):
    """A generator yielding the ball's position at time t."""
    x, y, vx, vy = x0, y0, vx0, vy0
❶   while x < XMAX:
        t += dt
        x += vx0 * dt
        y += vy * dt
        vy -= g * dt
        if y < 0:
            # bounce!
            y = 0
            vy = -vy * cor
        yield x, y

def init():
    """Initialize the animation figure."""
    ax.set_xlim(0, XMAX)
    ax.set_ylim(0, y0)
    ax.set_xlabel('$x$ /m')
    ax.set_ylabel('$y$ /m')
    line.set_data(xdata, ydata)
    ball.set_center((x0, y0))
    height_text.set_text(f'Height: {y0:.1f} m')
    return line, ball, height_text

def animate(pos):
    """For each frame, advance the animation to the new position, pos."""
    x, y = pos
    xdata.append(x)
    ydata.append(y)
    line.set_data(xdata, ydata)
    ball.set_center((x, y))
    height_text.set_text(f'Height: {y:.1f} m')
    return line, ball, height_text

# Set up a new Figure, with equal aspect ratio so the ball appears round.
fig, ax = plt.subplots()
ax.set_aspect('equal')

# These are the objects we need to keep track of.
line, = ax.plot([], [], lw=2)
ball = plt.Circle((x0, y0), 0.08)
height_text = ax.text(XMAX*0.5, y0*0.8, f'Height: {y0:.1f} m')
ax.add_patch(ball)
xdata, ydata = [], []

interval = 1000*dt
ani = animation.FuncAnimation(fig, animate, get_pos, blit=True,
                    interval=interval, repeat=False, init_func=init)
plt.show()
```

❶ The generator function will keep on producing the ball's position vector, (x, y) until the ball's x-coordinate reaches XMAX; then when the generator is exhausted and produces None, the animation stops.

The final frame of the animation is depicted in Figure 7.32.

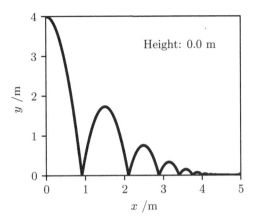

Figure 7.32 The final frame of an animation of a bouncing ball.

7.7.3   Exercises

Problems

P7.7.1   Use Matplotlib's `FuncAnimation` class to produce an animation of a swinging pendulum with an initial maximum angle from vertical and zero initial velocity. Integrate the equation of motion numerically, and repeat the animation after one period of the motion.

P7.7.2   Update Example E7.24 to display an animation of the evolution of temperature of the metal plate over time.

P7.7.3   NASA's Jet Propulsion Laboratory (JPL) maintains a web service and database called HORIZONS that can be used to calculate ephemerides (the trajectories of objects in the solar system over time). Use data from this resource, which has been pre-selected and can be downloaded from https://scipython.com/ex/bas, to produce an animation of the trajectory of the Voyager 2 space probe, between its launch in August 1977 and the end of 1999. This period includes several gravity-assist ("slingshot") maneuvers as the spacecraft flies past the larger planets. Only the (X, Y) coordinates of the relevant bodies need be considered.

8   SciPy

SciPy is a library of Python modules for scientific computing that provides more specific functionality than the generic data structures and mathematical algorithms of NumPy. For example, it contains modules for the evaluation of special functions frequently encountered in science and engineering, optimization, integration, interpolation and image manipulation. As with the NumPy library, many of SciPy's underlying algorithms are executed as compiled C code, so they are fast. Also like NumPy and Python itself, SciPy is free software.

There is little new syntax to learn in using the SciPy routines, so this chapter will focus on examples of the library's use in short programs of relevance to science and engineering.

8.1   Physical Constants and Special Functions

The useful `scipy.constants` package provides the internationally agreed standard values and uncertainties for physical constants. The `scipy.special` package also supplies a large number of algorithms for calculating functions that appear in science, mathematical analysis and engineering, including:

- Airy functions;
- elliptic functions and integrals;
- Bessel functions, their zeros, derivatives and integrals;
- spherical Bessel functions;
- a variety of statistical functions and distributions;
- gamma and beta functions;
- the error function;
- Fresnel integrals;
- Legendre functions and associated Legendre functions;
- a variety of orthogonal polynomials;
- hypergeometric functions;
- parabolic cylinder functions;
- Matheiu functions;
- spheroidal functions.

They are described in detail in the documentation;[1] this section focuses on a few representative examples.

Most of these special functions are implemented in SciPy as universal functions: that is, they support broadcasting and vectorization (automatic array-looping), and so work as expected with NumPy arrays.

8.1.1 Physical Constants

SciPy contains the 2018 CODATA internationally recommended values[2] of many physical constants. They are held, with their units and uncertainties, in a dictionary, `scipy.constants.physical_constants`, keyed by an identifying string. For example,

```
In [x]: import scipy.constants as pc
In [x]: pc.physical_constants['Avogadro constant']
Out[x]: (6.022140857e+23, 'mol^-1', 7400000000000000.0)
```

The convenience methods `value`, `unit` and `precision` retrieve the corresponding properties on their own:

```
In [x]: pc.value('electron mass')
Out[x]: 9.1093837015e-31
In [x]: pc.unit('electron mass')
Out[x]: 'kg'
In [x]: pc.precision('electron mass')
3.0737534961217373e-10
```

To save typing, it is usual to assign the value to a variable name at the start of a program, for example,

```
In [x]: muB = pc.value('Bohr magneton')
```

A full list of the constants and their names is given in the SciPy documentation,[3] but Table 8.1 lists the more important ones. Some particularly important constants have a direct variable assignment within `scipy.constants` (in SI units) and so can be imported directly:

```
In [x]: from scipy.constants import c, R, k
In [x]: c, R, k     # speed of light, gas constant, Boltzmann constant
Out[x]: (299792458.0, 8.314462618, 1.380649e-23)
```

Where this is the case, the variable name is given in the table. You will probably find it convenient to use the `scipy.constants` values, but should be aware that if and when newer values are released the package may be updated – this means that your code may produce slightly different results for different versions of SciPy. The values given below are from SciPy version 1.4, which includes the 2019 redefinition of the SI base units.

There are one or two useful conversion factors and methods defined within the `scipy.constants` package, which also includes representations of the SI prefixes. For example,

---

[1] https://docs.scipy.org/doc/scipy/reference/special.html.

[2] https://physics.nist.gov/cuu/Constants/.

[3] https://docs.scipy.org/doc/scipy/reference/constants.html.

Table 8.1 Physical constants in `scipy.constants`

Constant string	Variable	Value	Units
'atomic mass constant'	m_u	1.6605390666e-27	kg
'Avogadro constant'	N_A	6.02214076e+23	mol^{-1}
'Bohr magneton'		9.2740100783e-24	$\mathrm{J\,T}^{-1}$
'Bohr radius'		5.29177210903e-11	m
'Boltzmann constant'	k	1.380649e-23	$\mathrm{J\,K}^{-1}$
'electron mass'	m_e	9.1093837015e-31	kg
'elementary charge'	e	1.602176634e-19	C
'Faraday constant'		96485.33212	$\mathrm{C\,mol}^{-1}$
'fine-structure constant'	alpha	0.0072973525693	
'molar gas constant'	R	8.314462618	$\mathrm{J\,K}^{-1}\,\mathrm{mol}^{-1}$
'neutron mass'	m_n	1.67492749804e-27	kg
'Newtonian constant of gravitation'	G	6.6743e-11	$\mathrm{m}^3\,\mathrm{kg}^{-1}\,\mathrm{s}^{-2}$
'Planck constant'	h	6.62607015e-34	$\mathrm{J\,s}$
'proton mass'	m_p	1.67262192369e-27	kg
'Rydberg constant'	Rydberg	10973731.56816	m^{-1}
'speed of light in vacuum'	c	299792458.0	$\mathrm{m\,s}^{-1}$

```
In [x]: import scipy.constants as pc
In [x]: pc.atm
Out[x]: 101325.0                    # 1 atm in Pa
In [x]: pc.bar
Out[x]: 100000.0                    # 1 bar in Pa
In [x]: pc.torr
Out[x]: 133.32236842105263          # 1 torr in Pa
In [x]: pc.zero_Celsius
Out[x]: 273.15                      # 0 degC in K
In [x]: pc.micro                    # also nano, pico, mega, giga, etc.
Out[x]: 1e-06
```

Example E8.1 This example uses the `scipy.constants.physical_constants` dictionary to determine which are the least accurately known constants. To do this we need the *relative* uncertainties in the constants' values. The code mentioned here uses a structured array to calculate these and outputs the least-well-determined constants.

Listing 8.1 Least-well-defined physical constants

```
import numpy as np

from scipy.constants import physical_constants

def make_record(k, v):
    """

    Return the record for this constant from the key and value of its entry
    in the physical_constants dictionary.

    """
```

```
    name = k
    val, units, abs_unc = v
    # Calculate the relative uncertainty in ppm.
    rel_unc = abs_unc / abs(val) * 1.e6
    return name, val, units, abs_unc, rel_unc

dtype = [('name', 'S50'), ('val', 'f8'), ('units', 'S20'),
         ('abs_unc', 'f8'), ('rel_unc', 'f8')]
constants = np.array([make_record(k, v) for k, v in physical_constants.items()],
                     dtype=dtype)
constants.sort(order='rel_unc')

# List the 10 constants with the largest relative uncertainties.
for rec in constants[-10:]:
    print('{:.0f} ppm: {:s} = {:g} {:s}'.format(rec['rel_unc'],
            rec['name'].decode(), rec['val'], rec['units'].decode()))
```

The output is shown here:

```
90 ppm: tau Compton wavelength over 2 pi = 1.11056e-16 m
90 ppm: tau mass energy equivalent in MeV = 1776.82 MeV
193 ppm: W to Z mass ratio = 0.88153
348 ppm: deuteron rms charge radius = 2.12799e-15 m
428 ppm: proton mag. shielding correction = 2.5689e-05
428 ppm: proton magn. shielding correction = 2.5689e-05
829 ppm: shielding difference of t and p in HT = 2.414e-08
990 ppm: shielding difference of d and p in HD = 2.02e-08
1346 ppm: weak mixing angle = 0.2229
2258 ppm: proton rms charge radius = 8.414e-16 m
```

8.1.2   Airy and Bessel Functions

The Airy functions Ai(x) and Bi(x) are the linearly independent solutions to the Airy equation, $y'' - xy = 0$, which occurs in quantum mechanics, optics, electrodynamics and other areas of physics. The functions (Ai, Bi) and their derivatives (Aip, Bip) are returned by the function scipy.special.airy. The only required argument is x, which can be complex and can be a NumPy array:

```
In [x]: Ai, Aip, Bi, Bip = airy(0)
In [x]: Ai, Aip, Bi, Bip
(0.35502805388781722, -0.25881940379280682, 0.61492662744600068,
0.44828835735382638)
```

The first nt zeros of the Airy functions and their derivatives are returned by the function scipy.special.ai_zeros(nt):

```
In [x]: a, ap, ai, aip = ai_zeros(2)    # arrays for the first two zeros of Ai
In [x]: a[1], ap[1], ai[1], aip[1]      # look at the second zero:
Out[x]: (-4.0879494441309721, -3.248197582179837, -0.41901547803256406,
         -0.80311136965486463)
In [x]: airy(a[1])[0]                    # Ai(a) should = 0
Out[x]: 1.2774882441379295e-15           # close enough
In [x]: airy(ap[1])[1]                    # Aip(ap) should = 0
Out[x]: -3.2322209157744908e-16          # close enough
In [x]: airy(ap[1])[0]                    # Ai(ap) is returned as ai above
```

```
Out[x]: -0.41901547803256395
In [x]: airy(a[1])[1]                    # Aip(a) is returned as aip above
Out[x]: -0.80311136965486396
```

◇ **Example E8.2** Consider a particle of mass m moving in a constant gravitational field such that its potential energy at a height z above a surface is mgz. If the particle bounces elastically on the surface, the classical probability density corresponding to its position is

$$P_{cl}(z) = \frac{1}{\sqrt{z_{max}(z_{max} - z)}},$$

where z_{max} is the maximum height it reaches.

The quantum mechanical behavior of this system may described by the solution to the time-independent Schrödinger equation,

$$-\frac{\hbar^2}{2m}\frac{d^2\psi}{dz^2} + mgz\psi = E\psi,$$

which is simplified by the coordinate rescaling $q = z/\alpha$, where $\alpha = (\hbar^2/2m^2g)^{1/3}$:

$$\frac{d^2\psi}{dq^2} - (q - q_E)\psi = 0, \quad \text{where } q_E = \frac{E}{mg\alpha}.$$

The solutions to this differential equation are the Airy functions. The boundary condition $\psi(z) \to 0$ as $z \to \infty$ specifically gives:

$$\psi(q) = N_E \text{Ai}(q - q_E),$$

where N_E is a normalization constant.

The second boundary condition, $\psi(q = 0) = 0$, leads to quantization in terms of a quantum number $n = 1, 2, 3, \ldots$, with scaled energy values q_E found from the zeros of the Airy function: $\text{Ai}(-q_E) = 0$.

The following program plots the classical and quantum probability distributions, $P_{cl}(z)$ and $|\psi(z)|^2$, for $n = 1$ and $n = 16$ (Figure 8.1).

Listing 8.2 Probability densities for a particle in a uniform gravitational field

```
# eg8-qm-gravfield.py
import numpy as np
from scipy.special import airy, ai_zeros
import matplotlib.pyplot as plt


nmax = 16


# Find the first nmax zeros of Ai(x).
a, _, _, _ = ai_zeros(nmax)
# The actual boundary condition is Ai(-qE) = 0 at q = 0, so:
qE = -a


def prob_qm(n):
    """

    Return the quantum mechanical probability density for a particle moving
```

❶

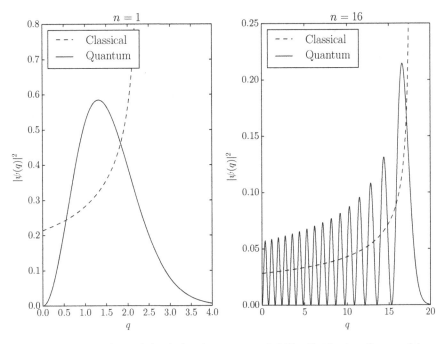

Figure 8.1 A comparison of classical and quantum probability distributions for a particle moving in a constant gravitational field at two different energies.

```
    in a uniform gravitational field.

    """
    # The quantum mechanical wavefunction is proportional to Ai(q - qE), where
    # the qE corresponding to quantum number n is indexed at n - 1.
    psi, _, _, _ = airy(q-qE[n-1])
    # Return the probability density, after rough-and-ready normalization.
    P = psi**2
    return P / (sum(P) * dq)

def prob_cl(n):
    """
    Return the classical probability density for a particle bouncing
    elastically in a uniform gravitational field.

    """
    # The classical probability density is already normalized.
    return 0.5/np.sqrt(qE[n-1]*(qE[n-1]-q))

# The ground state, n = 1.
q, dq = np.linspace(0, 4, 1000, retstep=True)
plt.plot(q, prob_cl(1), label='Classical')
plt.plot(q, prob_qm(1), label='Quantum')
plt.ylim(0, 0.8)
plt.legend()
plt.show()

# An excited state, n = 16.
```

```
q, dq = np.linspace(0, 20, 1000, retstep=True)
plt.plot(q, prob_cl(16), label='Classical')
plt.plot(q, prob_qm(16), label='Quantum')
plt.ylim(0, 0.25)
plt.legend(loc='upper left')
plt.show()
```

❶ We use scipy.special.ai_zeros to retrieve the $n = 1$ and $n = 16$ eigenvalues.

❷ scipy.special.airy finds the corresponding wavefunctions and hence probability densities.

❸ For the sake of illustration, these are normalized approximately by a very simple numerical integration.

Bessel functions are another important class of function with many applications to physics and engineering. SciPy provides several functions for evaluating them, their derivatives and their zeros.

- jn(v, x) and jv(v, x) return the *Bessel function of the first kind* at x for order v ($J_v(x)$). v can be real or integer.
- yn(n, x) and yv(v, x) return the *Bessel function of the second kind* at x for integer order n ($Y_n(x)$) and real order v ($Y_v(x)$), respectively.
- in(n, x) and iv(v, x) return the *modified Bessel function of the first kind* at x for integer order n ($I_n(x)$) and real order v ($I_v(x)$), respectively.
- kn(n, x) and kv(v, x) return the *modified Bessel function of the second kind* at x for integer order n ($K_n(x)$) and real order v ($K_v(x)$), respectively.
- The functions jvp(v, x), yvp(v, x), ivp(v, x) and kvp(v, x) return the *derivatives* of the earlier mentioned functions. By default, the first derivative is returned; to return the nth derivative, set the optional argument, n.
- Several functions can be used to obtain the *zeros* of the Bessel functions. Probably the most useful are jn_zeros(n, nt), jnp_zeros(n, nt), yn_zeros(n, nt) and ynp_zeros(n, nt), which return the first nt zeros of $J_n(x)$, $J_n'(x)$, $Y_n(x)$ and $Y_n'(x)$.

Example E8.3   The vibrations of a thin circular membrane stretched across a rigid circular frame (such as a drum head) can be described as normal modes written in terms of Bessel functions:

$$z(r, \theta; t) = A J_n(kr) \sin n\theta \cos kvt,$$

where (r, θ) describes a position in polar coordinates with the origin at the center of the membrane, t is time and v is a constant depending on the tension and surface density of the drum. The modes are labeled by integers $n = 0, 1, \ldots$ and $m = 1, 2, 3, \ldots$, where k is the mth zero of J_n.

The following program produces a plot of the displacement of the membrane in the $n = 3, m = 2$ normal mode at time $t = 0$ (Figure 8.2).

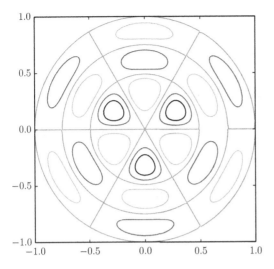

Figure 8.2 The $n = 3, m = 2$ normal mode of a vibrating circular drum.

Listing 8.3 Normal modes of a vibrating circular drum

```python
# eg8-drum-normal-modes.py
import numpy as np
from scipy.special import jn, jn_zeros
import matplotlib.pyplot as plt

# Allow calculations up to m = mmax.
mmax = 5

def displacement(n, m, r, theta):
    """

    Calculate the displacement of the drum membrane at (r, theta; t = 0)
    in the normal mode described by integers n >= 0, 0 < m <= mmax.

    """

    # Pick off the mth zero of Bessel function Jn.
    k = jn_zeros(n, mmax+1)[m]
    return np.sin(n*theta) * jn(n, r*k)

# Positions on the drum surface are specified in polar coordinates.
r = np.linspace(0, 1, 100)
theta = np.linspace(0, 2 * np.pi, 100)

# Create arrays of cartesian coordinates (x, y) ...
x = np.array([rr*np.cos(theta) for rr in r])
y = np.array([rr*np.sin(theta) for rr in r])
# ... and vertical displacement (z) for the required normal mode at
# time, t = 0.
n, m = 3, 2
z = np.array([displacement(n, m, rr, theta) for rr in r])

plt.contour(x, y, z)
```

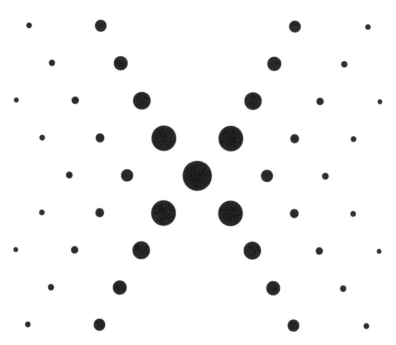

Figure 8.3 The diffraction pattern of a uniform, continuous helix.

```
plt.show()
```

Example E8.4 In an important paper in 1953[4] Rosalind Franklin published the X-ray diffraction pattern of DNA from calf thymus, which displays a characteristic X shape of diffraction spots, indicative of a helical structure.

The diffraction pattern of a uniform, continuous helix consists of a series of "layer lines" of spacing $1/p$ in reciprocal space, where p is the helix pitch (the height of one complete turn of the helix, measured parallel to its axis). The intensity distribution along the nth layer line is proportional to the square of the nth Bessel function, $J_n(2\pi rR)$, where r is the radius of the helix and R is the radial coordinate in reciprocal space.

Consider the diffraction pattern of a helix with $p = 34$ Å and $r = 10$ Å. The code listing here produces an SVG image of the diffraction pattern of a helix (Figure 8.3).

Listing 8.4 Generating an image of the diffraction pattern of a uniform, continuous helix

```
# eg8-dna-diffraction.py

import numpy as np
from scipy.special import jn
import matplotlib.pyplot as plt
```

[4] R. E. Franklin and R. G. Gosling, *Nature* **171**, 740 (1953).

```
# Vertical range of the diffraction pattern: plot nlayer line layers above and
# below the center horizontal.
nlayers = 5
ymin, ymax = -nlayers, nlayers

# Horizontal range of the diffraction pattern, x = 2pi.r.R.
xmin, xmax = -10, 10
npts = 4000
x = np.linspace(xmin, xmax, npts)

# Diffraction pattern along each line layer: |Jn(x)|^2
# for n = 0, 1, ..., nlayers - 1.
```
❶
```
layers = np.array([jn(i, x)**2 for i in range(nlayers)])

# Obtain the indexes of the maxima in each layer.
```
❷
```
maxi = [(np.diff(np.sign(np.diff(layers[i,:]))) < 0).nonzero()[0] + 1
                                            for i in range(nlayers)]

# Create the SVG image, using circles of different radii for diffraction spots.
svg_name='eg8-dna-diffraction.svg'
canvas_width = canvas_height = 500
fo = open(svg_name, 'w')
print("""<?xml version="1.0" encoding="utf-8"?>
        <svg xmlns="www.w3.org/2000/svg"
            xmlns:xlink="www.w3.org/1999/xlink"
            width="{}" height="{}" style="background: {}">""".format(
            canvas_width, canvas_height, '#ffffff'), file=fo)

def svg_circle(r, cx, cy):
    """ Return the SVG mark up for a circle of radius r centered at (cx, cy). """
    return r'<circle r="{}" cx="{}" cy="{}"/>'.format(r, cx, cy)

# For each spot in each layer, draw a circle on the canvas. The circle radius
# is the scaled value of the diffraction intensity maximum, with a ceiling
# value of spot_max_radius because the center spots are very intense.
spot_scaling, spot_max_radius = 50, 20
for i in range(nlayers):
    for j in maxi[i]:
```
❸
```
        sx = (x[j] - xmin)/(xmax - xmin) * canvas_width
        sy = (i - ymin)/(ymax - ymin) * canvas_height
        spot_radius = min(layers[i, j]*spot_scaling, spot_max_radius)
        print(svg_circle(spot_radius, sx, sy), file=fo)
        if i:
            # The pattern is symmetric about the center horizontal:
            # duplicate the layers with i > 0.
            sy = canvas_height - sy
            print(svg_circle(spot_radius, sx, sy), file=fo)

print(r'</svg>', file=fo)
```

❶ The two-dimensional array, `layers`, holds the diffraction intensity in each line layer, calculated as the square of a Bessel function.

❷ For plotting the pattern, we need to find the indexes of the maxima in the `layers` array: this line of code finds these maxima by determining where the *differences* between neighboring items go from positive to negative.

❸ Map the (`x`, `y`) coordinates in the reciprocal space of the diffraction pattern onto the canvas coordinates, (`sx`, `sy`).

8.1.3   The Gamma and Beta Functions; Elliptic Integrals

The gamma function is defined by the improper integral

$$\Gamma(x) = \int_0^\infty t^{x-1}e^{-t}\,dt,$$

for real $x > 0$, and extended to negative x and complex numbers by analytic continuation. It occurs frequently in integration problems, combinatorics and in expressions for other special functions.

The gamma function and its natural logarithm are returned by the functions `gamma(x)` and `gammaln(x)`. There are also methods for the evaluation of the incomplete gamma functions (obtained by replacing the lower or upper limits in the integral above with the parameter a) and their inverses; these will not be described in detail here.

Example E8.5   The gamma function is related to the factorial by $\Gamma(x) = (x - 1)!$ and both are plotted in the code mentioned later (see Figure 8.4). Note that $\Gamma(x)$ is not defined for negative integer x, which leads to discontinuities in the plot.

Listing 8.5 The Gamma function on the real line

```
# eg3-gamma.py
import numpy as np
from scipy.special import gamma
import matplotlib.pyplot as plt

# The Gamma function.
ax = plt.linspace(-5, 5, 1000)
plt.plot(ax, gamma(ax), ls='-', c='k', label='$\Gamma(x)$')

# (x - 1)! for x = 1, 2, ..., 6.
ax2 = plt.linspace(1, 6, 6)
xm1fac = np.array([1, 1, 2, 6, 24, 120])
plt.plot(ax2, xm1fac, marker='*', markersize=12, markeredgecolor='r',
         ls='', c='r', label='$(x-1)!$')

plt.ylim(-50, 50)
plt.xlim(-5, 5)
plt.xlabel('$x$')
plt.legend()
plt.show()
```

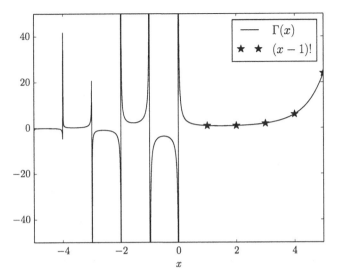

Figure 8.4 The gamma function on the real line, $\Gamma(x)$, and $(x - 1)!$ for integer $x > 0$.

The beta function is defined by the definite integral

$$B(a, b) = \int_0^1 t^{a-1}(1 - t)^{b-1}\, dt, \quad a > 0, b > 0.$$

It is closely related to the gamma function: $B(a, b) = \Gamma(a)\Gamma(b)/\Gamma(a + b)$. The scipy.special functions beta(a, b) and betaln(a, b) return the beta function and its natural logarithm, respectively. As with the gamma function, there is an *incomplete beta function*, $B(a, b; x)$, obtained by replacing the upper limit with x; the methods betainc(a, b, x) and betaincinv(a, b, y) return this function and its inverse.

Example E8.6   The exact classical mechanical description of a pendulum is quite complex, and the equations of motion are usually only solved in introductory texts for small displacements about equilibrium. In this case, the period $T \approx 2\pi \sqrt{L/g}$ and the motion is harmonic.

The general solution requires elliptic integrals, but the special case of a pendulum making 180° swings (i.e. ±90° about its equilibrium position) leads to the following expression for the period:

$$T = 2\sqrt{\frac{2l}{g}} \int_0^{\pi/2} \frac{d\theta}{\sqrt{\cos\theta}}.$$

The substitution $x = \sin^2 \theta$ transforms this integral into a beta function:

$$\int_0^{\pi/2} \frac{d\theta}{\sqrt{\cos \theta}} = \tfrac{1}{2} \int_0^1 x^{-1/2}(1-x)^{-3/4}\, dx = \tfrac{1}{2} B\left(\tfrac{1}{2}, \tfrac{1}{4}\right).$$

Therefore,

$$T = \sqrt{2} B\left(\tfrac{1}{2}, \tfrac{1}{4}\right) \sqrt{\frac{l}{g}}.$$

To find the period of the pendulum in units of $\sqrt{l/g}$:

```
In [x]: import numpy as np
In [x]: from scipy.special import beta
In [x]: np.sqrt(2) * beta(0.5, 0.25)
7.4162987092054875
```

(Compare with the harmonic approximation, $2\pi = 6.283185$.)

The group of elliptic integrals and related functions form an important class of mathematical objects and have been widely studied. They find application in geometry, cryptography, analysis and many areas of physics. The complete elliptic integrals of the first and second kind, $K(m)$ and $E(m)$, are defined for $0 \le m \le 1$ by

$$K(m) = \int_0^{\pi/2} \frac{d\theta}{\sqrt{1 - m \sin^2 \theta}},$$

$$E(m) = \int_0^{\pi/2} \sqrt{1 - m \sin^2 \theta}\, d\theta.$$

Their values for the parameter m are returned by the functions `ellipk(m)` and `ellipe(m)`. The incomplete elliptic integrals (defined by replacing the upper limit of $\pi/2$ with the variable ϕ) are returned by `ellipkinc(phi, m)` and `ellipeinc(phi, m)`, respectively:[5]

$$K(\phi, m) = \int_0^\phi \frac{d\theta}{\sqrt{1 - m \sin^2 \theta}},$$

$$E(\phi, m) = \int_0^\phi \sqrt{1 - m \sin^2 \theta}\, d\theta.$$

---

[5] It is necessary to be very careful with the notation of elliptic integrals; many sources use $F(\phi, m)$ instead of $K(\phi, m)$ for the first kind, define them with interchanged arguments (i.e. $F(m, \phi)$) or use the parameter k^2 instead of m

$$F(\phi, k) = F(\phi|k^2) = \int_0^\phi \frac{d\theta}{\sqrt{1 - k^2 \sin^2 \theta}},$$

$$E(\phi, k) = E(\phi|k^2) = \int_0^\phi \sqrt{1 - k^2 \sin^2 \theta}\, d\theta.$$

Example E8.7   The problem of finding an arc length of an ellipse is the origin of the name of the elliptic integrals. The equation of an ellipse with semi-major axis a and semi-minor axis b may be written in parametric form as

$$x = a \sin \phi,$$
$$y = b \cos \phi.$$

The element of length along the ellipse's perimeter is given by

$$ds = \sqrt{dx^2 + dy^2} = \sqrt{a^2 \cos^2 \phi + b^2 \sin^2 \phi} \, d\phi$$
$$= a \sqrt{1 - e^2 \sin^2 \phi} \, d\phi,$$

where $e = \sqrt{1 - b^2/a^2}$ is the *eccentricity*. The arc length may therefore be written in terms of incomplete elliptic integrals of the second kind:

$$\int ds = a \int_{\phi_1}^{\phi_2} \sqrt{1 - e^2 \sin^2 \phi} \, d\phi = a[E(e; \phi_2) - E(e; \phi_1)].$$

Earth's orbit is an ellipse with semi-major axis 149 598 261 km and eccentricity 0.01671123. We will find the distance traveled by the Earth in one orbit, and compare it with that obtained assuming a circular orbit of radius 1 AU $\equiv$ 149 597 870.7 km.

The perimeter of an ellipse may be written using the earlier expression with $\phi_1 = 0, \phi_2 = 2\pi$:

$$P = a[E(e, 2\pi) - E(e, 0)] = 4aE(e),$$

since the entire perimeter is four times the quarter-perimeters, which may be written in terms of the *complete* elliptic integral of the second kind. We have

```
In [x]: import numpy as np
In [x]: from scipy.special import ellipe
In [x]: a, e  = 149_598_261, 0.01671123    # semi-major axis (km), eccentricity
In [x]: pe = 4 * a * ellipe(e*e)
In [x]: print(pe)
939887967.974                   # "exact" answer
In [x]: AU = 149_597_870.7      # mean orbit radius, km
In [x]: pc = 2 * np.pi * AU
In [x]: print(pc)
939951143.1675915               # assuming circular orbit
In [x]: (pc - pe) / pe * 100
0.0067215663638305143
```

That is, the percentage error in the perimeter in treating the orbit as circular is about 0.0067%.

8.1.4   The Error Function and Related Integrals

The error function, defined by:

$$\mathrm{erf}(z) = \frac{2}{\sqrt{\pi}} \int_0^z e^{-t^2}\, dt$$

for real or complex z does not have a simple closed-form expression and so must be calculated numerically. `scipy.special` has several functions relating to the error function:

- `erf(z)`: the error function;
- `erfc(z)`: the complementary error function, $\mathrm{erfc}(z) = 1 - \mathrm{erf}(z)$; it is more accurate to use this function for large z than directly subtracting $\mathrm{erf}(z)$ from 1;
- `erfcx(z)`: the *scaled complementary error function*, $e^{z^2}\mathrm{erfc}(z)$;
- `erfinv(y)`: the inverse error function;
- `erfcinv(y)`: the inverse complementary error function;
- `wofz(z)`: the Faddeeva function, a scaled complementary error function with complex argument:

$$w(z) = e^{-z^2}\mathrm{erfc}(-iz) = \mathrm{erfcx}(-iz),$$

  which appears in problems related to plasma physics and radiative transfer;
- `dawsn(z)`: a related integral known as Dawson's integral:

$$D(z) = e^{-z^2} \int_0^z e^{t^2}\, dt.$$

Example E8.8   The wavefunction corresponding to the ground state of the one-dimensional quantum harmonic oscillator may be written as follows in terms of a parameter $\alpha = \sqrt{mk}/\hbar$, where m is the mass and k the oscillator force constant:

$$\psi_0(x) = \left(\frac{\alpha}{\pi}\right)^{1/4} \exp\left(-\alpha x^2/2\right).$$

The probability density of the oscillator's position is given by $P_0(x) = |\psi_0(x)|^2$ and is nonzero outside the classical turning points, $\pm\alpha^{-1/2}$, a phenomenon known as tunneling. We will calculate the probability of tunneling for an oscillator in the state ψ_0.

The wavefunction is symmetric about $x = 0$, so the probability of tunneling is

$$P(x < -\alpha) + P(x > \alpha) = 2P(x > \alpha) = 2\sqrt{\frac{\alpha}{\pi}} \int_{\alpha^{-1/2}}^{\infty} \exp\left(-\alpha x^2\right) dx$$

$$= \frac{2}{\sqrt{\pi}} \int_1^{\infty} e^{-y^2}\, dy = \mathrm{erfc}(1).$$

The complementary error function can be calculated directly:

```
In [x]: from scipy.special import erfc
In [x]: erfc(1)
0.15729920705028516
```

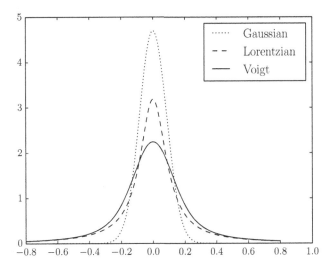

Figure 8.5 A comparison of the Lorentzian, Gaussian and Voigt line shapes with $\gamma = \alpha = 0.1$.

or about 16%.

Example E8.9 The *Voigt line profile* occurs in the modeling and analysis of radiative transfer in the atmosphere. It is the convolution of a Gaussian profile, $G(x; \sigma)$, and a Lorentzian profile, $L(x; \gamma)$:

$$V(x; \sigma, \gamma) = \int_{-\infty}^{\infty} G(x'; \sigma)L(x - x'; \gamma)\,dx', \quad \text{where}$$

$$G(x; \sigma) = \frac{1}{\sigma\sqrt{2\pi}}\exp\left(-\frac{x^2}{2\sigma^2}\right) \quad \text{and} \quad L(x; \gamma) = \frac{\gamma/\pi}{x^2 + \gamma^2}.$$

Here γ is the half-width at half-maximum (HWHM) of the Lorentzian profile and σ is the standard deviation of the Gaussian profile, related to its HWHM, α, by $\alpha = \sigma\sqrt{2\ln 2}$. In terms of frequency, ν, $x = \nu - \nu_0$, where ν_0 is the line center.

There is no closed form for the Voigt profile, but it is related to the real part of the Faddeeva function, $w(z)$, by

$$V(x; \sigma, \gamma) = \frac{\text{Re}\,[w(z)]}{\sigma\sqrt{2\pi}}, \quad \text{where } z = \frac{x + i\gamma}{\sigma\sqrt{2}}.$$

The program mentioned here plots the Voigt profile for $\gamma = 0.1, \alpha = 0.1$ and compares it with the corresponding Gaussian and Lorentzian profiles (Figure 8.5). The equations mentioned earlier are implemented in the three functions, G, L and V, defined in the code here.

Listing 8.6 A comparison of the Lorentzian, Gaussian and Voigt line shapes

```
# eg8-voigt.py
import numpy as np
```

```
from scipy.special import wofz
import matplotlib.pyplot as plt

def G(x, alpha):
    """ Return Gaussian line shape at x with HWHM alpha """
    return np.sqrt(np.log(2) / np.pi) / alpha\
                        * np.exp(-(x / alpha)**2 * np.log(2))

def L(x, gamma):
    """ Return Lorentzian line shape at x with HWHM gamma """
    return gamma / np.pi / (x**2 + gamma**2)

def V(x, alpha, gamma):
    """

    Return the Voigt line shape at x with Lorentzian component HWHM gamma
    and Gaussian component HWHM alpha.

    """
    sigma = alpha / np.sqrt(2 * np.log(2))

    return np.real(wofz((x + 1j*gamma)/sigma/np.sqrt(2))) / sigma\
                                                / np.sqrt(2*np.pi)

alpha, gamma = 0.1, 0.1
x = np.linspace(-0.8, 0.8, 1000)
plt.plot(x, G(x, alpha), ls=':', c='k', label='Gaussian')
plt.plot(x, L(x, gamma), ls='--', c='k', label='Lorentzian')
plt.plot(x, V(x, alpha, gamma), c='k', label='Voigt')
plt.legend()
plt.show()
```

8.1.5   Fresnel Integrals

The Fresnel integrals are encountered in optics and are defined by the equations

$$S(z) = \int_0^z \sin\left(\frac{\pi t^2}{2}\right) dt, C(z) = \int_0^z \cos\left(\frac{\pi t^2}{2}\right) dt.$$

Both are returned in a tuple for real or complex argument z by the `special.scipy` function `fresnel(z)`. The related function, `fresnel_zeros(nt)`, returns the first nt complex zeros of $S(z)$ and $C(z)$.

Example E8.10   As well as playing an important role in the description of diffraction effects in optics, the Fresnel integrals find an application in the design of motorway junctions (freeway intersections). The curve described by the parametric equations $(x, y) = (S(t), C(t))$ is called a *clothoid* (or Euler spiral) and has the property that its curvature is proportional to the distance along the path of the curve. Hence, a vehicle traveling at constant speed will experience a constant rate of angular acceleration as it travels around the curve – this means that the driver can turn the steering wheel at a constant rate, which makes the junction safer.

The following code plots the Euler spiral for $-10 \le t \le 10$ (Figure 8.6).

```
In [x]: import numpy as np
In [x]: from scipy.special import fresnel
In [x]: import matplotlib.pyplot as plt
In [x]: t = np.linspace(-10, 10, 1000)
In [x]: plt.plot(*fresnel(t), c='k')
In [x]: plt.show()
```

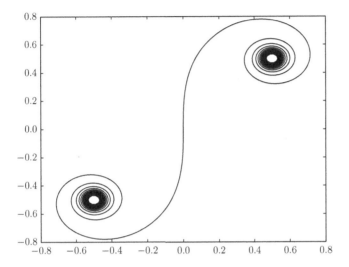

Figure 8.6 The Euler spiral.

8.1.6   Binomial Coefficients and Exponential Integrals

The binomial coefficient $\binom{n}{k} \equiv {}^nC_k$ is returned by the scipy.special function binom(n, k).

Various functions are supplied for the evaluation of different forms of the exponential integral. The standard form is returned by expi(z):

$$\mathrm{Ei}(z) = \int_{-\infty}^{z} \frac{e^t}{t}\, dt, \quad |\arg(-z)| < \pi.$$

expn(n, x) returns the value of

$$\int_{1}^{\infty} \frac{e^{-xt}}{t^n}\, dt.$$

For $n = 1$, it is faster and more accurate to use exp1(z):

$$\int_{1}^{\infty} \frac{e^{-zt}}{t}\, dt.$$

Example E8.11   Any integral of the form

$$\int f(z)e^z \, dz,$$

where $f(z) = P(z)/Q(z)$ is a rational function, can be reduced to the form

$$\int R(z)e^z \, dz + \sum_i \int \frac{e^z}{(z - a_i)^{n_i}} \, dz,$$

where $R(z)$ is a polynomial (which may be zero) by expansion in partial fractions. The first integral here can be evaluated by standard methods (repeated integration by parts). Provided the path of integration does not pass through any singular points of the integrand, the second term can be written in terms of exponential integrals.

For example, consider the integral

$$I = \int_{-\infty}^{-2} \frac{e^z}{z^2(z - 1)} \, dz.$$

It can easily be shown that

$$\frac{1}{z^2(z - 1)} = \frac{1}{z - 1} - \frac{1}{z} - \frac{1}{z^2}$$

and so the integral may be written as the three terms

$$I = \int_{-\infty}^{-2} \frac{e^z}{z - 1} \, dz - \int_{-\infty}^{-2} \frac{e^z}{z} \, dz - \int_{-\infty}^{-2} \frac{e^z}{z^2} \, dz.$$

The second integral is simply $-\text{Ei}(-2)$ and substitution $u = z - 1$ resolves the first integral to $e\text{Ei}(-3)$. The last integral may be written in terms of $En(z)$ or further reduced by integration by parts to

$$\int_{-\infty}^{-2} \frac{e^z}{z^2} \, dz = -\frac{e^{-2}}{2} + \text{Ei}(-2).$$

Therefore,

$$I = e\text{Ei}(-3) - 2\text{Ei}(-2) - \frac{e^{-2}}{2}.$$

In SciPy,

```
In [x]: import numpy as np
In [x]: from scipy.special import expi
In [x]: np.e * expi(-3) - 2*expi(-2) - np.exp(-2)/2
-0.0053357974213484663
```

8.1.7   Orthogonal Polynomials and Spherical Harmonics

There are a large number of functions in `scipy.special` for the evaluation of different sorts of orthogonal polynomials, including the Legendre, Jacobi, Laguerre, Hermite and different flavors of Chebyshev polynomials. They take the general name `eval_poly(n,`

Table 8.2 Some of the orthogonal polynomials in SciPy

Function	Description
eval_legendre(n, x)	Legendre polynomial, $P_n(x)$
eval_chebyt(n, x)	Chebyshev polynomial of the first kind, $T_n(x)$
eval_chebyu(n, x)	Chebyshev polynomial of the second kind, $U_n(x)$
eval_hermite(n, x)	(Physicists') Hermite polynomial, $H_n(x)$
eval_jacobi(n, alpha, beta, x)	Jacobi polynomial, $P_n^{(\alpha,\beta)}(x)$
eval_laguerre(n, x)	Laguerre polynomial of the first kind, $L_n(x)$
eval_genlaguerre(n, alpha x)	Generalized Laguerre polynomial of the first kind, $L_n^\alpha(x)$

x) where n is the order of the polynomial and x is an array-like sequence of values at which to evaluate the polynomial. Table 8.2 gives the names of some of these functions.

The spherical harmonics used in SciPy are defined by the formula

$$Y_n^m(\phi, \theta) = \sqrt{\frac{(2n+1)}{4\pi}\frac{(n-m)!}{(n+m)!}} P_n^m(\cos\phi)e^{im\theta},$$

where $n = 0, 1, 2, \ldots$ is called the *degree* and $m = -n, -n+1, \ldots n$ the *order* of the spherical harmonic. The functions $P_n^m(x)$ are the associated Legendre polynomials. As with so many special functions, different fields adopt different phase conventions and normalizations, so it is important to check these carefully and make the appropriate modifications when using them. In particular, many other fields use l for the degree of the harmonic and reverse the definition of θ and ϕ. To be clear, in SciPy, θ is the *azimuthal* (longitudinal) angle (taking values between 0 and 2π) and ϕ is the polar (colatitudinal) angle (between 0 and π).

The scipy.special.sph_harm method is called with the arguments:

```
scipy.special.sph_harm(m, n, theta, phi)
```

where theta and phi can be array-like objects.

Example E8.12 Visualizing the spherical harmonics is a little tricky because they are complex and defined in terms of angular coordinates, (θ, ϕ). One way is to plot the real part only on the unit sphere. Matplotlib provides a toolkit for such three-dimensional plots, mplot3d, as illustrated by the following code which produces Figure 8.7.[6]

Listing 8.7 The spherical harmonic defined by $l = 3, m = 2$

```
# eg8-spherical-harmonics.py

import matplotlib.pyplot as plt
from matplotlib import cm, colors
from mpl_toolkits.mplot3d import Axes3D
```

---

[6] See Section 7.6 and https://matplotlib.org/mpl_toolkits/mplot3d/.

Figure 8.7 A depiction of the spherical harmonic defined by $l = 3, m = 2$.

```
import numpy as np
from scipy.special import sph_harm

phi = np.linspace(0, np.pi, 100)
theta = np.linspace(0, 2*np.pi, 100)
phi, theta = np.meshgrid(phi, theta)

# The Cartesian coordinates of the unit sphere.
x = np.sin(phi) * np.cos(theta)
y = np.sin(phi) * np.sin(theta)
z = np.cos(phi)

m, l = 2, 3

# Calculate the spherical harmonic Y(l, m) and normalize to [0, 1].
fcolors = sph_harm(m, l, theta, phi).real
fmax, fmin = fcolors.max(), fcolors.min()
fcolors = (fcolors - fmin)/(fmax - fmin)

# Set the aspect ratio to 1 so our sphere looks spherical.
fig = plt.figure(figsize=plt.figaspect(1.))
ax = fig.add_subplot(projection='3d')
ax.plot_surface(x, y, z, rstride=1, cstride=1, facecolors=cm.jet(fcolors))
# Turn off the axis planes.
ax.set_axis_off()
plt.show()
```

8.1.8   Exercises

Questions

Q8.1.1   By changing a single line in the program of Example E8.1, output the 10 *most accurately* known constants (excluding those set to their values by definition).

Q8.1.2   Use SciPy's constants and conversion factors to calculate the number density, N/V, of ideal gas molecules at standard temperature and pressure ($T = 0$ °C, $p = 1$ atm). The ideal gas law is $pV = Nk_B T$.

Problems

P8.1.1 Use scipy.special.binom to create a depiction of Pascal's triangle of binomial coefficients $\binom{n}{k}$ up to $n = 8$.

P8.1.2 The *Airy pattern* is the circular diffraction pattern resulting from a uniformly illuminated circular aperture. It consists of a bright, central disc surrounded by fainter rings. Its mathematical description may be written in terms of the Bessel function of the first kind,

$$I(\theta) = I_0 \left(\frac{2J_1(x)}{x} \right)^2,$$

where θ is the observation angle and $x = ka \sin \theta$. a is the aperture radius and $k = 2\pi/\lambda$ is the *angular wavenumber* of the light with wavelength λ.

Plot the Airy pattern as $I(x)/I_0$ for $-10 \leq x \leq 10$ and deduce from the position of the first minimum in this function the maximum resolving power (in arcsec) of the human eye (pupil diameter 3 mm) at a wavelength of 500 nm.

P8.1.3 Write a function, get_wv, which takes a molar bond dissociation energy, D0, in kJ mol^{-1} and returns the wavelength of a photon corresponding to that energy *per molecule*, in nm. The energy of a photon with wavelength λ is $E = hc/\lambda$.

For example,

```
In [x]: get_wv(497)
Out[x]: 240.69731528286377
```

P8.1.4 An *ellipsoid* is the three-dimensional figure bounded by the surface described by the equation

$$\frac{x^2}{a^2} + \frac{y^2}{b^2} + \frac{z^2}{c^2} = 1,$$

where a, b and c are the *semi-principal axes*. If $a = b = c$, the ellipsoid is a sphere. The volume of an ellipsoid has a simple form,

$$V = \frac{4}{3}\pi abc.$$

There is no closed formula for the surface area of a general ellipsoid, but it may be expressed in terms of incomplete elliptic integrals of the first and second kinds, $K(\phi, k)$ and $E(\phi, k)$:

$$S = 2\pi c^2 + \frac{2\pi ab}{\sin \phi} \left(K(\phi, k^2) \cos^2 \phi + E(\phi, k^2) \sin^2 \phi \right),$$

where

$$\cos \phi = \frac{c}{a}, \quad k = \frac{a\sqrt{b^2 - c^2}}{b\sqrt{a^2 - c^2}}$$

and the coordinate system has been chosen such that $a \geq b \geq c$.

Define a function, `ellipsoid_surface`, to calculate the surface area of a general ellipsoid, and compare the results for different-shaped ellipsoids with the following approximate formula:

$$S \approx 2\pi c^2 + 2\pi abr \left(1 - \frac{b^2 - c^2}{6b^2}r^2\left(1 - \frac{3b^2 + 10c^2}{56b^2}r^2\right)\right), \quad \text{where } r = \frac{\phi}{\sin\phi}.$$

P8.1.5   The *drawdown* or change in hydraulic head, s (a measure of the water pressure above some geodetic datum), a distance r from a well at time t, from which water is being pumped at a constant rate, Q, can be modeled using the *Theis* equation,

$$s(r, t) = H_0 - H(r, t) = \frac{Q}{4\pi T}W(u), \quad \text{where} \quad u = \frac{r^2 S}{4Tt}.$$

Here, H_0 is the hydraulic head in the absence of the well, S is the aquifer storage coefficient (volume of water released per unit decrease in H per unit area) and T is the transmissivity (a measure of how much water is transported horizontally per unit time). The *well function*, $W(u)$, is simply the exponential integral, $E_1(u)$.

For a well being pumped at a rate of $Q = 1000$ m^3 d^{-1} from an aquifer described by the parameters $H_0 = 20$ m, $S = 0.0003$, $T = 1000$ m^2 d^{-1}, determine the height of the hydraulic head as a function of r after $t = 1$ d of pumping.

Compare your answer with the approximate version of the Theis equation known as the Jacob equation, in which the well function is taken to be appoximately $W(u) \approx -\gamma - \ln u$, where $\gamma = 0.577215664\ldots$ is the Euler–Mascheroni constant.

P8.1.6   Some electronic components are cooled by annular fins (heatsinks), which conduct heat away from the component and provide a larger surface area for that heat to dissipate to the surroundings.

The cooling efficiency of an annular fin of width $2w$ and inner and outer radii r_0 and r_1 may be written in terms of modified Bessel functions of the first and second kinds:

$$\eta = \frac{2r_0}{\beta(r_1^2 - r_0^2)} \frac{K_1(u_0)I_1(u_1) - I_1(u_0)K_1(u_1)}{K_0(u_0)I_1(u_1) + I_0(u_0)K_1(u_1)},$$

where $u_0 = \beta r_0$, $u_1 = \beta r_1$ and

$$\beta = \sqrt{\frac{h_c}{\kappa w}}.$$

h_c is the heat transfer coefficient (which is taken to be constant over the fin's surface) and κ is the thermal conductivity of the fin material.

What is the cooling efficiency of an aluminium annular fin with dimensions $r_0 = 5$ mm, $r_1 = 10$ mm, $w = 0.1$ mm? Take $h_c = 10$ W m^{-2} K^{-1} and $\kappa = 200$ W m^{-1} K^{-1}.

Calculate the heat dissipation, $\dot{Q}$ (the product of the efficiency, the fin area and the temperature difference), for a component temperature $T_0 = 400$ K and ambient temperature $T_e = 300$ K.

8.2   Integration and Ordinary Differential Equations

The `scipy.integrate` package contains functions for computing definite integrals. It can evaluate both proper (with finite limits) and improper (infinite limits) integrals. It can also perform numerical integration of systems of ordinary differential equations.

8.2.1   Definite Integrals of a Single Variable

The basic numerical integration routine is `scipy.integrate.quad`, which is based on the venerable FORTRAN 77 QUADPACK library. It uses adaptive quadrature to approximate the value of an integral by dividing its domain into subintervals that are chosen iteratively to meet a particular tolerance (that is, estimated absolute or relative error). In its simplest form, it takes three arguments: a Python function object corresponding to the function to integrate, `func`, and the limits of integration, a and b. `func` must take at least one argument; if it takes more than one it is integrated along the coordinate corresponding to the first argument. In simple usage, `lambda` expressions are a convenient way to define `func`. For example, to evaluate $\int_1^4 x^{-2}\, dx = \frac{3}{4}$ numerically:

```
In [x]: from scipy.integrate import quad
In [x]: f = lambda x: 1/x**2
Out[x]: quad(f, 1, 4)
(0.7500000000000002, 1.913234548258995e-09)
```

quad returns two values in a tuple – the value of the integral and an estimate of the absolute error in the result.

Use `np.inf` to evaluate improper integrals:

```
In [x]: quad(lambda x: np.exp(-x**2), 0, np.inf)
Out[x]: (0.8862269254527579, 7.101318390472462e-09)
In [x]: np.sqrt(np.pi)/2     # analytical result
Out[x]: 0.88622692545275794
```

Note that in this call to quad we didn't even give the function a name but simply passed it as an anonymous `lambda` object.

More complicated functions require a Python function object defined with `def`:

```
In [x]: def g(x):
   ...:       if abs(x) < 0.5:
   ...:           return -x
   ...:       return x - np.sign(x)
   ...:
In [x]: quad(g, -0.6, 0.8)
Out[x]: (-0.06000000000000002, 6.661338147750941e-17)
```

Functions with singularities or discontinuities can cause problems for the numerical quadrature routine even if the required integral is well defined. For example, the sinc function $f(x) = \sin(x)/x$ has a removable singularity at $x = 0$, which causes the following simple application of quad to fail:

```
In [x]: sinc = lambda x: np.sin(x)/x
```

```
In [x]: quad(sinc, -2, 2)
...: RuntimeWarning: invalid value encountered in double_scalars
Out[37]: (nan, nan)
```

The solution is to configure quad by passing a list of such *break points* to the points argument (the list does not have to be ordered):

```
In [x] quad(sinc, -2, 2, points=[0,])
(3.210825953605389, 3.5647329017567276e-14)
```

Note that break points cannot be specified with infinite limits.

The arguments epsrel and epsabs allow the specification of a desired accuracy of the quadrature as a relative or absolute tolerance. The default values are both 1.49e-8, but the integration can be done faster if a less-accurate answer is required. As an example, consider integrating the rapidly varying function $f(x) = e^{-|x|}\sin^2 x^2$:

```
In [x]: f = lambda x: np.sin(x**2)**2 * np.exp(-np.abs(x))
In [x]: quad(f, -1, 2, epsabs=0.1)
Out[x]: (0.29551455828969975, 0.0015295718279111671)
In [x]: quad(f, -1, 2, epsabs=1.49e-8)    # (the default absolute tolerance)
Out[x]: (0.29551455505239044, 4.449763315720537e-10)
```

Note that epsabs is only a requested upper bound: the actual estimated accuracy in the result may be much better, and in fact the actual result may be more accurate than this estimate.

If a function takes one or more parameters in addition to its principal argument, these need to be passed to quad as a tuple in args. For example, the integral

$$I_{n,m} = \int_{-\pi/2}^{\pi/2} \sin^n x \cos^m x \, \mathrm{d}x$$

can be evaluated numerically with

```
In [x]: def f(x, n, m):
....:        return np.sin(x)**n * np.cos(x)**m
....:
In [x]: n, m = 2, 1
In [x]: quad(f, -np.pi/2, np.pi/2, args=(n, m))
(0.6666666666666666, 1.6257468410185671e-13)
```

Note that the additional parameters, n and m here, appear as arguments to our function *after* the coordinate to be integrated over (x).

Example E8.13   Consider a torus of average radius R and cross-sectional radius r. The volume of this shape may be evaluated analytically in Cartesian coordinates as a volume of revolution:

$$V = 2 \int_{R-r}^{R+r} 2\pi xz \, \mathrm{d}x, \quad \text{where } z = \sqrt{r^2 - (x - R)^2}.$$

The center of the torus is at the origin and the z axis is taken to be its symmetry axis.

The integral is tedious but yields to standard methods: $V = 2\pi^2 R r^2$. Here we take a numerical approach with the values $R = 4$, $r = 1$:

```
In [x]: R, r = 4, 1
In [x]: f = lambda x, R, r: x * np.sqrt(r**2 - (x-R)**2)
In [x]: V, _ = quad(f, R-r, R+r, args=(R, r))
In [x]: V *= 4 * np.pi
In [x]: Vexact = 2 * np.pi**2 * R * r**2
In [x]: print('V = {} (exact: {})'.format(V, Vexact))
Out[x]: V = 78.95683520871499 (exact: 78.95683520871486)
```

8.2.2 Integrals of Two and More Variables

The scipy.integrate functions dblquad, tplquad and nquad evaluate double, triple and multiple integrals, respectively. Because, in general, the limits on one coordinate may depend on another coordinate, the syntax for calling these functions is a little more complicated.

dblquad evaluates the double integral:

$$\int_a^b \int_{g(x)}^{h(x)} f(x, y)\, dy\, dx.$$

It is passed $f(x, y)$ as a function of at least two variables, func(y, x, ...). The function must take y as its first argument and x as its second argument. The integral limits are passed to dblquad in four further arguments. First, the two arguments, a and b, specify the lower and upper limits on the x-integral, respectively, as for quad. The next two arguments, gfun and hfun, are the lower and upper limits on the y-integral and they must be *callable objects* taking a single floating-point argument, the value of x at which the limit applies (i.e. they must themselves be functions of x). If either of the y-integral limits does not depend on x, gfun or hfun can return a constant value.

As a simple example, the integral

$$\int_1^4 \int_0^2 x^2 y\, dy dx$$

can be evaluated with

```
In [x]: f = lambda y, x: x**2 * y
In [x]: a, b = 1, 4
In [x]: gfun = lambda x: 0
In [x]: hfun = lambda x: 2
In [x]: dblquad(f, a, b, gfun, hfun)
Out[x]: (42.00000000000001, 4.662936703425658e-13)
```

Here, gfun and hfun are each called with a value of x, but they return a constant (0 and 2, respectively) no matter what this value is.

Of course, it is possible to wrap all of this into a single line:

```
In [x]: dblquad(lambda y, x: x**2 * y, 1, 4, lambda x: 0, lambda x: 2)
Out[x]: (42.00000000000001, 4.662936703425658e-13)
```

A double integral can be used to find the area of some two-dimensional shape bounded by one or more functions. For an example in polar coordinates, consider the

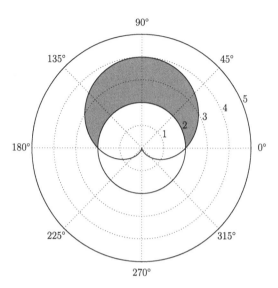

Figure 8.8 The region defined as the area inside $r = 2 + 2 \sin \theta$ but outside the circle $r = 2$.

area inside the curve $r = 2 + 2 \sin \theta$ but outside the circle defined by $r = 2$ for θ in $[0, 2\pi]$ (see Figure 8.8). These curves intersect at $\theta = 0, \pi$ so the required integral is

$$A = \int_0^\pi \int_2^{2+2\sin\theta} r \, dr \, d\theta,$$

where $r \, dr \, d\theta$ is the infinitesimal area element in polar coordinates. This particular integral is fairly straightforward to evaluate analytically ($A = 8 + \pi$), so the numerical result is easy to check:

```
In [x]: r1, r2 = lambda theta: 2, lambda theta: 2 + 2*np.sin(theta)
In [x]: A, _ = dblquad(lambda r, theta: r, 0, np.pi, r1, r2)
Out[x]: 11.141592653589791
In [x]: 8 + np.pi        # exact answer
Out[x]: 11.141592653589793
```

The function to evaluate is simply r, defined by `lambda r, theta: r`; in the inner integral the limits on r are 2 and $2 + 2 \sin \theta$; for the outer integral θ ranges from 0 to π.

The method `tplquad` evaluates triple integrals and takes a function of three variables, `func(z, y, x)` and *six* further arguments: constant x-limits, a and b; y-limits, `gfun(x)` and `hfun(x)` (which are functions, as for dblquad); and z-limits, `qfun(x, y)` and `rfun(x, y)` (functions of x and y in that order).

Higher-dimensional integrations are handled by the `scipy.integrate.nquad` method, which will not be discussed here (documentation and examples are available online).[7]

---

[7] https://docs.scipy.org/doc/scipy/reference/generated/scipy.integrate.nquad.html.

Example E8.14 The volume of the unit sphere, $4\pi/3$, can be expressed as a triple integral in spherical polar coordinates with constant limits:

$$\int_0^{2\pi} \int_0^{\pi} \int_0^1 r^2 \sin\theta \, dr d\theta d\phi.$$

```
In [x]: from scipy.integrate import tplquad
In [x]: tplquad(lambda phi, theta, r: r**2 * np.sin(theta),
                 0, 1,
                 lambda theta: 0, lambda theta: np.pi,
                 lambda theta, phi: 0, lambda theta, phi: 2*np.pi)
Out[x]: (4.18879020478639, 4.650491330678174e-14)
```

Alternatively, it can be expressed in Cartesian coordinates with limits as functions:

$$8 \int_0^1 \int_0^{\sqrt{1-x^2}} \int_0^{\sqrt{1-x^2-y^2}} dz \, dy \, dx,$$

where the integral is in the positive octant of the three-dimensonal Cartesian axes.

```
In [x]: A, _ = tplquad(lambda z, y, x: 1,
                 0, 1,
                 lambda x: 0, lambda x: np.sqrt(1 - x**2),
                 lambda x, vy: 0, lambda x, y: np.sqrt(1 - x**2 - y**2))
In [x]: 8*A
Out[x]: 4.188790204786391
```

Example E8.15 This example finds the mass and center of mass of the tetrahedron bounded by the coordinate axes and the plane $x + y + z = 1$ with density $\rho = \rho(x, y, z)$, where $\rho(x, y, z)$ is provided as a lambda function. We test it with the functions $\rho = 1$, $\rho = x$ and $\rho = x^2 + y^2 + z^2$.

The mass may be written as a triple integral of the density over the volume of the tetrahedron:

$$m = \int_V \rho(x, y, z) \, dV = \int_0^1 \int_0^{1-x} \int_0^{1-x-y} \rho(x, y, z) \, dz \, dy \, dx,$$

and the coordinates of the center of mass are given by

$$m\bar{x} = \int_V x\rho(x, y, z) \, dV, \quad m\bar{y} = \int_V y\rho(x, y, z) \, dV, \quad m\bar{z} = \int_V z\rho(x, y, z) \, dV.$$

The following program uses `scipy.integrate.tplquad` to perform the necessary integrations (which can also be solved analytically).

Listing 8.8 Calculating the mass and center of mass of a tetrahedron given three different densities

```
# eg8-tetrahedron-cofm.py

import numpy as np
from scipy.integrate import tplquad
```

```
# The integration limits on x, y, z.
a, b = 0, 1
gfun, hfun = lambda x: 0, lambda x: 1 - x
qfun, rfun = lambda x, y: 0, lambda x, y: 1 - x - y
```
❶ `lims = (a, b, gfun, hfun, qfun, rfun)`

```
# The three different density functions.
rhos = [lambda x, y, z: 1,
        lambda x, y, z: x,
        lambda x, y, z: x**2 + y**2 + z**2]

for rho in rhos:
    # The mass as a triple integral of rho over the volume.
    m, _ = tplquad(rho, *lims)
    # The center of mass (xbar, ybar, zbar).
    mxbar, _ = tplquad(lambda x, y, z: x * rho(x, y, z), *lims)
    mybar, _ = tplquad(lambda x, y, z: y * rho(x, y, z), *lims)
    mzbar, _ = tplquad(lambda x, y, z: z * rho(x, y, z), *lims)
    xbar, ybar, zbar = mxbar / m, mybar / m, mzbar / m

    print('mass = {:g}, CofM = ({:g}, {:g}, {:g})'.format(m, xbar, ybar, zbar))
```

❶ Note that the six arguments representing the limits on the triple integral (two constants and two pairs of `lambda` functions) have been packed into a tuple, `lims` (the parentheses are optional here).

The output is:

```
mass = 0.166667, CofM = (0.25, 0.25, 0.25)
mass = 0.0416667, CofM = (0.4, 0.2, 0.2)
mass = 0.05, CofM = (0.277778, 0.277778, 0.277778)
```

8.2.3   Ordinary Differential Equations

Ordinary differential equations (ODEs) can be solved numerically with `scipy.integrate.odeint` or `scipy.integrate.solve_ivp` ("solve an initial value problem"). The latter function was introduced in version 1.0 of the SciPy library and is the recommended approach. However, much legacy code still uses `odeint` and this function is described in Appendix C. `solve_ivp`, solves first-order differential equations – *to solve a higher-order equation, it must be decomposed into a system of first-order equations first*, as explained later.

A Single First-Order ODE
In its simplest use for the solution of a single first-order ODE,

$$\frac{dy}{dt} = f(t, y),$$

`solve_ivp` takes three arguments: a function object returning dy/dt, the initial and final time points for the integration, and a set of initial conditions, y_0. If not specified, the ODE solver selects and returns a sequence of suitable time points on which the integration is carried out.

For example, consider the first-order differential equation describing the rate of the reaction A → P in terms of the concentration of the reactant, A:

$$\frac{d[A]}{dt} = -k[A].$$

This example has an easily obtainable analytical solution:

$$[A] = [A]_0 e^{-kt},$$

where $[A]_0$ is the initial concentration of $[A]$.

To solve the equation numerically with solve_ivp, write it in the form as shown above, with a single dependent variable, $y(t) \equiv [A]$, which is a function of the independent variable, t (time). We have:

$$\frac{dy}{dt} = -ky.$$

We need to provide a function returning dy/dt as $f(t, y)$ (in general a function of both t and y), which here is simply:

```
def dydt(t, y):
    return -k * y
```

(the order of the arguments is important). The initial and final time points, t_span, should be provided as a (t0, tf) tuple, and the initial conditions must be an array-like object even if, as here, there is only one value. We have:

```
soln = solve_ivp(dydt, (t0, tf), [y0])
```

The returned object, soln, is an instance of the OdeResult class, which defines a number of relevant properties including arrays soln.t for the time points used in the integration, soln.y, the values of the solution at these time points, and soln.success, a boolean flag indicating whether or not the solver successfully reached the final time point requested.

A program comparing the numerical and analytical results for a reaction with $k = 0.2$ s^{-1} and $y(0) \equiv [A]_0 = 100$ is given below; the resulting plot is Figure 8.9.

Listing 8.9 First-order reaction kinetics

```
import numpy as np
from scipy.integrate import solve_ivp
import matplotlib.pyplot as plt

# First-order reaction rate constant, s-1.
k = 0.2
# Initial condition on y: 100% of reactant is present at t = 0.
y0 = 100

# Initial and final time points for the integration.
t0, tf = 0, 20

def dydt(t, y):
    """ Return dy/dt = f(t, y) at time t. """
```

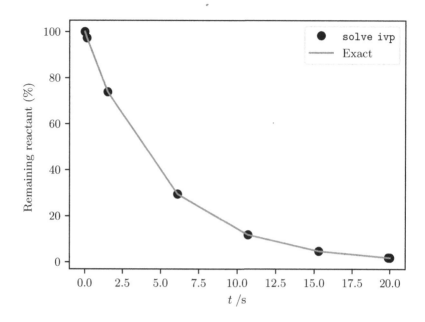

Figure 8.9 Exponential decay of a reactant in a first-order reaction: exact solution and numerical solution with time points selected by the ODE solver.

```
    return -k * y

# Integrate the differential equation.
soln = solve_ivp(dydt, (t0, tf), [y0])
t, y = soln.t, soln.y[0]

# Plot and compare the numerical and exact solutions.
plt.plot(t, y, 'o', color='k', label=r'\texttt{solve\_ivp}')
plt.plot(t, y0 * np.exp(-k*t), color='gray', label='Exact')
plt.xlabel(r'$t\;/\mathrm{s}$')
plt.ylabel('Remaining reactant (\%)')
plt.legend()
plt.show()
```

This approach is certainly suitable if all that is required is the final reactant concentration, but to follow the change in concentration at a higher time resolution, a specific sequence of time points can be provided to the argument t_eval:

```
# A suitable grid of 21 time points over 0-20s for following the reaction.
t0, tf = 0, 20
t_eval = np.linspace(t0, tf, 21)
# Integrate the differential equation.
soln = solve_ivp(dydt, (t0, tf), [y0], t_eval=t_eval)
t, y = soln.t, soln.y[0]
```

Better still, setting the dense_output argument to True defines an OdeSolution object called sol as one of the returned objects. This can be used to generate interpolated values of the solution for intermediate values of the time points:

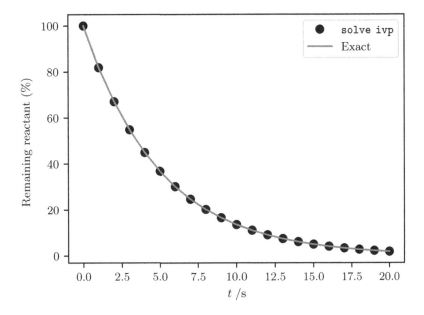

Figure 8.10 Exponential decay of a reactant in a first-order reaction: exact solution and numerical solution with pre-determined time points.

```
# Initial and final time points for the integration.
t0, tf = 0, 20

# Integrate the differential equation
soln = solve_ivp(dydt, (t0, tf), [y0], dense_output=True)
t = np.linspace(t0, tf, 20)
y = soln.sol(t)[0]
```

Note that the soln.sol object is *callable*: a value of the independent variable, time, is passed to it and the solution array at that time is returned. Here there is only one dependent variable, y, so we index this array at [0].

Plotting the solution points, y, as before yields Figure 8.10.

As with the quad family of routines, if the function returning the derivative requires further arguments, they can be passed to solve_ivp in the args parameter. In the earlier mentioned example, k is resolved in global scope, but we could pass it with:

```
def dydt(t, y, k):
    return -k * y
```

(note that additional parameters must appear after the independent and dependent variables). The call to solve_ivp would then be:

```
soln = solve_ivp(dydt, (t0, tf), [y0], args=(k,))
```

Oddly, the ability to pass additional arguments in args was only added in SciPy version 1.4. An alternative way to pass arguments from the calling scope into the derivative function, dydt, is to wrap this function in a lambda expression:

```
soln = solve_ivp(lambda t, y: dydt(t, y, k), (t0, tf), y0, t_eval=t)
```

Coupled First-Order ODEs

`solve_ivp` can also solve a set of coupled first-order ODEs in more than one dependent variable: $y_1(t), y_2(t), \ldots, y_n(t)$:

$$\frac{dy_1}{dt} = f_1(y_1, y_2, \ldots, y_n; t),$$

$$\frac{dy_2}{dt} = f_2(y_1, y_2, \ldots, y_n; t),$$

$$\cdots$$

$$\frac{dy_n}{dt} = f_n(y_1, y_2, \ldots, y_n; t).$$

In this case, the function passed to `solve_ivp()` must return a sequence of derivatives, $dy_1/dt, dy_2/dt, \ldots, dy_n/dt$ for each of the dependent variables; that is, it evaluates the earlier mentioned functions $f_i(y_1, y_2, \ldots, y_n; t)$ for each of the y_i passed to it in a sequence, y. The form of this function is:

```
def deriv(t, y):
    # y = [y1, y2, y3, ...] is a sequence of dependent variables.
    dy1dt = f1(y, t)    # calculate dy1/dt as f1(y1, y2, ..., yn; t)
    dy2dt = f2(y, t)    # calculate dy2/dt as f2(y1, y2, ..., yn; t)
    # ... etc
    # Return the derivatives in a sequence such as a tuple:
    return dy1dt, dy2dt, ..., dyndt
```

For a concrete example, suppose a reaction proceeds via two first-order reaction steps: A → B → P, with rate constants k_1 and k_2. The equations governing the rate of change of A and B are

$$\frac{d[A]}{dt} = -k_1[A],$$

$$\frac{d[B]}{dt} = k_1[A] - k_2[B].$$

Again, we can solve this pair of coupled equations analytically, but in our numerical solution, let $y_1 \equiv [A]$ and $y_2 \equiv [B]$:

$$\frac{dy_1}{dt} = -k_1 y_1,$$

$$\frac{dy_2}{dt} = k_1 y_1 - k_2 y_2.$$

The code presented below integrates these equations for $k_1 = 0.2 \ \text{s}^{-1}, k_2 = 0.8 \ \text{s}^{-1}$ and initial conditions $y_1(0) = 100, y_2(0) = 0$, and compares the result with the analytical solution (Figure 8.11).

Listing 8.10 Two coupled first-order reactions

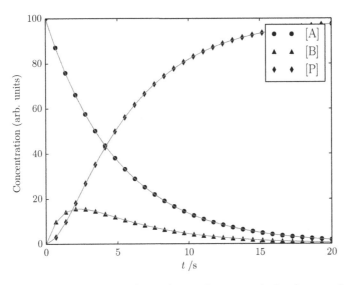

Figure 8.11 Two coupled first-order reactions: numerical and exact solutions.

```python
import numpy as np
from scipy.integrate import solve_ivp
import matplotlib.pyplot as plt

# First-order reaction rate constants, s-1.
k1, k2 = 0.2, 0.8
# Initial condition on y1, y2: [A](t=0) = 100, [B](t=0) = 0.
A0, B0 = 100, 0

# A suitable grid of time points for the reaction.
t0, tf = 0, 20

def dydt(t, y, k1, k2):
    """ Return dy_i/dt = f(y_i, t) at time t. """
    y1, y2 = y
    dy1dt = -k1 * y1
    dy2dt = k1 * y1 - k2 * y2
    return dy1dt, dy2dt

# Integrate the differential equation.
y0 = A0, B0
soln = solve_ivp(dydt, (t0, tf), y0, dense_output=True, args=(k1, k2))

t = np.linspace(t0, tf, 100)
A, B = soln.sol(t)

# [P] is determined by conservation.
P = A0 - A - B

# Analytical result.
Aexact = A0 * np.exp(-k1*t)
Bexact = A0 * k1/(k2-k1) * (np.exp(-k1*t) - np.exp(-k2*t))
```

```
Pexact = A0 - Aexact - Bexact

plt.plot(t, A, 'o', label='[A]')
plt.plot(t, B, '^', label='[B]')
plt.plot(t, P, 'd', label='[P]')
plt.plot(t, Aexact)
plt.plot(t, Bexact)
plt.plot(t, Pexact)
plt.xlabel(r'$t\;/\mathrm{s}$')
plt.ylabel('Concentration (arb. units)')
plt.legend()
plt.show()
```

❶ Again, if you are using `solve_ivp` from a version of SciPy prior to version 1.4, there is no way to directly pass the additional arguments k1 and k2 to the derivative function. The work-around is to wrap the function in a `lambda` expression:

```
soln = solve_ivp(lambda t, y: dydt(t, y, k1, k2), (t0, tf), y0, dense_output=True)
```

A Single Second-Order ODE

To solve an ODE of higher than first order, it must first be reduced into a system of first-order differential equations. In general, any differential equation with a single dependent variable of order n can be written as a system of n first-order differential equations in n dependent variables.

For example, the equation of motion for a harmonic oscillator is a second-order differential equation:

$$\frac{d^2x}{dt^2} = -\omega^2 x,$$

where x is the displacement from equilibrium and ω is the angular frequency. This equation may be decomposed into two first-order equations as follows:

$$\frac{dx_1}{dt} = x_2,$$

$$\frac{dx_2}{dt} = -\omega^2 x_1,$$

where x_1 is identified with x and x_2 with dx/dt.

This pair of coupled first-order equations may be solved as before:

Listing 8.11 Solution of the harmonic oscillator equation of motion

```
import numpy as np
from scipy.integrate import solve_ivp
import matplotlib.pyplot as plt

# Harmonic oscillator frequency (s-1).
omega = 0.9
# Initial conditions on x1 = x and x2 = dx/dt at t = 0.
A, v0 = 3, 0          # cm, cm.s-1
x0 = A, v0

# A suitable grid of time points.
```

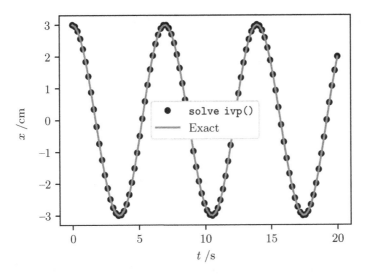

Figure 8.12 The harmonic oscillator: numerical and exact solutions.

```
t0, tf = 0 , 20

def dxdt(t, x):
    """ Return dx/dt = f(t, x) at time t. """
    x1, x2 = x
    dx1dt = x2
    dx2dt = -omega**2 * x1
    return dx1dt, dx2dt

# Integrate the differential equation.
soln = solve_ivp(dxdt, (t0, tf), x0, dense_output=True)

t = np.linspace(t0, tf, 100)
x1, x2 = soln.sol(t)

# Plot and compare the numerical and exact solutions.
plt.plot(t, x1, 'o', color='k', label=r'\texttt{solve_ivp()}')
plt.plot(t, A * np.cos(omega * t), color='gray', label='Exact')
plt.xlabel(r'$t\;/\mathrm{s}$')
plt.ylabel(r'$x\;/\mathrm{cm}$')
plt.legend()
plt.show()
```

The plot produced by this code is given in Figure 8.12.

Example E8.16 An object falling slowly in a viscous fluid under the influence of gravity is subject to a drag force (*Stokes' drag*), which varies linearly with its velocity.

Its equation of motion may be written as the second-order differential equation:

$$m\frac{d^2z}{dt^2} = -c\frac{dz}{dt} + mg',$$

where z is the object's position as a function of time, t, c is a drag constant which depends on the shape of the object and the fluid viscosity and

$$g' = g\left(1 - \frac{\rho_{fluid}}{\rho_{obj}}\right)$$

is the effective gravitational acceleration, which accounts for the buoyant force due to the fluid (density ρ_{fluid}) displaced by the object (density ρ_{obj}). For a small sphere of radius r in a fluid of viscosity η, Stokes' law predicts $c = 6\pi\eta r$.

Consider a sphere of platinum ($\rho = 21.45$ g cm^{-3}) with radius 1 mm, initially at rest, falling in mercury ($\rho = 13.53$ g cm^{-3}, $\eta = 1.53 \times 10^{-3}$ Pa s). The earlier mentioned second-order differential equation can be solved analytically, but to integrate it numerically using `solve_ivp`, it must be treated as two first-order ODEs:

$$\frac{dz}{dt} = \dot{z},$$

$$\frac{d^2z}{dt^2} = \frac{d\dot{z}}{dt} = g' - \frac{c}{m}\dot{z}.$$

In the code presented here, the function `deriv` calculates these derivatives and is passed to `solve_ivp` with the intial conditions ($z = 0$, $\dot{z} = 0$) and a grid of time points.

Listing 8.12 Calculating the motion of a sphere falling under the influence of gravity and Stokes' drag

```
# eg8-stokes-drag.py
import numpy as np
from scipy.integrate import solve_ivp
import matplotlib.pyplot as plt

# Platinum sphere falling from rest in mercury.

# Acceleration due to gravity (m.s-2).
g = 9.81
# Densities (kg.m-3).
rho_Pt, rho_Hg = 21450, 13530
# Viscosity  of mercury (Pa.s).
eta = 1.53e-3

# Radius and mass of the sphere.
r = 1.e-3    # radius (m)
m = 4*np.pi/3 * r**3 * rho_Pt
# Drag constant from Stokes' law.
c = 6 * np.pi * eta * r
# Effective gravitational acceleration.
gp = g * (1 - rho_Hg/rho_Pt)

def deriv(t, z):
    """ Return the dz/dt and d2z/dt2. """
    dz0 = z[1]
```

```
        dz1 = gp - c/m * z[1]
        return dz0, dz1

t0, tf = 0, 20
t_eval = np.linspace(t0, tf, 50)
# Initial conditions: z = 0, dz/dt = 0 at t = 0.
z0 = (0, 0)

# Integrate the pair of differential equations.
sol = solve_ivp(deriv, (t0, tf), z0, t_eval=t_eval)
t = sol.t
z, zdot = sol.y
plt.plot(t, zdot)

print('Estimate of terminal velocity = {:.3f} m.s-1'.format(zdot[-1]))

# Exact solution: terminal velocity vt (m.s-1) and characteristic time tau (s).
v0, vt, tau = 0, m*gp/c, m/c
print('Exact terminal velocity = {:.3f} m.s-1'.format(vt))
z = vt*t + v0*tau*(1-np.exp(-t/tau)) + vt*tau*(np.exp(-t/tau)-1)
zdot_exact = vt + (v0-vt)*np.exp(-t/tau)
plt.plot(t, zdot_exact)
plt.xlabel('$t$ /s')
plt.ylabel('$\dot{z}\;/\mathrm{m\, s^{-1}}$')
plt.show()
```

The plot produced by this program is shown in Figure 8.13: the numerical and analytical results are indistinguishable at this scale but are reported to three decimal places in the output:

```
Estimate of terminal velocity = 11.266 m.s-1
Exact terminal velocity = 11.285 m.s-1
```

The `solve_ivp` function can be configured to use different algorithms to solve a system of ODEs by setting its `method` attribute; a list of options is given in Table 8.3. The default, `'RK45'`, is an explicit Runge–Kutta method of order 5(4) and is a good general-purpose approach for non-stiff problems.

A problem is said to be *stiff* if a numerical method is required to take excessively small steps in its intervals of integration in relation to the smoothness of the exact underlying solution. Stiff problems frequently occur when terms in the ODE represent a variable changing in magnitude with very different timescales. The methods `'Radau'`, `'BDF'` and `'LSODA'` are worth trying if you suspect your ODE is stiff, as in the following example.

Example E8.17   A classic example of a stiff system of ODEs is the kinetic analysis of Robertson's autocatalytic chemical reaction[8] involving three species, $x = [X]$, $y = [Y]$ and $z = [Z]$ with initial conditions $x = 1$, $y = z = 0$:

---

[8] H. H. Robertson, The solution of a set of reaction rate equations, in J. Walsh (Ed.), *Numerical Analysis: An Introduction*, pp. 178–182, Academic Press, London (1966).

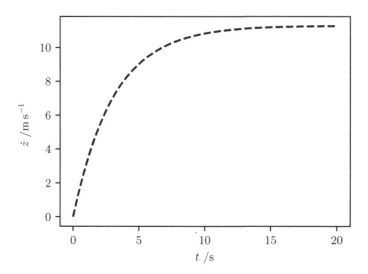

Figure 8.13 The velocity of a platinum sphere falling in mercury as a function of time, modeled with Stokes' law.

Table 8.3 ODE integration methods defined within `scipy.integrate.solve_ivp`

method	Description
'RK45'	Explicit Runge–Kutta method of order 5(4)
'RK23'	Explicit Runge–Kutta method of order 3(2)
'Radau'	Implicit Runge–Kutta method of the Radau IIA family of order 5: suitable for stiff problems
'BDF'	Backward differentiation formula approximation, an implicit method suitable for stiff problems
'LSODA'	A flexible method that can automaticlly detect stiffness and switch between the Adams (for non-stiff problems) and BDF (for stiff problems) algorithms

$$\dot{x} \equiv \frac{dx}{dt} = -0.04x + 10^4 yz,$$

$$\dot{y} \equiv \frac{dy}{dt} = 0.04x - 10^4 yz - 3 \times 10^7 y^2,$$

$$\dot{z} \equiv \frac{dz}{dt} = 3 \times 10^7 y^2.$$

(Note the very different timescales of the reactions, particularly for [Y].)

With the default Runge–Kutta algorithm on the time interval $[0, 500]$:

```
def deriv(t, y):
    x, y, z = y
    xdot = -0.04 * x + 1.e4 * y * z
    ydot = 0.04 * x - 1.e4 * y * z - 3.e7 * y**2
    zdot = 3.e7 * y**2
    return xdot, ydot, zdot

t0, tf = 0, 500
y0 = 1, 0, 0
soln = solve_ivp(deriv, (t0, tf), y0)
print(soln)
```

We get there eventually:

```
 message: 'The solver successfully reached the end of the integration interval.'
    nfev: 6123410
    njev: 0
     nlu: 0
     sol: None
  status: 0
 success: True
       t: array([0.00000000e+00, 6.36669332e-04, 1.06518798e-03, ...,
          4.99999288e+02, 4.99999819e+02, 5.00000000e+02])
t_events: None
       y: array([[1.00000000e+00, 9.99974534e-01, 9.99957394e-01, ...,
          4.19780946e-01, 4.19780771e-01, 4.19780487e-01],
         [0.00000000e+00, 2.20107324e-05, 3.00616449e-05, ...,
          2.41400796e-06, 2.47838908e-06, 2.72514279e-06],
         [0.00000000e+00, 3.45561028e-06, 1.25439771e-05, ...,
          5.80216640e-01, 5.80216750e-01, 5.80216788e-01]])
```

but at the cost of more than 6 million function evaluations (soln.nfev). The 'Radau' method fares much better:

Listing 8.13 Solution of the Robertson system of chemical reactions.

```
import numpy as np
from scipy.integrate import solve_ivp
import matplotlib.pyplot as plt

def deriv(t, y):
    """ODEs for Robertson's chemical reaction system."""
    x, y, z = y
    xdot = -0.04 * x + 1.e4 * y * z
    ydot = 0.04 * x - 1.e4 * y * z - 3.e7 * y**2
    zdot = 3.e7 * y**2
    return xdot, ydot, zdot

# Initial and final times.
t0, tf = 0, 500
# Initial conditions: [X] = 1; [Y] = [Z] = 0.
y0 = 1, 0, 0
# Solve, using a method resilient to stiff ODEs.
soln = solve_ivp(deriv, (t0, tf), y0, method='Radau')
print(soln.nfev, 'evaluations required.')

# Plot the concentrations as a function of time. Scale [Y] by 10**YFAC
# so its variation is visible on the same axis used for [X] and [Z].
```

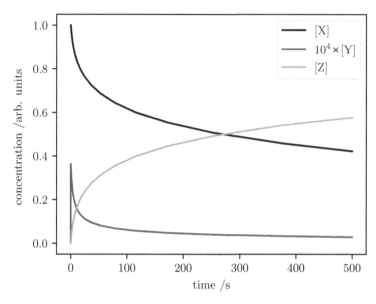

Figure 8.14 The Robertson chemical equation system, numerically integrated with the Radau IIA method.

```
YFAC = 4
plt.plot(soln.t, soln.y[0], label='[X]')
plt.plot(soln.t, 10**YFAC*soln.y[1], label=r'$10^{}\times$[Y]'.format(YFAC))
plt.plot(soln.t, soln.y[2], label='[Z]')
plt.xlabel('time /s')
plt.ylabel('concentration /arb. units')
plt.legend()
plt.show()
```

The output indicates only 248 evaluations required. The concentrations of [X], [Y] and [Z] are plotted against time in Figure 8.14

The `solve_ivp` function can also detect and respond to *events* in the integration of a set of differential equations. One or more functions can be provided in the `events` argument, which should return zero when the state of the system corresponds to the event to be triggered. For a very simple example consider a car traveling with velocity $v = 20$ m s^{-1} subject to a braking force which decelerates it at a constant rate, $a = dv/dt = -3$ m s^{-2}. How long will it take the car to stop? The analytical answer is obviously $20/3 = 6.67$ s. To analyze the problem numerically, we can define an event function that will return the car's speed: the time at which this function returns 0 is assigned to the solution object's `t_events` variable:

```
In [x]: def car_stopped(t, y):
   ...:     return y[0]
```

```
   ...:
In [x]: t0, tf = 0, 100     # a generous time interval to consider, s
In [x]: v0 = 20             # initial speed, m.s-1
In [x]: solve_ivp(lambda t, y: -3, (t0, tf), [v0], events=car_stopped)
Out[x]:
  message: 'The solver successfully reached the end of the integration interval.'
     nfev: 26
     njev: 0
      nlu: 0
      sol: None
   status: 0
  success: True
        t: array([  0.       ,   0.14614572,   1.60760288,   16.22217454,
           100.        ])
 t_events: [array([6.66666667])]
        y: array([[  20.       ,   19.56156285,   15.17719135,   -28.66652362,
           -280.      ]])
 y_events: [array([[-1.77635684e-15]])]
```

Note that the stopping time, given in the t_events array, is accurate: it is obtained using a root-finding algorithm once the ODE integration has detected a change in sign of the return value from the car_stopped event function. It has continued to integrate for unphysical solutions beyond this time though (the car does not go backwards after stopping). We can force solve_ivp to terminate after an event by attaching the boolean object terminal = True to it:

```
In [x]: car_stopped.terminal = True
In [x]: solve_ivp(lambda t, y: -3, (t0, tf), [v0], events=car_stopped)
Out[x]:
  message: 'A termination event occurred.'
     nfev: 20
     njev: 0
      nlu: 0
      sol: None
   status: 1
  success: True
        t: array([0.       ,   0.14614572,   1.60760288,   6.66666667])
 t_events: [array([6.66666667])]
        y: array([[ 2.00000000e+01,   1.95615629e+01,   1.51771914e+01,
           -1.77635684e-15]])
```

The terminal = True attribute must be attached to the function car_stopped *after* its definition or it won't be accessible to solve_ivp.[9]

Example E8.18   A spherical projectile of mass m launched with some initial velocity moves under the influence of two forces: gravity, $F_g = -mg\hat{z}$, and air resistance (drag), $F_D = -\frac{1}{2}c\rho Av^2v/|v| = -\frac{1}{2}c\rho Avv$, acting in the opposite direction to the projectile's velocity and proportional to the square of that velocity (under most realistic conditions).

---

[9] This sort of post-definition modification of a function has been available in Python since version 2.1 (PEP 232) and is an example of *monkey-patching*. Some consider it to be poor style since it separates the method's definition from its functionality.

Here, c is the drag coefficient, ρ the air density, and A the projectile's cross-sectional area.

The relevant equations of motion are therefore:

$$m\ddot{x} = -k\sqrt{\dot{x}^2 + \dot{z}^2}\,\dot{x},$$

$$m\ddot{z} = -k\sqrt{\dot{x}^2 + \dot{z}^2}\,\dot{z} - mg,$$

where $v = |\boldsymbol{v}| = \sqrt{\dot{x}^2 + \dot{z}^2}$ and $k = \frac{1}{2}c\rho A$. These can be decomposed into the following four first-order ODEs with $u_1 \equiv x, u_2 \equiv \dot{x}, u_3 \equiv z, u_4 \equiv \dot{z}$:

$$\dot{u}_1 = u_2,$$

$$\dot{u}_2 = -\frac{k}{m}\sqrt{u_2^2 + u_4^2}u_2,$$

$$\dot{u}_3 = u_4,$$

$$\dot{u}_4 = -\frac{k}{m}\sqrt{u_2^2 + u_4^2}u_4 - mg.$$

The following code integrates this system and identifies two events: hitting the target (the projectile returning to the ground at $z = 0$) and reaching its maximum height (at which the z-component of its velocity is zero). We set the additional attribute `hit_target.direction = -1` to ensure that `hit_target` only triggers the event when its return value (the projectile's elevation) goes from positive to negative; otherwise the event would be triggered at launch since $z_0 = 0$. Other possibilities are `direction = 1`: trigger the event when the return value changes from negative to positive or `direction = 0` (the default): the event is triggered when the return value is zero from either direction.

Listing 8.14 Calculating and plotting the trajectory of a spherical projectile including air resistance.

```
import numpy as np
from scipy.integrate import solve_ivp
import matplotlib.pyplot as plt

# Drag coefficient, projectile radius (m), area (m2) and mass (kg).
c = 0.47
r = 0.05
A = np.pi * r**2
m = 0.2
# Air density (kg.m-3), acceleration due to gravity (m.s-2).
rho_air = 1.28
g = 9.81
# For convenience, define  this constant.
k = 0.5 * c * rho_air * A

# Initial speed and launch angle (from the horizontal).
v0 = 50
phi0 = np.radians(65)

def deriv(t, u):
```

```
    x, xdot, z, zdot = u
    speed = np.hypot(xdot, zdot)
    xdotdot = -k/m * speed * xdot
    zdotdot = -k/m * speed * zdot - g
    return xdot, xdotdot, zdot, zdotdot

# Initial conditions: x0, v0_x, z0, v0_z.
u0 = 0, v0 * np.cos(phi0), 0., v0 * np.sin(phi0)
# Integrate up to tf unless we hit the target sooner.
t0, tf = 0, 50

def hit_target(t, u):
    # We've hit the target if the z-coordinate is 0.
    return u[2]
# Stop the integration when we hit the target.
hit_target.terminal = True
# We must be moving downwards (don't stop before we begin moving upwards!)
hit_target.direction = -1

def max_height(t, u):
    # The maximum height is obtained when the z-velocity is zero.
    return u[3]

soln = solve_ivp(deriv, (t0, tf), u0, dense_output=True,
                 events=(hit_target, max_height))
print(soln)
print('Time to target = {:.2f} s'.format(soln.t_events[0][0]))
print('Time to highest point = {:.2f} s'.format(soln.t_events[1][0]))

# A fine grid of time points from 0 until impact time.
t = np.linspace(0, soln.t_events[0][0], 100)

# Retrieve the solution for the time grid and plot the trajectory.
sol = soln.sol(t)
x, z = sol[0], sol[2]
print('Range to target, xmax = {:.2f} m'.format(x[-1]))
print('Maximum height, zmax = {:.2f} m'.format(max(z)))
plt.plot(x, z)
plt.xlabel('x /m')
plt.ylabel('z /m')
plt.show()
```

The output is:

```
Time to target = 6.34 s
Time to highest point = 2.79 s
Range to target, xmax = 64.12 m
Maximum height, zmax = 49.42 m
```

and the plot created is shown in Figure 8.15.

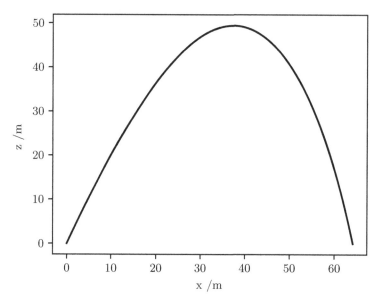

Figure 8.15 The trajectory of a spherical projectile launched with $v_0 = 50$ m s^{-1} at $\phi_0 = 50°$ including air resistance.

8.2.4   Exercises

Questions

Q8.2.1   Use scipy.integrate.quad to evaluate the following integral:

$$\int_0^6 \lfloor x \rfloor - 2 \left\lfloor \frac{x}{2} \right\rfloor \, dx.$$

Q8.2.2   Use scipy.integrate.quad to evaluate the following definite integrals (most of which can also be expressed in closed form over the range given but are awkward).

(a)

$$\int_0^1 \frac{x^4(1-x)^4}{1+x^2} \, dx.$$

(Compare with $22/7 - \pi$.)

(b)   The following integral appears in the Debye theory of the heat capacity of crystals at low temperature

$$\int_0^\infty \frac{x^3}{e^x - 1} \, dx.$$

(Compare with $\pi^4/15$.)

(c)   The integral sometimes known as the *Sophomore's dream*:

$$\int_0^1 x^{-x}\, dx.$$

(Compare the value you obtain from the summation $\sum_{n=1}^{\infty} n^{-n}$.)

(d)

$$\int_0^1 [\ln(1/x)]^p\, dx.$$

(Compare with $p!$ for integer $0 \le p \le 10$.)

(e)

$$\int_0^{2\pi} e^{z\cos\theta}\, d\theta.$$

(Compare with $2\pi I_0(z)$, where $I_0(z)$ is a modified Bessel function of the first kind, for $0 \le z \le 2$.)

Q8.2.3   Use `scipy.integrate.dblquad` to evaluate π by integration of the constant function $f(x, y) = 4$ over the quarter circle with unit radius in the quadrant $x > 0, y > 0$.

Q8.2.4   What is wrong with the following attempt to calculate the area of the unit circle (π) as a double integral in polar coordinates?

```
In [x]: dblquad(lambda r, theta: r, 0, 1, lambda r: 0, lambda r: 2*np.pi)
Out[x]: (19.739208802178712, 2.1914924100062363e-13)
```

Problems

P8.2.1   The area of the surface of revolution about the x-axis between a and b of the function $y = f(x)$ is given by the integral

$$S = 2\pi \int_a^b y\, ds, \quad \text{where } ds = \sqrt{1 + \left(\frac{dy}{dx}\right)^2}\, dx.$$

Use this equation to write a function to determine the surface area of revolution of a function $y = f(x)$ about the x-axis, given Python function objects that return y and dy/dx, and test it for the paraboloid obtained by rotation of the function $f(x) = \sqrt{x}$ about the x-axis between $a = 0$ and $b = 1$. Compare with the exact result, $\pi(5^{3/2} - 1)/6$.

P8.2.2   The integral of the secant function,

$$\int_0^\theta \sec\phi\, d\phi$$

for $-\pi/2 < \theta < \pi/2$ is important in navigation and the theory of map projections. It can be expressed in closed form as the inverse Gudermannian function,

$$\mathrm{gd}^{-1}(\theta) = \ln|\sec\theta + \tan\theta|.$$

Use `scipy.integrate.quad` to calculate values for the integral across the relevant range for θ given earlier and compare graphically with the exact answer.

P8.2.3 Consider a torus of uniform density, unit mass, average radius R and cross-sectional radius r. The volume and moments of inertia of such a torus may be evaluated analytically and give the results:

$$V = 2\pi^2 R r^2,$$
$$I_z = R^2 + \tfrac{3}{4} r^2,$$
$$I_x = I_y = \tfrac{1}{2} R^2 + \tfrac{5}{8} r^2,$$

where the center of mass of the torus is at the origin and the z axis is taken to be its symmetry axis.

Here, we take a numerical approach. In cylindrical coordinates (ρ, θ, z), it may be shown that:

$$V = 2 \int_0^{2\pi} \int_{R-r}^{R+r} \int_0^{\sqrt{r^2-(\rho-R)^2}} \rho \, dz \, d\rho \, d\theta,$$

$$I_z = \frac{2}{V} \int_0^{2\pi} \int_{R-r}^{R+r} \int_0^{\sqrt{r^2-(\rho-R)^2}} \rho^3 \, dz \, d\rho \, d\theta,$$

$$I_x = I_y = \frac{2}{V} \int_0^{2\pi} \int_{R-r}^{R+r} \int_0^{\sqrt{r^2-(\rho-R)^2}} (\rho^2 \sin^2 \theta + z^2) \rho \, dz \, d\rho \, d\theta.$$

Evaluate these integrals for the torus with dimensions $R = 4$, $r = 1$ and compare with the exact values.

P8.2.4 The *Brusselator* is a theoretical model for an autocatalytic reaction. It assumes the following reaction sequence, in which species A and B are taken to be in excess with constant concentration and species D and E are removed as they are produced. The concentrations of species X and Y can show oscillatory behavior under certain conditions.

$$\begin{aligned} A &\rightarrow X & k_1, \\ 2X + Y &\rightarrow 3X & k_2, \\ B + X &\rightarrow Y + D & k_3, \\ X &\rightarrow E & k_4. \end{aligned}$$

It is convenient to introduce the scaled quantities

$$x = [X] \sqrt{\frac{k_2}{k_4}}, \quad y = [Y] \sqrt{\frac{k_2}{k_4}},$$

$$a = [A] \frac{k_1}{k_4} \sqrt{\frac{k_2}{k_4}}, \quad b = [B] \frac{k_3}{k_4},$$

and to scale the time by the factor k_4, which gives rise to the dimensionless equations

$$\frac{dx}{dt} = a - (1 + b)x + x^2 y,$$

$$\frac{dy}{dt} = bx - x^2 y.$$

Show how these equations predict x and y to vary for (a) $a = 1, b = 1.8$ and (b) $a = 1, b = 2.02$ by plotting in each case (i) x, y as functions of (dimensionless) time and (ii) y as a function of x.

P8.2.5 The equation governing the motion of a pendulum consisting of a mass at the end of a light, rigid rod of length l may be written

$$\frac{d^2\theta}{dt^2} = -\frac{g}{l} \sin \theta,$$

where θ is the angle the pendulum makes with the vertical.

Taking $l = 1$ m and $g = 9.81$ m s^{-2}, determine the subsequent motion of the pendulum if it is started at rest with an initial angle $\theta_0 = 30°$. Compare the motion with the harmonic approximation reached by assuming θ is small, which has the analytical solution $\theta = \theta_0 \cos(\omega t)$ with $\omega = \sqrt{g/l}$.

P8.2.6 A simple mechanism for the formation of ozone in the stratosphere consists of the following four reactions (known as the *Chapman cycle*):

$$O_2 + hv \rightarrow 2O \qquad\qquad k_1 = 3 \times 10^{-12} \text{ s}^{-1},$$
$$O_2 + O + M \rightarrow O_3 + M \qquad k_2 = 1.2 \times 10^{-33} \text{ cm}^6 \text{ molec}^{-2} \text{ s}^{-1},$$
$$O_3 + hv' \rightarrow O + O_2 \qquad\qquad k_3 = 5.5 \times 10^{-4} \text{ s}^{-1},$$
$$O + O_3 \rightarrow 2O_2 \qquad\qquad k_4 = 6.9 \times 10^{-16} \text{ cm}^3 \text{ molec}^{-1} \text{ s}^{-1},$$

where M is a nonreacting third body taken to be at the total air molecule concentration for the altitude being considered. These reactions lead to the following rate equations for [O], [O$_3$] and [O$_2$]:

$$\frac{d[O_2]}{dt} = -k_1[O_2] - k_2[O_2][O][M] + k_3[O_3] + 2k_4[O][O_3],$$

$$\frac{d[O]}{dt} = 2k_1[O_2] - k_2[O_2][O][M] + k_3[O_3] - k_4[O][O_3],$$

$$\frac{d[O_3]}{dt} = k_2[O_2][O][M] - k_3[O_3] - k_4[O][O_3].$$

The rate constants apply at an altitude of 25 km, where $[M] = 9 \times 10^{17}$ molec cm^{-3}. Write a program to determine the concentrations of O$_3$ and O as a function of time at this altitude (you should find the [O$_2$] remains pretty much constant). Start with initial conditions $[O_2]_0 = 0.21[M]$, $[O]_0 = [O_3]_0 = 0$ and integrate for 10^8 s (starting from scratch it takes about three years to build an ozone layer with this mechanism). Compare the equilibrium concentrations with the approximate analytical result obtained using the *steady-state approximation*:

$$[O_3] = \sqrt{\frac{k_1 k_2}{k_3 k_4}} [O_2][M]^{\frac{1}{2}}, \quad \frac{[O]}{[O_3]} = \frac{k_3}{k_2[O_2][M]}.$$

P8.2.7 Hyperion is an irregularly shaped moon of Saturn notable for its chaotic rotation. Its motion may be modeled as follows.

The orbit of Hyperion (H) about Saturn (S) is an ellipse with semi-major axis a and eccentricity e. Let its point of closest approach (*periapsis*) be P. Its distance from the planet, SH, as a function of its *true anomaly* (orbital angle, ϕ, measured from the line SP) is therefore

$$r = \frac{a(1 - e^2)}{1 + e \cos \phi}.$$

Define the angle θ to be that between the axis of the smallest principal moment of inertia (loosely, the longest axis of the moon) and SP, and the quantity Ω to be a scaled rate of change of θ with ϕ (i.e. the rate at which Hyperion spins as it orbits Saturn) as follows:

$$\Omega = \frac{a^2}{r^2} \frac{d\theta}{d\phi}.$$

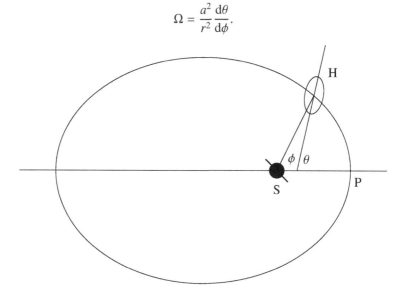

Now, it can be shown that

$$\frac{d\Omega}{d\phi} = -\frac{B - A}{C} \frac{3}{2(1 - e^2)} \frac{a}{r} \sin[2(\theta - \phi)],$$

where A, B and C are the principal moments of inertia.

Use `scipy.integrate.solve_ivp` to find and plot the spin rate, Ω, as a function of ϕ for the initial conditions (a) $\theta = \Omega = 0$ at $\phi = 0$, and (b) $\theta = 0$, $\Omega = 2$ at $\phi = 0$. Take $e = 0.1$ and $(B - A)/C = 0.265$.

P8.2.8 The radioactive decay chain of ^{212}Pb to the stable isotope ^{208}Pb may be considered as the following sequence of steps with the given rate constants, k_i:

$$^{212}\text{Pb} \rightarrow {}^{212}\text{Bi} + \beta^- \qquad k_1 = 1.816 \times 10^{-5} \text{ s}^{-1},$$

$$^{212}\text{Bi} \rightarrow {}^{208}\text{Tl} + \alpha \qquad k_2 = 6.931 \times 10^{-5} \text{ s}^{-1},$$

$$^{212}\text{Bi} \rightarrow {}^{212}\text{Po} + \beta^- \qquad k_3 = 1.232 \times 10^{-4} \text{ s}^{-1},$$

$$^{208}\text{Tl} \rightarrow {}^{208}\text{Pb} + \beta^- \qquad k_4 = 3.851 \times 10^{-3} \text{ s}^{-1},$$

$$^{212}\text{Po} \rightarrow {}^{208}\text{Pb} + \alpha \qquad k_5 = 2.310 \text{ s}^{-1}.$$

By considering the following first-order differential equations giving the rates of change for each species, plot their concentrations as a function of time:

$$\frac{d[^{212}\text{Pb}]}{dt} = -k_1[^{212}\text{Pb}],$$

$$\frac{d[^{212}\text{Bi}]}{dt} = k_1[^{212}\text{Pb}] - k_2[^{212}\text{Bi}] - k_3[^{212}\text{Bi}],$$

$$\frac{d[^{208}\text{Tl}]}{dt} = k_2[^{212}\text{Bi}] - k_4[^{208}\text{Tl}],$$

$$\frac{d[^{212}\text{Po}]}{dt} = k_3[^{212}\text{Bi}] - k_5[^{212}\text{Po}],$$

$$\frac{d[^{208}\text{Pb}]}{dt} = k_4[^{208}\text{Tl}] + k_5[^{212}\text{Po}].$$

If all the intermediate species, J, are treated in "steady state" (i.e. $d[J]/dt = 0$), the approximate expression for the ^{208}Pb concentration as a function of time is

$$[^{208}\text{Pb}] = [^{212}\text{Pb}]_0 \left(1 - e^{-k_1 t}\right).$$

Compare the "exact" result obtained by numerical integration of the differential equations with this approximate answer.

P8.2.9 A simple model of the evolution of a match flame considers the flame radius, y, to change in time as

$$\frac{dy}{dt} = \alpha y^2 - \beta y^3,$$

where α and β are some constants relating to the transport of oxygen through the surface of the flame and the rate of its consumption inside it. The flame is initally small, $y(0) \ll \alpha/\beta$, but at some point rapidly grows until it reaches a steady state of constant radius (assuming an unlimited supply of fuel).

Taking $\alpha = \beta = 1$, solve this ODE numerically using `scipy.integrate.solve_ivp` using a suitable integration method over a time interval $(0, 5/y(0))$ for (a) $y(0) = 0.01$, (b) $y(0) = 0.0001$. How many time steps must be taken in each case?

The exact solution may be written as

$$y(t) = \frac{\alpha}{\beta\left[1 + W\left(ae^{a-\alpha^2 t/\beta}\right)\right]},$$

Table 8.4 Interpolation methods specified by the kind argument to
`scipy.interpolate.interp1d`

kind	Description
`'linear'`	The default, linear interpolation using only the values from the original data arrays bracketing the desired point
`'nearest'`	"Snap" to the nearest data point
`'zero'`	A zeroth-order spline: interpolates to the last value seen in its traversal of the data arrays
`'slinear'`	First-order spline interpolation (in practice, the same as `'linear'`)
`'quadratic'`	Second-order spline interpolation
`'cubic'`	Cubic spline interpolation
`'previous'`	Use the previous data point
`'next'`	Use the next data point

where $a = \alpha/(\beta y(0))$ and $W(x)$ is the Lambert W function, implemented in SciPy as `scipy.special.lambertw`. Compare the accuracy of the various numerical solutions with this expression.

8.3   Interpolation

The package `scipy.interpolate` contains a large variety of functions and classes for interpolation and splines in one and more dimensions. Some of the more important are described in this section.

8.3.1   Univariate Interpolation

The most straightforward one-dimensional interpolation functionality is provided by `scipy.interpolate.interp1d`. Given arrays of points x and y, a function is returned, which can be called to generate interpolated values at intermediate values of x. The default interpolation scheme is linear, but other options (see Table 8.4) allow for different schemes, as shown in the following example.

Example E8.19   This example demonstrates some of the different interpolation methods available in `scipy.interpolation.interp1d` (see Figure 8.16).

Listing 8.15 A comparison of one-dimensional interpolation types using
`scipy.interpolate.interp1d`

```
# eg8-interp1d.py
import numpy as np
from scipy.interpolate import interp1d
import matplotlib.pyplot as plt

A, nu, k = 10, 4, 2
```

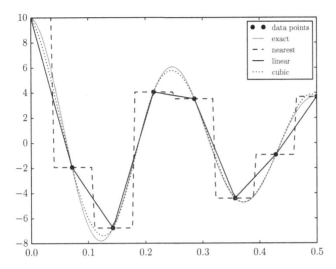

Figure 8.16 An illustration of different one-dimensional interpolation methods with
`scipy.interpolation.interp1d`.

```
def f(x, A, nu, k):
    return A * np.exp(-k*x) * np.cos(2*np.pi * nu * x)

xmax, nx = 0.5, 8
x = np.linspace(0, xmax, nx)
y = f(x, A, nu, k)

f_nearest = interp1d(x, y, kind='nearest')
f_linear  = interp1d(x, y)
f_cubic   = interp1d(x, y, kind='cubic')

x2 = np.linspace(0, xmax, 100)
plt.plot(x, y, 'o', label='data points')
plt.plot(x2, f(x2, A, nu, k), label='exact')
plt.plot(x2, f_nearest(x2), label='nearest')
plt.plot(x2, f_linear(x2), label='linear')
plt.plot(x2, f_cubic(x2), label='cubic')
plt.legend()
plt.show()
```

8.3.2   Multivariate Interpolation

We shall consider two kinds of multivariate interpolation corresponding to whether or
not the source data are structured (arranged on some kind of grid) or not.

Interpolation from a Rectangular Grid

The simplest two-dimensional interpolation routine is `scipy.interpolate.interp2d`. It
requires a two-dimensional array of values, z, and the two (one-dimensional) coordinate

arrays x and y to which they correspond. These arrays need not have constant spacing. Three kinds of interpolation spline are supported through the kind argument: `'linear'` (the default), `'cubic'` and `'quintic'`.

Example E8.20 In the following example, we calculate the function

$$z(x, y) = \sin\left(\frac{\pi x}{2}\right) e^{y/2}$$

on a grid of points (x, y) which is not evenly spaced in the y-direction. We then use `scipy.interpolate.interp2d` to interpolate these values onto a finer, evenly spaced (x, y) grid: see Figure 8.17.

Listing 8.16 Two-dimensional interpolation with `scipy.interpolate.interp2d`

```
# eg8-interp2d.py
import numpy as np
from scipy.interpolate import interp2d
import matplotlib.pyplot as plt

x = np.linspace(0, 4, 13)
y = np.array([0, 2, 3, 3.5, 3.75, 3.875, 3.9375, 4])
X, Y = np.meshgrid(x, y)
Z = np.sin(np.pi*X/2) * np.exp(Y/2)

x2 = np.linspace(0, 4, 65)
y2 = np.linspace(0, 4, 65)
f = interp2d(x, y, Z, kind='cubic')
Z2 = f(x2, y2)

fig, ax = plt.subplots(nrows=1, ncols=2)
ax[0].pcolormesh(X, Y, Z)

X2, Y2 = np.meshgrid(x2, y2)
ax[1].pcolormesh(X2, Y2, Z2)

plt.show()
```

❶ is placed next to `f = interp2d(x, y, Z, kind='cubic')`

❶ Note that `interp2d` requires the *one-dimensional* arrays, x and y.

If the mesh of (x, y) coordinates form a *regularly spaced* grid, the fastest way to interpolate values from values of z is to use a `scipy.interpolate.RectBivariateSpline` object as in the following example.

Example E8.21 In the following code, the function

$$z(x, y) = e^{-4x^2} e^{-y^2/4}$$

is calculated on a regular, coarse grid and then interpolated onto a finer one (Figure 8.18).

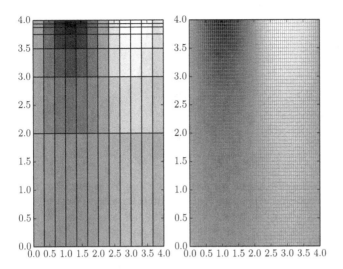

Figure 8.17 Two-dimensional interpolation with `scipy.interpolate.interp2d`.

Listing 8.17 Interpolation onto a regular two-dimensional grid with
`scipy.interpolate.RectBivariateSpline`

```
# eg8-RectBivariateSpline.py
import numpy as np
from scipy.interpolate import RectBivariateSpline
import matplotlib.pyplot as plt
from mpl_toolkits.mplot3d import Axes3D

# Regularly spaced, coarse grid.
dx, dy = 0.4, 0.4
xmax, ymax = 2, 4
x = np.arange(-xmax, xmax, dx)
y = np.arange(-ymax, ymax, dy)
X, Y = np.meshgrid(x, y)
Z = np.exp(-(2*X)**2 - (Y/2)**2)

interp_spline = RectBivariateSpline(y, x, Z)

# Regularly spaced, fine grid.
dx2, dy2 = 0.16, 0.16
x2 = np.arange(-xmax, xmax, dx2)
y2 = np.arange(-ymax, ymax, dy2)
X2, Y2 = np.meshgrid(x2, y2)
Z2 = interp_spline(y2, x2)

fig, ax = plt.subplots(nrows=1, ncols=2, subplot_kw={'projection': '3d'})
ax[0].plot_wireframe(X, Y, Z, color='k')

ax[1].plot_wireframe(X2, Y2, Z2, color='k')
for axes in ax:
    axes.set_zlim(-0.2, 1)
    axes.set_axis_off()
```

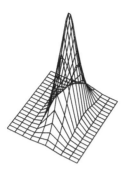

Figure 8.18 Two-dimensional interpolation from a coarse rectangular grid (left-hand plot) to a finer one (right-hand plot) with `scipy.interpolate.RectBivariateSpline`.

```
fig.tight_layout()
plt.show()
```

❶ Note that for our function, Z, defined using the `meshgrid` set up here, the RectBivariateSpline method expects the corresponding one-dimensional arrays y and x to be passed in this order (opposite to that of interp2d).[10]

Interpolation of Unstructured Data

To interpolate unstructured data, that is, data points provided at *arbitrary* coordinates (x, y), onto a grid, the method `scipy.interpolate.griddata` can be used. Its basic usage for two dimensions is:

```
scipy.interpolate.griddata(points, values, xi, method='linear')
```

where the provided data are given as the one-dimensional array, values, at the coordinates, points, which is provided as a tuple of arrays x and y or as a single array of shape (n, 2), where n is the length of the values array. xi is an array of the coordinate grid to by interpolated onto (of shape (m, 2).) The methods available are 'linear' (the default), 'nearest' and 'cubic'.

Example E8.22   The code mentioned here illustrates the different kinds of interpolation method available for `scipy.interpolate.griddata` using 400 points chosen randomly from an interesting function. The results can be compared in Figure 8.19.

Listing 8.18 Interpolation from an unstructured array of two-dimensional points with `scipy.interpolate.griddata`

```
# eg8-gridinterp.py
import numpy as np
from scipy.interpolate import griddata
import matplotlib.pyplot as plt

x = np.linspace(-1, 1, 100)
```

---

[10] This issue is related to the way that meshgrid is indexed, which is based on the conventions of MATLAB.

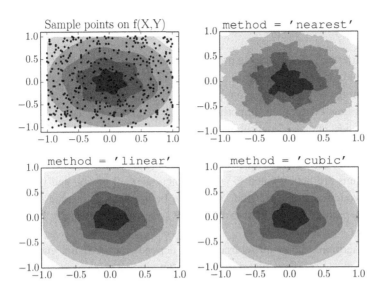

Figure 8.19 Some different interpolation schemes for `scipy.interpolate.griddata`.

```
y =  np.linspace(-1, 1, 100)
X, Y = np.meshgrid(x, y)

def f(x, y):
    s = np.hypot(x, y)
    phi = np.arctan2(y, x)
    tau = s + s * (1 - s) / 5 * np.sin(6 * phi)
    return 5 * (1 - tau) + tau

T = f(X, Y)
# Choose npts random points from the discrete domain of our model function.
npts = 400
px, py = np.random.choice(x, npts), np.random.choice(y, npts)

fig, ax = plt.subplots(nrows=2, ncols=2)
# Plot the model function and the randomly selected sample points.
ax[0, 0].contourf(X, Y, T)
ax[0, 0].scatter(px, py, c='k', alpha=0.2, marker='.')
ax[0, 0].set_title('Sample points on f(X, Y)')

# Interpolate using three different methods and plot.
for i, method in enumerate(('nearest', 'linear', 'cubic')):
    Ti = griddata((px, py), f(px, py), (X, Y), method=method)
    r, c = (i + 1) // 2, (i + 1) % 2
    ax[r, c].contourf(X, Y, Ti)
    ax[r, c].set_title("method = '{}'".format(method))

fig.tight_layout()
plt.show()
```

8.4   Optimization, Data-Fitting and Root-Finding

The `scipy.optimize` package provides a range of popular algorithms for minimization of multidimensional functions (with or without additional constraints), least-squares data-fitting and multidimensional equation solving (root-finding). This section will give an overview of the more important options available, but it should be borne in mind that the best choice of algorithm will depend on the individual function being analyzed. For an arbitrary function, there is no guarantee that a particular method will converge on the desired minimum (or root, etc.), or that if it does it will do so quickly. Some algorithms are better suited to certain functions than others, and the more you know about your function the better. SciPy can be configured to issue a warning message when a particular algorithm fails, and this message can usually help to analyze the problem.

Furthermore, the result returned often depends on the initial guess provided to the algorithm – consider a two-dimensional function as a landscape with several valleys separated by steep ridges: an initial guess placed within one valley is likely to lead most algorithms to wander downhill and find the minimum in that valley (even if it isn't the *global* minimum) without climbing the ridges. Similarly, you might expect (but cannot *guarantee*) that most numerical root-finders return the "nearest" root to the initial guess.

8.4.1   Minimization

SciPy's optimization routines *minimize* a function of one or more variables, $f(x_1, x_2, \ldots, x_n)$. To find the *maximum*, one determines the minimum of $-f(x_1, x_2, \ldots, x_n)$.

Some of the minimization algorithms only require the function itself to be evaluated; others require its first derivative with respect to each of the variables in an array known as the *Jacobian*:

$$J(f) = \left(\frac{\partial f}{\partial x_1}, \frac{\partial f}{\partial x_2}, \ldots, \frac{\partial f}{\partial x_n} \right).$$

Some algorithms will attempt to estimate the Jacobian numerically if it cannot be provided as a separate function.

Furthermore, some sophisticated optimization algorithms require information about the second derivatives of the function, a symmetric matrix of values called the *Hessian*:

$$H(f) = \begin{pmatrix} \frac{\partial^2 f}{\partial x_1^2} & \frac{\partial^2 f}{\partial x_2 \partial x_1} & \cdots & \frac{\partial^2 f}{\partial x_n \partial x_1} \\ \frac{\partial^2 f}{\partial x_1 \partial x_2} & \frac{\partial^2 f}{\partial x_2^2} & \cdots & \frac{\partial^2 f}{\partial x_n \partial x_2} \\ \vdots & \vdots & \ddots & \vdots \\ \frac{\partial^2 f}{\partial x_1 \partial x_n} & \frac{\partial^2 f}{\partial x_2 \partial x_n} & \cdots & \frac{\partial^2 f}{\partial x_n^2} \end{pmatrix}.$$

Just as the Jacobian represents the local *gradient* of a function of several variables, the Hessian represents the local *curvature*.

Unconstrained Minimization

The general algorithm for the minimization of multivariate scalar functions is `scipy.optimize.minimize`, which takes two mandatory arguments:

```
minimize(fun, x0, ...)
```

The first is a function object, `fun`, for evaluating the function to be minimized: this function should take an array of values, `x`, defining the point at which it is to be evaluated $(x_1, x_2, \ldots, x_n)$ followed by any further arguments it requires. The second required argument, `x0`, is an array of values representing the initial guess for the minimization algorithm to start at.

In this section we will demonstrate the use of `minimize` with *Himmelblau's function*, a simple two-dimensional function with some awkward features that make it a good test-function for optimization algorithms. Himmelblau's function is

$$f(x, y) = (x^2 + y - 11)^2 + (x + y^2 - 7)^2.$$

The region $-5 \le x \le 5$, $-5 \le y \le 5$ contains one local maximum,

$$f(-0.270845, -0.923039) = 181.617$$

(though the function climbs steeply outside of this region). There are four minima:

$$f(3, 2) = 0,$$
$$f(-2.805118, 3.131312) = 0,$$
$$f(-3.779310, -3.283186) = 0,$$
$$f(3.584428, -1.848126) = 0,$$

and four saddle points. Figure 8.20 shows a contour plot of the function.

The function may be defined in Python in the usual way:

```
In [x]: def f(X):
   ...:     x, y = X
   ...:     return (x**2 + y - 11)**2 + (x + y**2 - 7)**2
```

where for clarity we have unpacked the array, `X`, holding (x_1, x_2) into the named values $x_1 \equiv x$ and $x_2 \equiv y$.

To find a minimum, call `minimize` with some initial guess, say $(x, y) = (0, 0)$:

```
In [x]: from scipy.optimize import minimize
In [x]: minimize(f, (0, 0))
      jac: array([ -8.77780211e-06,  -3.52519449e-06])
  message: 'Optimization terminated successfully.'
      fun: 6.15694370233122e-13
     njev: 16
 hess_inv: array([[ 0.01575433, -0.00956965],
          [-0.00956965,  0.03491686]])
   status: 0
     nfev: 64
  success: True
        x: array([ 2.99999989,  1.99999996])
```

Table 8.5 Minimization information dictonary returned by `scipy.optimize.minimize`

Key	Description
success	A boolean value indicating whether or not the minimization was successful
x	If successful, the solution: the values of $(x_1, x_2, \ldots, x_n)$ at which the function is a minimum; if the algorithm was not successful, x indicates the point at which it gave up
fun	If successful, the value of the function at the minimum identified as x
message	A string describing of the outcome of the minimization
jac	The value of the Jacobian: if the minimization is successful the values in this array should be close to zero
hess, hess_inv	The Hessian and its inverse (if used)
nfev, njev, nhev	The number of evaluations of the function, its Jacobian and its Hessian

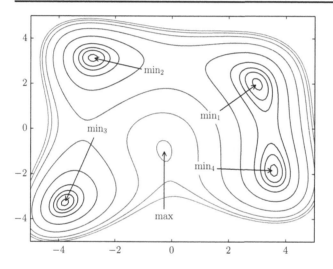

Figure 8.20 Contour plot of Himmelblau's function.

`minimize` returns a dictionary-like object with information about the minimization. The important fields are described in Table 8.5: if the minimization is successful, the minimum appears as x in this object – here we have converged close to the minimum $f(3, 2) = 0$.

The algorithm to be used by `minimize` is specified by setting its `method` argument to one of the strings given in Table 8.6. The default algorithm, BFGS, is a good general-purpose quasi-Newton method that can approximate the Jacobian if it is not provided and does not use the Hessian. However, it struggles to find the maximum of Himmelblau's function:

```
In [x]: mf = lambda X: -f(X)      # to find the maximum, minimize -f(x, y)
In [x]: minimize(mf, (0.1, -0.2))
Out[x]:
        fun: -1.2100579056485772e+35
```

Table 8.6 Some of the minimization methods used by `scipy.optimize.minimize`

method	Description
BFGS	Broyden–Fletcher–Goldfarb–Shanno (BFGS) algorithm, the default for minimization without constraints or bounds
Nelder-Mead	Nelder–Mead algorithm, also known as the downhill simplex or amoeba method; no derivatives are needed
CG	Conjugate gradient method
Powell	Powell's method (no derivatives are needed with this algorithm)
dogleg	Dog-leg trust-region algorithm (unconstrained minimization); requires the Jacobian and the Hessian (which must be positive-definite)
TNC	Truncated Newton algorithm for minimization within bounds
l-bfgs-b	Bound-constrained minimization with the L-BFGS-B algorithm
slsqp	"Sequential least-squares programming" method for minimization with bounds and equality and inequality constraints
cobyla	"Constrained optimization by linear approximation" method for constrained minimization

```
 hess_inv: array([[ 0.254751  , -0.43222419],
       [-0.43222419,  0.83976276]])
      jac: array([0., 0.])
  message: 'Optimization terminated successfully.'
     nfev: 68
      nit: 2
     njev: 17
   status: 0
  success: True
        x: array([ 3.45579856e+08, -5.71590777e+08])
```

Starting at $(0.1, -0.2)$, the BFGS algorithm has wandered up one of the steep sides of the Himmelblau function and failed to converge. Unfortunately, in this case it doesn't know it failed and returned `True` in the `success` flag (you may or may not see an error message: `'Desired error not necessarily achieved due to precision loss.'`, depending on the precise setup of your system). In fact, we need to start quite close to the maximum to be sure of success:

```
In [x]: minimize(mf, (-0.2,-1))
Out[x]:
      jac: array([ 3.81469727e-06,  1.90734863e-06])
  message: 'Optimization terminated successfully.'
      fun: -181.61652152258262
     njev: 8
 hess_inv: array([[ 0.0232834 , -0.00626945],
       [-0.00626945,  0.06137267]])
   status: 0
     nfev: 32
  success: True
        x: array([-0.27084453, -0.92303852])
```

This is, of course, not much help if we don't know in advance where the maximum is! Let's try a different minimization algorithm, starting at our arbitrary guess, $(0, 0)$:

```
In [x]: minimize(mf, (0, 0), method='Nelder-Mead')
Out[x]:
  status: 0
    nfev: 115
 success: True
 message: 'Optimization terminated successfully.'
     fun: -181.61652150549165
     nit: 59
       x: array([-0.27086815, -0.92300745])
```

The Nelder–Mead algorithm is a simplex method that does not need or estimate the derivatives of the function, so it isn't tempted up the steep sides of the function. However, it has taken 115 function evaluations to converge on the local maximum.

As a final example, consider the dogleg method, which requires minimize to be passed functions evaluating the Jacobian and the Hessian. The necessary derivatives have simple analytical forms for Himmelblau's function:

$$\frac{\partial f}{\partial x} = 4x(x^2 + y - 11) + 2(x + y^2 - 7),$$

$$\frac{\partial f}{\partial y} = 2(x^2 + y - 11) + 4y(x + y^2 - 7),$$

$$\frac{\partial^2 f}{\partial x^2} = 12x^2 + 4y - 42,$$

$$\frac{\partial^2 f}{\partial y^2} = 12y^2 + 4x - 26,$$

$$\frac{\partial^2 f}{\partial y \partial x} = \frac{\partial^2 f}{\partial x \partial y} = 4x + 4y.$$

The Jacobian and Hessian can be coded up as follows:

```
In [x]: def df(X):
   ...:     x, y = X
   ...:     f1, f2  = x**2 + y - 11, x + y**2 - 7
   ...:     dfdx = 4*x*f1 + 2*f2
   ...:     dfdy = 2*f1 + 4*y*f2
   ...:     return np.array([dfdx, dfdy])
   ...:
In [x]: def ddf(X):
   ...:     x, y = X
   ...:     d2fdx2 = 12*x**2 + 4*y - 42
   ...:     d2fdy2 = 12*y**2 + 4*x - 26
   ...:     d2fdxdy = 4*(x + y)
   ...:     return np.array([[d2fdx2, d2fdxdy], [d2fdxdy, d2fdy2]])
   ...:
❶ In [x]: mdf = lambda X: -df(X)
In [x]: mddf = lambda X: -ddf(X)
```

❶ Note that as with the function itself, we need to use the negative of the Jacobian and Hessian if we seek the maximum: these are defined as lambda functions mdf and mddf.

```
In [x]: minimize(mf, (0, 0), jac=mdf, hess=mddf, method='dogleg')
Out[x]:
    jac: array([ -1.26922473e-10,   1.23685240e-09])
```

```
message: 'Optimization terminated successfully.'
    fun: -181.6165215225827
   hess: array([[ 44.81187272,   4.77553259],
         [  4.77553259,  16.85937624]])
    nit: 4
   njev: 5
      x: array([-0.27084459, -0.92303856])
 status: 0
   nfev: 5
success: True
   nhev: 4
```

The algorithm has converged successfully on the local maximum in five function evaluations, five Jacobian evaluations and four Hessian evaluations.

◊ **Constrained Optimization**

Sometimes it is necessary to find the maximum or minimum of a function subject to
one or more constraints. To use the earlier mentioned function as an example, you may
wish for the single minimum of $f(x, y)$ that satisfies $x > 0, y > 0$; or the minimum value
of the function along the line $x = y$.

The algorithms l-bfgs-b, tnc and slsqp support the bounds argument to minimize.
bounds is a sequence of tuples, each giving the (min, max) pairs for each variable of the
function defining the bounds on that variable to the minimization. If there is no bound
in either direction, use None.

For example, if we try to find a minimum in $f(x, y)$ starting at $(-\frac{1}{2}, -\frac{1}{2})$ without specifying any bounds, the slsqp method converges (just about) on the one at $(-2.805118,$
$3.131312)$:

```
In [x]: minimize(f, (-0.5,-0.5), method='slsqp')
Out[x]:
     jac: array([-0.00721077,  0.00037714,  0.        ])
 message: 'Optimization terminated successfully.'
     fun: 4.0198760213901536e-07
     nit: 10
    njev: 10
       x: array([-2.80522924,  3.131319  ])
  status: 0
    nfev: 46
 success: True
```

To stay in the quadrant $x < 0, y < 0$, set bounds with no minimum on x or y and a
maximum bound at $x = 0$ and $y = 0$:

```
In [x]: xbounds = (None, 0)
In [x]: ybounds = (None, 0)
In [x]: bounds = (xbounds, ybounds)
In [x]: minimize(f, (-0.5,-0.5), bounds=bounds, method='slsqp')
Out[x]:
     jac: array([-0.00283595, -0.00034243,  0.        ])
 message: 'Optimization terminated successfully.'
     fun: 4.115667606325133e-08
     nit: 11
    njev: 11
```

```
        x: array([-3.77933774, -3.28319868])
   status: 0
     nfev: 50
  success: True
```

Suppose we wish to find the extrema of Himmelblau's function that also satisfy the condition $x = y$ (that is, they lie along the diagonal of Figure 8.20). Two of the minimization methods listed in Table 8.6 allow for constraints, cobyla and slsqp, so we must use one of these.

Constraints are specified as the argument constraints to the minimize function as a sequence of dictionaries defining string keys: 'type', the *type* of constraint; and 'fun', a callable object implementing the constraint. 'type' may be 'eq' or 'ineq' for a constraint based on an equality (such as $x = y$) or an inequality (e.g. $x > 2y - 1$). *Note that cobyla does not support equality constraints.*

An equality constraint function should return zero if the constraint function is met; an inequality constraint function should return a non-negative value if the inequality is met.

To find the minima in $f(x, y)$ subject to the constraint $x = y$, we can use the slsqp method with an equality constraint function returning $x - y$:

```
In [x]: con = {'type': 'eq', 'fun': lambda X: X[0] - X[1]}
In [x]: minimize(f, (0, 0), constraints=con, method='slsqp')
     jac: array([-16.33084416,   16.33130538,    0.      ])
 message: 'Optimization terminated successfully.'
     fun: 8.0000000007160867
     nit: 7
    njev: 7
       x: array([ 2.54138438,   2.54138438])
  status: 0
    nfev: 32
 success: True
```

The method converged on one of the minima (there is another: start at, for example, $(-2, -2)$ to find it). What about the maximum?

```
In [x]: minimize(mf, (0, 0), constraints=con, method='slsqp')
Out[x]:
     jac: array([ 0.,   0.,   0.])
 message: 'Singular matrix C in LSQ subproblem'
     fun: -3.1826053300603689e+68
     nit: 4
    njev: 4
       x: array([ -1.12315113e+17,   -1.12315113e+17])
  status: 6
    nfev: 16
 success: False
```

That didn't go so well – the algorithm wandered up the side of a valley. A better choice of algorithm here is cobyla, but this method doesn't support equality constraints, so we will build one from a pair of inequalities: $x = y$ if both of $x > y$ and $x < y$ are not satisified:

```
In [x]: con1 = {'type': 'ineq', 'fun': lambda X: X[0] - X[1]}
```

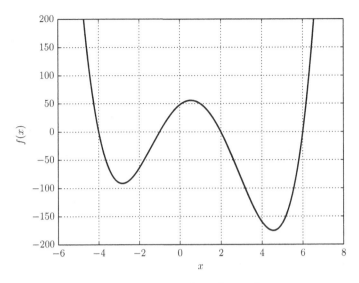

Figure 8.21 The polynomial $f(x) = x^4 - 3x^3 - 24x^2 + 28x + 48$.

```
In [x]: con2 = {'type': 'ineq', 'fun': lambda X: X[1] - X[0]}
In [x]: minimize(mf, (0, 0), constraints=(con1, con2), method='cobyla')
Out[x]:
  status: 1
    nfev: 34
 success: True
 message: 'Optimization terminated successfully.'
     fun: -179.12499987327624
   maxcv: 0.0
       x: array([-0.49994148, -0.49994148])
```

Here, the constraint function defined in con1 returns a non-negative value if $x > y$ and that defined in con2 returns a non-negative value if $x < y$. The only way both can be satisfied is if $x = y$.

Minimizing a Function of One Variable

If the function to be minimized is *univariate* (i.e. takes only one, scalar, variable), a faster algorithm is provided by scipy.optimize.minimize_scalar. To simply return a minimum, this function can be called with method='brent', which implements Brent's method for locating a minimum.

Ideally, one should "bracket" the minimum first by providing values for x, (a, b, c) such that $f(a) > f(b)$ and $f(c) > f(b)$. This can be done with the bracket argument which takes the tuple (a, b, c). If this isn't possible or feasible, provide an interval of two values of x on which to start a search for such a bracket (in the downhill direction). If no bracket argument is specified, this search is initiated from the interval $(0, 1)$.

Figure 8.21 gives an example polynomial with two minima and a maximum.

With no bracket, minimize_scalar converges on the minimum at -2.841 for this function:

```
In [x]: Polynomial = np.polynomial.Polynomial
In [x]: from scipy.optimize import minimize_scalar
In [x]: f = Polynomial( (48., 28., -24., -3., 1.))
In [x]: minimize_scalar(f)
Out[x]:
   fun: -91.32163915433344
  nfev: 11
     x: -2.8410443265958261
   nit: 10
```

If we bracket the other minimum by providing values (a, b, c)=(3, 4, 6), which can be seen from Figure 8.21 to satisfy $f(a) > f(b) < f(c)$, the algorithm converges on 4.549:

```
In [x]: minimize_scalar(f, bracket=(3, 4, 6))
Out[x]:
   fun: -175.45563549487974
  nfev: 11
     x: 4.5494683642571934
   nit: 10
```

Finally, to find the maximum, call `minimize_scalar` with $-f(x)$. This time, we will initialize a search for a bracket to the minimum of $-f(x)$ with the pair of values $(-1, 0)$:

```
In [x]: minimize_scalar(-f, bracket=(-1, 0))
Out[x]:
   fun: -55.734305899213226
  nfev: 9
     x: 0.54157595897344157
   nit: 8
```

Example E8.23 A simple model for the envelope of an airship treats it as the volume of revolution obtained from a pair of quarter-ellipses joined at their (equal) semi-minor axes. The semi-major axis of the aft ellipse is taken to be longer than that representing the bow by a factor $\alpha = 6$. Equations describing the cross section (in the vertical plane) of the airship envelope may be written

$$y = \begin{cases} \frac{b}{a} \sqrt{x(2a - x)} & (x \le a), \\ \frac{b}{a} \sqrt{a^2 - \frac{(x-a)^2}{\alpha^2}} & (a < x \le \alpha(a + 1)). \end{cases}$$

The drag on the envelope is given by the formula

$$D = \tfrac{1}{2}\rho_{air}v^2 V^{2/3} C_{DV},$$

where ρ_{air} is the air density, v the speed of the airship, V the envelope volume and the drag coefficient, C_{DV}, is estimated using the following empirical formula:[11]

$$C_{DV} = Re^{-1/6}[0.172(l/d)^{1/3} + 0.252(d/l)^{1.2} + 1.032(d/l)^{2.7}].$$

Here, $Re = \rho_{air}vl/\mu$ is the Reynold's number and μ the dynamic viscosity of the air. l and d are the airship length and maximum diameter ($= 2b$), respectively.

---

[11] S. F. Hoerner, *Fluid Dynamic Drag*, Hoerner Fluid Dynamics (1965).

Suppose we want to minimize the drag with respect to the parameters a and b but fix the total volume of the airship envelope, $V = \frac{2}{3}\pi ab^2(1+\alpha)$. The following program does this using the `slsqp` algorithm, for a volume of $200\,000$ m^3, that of the Hindenburg.

Listing 8.19 Minimizing the drag on an airship envelope

```python
# eg8-airship.py
import numpy as np
from scipy.optimize import minimize

# Air density (kg.m-3) and dynamic viscosity (Pa.s) at cruise altitude.
rho, mu = 1.1, 1.5e-5
# Air speed (m.s-1) at cruise altitude.
v = 30

def CDV(L, d):
    """ Calculate the drag coefficient. """
    Re = rho * v * L / mu        # Reynold's number
    r = L / d                    # "fineness" ratio
    return (0.172 * r**(1/3) + 0.252 / r**1.2 + 1.032 / r**2.7) / Re**(1/6)

def D(X):
    """ Return the total drag on the airship envelope. """
    a, b = X
    L = a * (1+alpha)
    return 0.5 * rho * v**2 * V(X)**(2/3) * CDV(L, 2*b)

# Fixed total volume of the airship envelope (m3).
V0 = 2.e5
# Parameter describing the tapering of the stern of the envelope.
alpha = 6

def V(X):
    """ Return the volume of the envelope. """
    a, b = X
    return 2 * np.pi * a * b**2 * (1+alpha) / 3

# Minimize the drag, constraining the volume to be equal to V0.
a0, b0 = 70, 45        # initial guesses for a, b
con = {'type': 'eq', 'fun': lambda X: V(X)-V0}
res = minimize(D, (a0, b0), method='slsqp', constraints=con)
if res['success']:
    a, b = res['x']
    L, d = a * (1+alpha), 2*b        # length, greatest diameter
    print('Optimum parameters: a = {:g} m, b = {:g} m'.format(a, b))
    print('V = {:g} m3'.format(V(res['x'])))
    print('Drag, D = {:g} N'.format(res['fun']))
    print('Total length, L = {:g} m'.format(L))
    print('Greatest diameter, d = {:g} m'.format(d))
    print('Fineness ratio, L/d = {:g}'.format(L/d))
else:
    # We failed to converge: output the results dictionary.
    print('Failed to minimize D!', res, sep='\n')
```

This example is a little contrived, since for fixed α the requirement that V be constant means that a and b are not independent, but a solution is found readily enough:

```
Optimum parameters: a = 32.9301 m, b = 20.3536 m
V = 200000 m3
Drag, D = 20837.6 N
Total length, L = 230.51 m
Greatest diameter, d = 40.7071 m
Fineness ratio, L/d = 5.66266
```

The actual dimensions of the Hindenburg were $l = 245$ m, $d = 41$ m giving the ratio $l/d = 5.98$; so we didn't do too badly.

8.4.2   Nonlinear Least-Squares Fitting

SciPy's general *nonlinear* least-squares fitting routine is `scipy.optimize.leastsq`, which has as its most basic call signature:

    `scipy.optimize.leastsq(func, x0, args=())`.

This will attempt to fit a sequence of data points, y, to a model function, f, which depends on one or more fit parameters. `leastsq` is passed a related function object, func, which returns the *difference* between y and f (the *residuals*). `leastsq` also requires an initial guess for the fitted parameters, x0. If func requires any other arguments (typically, arrays of the data, y, and one or more independent variables), pass them in the sequence args. For example, consider fitting the artificial noisy decaying cosine function, $f(t) = Ae^{-t/\tau} \cos 2\pi vt$ (Figure 8.22).

```
In [x]: import numpy as np
In [x]: import matplotlib.pyplot as plt

In [x]: A, freq, tau = 10, 4, 0.5
In [x]: def f(t, A, freq, tau):
   ...:         return A * np.exp(-t/tau) * np.cos(2*np.pi * freq * t)
   ...:
In [x]: tmax, dt = 1, 0.01
In [x]: t = np.arange(0, tmax, dt)
In [x]: yexact = f(t, A, freq, tau)
In [x]: y = yexact + np.random.randn(len(yexact))*2
In [x]: plt.plot(t, yexact)
In [x]: plt.plot(t, y)
In [x]: plt.show()
```

To fit this noisy data, y, to the parameters A, freq and tau (pretending we don't know them), we first define our residuals function:

```
In [x]: def residuals(p, y, t):
   ...:         A, freq, tau = p
   ...:         return y - f(t, A, freq, tau)
```

The first argument is the sequence of parameters, p, which we unpack into named variables for clarity. The additional arguments needed are the data points, y, and the independent variable, t. Now make some initial guesses for the parameters that aren't too wildly off and call `leastsq`:

```
In [x]: from scipy.optimize import leastsq
In [x]: p0 = 5, 5, 1
```

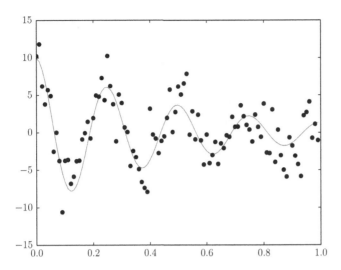

Figure 8.22 A synthetic noisy decaying cosine function.

```
In [x]: plsq = leastsq(residuals, p0, args=(y, t))
In [x]: plsq[0]
Out[x]: [ 9.33962672  4.04958427  0.48637434]
```

As with SciPy's other optimization routines, `leastsq` can be configured to return more information about its working, but here we report only the solution (best-fit parameters), which is always the first item in the `plsq` tuple.

The true values are `A, freq, tau = 10, 4, 0.5`, so given the noise we haven't done badly. Graphically,

```
In [x]: plt.plot(t, y, 'o', c='k', label='Data')
In [x]: plt.plot(t, yexact, c='gray', label='Exact')
In [x]: pfit = plsq[0]
In [x]: plt.plot(t, f(t, *pfit), c='k', label='Fit')
In [x]: plt.legend()
In [x]: plt.show()
```

The fit is illustrated in Figure 8.23.

If it is known, it is also possible to pass the Jacobian to `leastsq`, as the following example demonstrates.

Example E8.24 In this example, we are given a noisy series of data points that we want to fit to an ellipse. The equation for an ellipse may be written as a nonlinear function of angle θ ($0 \leq \theta < 2\pi$), which depends on the parameters a (the semi-major axis) and e (the eccentricity):

$$r(\theta; a, e) = \frac{a(1 - e^2)}{1 - e \cos \theta}.$$

To fit a sequence of data points (θ, r) to this function, we first code it as a Python function taking two arguments: the independent variable, `theta`, and a tuple of the parameters,

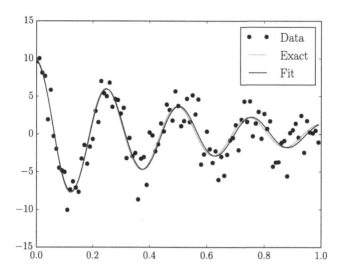

Figure 8.23 Non-linear least-squares fit to a noisy decaying cosine function.

p = (a, e). The function we wish to minimize is the difference between this model function and the data, r, defined as the method `residuals`:

```python
def f(theta, p):
    a, e = p
    return a * (1 - e**2)/(1 - e*np.cos(theta))

def residuals(p, r, theta):
    return r - f(theta, p)
```

We also need to give `leastsq` an initial guess for the fit parameters, say p0 = (1, 0.5). The simplest call to fit the function would then pass to `leastsq` the function object, `residuals`; the initial guesses, p0; and args=(r, theta), the additional arguments needed by the `residuals` function:

```python
plsq = leastsq(residuals, p0, args=(r, theta))
```

If at all possible, however, it is better to also provide the Jacobian (the first derivative of the fit function with respect to the parameters to be fitted). Expressions for these are straightforward to calculate and implement:

$$\frac{\partial f}{\partial a} = \frac{(1 - e^2)}{1 - e\cos\theta},$$
$$\frac{\partial f}{\partial e} = \frac{a[\cos\theta(1 + e^2) - 2e]}{(1 - e\cos\theta)^2}.$$

However, the function we wish to minimize is the residuals function, $r-f$, so we need the negatives of these derivatives. Here is the working code and the fit result (Figure 8.24).

Listing 8.20 Nonlinear least squares-fit to an ellipse

```python
# eg8-leastsq.py
```

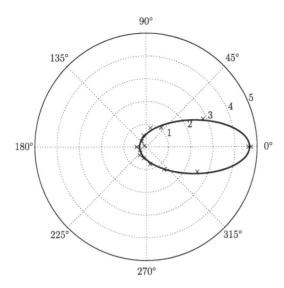

Figure 8.24 Nonlinear least-squares fitting of data to the equation of an ellipse in polar coordinates.

```python
import numpy as np
from scipy import optimize
import matplotlib.pyplot as plt

def f(theta, p):
    a, e = p
    return a * (1 - e**2)/(1 - e*np.cos(theta))

# The data to fit.
theta = np.array([0.0000, 0.4488, 0.8976, 1.3464, 1.7952, 2.2440, 2.6928,
                  3.1416, 3.5904, 4.0392, 4.4880, 4.9368, 5.3856, 5.8344, 6.2832])
r = np.array([4.6073, 2.8383, 1.0795, 0.8545, 0.5177, 0.3130, 0.0945, 0.4303,
              0.3165, 0.4654, 0.5159, 0.7807, 1.2683, 2.5384, 4.7271])

def residuals(p, r, theta):
    """ Return the observed - calculated residuals using f(theta, p). """
    return r - f(theta, p)

def jac(p, r, theta):
    """ Calculate and return the Jacobian of residuals. """
    a, e = p
    da = (1 - e**2)/(1 - e*np.cos(theta))
    de = a * (np.cos(theta) * (1 + e**2) - 2*e) / (1 - e*np.cos(theta))**2
    return -da, -de

# Initial guesses for a, e.
p0 = (1, 0.5)
plsq = optimize.leastsq(residuals, p0, Dfun=jac, args=(r, theta), col_deriv=True)
print(plsq)

plt.polar(theta, r, 'x')
```

```
theta_grid = np.linspace(0, 2*np.pi, 200)
plt.polar(theta_grid, f(theta_grid, plsq[0]), lw=2)
plt.show()
```

SciPy also includes a curve-fitting function, `scipy.optimize.curve_fit`, that can fit data to a function directly (without the need for an additional function to calculate the residuals) and supports weighted least-squares fitting. The call signature is

```
curve_fit(f, xdata, ydata, p0, sigma, absolute_sigma)
```

where f is the function to fit to the data (xdata, ydata). p0 is the initial guess for the parameters, and sigma, if provided, gives the weights of the ydata values. If absolute_sigma is True, these are treated as one standard deviation error (that is, *absolute* weights); the default, absolute_sigma=False, treats them as *relative* weights.

The curve_fit function returns popt, the best-fit values of the parameters, and pcov, the covariance matrix of the parameters.

Example E8.25   To illustrate the use of `curve_fit` in weighted and unweighted least-squares fitting, the following program fits the Lorentzian line shape function centered at x_0 with a HWHM of γ and an amplitude, A:

$$f(x) = \frac{A\gamma^2}{\gamma^2 + (x - x_0)^2},$$

to some artificial noisy data. The fit parameters are A, γ and x_0. The data close to the line center are simulated as being much noisier than the rest.

Listing 8.21 Weighted and unweighted least-squares fitting with `curve_fit`

```
# eg8-curve-fit.py
import numpy as np
from scipy.optimize import curve_fit
import matplotlib.pyplot as plt


x0, A, gamma = 12, 3, 5


n = 200
x = np.linspace(1, 20, n)
yexact = A * gamma**2 / (gamma**2 + (x-x0)**2)


# Add some noise with a sigma of 0.5 apart from a particularly noisy region
# near x0 where sigma is 3.
sigma = np.ones(n)*0.5
sigma[np.abs(x-x0+1)<1] = 3
noise = np.random.randn(n) * sigma
y = yexact + noise


def f(x, x0, A, gamma):
    """ The Lorentzian entered at x0 with amplitude A and HWHM gamma. """
    return A *gamma**2 / (gamma**2 + (x-x0)**2)
```

```python
def rms(y, yfit):
    return np.sqrt(np.sum((y-yfit)**2))

# Unweighted fit.
p0 = 10, 4, 2
popt, pcov = curve_fit(f, x, y, p0)
yfit = f(x, *popt)
print('Unweighted fit parameters:', popt)
print('Covariance matrix:'); print(pcov)
print('rms error in fit:', rms(yexact, yfit))
print()

# Weighted fit.
popt2, pcov2 = curve_fit(f, x, y, p0, sigma=sigma, absolute_sigma=True)
yfit2 = f(x, *popt2)
print('Weighted fit parameters:', popt2)
print('Covariance matrix:'); print(pcov2)
print('rms error in fit:', rms(yexact, yfit2))

plt.plot(x, yexact, label='Exact')
plt.errorbar(x, y, yerr=noise, elinewidth=0.5, c='0.5', marker='+',
             lw=0, label='Noisy data')
plt.plot(x, yfit, label='Unweighted fit')
plt.plot(x, yfit2, label='Weighted fit')
plt.ylim(-1, 4)
plt.legend(loc='lower center')
plt.show()
```

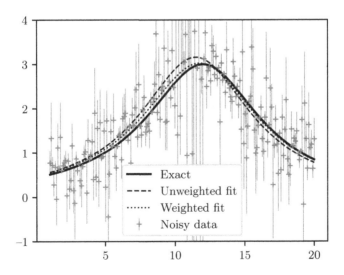

Figure 8.25 Example of least-squares fit with `scipy.optimize.curve_fit`.

As Figure 8.25 shows, the unweighted fit is thrown off by the noisy region. Data in this region are given a lower weight in the weighted fit and so the parameters are closer to their true values and the fit is better. The output is

```
Unweighted fit parameters: [ 11.61282984    3.64158981    3.93175714]
Covariance matrix:
[[ 0.0686249   -0.00063262   0.00231442]
 [-0.00063262   0.06031262  -0.07116127]
 [ 0.00231442  -0.07116127   0.16527925]]
rms error in fit: 4.10434012348

Weighted fit parameters: [ 11.90782988    3.0154818    4.7861561 ]
Covariance matrix:
[[ 0.01893474  -0.00333361   0.00639714]
 [-0.00333361   0.01233797  -0.02183039]
 [ 0.00639714  -0.02183039   0.06062533]]
rms error in fit: 0.694013741786
```

8.4.3   Root-Finding

`scipy.optimize` provides several methods for obtaining the roots of both univariate and multivariate functions. Only the algorithms relating to functions of a single variable, `brentq`, `brenth`, `ridder` and `bisect`, are described here. Each of these methods requires a continuous function, $f(x)$, and a pair of numbers defining a *bracketing interval* for the root to find; that is, values a and b such that the root lies in the interval $[a, b]$ and $\text{sgn}[f(a)] = -\text{sgn}[f(b)]$. Details of the algorithms behind these root-finding methods can be found in standard textbooks on numerical analysis.[12]

In general, the method of choice for finding the root of a well-behaved function is `scipy.optimize.brentq`, which implements a version of Brent's method with inverse quadratic extrapolation (`scipy.optimize.brenth` is a similar algorithm but with hyperbolic extrapolation). As an example, consider the following function for $-1 \leq x \leq 1$:

$$f(x) = \frac{1}{5} + x \cos\left(\frac{3}{x}\right).$$

A plot of this function (Figure 8.26) suggests there is a root between -0.7 and -0.5:

```
In [x]: f = lambda x: 0.2 + x*np.cos(3/x)
In [x]: x = np.linspace(-1, 1, 1000)
In [x]: plt.plot(x, f(x))
In [x]: plt.axhline(0, color='k')
In [x]: plt.show()

In [x]: from scipy.optimize import brentq
In [x]: brentq(f, -0.7, -0.5)
Out[x]: -0.5933306271014237
```

---

[12] For example, Press et al., *Numerical Recipes. The Art of Scientific Computing*, 3rd edn., Cambridge University Press, Cambridge (2007).

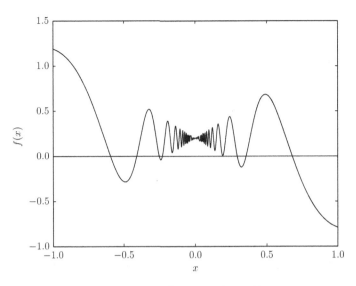

Figure 8.26 The function $f(x) = \frac{1}{5} + x\cos(3/x)$ and its roots.

The algorithm for root-finding known as *Ridder's method* is implemented in the function `scipy.optimize.ridder` and the slower but very reliable (for continuous functions) method of bisection is `scipy.optimize.bisect`.

Finally, root-finding by the Newton–Raphson algorithm can be very fast (quadratic) for many continuous functions, provided the first derivative, $f'(x)$, can be calculated. For functions for which an analytical expression for $f'(x)$ can be coded, this is passed to the method `scipy.optimize.newton` as the argument `fprime`, along with a starting point, `x0`, which should (in general) be as near to the root as possible. It is not necessary to bracket the root. If the $f'(x)$ cannot be provided, the secant method is used by `newton`. If you are in the happy position of being able to provide the second derivative, $f''(x)$, as `fprime2` as well as the first, Halley's method (which converges even faster than the basic Newton–Raphson algorithm) is used instead.

Note that the stopping condition within the iterative algorithm used by `newton` is the step size so there is no guarantee that it has converged on the desired root: the result should be verified by evaluating the function at the returned value to check that it is (close to) zero.

Example E8.26   In ecology, the *Euler–Lotka equation* describes the growth of a population in terms of $P(x)$, the fraction of individuals alive at age x, and $m(x)$, the mean number of live females born per time period per female alive during that time period:

$$\sum_{x=\alpha}^{\beta} P(x)m(x)e^{-rx} = 1,$$

where α and β are the boundary ages for reproduction defining the discrete growth rate, $\lambda = e^r$. $r = \ln \lambda$ is known as *Lotka's intrinsic rate of natural increase*.

Table 8.7 Population data for voles
measured by Leslie and Ranson

x /weeks	$m(x)$	$P(x)$
8	0.6504	0.83349
16	2.3939	0.73132
24	2.9727	0.58809
32	2.4662	0.43343
40	1.7043	0.29277
48	1.0815	0.18126
56	0.6683	0.10285
64	0.4286	0.05348
72	0.3000	0.02549

In a paper by Leslie and Ranson,[13] $P(x)$ and $m(x)$ were measured for a population of voles (*Microtus agrestis*) using a time period of eight weeks. The data are given in Table 8.7.

The sum $R_0 = \sum_{x=\alpha}^{\beta} P(x)m(x)$ gives the ratio between the total number of female births in successive generations; a population grows if $R_0 > 1$ and r determines how fast this growth is. In order to find r, Leslie and Ranson used an approximate numerical method; the code mentioned here determines r by finding the real root of the Lotka-Euler equation directly (it can be shown that there is only one root).

Listing 8.22 Solution of the Euler–Lotka equation

```
# eg8-euler-lotka.py
import numpy as np
from scipy.optimize import brentq

# The data, from Table 6 of:
# P. H. Leslie and R. M. Ranson, J. Anim. Ecol. 9, 27 (1940).
x = np.linspace(8, 72, 9)
m = np.array( [0.6504, 2.3939, 2.9727, 2.4662, 1.7043,
               1.0815, 0.6683, 0.4286, 0.3000] )
P = np.array( [0.83349, 0.73132, 0.58809, 0.43343, 0.29277,
               0.18126, 0.10285, 0.05348, 0.02549] )

# Calculate the product sequence f and R0, the ratio between the number of
# female births in successive generations.
f = P * m
R0 = np.sum(f)
if R0 > 1:
    msg = 'R0 > 1: population grows'
else:
    msg = 'Population does not grow'

# The Euler-Lotka equation: we seek the one real root in r.
def func(r):
```

[13] P. H. Leslie and R. M. Ranson, The mortality, fertility and rate of natural increase of the vole (*Microtus agrestis*) as observed in the laboratory, *J. Anim. Ecol.* **9**, 27 (1940).

```
    return np.sum(f * np.exp(-r * x)) - 1

# Bracket the root and solve with scipy.optimize.brentq.
a, b = 0, 10
r = brentq(func, a, b)
print('R0 = {:.3f} ({})'.format(R0, msg))
print('r = {:.5f} (lambda = {:.5f})'.format(r, np.exp(r)))
```

The output of this program is as follows:

```
R0 = 5.904 (R0 > 1: population grows)
r = 0.08742 (lambda = 1.09135)
```

This value of r may be compared with the approximate value obtained by Leslie and Ranson, who comment:

The required root is 0.087703 which slightly overestimates the value of r, to which the series is approaching. This lies between 0.0861 (the third degree approximation) and 0.0877, but nearer the latter than the former, the error being probably in the last decimal place.

Example E8.27 The Newton–Raphson method for finding the roots of a function takes an initial guess to a root, x_0, and seeks successively better approximations of it as:

$$x_{n+1} = x_n - \frac{f(x_n)}{f'(x_n)}.$$

That is, at each iteration, the root is approximated as x_{n+1}, the x-axis intercept of the tangent to the graph at $f(x_n)$. When applied to functions of a complex variable z, the method can be used to create interesting fractals by considering which root it converges to for a set of numbers in the complex plane. The code below generates a fractal image (Figure 8.27) by coloring those points in the complex plane used as the initial guess by the root found.

Listing 8.23 Generating a Newton fractal image

```
import numpy as np
import matplotlib.pyplot as plt
from matplotlib.colors import ListedColormap

# A list of colors to distinguish the roots.
colors = ['b', 'r', 'g', 'y']

TOL = 1.e-8

def newton(z0, f, fprime, MAX_IT=1000):
    """The Newton-Raphson method applied to f(z).

    Returns the root found, starting with an initial guess, z0, or False
    if no convergence to tolerance TOL was reached within MAX_IT iterations.

    """
```

```
        z = z0
        for i in range(MAX_IT):
            dz = f(z)/fprime(z)
            if abs(dz) < TOL:
                return z
            z -= dz
        return False

def plot_newton_fractal(f, fprime, n=200, domain=(-1, 1, -1, 1)):
    """Plot a Newton Fractal by finding the roots of f(z).

    The domain used for the fractal image is the region of the complex plane
    (xmin, xmax, ymin, ymax) where z = x + iy, discretized into n values along
    each axis.

    """

    roots = []
    m = np.zeros((n, n))

    def get_root_index(roots, r):
        """Get the index of r in the list roots.

        If r is not in roots, append it to the list.

        """

        try:
            return np.where(np.isclose(roots, r, atol=TOL))[0][0]
        except IndexError:
            roots.append(r)
            return len(roots) - 1

    xmin, xmax, ymin, ymax = domain
    for ix, x in enumerate(np.linspace(xmin, xmax, n)):
        for iy, y in enumerate(np.linspace(ymin, ymax, n)):
            z0 = x + y*1j
            r = newton(z0, f, fprime)
            if r is not False:
                ir = get_root_index(roots, r)
                m[iy, ix] = ir
    nroots = len(roots)
    if nroots > len(colors):
        # Use a "continuous" colormap if there are too many roots.
        cmap = 'hsv'
    else:
        # Use a list of colors for the colormap: one for each root.
        cmap = ListedColormap(colors[:nroots])
    plt.imshow(m, cmap=cmap, origin='lower')
    plt.axis('off')
    plt.show()

f = lambda z: z**4 - 1
fprime = lambda z: 4*z**3
```

Figure 8.27 The Newton fractal for the function $f(z) = z^4 - 1$. Intricate, self-similar structures are seen for initial guesses, z_0, in between the roots $(-1, 1, -i, i)$.

```
plot_newton_fractal(f, fprime, n=500)
```

8.4.4   Exercises

Questions

Q8.4.1   Use `scipy.optimize.brentq` to find the solutions to the equation

$$x + 1 = -\frac{1}{(x - 3)^3}.$$

Q8.4.2   The `scipy.optimize.newton` method fails to find a root of the following functions with the given starting point, x_0. Explain why and find the roots either by modifying the call to `newton` or by using a different method.

(a)

$$f(x) = x^3 - 5x, \quad x_0 = 1.$$

(b)

$$f(x) = x^3 - 3x + 1, \quad x_0 = 1.$$

(c)

$$f(x) = 2 - x^5, \quad x_0 = 0.01.$$

(d)

$$f(x) = x^4 - (4.29)x^2 - 5.29, \quad x_0 = 0.8.$$

Q8.4.3 The trajectory of a projectile in the xz-plane launched from the origin at an angle θ_0 with speed $v_0 = 25 \text{ m s}^{-1}$ is

$$z = x \tan \theta_0 - \frac{g}{2v_0^2 \cos^2 \theta_0} x^2.$$

If the projectile passes through the point $(5, 15)$, use Brent's method to determine the possible values of θ_0.

Problems

P8.4.1 A rectangular field with area $A = 10\,000 \text{ m}^3$ is to be fenced off beside a straight river (the boundary with the river does not need to be fenced). What dimensions a, b minimize the amount of fencing required? Verify that a constrained minimization algorithm gives the same answer as the algebraic analysis.

P8.4.2 Find all of the roots of

$$f(x) = \frac{1}{5} + x \cos\left(\frac{3}{x}\right)$$

using (a) `scipy.optimize.brentq` and (b) `scipy.optimize.newton`.

P8.4.3 The *Wien displacement law* predicts that the wavelength of maximum emission from a black body described by Planck's law is proportional to $1/T$:

$$\lambda_{\max} T = b,$$

where b is a constant known as *Wien's displacement constant*. Given the Planck distribution of emitted energy density as a function of wavelength,

$$u(\lambda, T) = \frac{8\pi^2 hc}{\lambda^5} \frac{1}{e^{hc/\lambda k_B T} - 1},$$

determine the constant b by using `scipy.optimize.minimize_scalar` to find the maximum in $u(\lambda, T)$ for temperatures in the range $500 \text{ K} \leq T \leq 6000 \text{ K}$ and fitting $\lambda_{\max}$ to a straight line against $1/T$. Compare with the "exact" value of b, which is available within `scipy.constants` (see Section 8.1.1).

◇   **P8.4.4** Consider a one-dimensional quantum mechanical particle in a box $(-1 \leq x \leq 1)$ described by the Schrödinger equation:

$$-\frac{d^2\psi}{dx^2} = E\psi,$$

in energy units for which $\hbar^2/(2m) = 1$, with m the mass of the particle. The exact solution for the ground state of this system is given by

$$\psi = \cos\left(\frac{\pi x}{2}\right), \quad E = \frac{\pi^2}{4}.$$

An approximate solution may be arrived at using the *variational principle* by minimizing the expectation value of the energy of a trial wavefunction,

$$\psi_{\text{trial}} = \sum_{n=0}^{N} a_n \phi_n(x)$$

with respect to the coefficients a_n. Taking the basis functions to have the following symmetrized polynomial form,

$$\phi_n = (1 - x)^{N-n+1}(x + 1)^{n+1},$$

use `scipy.optimize.minimize` and `scipy.integrate.quad` to find the optimum value of the expectation value (Rayleigh–Ritz ratio):

$$\mathcal{E} = \frac{\langle \psi_{\text{trial}} | \hat{H} | \psi_{\text{trial}} \rangle}{\langle \psi_{\text{trial}} | \psi_{\text{trial}} \rangle} = -\frac{\int_{-1}^{1} \psi_{\text{trial}} \frac{d^2}{dx^2} \psi_{\text{trial}} \, dx}{\int_{-1}^{1} \psi_{\text{trial}} \psi_{\text{trial}} \, dx}.$$

Compare the estimated energy, $\mathcal{E}$, with the exact answer for $N = 1, 2, 3, 4$. (*Hint*: use `np.polynomial.Polynomial` objects to represent the basis and trial wavefunctions.)

9 Data Analysis with pandas

9.1 Introduction to pandas

9.1.1 What is pandas?

pandas is a widely-used, open-source Python library for data manipulation and analysis. Unlike NumPy, its basic array-like data structure, the `DataFrame` object, can contain heterogeneous data types (floats, integers, strings, dates, etc.) that may be structured in a hierarchy and indexed. It provides a large number of vectorized functions for cleaning, transforming and aggregating data efficiently using similar idioms to those used by NumPy. Its name derives from the term "panel data" (otherwise known as "longitudinal data"), which refers to data sets of several variables followed over multiple time periods for the same individual.

The pandas homepage, https://pandas.pydata.org/, contains details of the latest release and how to download and install pandas. In this chapter we follow the common convention of importing the library as the alias pd:

```
import pandas as pd
```

The key pandas data structures are `Series` and `DataFrame`, representing a one-dimensional sequence of values and a data table, respectively. Their basic properties and use will be described in this section, followed by more advanced features and applications to examples in subsequent sections. pandas is a large, complex library with a lot of functionality; this chapter aims to cover the basics: more detailed examples are provided on the website accompanying the book.

9.1.2 Series

In its simplest form, a `Series` may be created in the same way as a one-dimensional NumPy array:

```
In [x]: river_lengths = pd.Series([6300, 6650, 6275, 6400])
In [x]: river_lengths
Out[x]:
0    6300
1    6650
2    6275
3    6400
dtype: int64
```

The Series can be given a name and dtype:

```
In [x]: river_lengths = pd.Series([6300, 6650, 6275, 6400], name='Length /km',
                                   dtype=float)
Out[x]:
0    6300.0
1    6650.0
2    6275.0
3    6400.0
Name: Length /km, dtype: float64
```

Unlike a NumPy array, however, each element in a pandas Series is associated with an *index*. Since we did not set the index explicitly here, a default integer sequence (starting at 0) is used for the index:

```
In [x]: river_lengths.index
Out[x]: RangeIndex(start=0, stop=4, step=1)
```

RangeIndex is a pandas object that works in a memory-efficient way like Python's range built-in to provide a monotonic integer sequence. It is often useful to refer to the rows of a Series with some other label than an integer index. Explicit indexing of the entries can be achieved by passing a sequence as the index argument or by creating the Series from a dictionary:

```
In [x]: river_lengths = pd.Series(data=[6300, 6650, 6275, 6400],
   ...:                           index=['Yangtze', 'Nile', 'Mississippi', 'Amazon'],
   ...:                           name='Length /km')
```

or:

```
In [x]: river_lengths = pd.Series(data={'Yangtze': 6300, 'Nile': 6650,
   ...:                                  'Mississippi': 6275, 'Amazon': 6400},
   ...:                            name='Length /km')
In [x]: river_lengths
Out[x]:
Yangtze        6300
Nile           6650
Mississippi    6275
Amazon         6400
Name: Length /km, dtype: int64
```

This allows a nicely expressive way of referring to Series entries using the index labels instead of integers; either individually:

```
In [x]: river_lengths['Nile']
Out[x]: 6650
```

instead of river_lengths[1]; or from another sequence:

```
In [x]: river_lengths[['Amazon', 'Nile', 'Yangtze']]
Out[x]:
Amazon     6400
Nile       6650
Yangtze    6300
Name: Length /km, dtype: int64
```

instead of river_lengths[[3, 1, 0]]. Python-style slicing also works as expected:

```
In [x]: river_lengths[2::-1]
Out[x]:
Mississippi    6275
Nile           6650
Yangtze        6300
Name: Length /km, dtype: int64
```

It is even possible to use a slice-like notation for the index labels, but note that in this case the endpoint is *inclusive*:

```
In [x]: river_lengths['Nile':'Amazon']
Out[x]:
Nile           6650
Mississippi    6275
Amazon         6400
Name: Length /km, dtype: int64
```

Providing the index label is a valid Python identifier, one can refer to a row as an *attribute* of the Series:

```
In [x]: river_lengths.Mississippi
Out[x]: 6275
```

It is, of course, possible to do numerical operations on Series data, in a vectorized fashion, as for NumPy arrays:

```
In [x]: KM_TO_MILES = 0.621371
In [x]: river_lengths *= KM_TO_MILES
In [x]: river_lengths.name = 'Length /miles'
In [x]: river_lengths
Out[x]:
Yangtze        3914.637300
Nile           4132.117150
Mississippi    3899.103025
Amazon         3976.774400
Name: Length /miles, dtype: float64
```

In the above we have also chosen to update the Series object's name attribute. Note that the dtype has also changed appropriately from int64 to float64 to accommodate the new values.

Comparison operations and filtering a Series with a boolean operation creates a new Series:

```
In [x]: river_lengths > 4000
Out[x]:
Nile           True
Amazon         False
Yangtze        False
Mississippi    False
Name: Length /miles, dtype: bool
```

```
In [x]: river_lengths[river_lengths <= 4000]
Out[x]:
Amazon         3976.774400
Yangtze        3914.637300
Mississippi    3899.103025
Name: Length /miles, dtype: float64
```

Tests for membership of a Series examine the *index*, not the *values*:

```
In [x]: 'Yangtze' in river_lengths
Out[x]: True
```

Series can be *sorted*, either by their index or their values, using Series.sort_index and Series.sort_values, respectively. By default these methods return a new Series, but they can also be used to update the original Series with the argument inplace=True. A further argument, ascending, can be True (the default) or False to set the ordering:

```
In [x]: river_lengths.sort_index()
Out[x]:
Amazon          3976.774400
Mississippi     3899.103025
Nile            4132.117150
Yangtze         3914.637300
Name: Length /miles, dtype: float64


In [x]: river_lengths.sort_values(ascending=False, inplace=True)
In [x]: river_lengths
Out[x]:
Nile            4132.117150
Amazon          3976.774400
Yangtze         3914.637300
Mississippi     3899.103025
Name: Length /miles, dtype: float64
```

When two series are combined, they are aligned by index label.

```
In [x]: masses = pd.Series({'Ganymede': 1.482e23,
                            'Callisto': 1.076e23,
                            'Io': 8.932e22,
                            'Europa': 4.800e22,
                            'Moon': 7.342e22,
                            'Earth': 5.972e24}, name='mass /kg')
In [x]: radii = pd.Series({'Ganymede': 2.634e6,
                          'Io': 1.822e6,
                          'Moon': 1.737e6,
                          'Earth': 6.371e6}, name='radius /m')
In [x]: from scipy.constants import G
In [x]: surface_g = G * masses / radii**2
In [x]: surface_g.name = 'surface gravity /m.s-2'
In [x]: surface_g.index.name = 'Body'
In [x]: surface_g
Body
Callisto            NaN
Earth          9.819650
Europa              NaN
Ganymede       1.425634
Io             1.795740
Moon           1.624075
Name: surface gravity /m.s-2, dtype: float64
```

Note that where no correspondence can be made within the indexes (an index label in one Series that is missing from the other), the result is "Not a Number" (NaN). The methods isnull and notnull test for this:

```
In [x]: surface_g.isnull()
Out[x]:
Body
Callisto      True
Earth         False
Europa        True
Ganymede      False
Io            False
Moon          False
Name: surface gravity /m.s-2, dtype: bool
```

To return a list without any missing values, either filter with surface_g[surface_g.notnull]
or use the dropna method:

```
In [x]: surface_g.dropna()
Out[x]:
Body
Earth         9.819650
Ganymede      1.425634
Io            1.795740
Moon          1.624075
Name: surface gravity /m.s-2, dtype: float64
```

Finally, to convert a Series into a NumPy ndarray (dropping the index and other metadata), use the values property:

```
In [x]: surface_g.values
Out[x]: array([      nan, 9.81964974,      nan, 1.42563409, 1.79573967,
          1.62407526])
```

Example E9.1   NaN entries can be replaced in a pandas Series with a specified value using the fillna method:

```
In [x]: ser1 = pd.Series({'b': 2, 'c': -5, 'd': 6.5}, index=list('abcd'))
In [x]: ser1
Out[x]:
a     NaN
b     2.0
c    -5.0
d     6.5
dtype: float64
```

```
In [x]: ser1.fillna(1, inplace=True)
In [x]: ser1
Out[x]:
a     1.0
b     2.0
c    -5.0
d     6.5
dtype: float64
```

Infinities (represented by the floating-point inf value) can be replaced with the replace method, which takes a scalar or sequence of values and substitutes them with another, single value:

```
In [x]: ser2 = pd.Series([-3.4, 0, 0, 1], index=ser1.index)
In [x]: ser2
Out[x]:
a   -3.4
b    0.0
c    0.0
d    1.0
dtype: float64

In [x]: ser3 = ser1 / ser2
In [x]: ser3
Out[x]:
a   -0.294118
b         inf
c        -inf
d    6.500000
dtype: float64

In [x]: ser3.replace([np.inf, -np.inf], 0)
Out[x]:
a   -0.294118
b    0.000000
c    0.000000
d    6.500000
dtype: float64
```

(Assuming NumPy has been imported with import numpy as np.)

9.1.3   DataFrame

Creating a DataFrame

A DataFrame is a two-dimensional table of data that can be thought of as an ordered set of Series columns, which all have the same index. To create a simple DataFrame from a dictionary, assign value sequences[1] to column name keys:

```
In [x]: data = {'mass': [1.482e23, 1.076e23, 8.932e22, 4.800e22, 7.342e22],
                'radius': [2.634e6, None, 1.822e6, None, 1.737e6],
                'parent': ['Jupiter', 'Jupiter', 'Jupiter', 'Jupiter', 'Earth']
               }
In [x]: index = ['Ganymede', 'Callisto', 'Io', 'Europa', 'Moon']
In [x]: df = pd.DataFrame(data, index=index)
In [x]: df
Out[x]:
                  mass      radius    parent
Ganymede  1.482000e+23  2634000.0   Jupiter
Callisto  1.076000e+23        NaN   Jupiter
Io        8.932000e+22  1822000.0   Jupiter
Europa    4.800000e+22        NaN   Jupiter
Moon      7.342000e+22  1737000.0     Earth
```

Values which were None in the data have been assigned to NaN in the DataFrame. We may wish to rename a column or index row: to do this, call rename, declaring which

---

[1] The (unspecified) units here are SI units: kg and m.

axis ('index' [the same as 'rows', and the default] or 'columns'[2]) contains the label(s) to be renamed, and passing a dictionary mapping each original label to its replacement. Remember to set inplace=True if you want the orginal DataFrame modified rather than a new copy returned. For example,

```
In [x]: df.rename({'parent': 'planet'}, axis='columns', inplace=True)
In [x]: df.rename({'Moon': 'The Moon'}) # change a row index label
Out[x]:
                    mass       radius   planet
Ganymede   1.482000e+23   2634000.0   Jupiter
Callisto   1.076000e+23         NaN   Jupiter
Io         8.932000e+22   1822000.0   Jupiter
Europa     4.800000e+22         NaN   Jupiter
The Moon   7.342000e+22   1737000.0     Earth
```

This last statement has returned a new DataFrame but not altered the original one, df.

Accessing Rows, Columns and Cells

An individual column can be obtained by indexing or by attribute (if its name is a valid Python identifier):

```
In [x]: df['mass']        # or df.mass
Out[x]:
Ganymede      1.482000e+23
Callisto      1.076000e+23
Io            8.932000e+22
Europa        4.800000e+22
Moon          7.342000e+22
Name: mass, dtype: float64
```

Since this column is just a pandas Series, individual values can be retrieved by position or reference to the index label:

```
In [x]: df['mass'][2]
Out[x]: 8.932e+22
In [x]: df['mass']['Io']     # or df['mass'].Io or df.mass.Io
Out[x]: 8.932e+22
```

Now, for retrieving columns and individual values this is fine, but assignment raises a warning:

```
In [x]: df['radius']['Callisto'] = 2.410e6
/Users/christian/envs/py37/bin/ipython:1: SettingWithCopyWarning:
A value is trying to be set on a copy of a slice from a DataFrame
...
```

In this case, it has worked:

```
In [x]: df['radius']['Callisto']
Out[x]: 2410000.0
```

but the message is a general warning that "chained indexing" ([..][..]) can lead to unpredictable results when used for assignment: depending on how the data are stored

---

[2] One can also refer to the rows and columns of a DataFrame with axis=0 and axis=1, respectively.

in memory, it is possible for the indexing expression to yield a *copy* of the data rather than a view. Assigning to the copy rather than modifying the data in-place is what the `SettingWithCopyWarning` is warning could happen. *Chained indexing for assignment operations should be avoided.*

Two `DataFrame` methods, `loc` and `iloc`, can be used to reliably access and assign to columns, rows and cells; using them is strongly recommended. `loc` selects by row and column *labels*:

```
In [x]: df.loc['Europa']
Out[x]:
mass       4.8e+22
radius         NaN
planet     Jupiter
Name: Europa, dtype: object
```

The single row of data indexed by the label Europa is returned as a `Series`. If only a subset of the columns are required, pass their names in a sequence to the second axis:[3]

```
In [x]: df.loc['Europa', ['mass', 'planet']]
Out[x]:
mass       4.8e+22
planet     Jupiter
Name: Europa, dtype: object
```

Slicing, "fancy" indexing and boolean indexing are all supported by `loc`:

```
In [x]: df.loc[:, 'mass']   # the same as df['mass'] - returns a Series
Out[x]:
Ganymede    1.482000e+23
Callisto    1.076000e+23
Io          8.932000e+22
Europa      4.800000e+22
Moon        7.342000e+22
Name: mass, dtype: float64


In [x]: df.loc['Ganymede':'Io', ['mass', 'radius']]
Out[x]:
                  mass      radius
Ganymede    1.482000e+23   2634000.0
Callisto    1.076000e+23   2410000.0
Io          8.932000e+22   1822000.0


In [x]: df.loc[['Moon', 'Europa'], 'planet']
Out[x]:
Moon          Earth
Europa      Jupiter
Name: planet, dtype: object


In [x]: df.loc[df.planet=='Jupiter', 'radius']
Out[x]:
Ganymede    2634000.0
Callisto    2410000.0
```

---

[3] Note that whilst chained indexing refers to a cell in column, row order: `df[col][row]`, `loc` locates cells the other way round: `df.loc[row, col]` or `df.loc[row][col]`.

```
Io              1822000.0
Europa          NaN
Name: radius, dtype: float64
```

The value of a single cell can therefore be retrieved from the row and column labels:

```
In [x]: df.loc['Europa', 'mass']
Out[x]: 4.8e+22
```

This is the safe way to modify data in a DataFrame:

```
In [x]: df.loc['Europa', 'radius'] = 1.561e6    # no warning, data changed in place
In [x]: df.loc['Europa']
Out[x]:
mass        4.8e+22
radius      1.561e+06
parent      Jupiter
Name: Europa, dtype: object
```

It is common to use loc in combination with boolean indexing to filter rows by column values. For example, the masses of Jupiter's moons:

```
In [x]: df.loc[df.planet=='Jupiter', 'mass']
Out[x]:
Ganymede     1.482000e+23
Callisto     1.076000e+23
Io           8.932000e+22
Europa       4.800000e+22
Name: mass, dtype: float64
```

The rows corresponding to moons with radii less than 2000 km:

```
In [x]: df.loc[df.radius < 2.e6]
Out[x]:
                 mass        radius      planet
Io       8.932000e+22     1822000.0    Jupiter
Europa   4.800000e+22     1561000.0    Jupiter
Moon     7.342000e+22     1737000.0      Earth
```

The second method, iloc, retrieves data by numerical index position:

```
In [x]: df.iloc[1]              # the second row
Out[x]:
mass        1.076e+23
radius      2.41e+06
parent      Jupiter
Name: Callisto, dtype: object
```

```
In [x]: df.iloc[:, [1, 2]]     # all rows, second and third columns
Out[x]:
              radius     planet
Ganymede    2634000.0   Jupiter
Callisto    2410000.0   Jupiter
Io          1822000.0   Jupiter
Europa      1561000.0   Jupiter
Moon        1737000.0     Earth
```

```
In [x]: df.iloc[-1, 1]         # last row, second column
Out[x]: 1737000.0
```

For single scalar values, there are also `at` and `iat`:

```
In [x]: df.at['Moon', 'mass']    # same as df.loc['Moon', 'mass']
Out[x]: 7.342e+22
In [x]: df.iat[-1, 0]            # same as df.iloc[-1, 0]
Out[x]: 7.342e+22
```

Example E9.2   There is a potential source of confusion when using `loc` for a `Series` or `DataFrame` with an integer index: it is important to remember that `loc` *always refers to the index labels*, whereas `iloc` takes a (zero-based) integer *location index*:

```
In [x]: df = pd.DataFrame(np.arange(12).reshape(4, 3) + 10,
                          index=[1, 2, 3, 4], columns=list('abc'))
In [x]: df
Out[x]:
    a    b    c
1   10   11   12
2   13   14   15
3   16   17   18
4   19   20   21

In [x]: df.loc[1]    # the row with index *label* 1 (the first row)
Out[x]:
a    10
b    11
c    12
Name: 1, dtype: int64

In [x]: df.iloc[1]   # the row with index *location* 1 (the row labeled 2)
a    13
b    14
c    15
Name: 2, dtype: int64
```

Note also that index labels do not have to be unique:

```
In [x]: df.index = [1, 2, 2, 3]     # change the index labels
In [x]: df
Out[x]:
    a    b    c
1   10   11   12
2   13   14   15
2   16   17   18
3   19   20   21

In [x]: df.loc[2]    # a DataFrame: all rows labeled 2
Out[x]:
    a    b    c
2   13   14   15
2   16   17   18

In [x]: df.iloc[2]    # a Series: there is only one row located at index 2
Out[x]:
a    16
b    17
```

```
c    18
Name: 2, dtype: int64
```

Combining `Series` and `DataFrames`

Another way to create a `DataFrame` is from a nested dictionary or from a dictionary of `Series`. In each case, the outer dictionary keys contain the *column* names; `Series` and inner dictionaries end up as rows:

```
boeing_wingspan = pd.Series({'B747-8': 68.4, 'B777-9': 64.8, 'B787-10': 60.12},
                            name='wingspan')
boeing_length = pd.Series({'B747-8': 76.3, 'B777-9': 76.7, 'B787-10': 68.28},
                          name='length')
boeing_range = pd.Series({'B777-9': 13940, 'B787-10': 11910},
                         name='range', dtype=float)

# Create a DataFrame from a dictionary of Series.
df_boeing = pd.DataFrame({'wingspan': boeing_wingspan, 'length': boeing_length,
                          'range': boeing_range})

# Create a DataFrame from a dictionary of dictionaries.
df_airbus = pd.DataFrame({'range': {'A350-1000': 16100, 'A380-800': 14800},
                          'wingspan': {'A350-1000': 64.75, 'A380-800': 79.75},
                          'length': {'A350-1000': 73.8, 'A380-800': 72.72} })
```

```
In [x]: df_boeing
Out[x]:
          wingspan  length    range
B747-8       68.40   76.30      NaN
B777-9       64.80   76.70  13940.0
B787-10      60.12   68.28  11910.0
```

```
In [x]: df_airbus
Out[x]:
            range  wingspan  length
A350-1000   16100     64.75   73.80
A380-800    14800     79.75   72.72
```

Note that missing values in the columns become NaN in the `DataFrame`. To concatenate two `DataFrames`, use `pd.concat`:[4]

```
In [x]: pd.concat((df_airbus, df_boeing))
Out[x]:
           length    range  wingspan
A350-1000   73.80  16100.0     64.75
A380-800    72.72  14800.0     79.75
B747-8      76.30      NaN     68.40
B777-9      76.70  13940.0     64.80
B787-10     68.28  11910.0     60.12
```

---

[4] Note that the concat and append functions require data to be copied into a new `DataFrame` and for large data sets can be slow and memory-inefficient. In this case, if at all possible, it is better to pre-allocate an empty `DataFrame` of the right size and to insert data directly into it.

(df_airbus.append(df_boeing) would give the same result.)

To add a single column to a DataFrame, assign a sequence of values or a Series object:

```
In [x]: df_airbus['speed'] = [950, 903]
In [x]: df_airbus

Out[x]:
          range  wingspan  length  speed
A350-1000 16100     64.75   73.80    950
A380-800  14800     79.75   72.72    903
```

Concatenating DataFrames with different columns fills the unknown values with NaN:

```
In [x]: df_aircraft = pd.concat((df_airbus, df_boeing))
In [x]: df_aircraft

Out[x]:
          length    range  speed  wingspan
A350-1000  73.80  16100.0  950.0     64.75
A380-800   72.72  14800.0  903.0     79.75
B747-8     76.30      NaN    NaN     68.40
B777-9     76.70  13940.0    NaN     64.80
B787-10    68.28  11910.0    NaN     60.12
```

Note that retrieving a Series as a row or column returns a *view* on the DataFrame, so changes to this Series will be reflected in it:

```
In [x]: speeds = df_aircraft['speed']
In [x]: speeds['B747-8','B787-10'] = 903, 956   # changes df_aircraft data
In [x]: jumbo = df_aircraft.loc['B747-8']
In [x]: jumbo.range = 15000                      # changes df_aircraft data
In [x]: df_aircraft
Out[x]:
          length    range  speed  wingspan
A350-1000  73.80  16100.0  950.0     64.75
A380-800   72.72  14800.0  903.0     79.75
B747-8     76.30  15000.0  903.0     68.40
B777-9     76.70  13940.0    NaN     64.80
B787-10    68.28  11910.0  956.0     60.12
```

To remove a column from a DataFrame, call Python's del keyword:

```
In [x]: del df_aircraft['speed']         # NB but not del df_aircraft.speed
In [x]: df_aircraft

Out[x]:
          length    range  wingspan
A350-1000  73.80  16100.0     64.75
A380-800   72.72  14800.0     79.75
B747-8     76.30  15000.0     68.40
B777-9     76.70  13940.0     64.80
B787-10    68.28  11910.0     60.12
```

The drop function can be used to selectively remove rows and columns from a DataFrame. A new object is returned unless inplace=True is specified:

```
In [x]: df_aircraft.drop(['A350-1000', 'A380-800'])    # drop rows by default
Out[x]:
          length    range   wingspan
B747-8     76.30   15000.0     68.40
B777-9     76.70   13940.0     64.80
B787-10    68.28   11910.0     60.12

In [x]: df_aircraft.drop(['length', 'wingspan'], axis='columns', inplace=True)
In [x]: df_aircraft
Out[x]:
              range
A350-1000   16100.0
A380-800    14800.0
B747-8      15000.0
B777-9      13940.0
B787-10     11910.0
```

9.1.4   Sorting, Arithmetic and Statistics

As might be expected, many of the most useful functions for data analysis are available from within pandas.

Example E9.3   The file india-data.csv, available at https://scipython.com/eg/bak, contains columns of demographic data on the 36 states and union territories (UTs) of India. When read in with:

```
In [x]: df = pd.read_csv('india-data.csv', index_col=0)
```

(more on this method in the next section), the DataFrame produced contains an Index of State/UT name and columns:

```
In [x]: df.index
Out[x]:
Index(['Uttar Pradesh', 'Maharashtra', 'Bihar', 'West Bengal',
       ...
       'Dadra and Nagar Haveli', 'Daman and Diu', 'Lakshadweep'],
      dtype='object', name='State/UT')

In [x]: df.columns
Out[x]:
Index(['Male Population', 'Female Population', 'Area (km2)',
       'Male Literacy (%)', 'Female Literacy (%)', 'Fertility Rate'],
      dtype='object')
```

We can quickly inspect the DataFrame with df.head(n), which outputs the first n rows (or five rows if n is not specified):

```
In [x]: df.head()
Out[x]:
                 Male Population  ...  Female Literacy (%)
State/UT                         ...
Uttar Pradesh         104480510  ...                59.26
Maharashtra            58243056  ...                75.48
Bihar                  54278157  ...                53.33
```

```
West Bengal               46809027  ...                  71.16
Madhya Pradesh            37612306  ...                  60.02

[5 rows x 5 columns]
```

pandas makes it straightforward to compute new columns for our DataFrame:

```
In [x]: df['Population'] = df['Male Population'] + df['Female Population']
In [x]: total_pop = df['Population'].sum()
In [x]: print(f'Total population: {total_pop:,d}')
Total population: 1,210,754,977


In [x]: df['Population Density (km-2)'] = df['Population'] / df['Area (km2)']
In [x]: df.loc['West Bengal', 'Population Density (km-2)']
Out[x]: 1028.440091490896      # population density of West Bengal


In [x]: total_pop / df['Area (km2)'].sum()
Out[x]: 368.3195047153525      # mean population density
```

Maximum and minimum values are obtained in the same way as in NumPy, for example:

```
In [x]: df['Male Literacy (%)'].min()
Out[x]: 73.39
```

Perhaps more usefully, idxmin and idxmax return the index *label(s)* of the minimum and maximum values, respectively:

```
In [x]: df['Area (km2)'].idxmax()    # largest state/UT by area
Out[x]: 'Rajasthan'
```

Naturally, the value returned can be passed to df.loc to obtain the entire row. For example, the row corresponding to the most densely populated state / UT:

```
In [x]: df.loc[df['Population Density (km-2)'].idxmax()]
Out[x]:
Male Population           8887326
Female Population         7800615
Area (km2)                1484
Male Literacy (%)         91.03
Female Literacy (%)       80.93
Population                16687940
Population Density (km-2) 1.124524e+04
Name: Delhi, dtype: float64
```

Correlation statistics between DataFrames or Series can be calculated with the corr function:

```
In [x]: df['Female Literacy (%)'].corr(df['Fertility Rate'])
Out[x]: -0.7361949271996956
```

In this case (two columns of data being compared), a single correlation coefficient is produced. More generally, the correlation *matrix* is returned as a new DataFrame. pandas can be used to quickly produce a variety of simple, labeled plots and charts from a DataFrame with a family of df.plot methods. By default, these use the Matplotlib backend, so the syntax is the same as presented in Chapter 7. For example,

```
In [x]: df.plot.scatter('Female Literacy (%)', 'Fertility Rate')
```

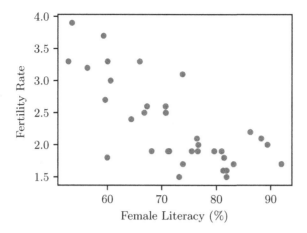

Figure 9.1 Scatter plot of fertility rate against female literacy for the 36 states and UTs of India.

Figure 9.1 shows the resulting plot.

9.2   Reading and Writing `Series` and `DataFrames`

9.2.1   Reading Text Files

Delimited Text Files

The core method for reading text files of data into a `DataFrame` is `pd.read_csv`. This works in much the same way as NumPy's `genfromtxt` method, but with additional functionality for naming columns and setting the `DataFrame` index. It takes no fewer than 49 possible arguments, but the most important are described below.[5]

- `filepath_or_buffer` (required): The path to the file to read: this can be a local file or a URL for fetching data across the internet.
- `sep`: The column delimiter; by default `','`, but use `'\s+'` for whitespace-delimited columns, `'\t'` for tab-delimiters, or `None` to force pandas to try to infer the delimiter. See also `delim_whitespace`.
- `delimiter`: An alias for `sep`.
- `header`: The row numbers (indices) to use for the column names. The default is `header=0`: use the first row for the column names. *Note*: if the file does not have a header, specify `header=None` and set the column names with the `names` argument.

---

[5] See the documentation at https://pandas.pydata.org/pandas-docs/stable/reference/api/pandas.read_csv.html for a complete description.

- names: A sequence of unique column names to use. If the file contains no header, set header=None in addition to setting names.
- index_col: The column(s) to use as the row labels in the DataFrame.
- usecols: A sequence of column indices (as for NumPy's loadtxt method) or column names identifying the columns to be read into the DataFrame.
- squeeze: If the data required consist of a single column, then squeeze=True will return a Series instead of the default, a DataFrame.
- converters: A dictionary of functions for converting the values in specified columns in the input file into data values for the DataFrame. The dictionary keys can be column indices or column names.
- skiprows: An integer giving the number of lines at the start of the file to skip over before reading the data or a sequence giving the indices of rows to skip.
- skipfooter: The number of rows at the bottom of the file to skip (by default, 0).
- nrows: The number of rows of the file to read: this is useful for reading a subset of lines from a very large file for testing or exploring its data.
- na_values: A string or sequence of strings to treat as NaN values, in addition to the default values which include 'NaN', 'NA', 'NULL' and '#N/A' (see the documentation for a full list).
- parse_dates: Set to True to parse the index column(s) as a sequence of datetime objects (see Section 9.3.2); other options are available for this argument (see the online documentation).
- comment: Specify a single character, such as '#', which, when found at the start of a line, signals that the whole line is to be ignored.
- skip_blank_lines: The default, True, skips over blank lines in the input file; set to False to interpret these as a row of NaN values instead.
- delim_whitespace: Can be set to True instead of specifying sep='\s+' to indicate that the data columns are separated by whitespace.

Example E9.4 The file ionization-energies.csv, available to download at https://scipython.com/eg/baq, contains the ionization energies (in eV) of some of the elements of the periodic table:

```
Ionization Energies (eV) of the first few elements of the periodic table
Element, IE1, IE2, IE3, IE4, IE5
H, 13.59844
He, 24.58741, 54.41778
Li, 5.39172, 75.64018, 122.45429
Be, 9.3227, 18.21116, 153.89661, 217.71865
B, 8.29803, 25.15484, 37.93064, 259.37521, 340.22580
C, 11.26030, 24.38332, 47.8878, 64.4939, 392.087
N, 14.53414, 29.6013, 47.44924, 77.4735, 97.8902
O, 13.61806, 35.11730, 54.9355, 77.41353, 113.8990
F, 17.42282, 34.97082, 62.7084, 87.1398, 114.2428
Ne, 21.5646, 40.96328, 63.45, 97.12, 126.21
Na, 5.13908, 47.2864, 71.6200, 98.91, 138.40
```

These data can be read into a `DataFrame` as follows. Here, we suppose that we are only interested in the first two periods of the periodic table and the first four ionization energies:

❶ ```
In [x]: df = pd.read_csv('ionization-energies.csv', skiprows=1, index_col=0,
   ...:   usecols=range(5), nrows=11)
```
❷ ```
In [x]: df.columns = df.columns.str.strip()
In [x]: print('Second ionization energy of Li: {} eV'.format(df.loc['Li'].IE2))
```
```
Second ionization energy of Li: 75.64018 eV
```

❶ Note that the `usecols` argument includes the column we want to set to the `DataFrame` index and `nrows` includes the column headers (but not the skipped rows).

❷ The whitespace around the column names is not automatically removed. pandas provides a variety of methods for manipulating strings within the `str` "accessor" namespace, which can be applied to all the column names in one statement; this is faster than using rename:

```
df.rename(columns=lambda s: s.strip(), inplace=True)
```

Example E9.5 The following text file, available at https://scipython.com/eg/bao, contains data concerning 13 vitamins important for human health.

```
List of vitamins, their solubility (in fat or water) and recommended dietary
allowances for men / women.
Data from the US Food and Nutrition Board, Institute of Medicine, National
Academies

Vitamin A    Fat     900ug/700ug

Vitamin B1   Water   1.2mg/1.1mg
Vitamin B2   Water   1.3mg/1.1mg
Vitamin B3   Water   16mg/14mg
Vitamin B5   Water   5mg
Vitamin B6   Water   1.5mg/1.4mg
Vitamin B7   Water   30ug
Vitamin B9   Water   400ug
Vitamin B12  Water   2.4ug

Vitamin C    Water   90mg/75mg
Vitamin D    Fat     15ug
Vitamin E    Fat     15mg
Vitamin K    Fat     110ug/120ug
--- Data for guidance only, consult your physician ---
```

The recommended (daily) dietary allowances are listed in either of two units in the final column; sometimes these are different for men and women. If we wish to parse this column into an average value in µg, we can use a converter function as in the following code.

Listing 9.1 Reading in a text table of vitamin data

```
import pandas as pd

def average_rda_in_micrograms(col):
    def ensure_micrograms(s):
        if s.endswith('ug'):
            return float(s[:-2])
        elif s.endswith('mg'):
            return float(s[:-2]) * 1000
        raise ValueError(f'Unrecognised units in {s}')
    fields = col.split('/')
    return sum([ensure_micrograms(s) for s in fields]) / len(fields)

df = pd.read_csv('vitamins.txt', delim_whitespace=True, skiprows=4,
                 skipfooter=1, header=None, usecols=(1, 2, 3),
                 converters={'RDA': average_rda_in_micrograms},
                 names=['Vitamin', 'Solubility', 'RDA'],
                 index_col=0
                 )
```

In this code, the four header rows and one footer row are skipped (blank lines are skipped automatically); the Index is set to the first *used* column (index_col=0, identifying the vitamin). The converter function averages the numerical values encountered (after conversion to µg), where multiple values are assumed to be separated by a solidus character (/).

Fixed-Width Text Files

The method read_fwf reads fixed-width formatted files. The field widths are passed as a list of tuples to the argument colspecs, giving the half-open intervals of the fields to read in from each line; i.e. (i, j) refers to the field from index i to index j - 1. Alternatively, if the intervals are contiguous, a list of field widths can be passed to the argument widths.

We return to the np.genfromtxt example of Section 6.2.3. The following short file, data.txt, consists of four columns with widths 2, 1, 9 and 3 characters (spaces are indicated with '␣'):

```
␣12␣␣100.231.03
␣11␣1201.842.04
␣11␣␣␣99.324.02
```

To read in this file with pandas, use either:

```
df = pd.read_fwf('data.txt',
                 colspecs=[(0, 2), (2, 3), (3, 12), (12, 15)], header=None)
```

or, since the intervals are contiguous:

```
df = pd.read_fwf('data.txt', widths=(2, 1, 9, 3), header=None)
```

to give the DataFrame:

```
     0  1        2     3
0    1  2   100.231  0.03
1    1  1  1201.842  0.04
2    1  1    99.324  0.02
```

9.2.2   Writing Text Files

The DataFrame method to_csv outputs its data to a text file, formatted according to the arguments summarized below.[6]

- path_or_buf: A file path or file object to output to; if None, the DataFrame is returned as string.
- sep: The single-character field-delimiter (defaults to ',').
- na_rep: The string to use to represent missing data (defaults to the empty string, '').
- float_format: The C-style format specifier (see Section 2.3.7) for floating-point numbers.
- columns: A sequence identifying the columns to output.
- header: By default, True, indicating that column names should be output; can be set to False or a list of column names.
- index: By default, True, indicating that row names should be output.
- compression: One of 'infer', 'gzip', 'bz2', 'zip', 'xz', None to specify whether and how to compress the output file. The default is 'infer': pandas determines the intended compression method from the filename extension.

Example E9.6   To write a comma-separated file containing data on vitamins from the DataFrame created in Example E9.5 to_csv can be used as follows:

```
df.to_csv('vitamins.csv', float_format='%.1f', columns=['Solubility', 'RDA'])
```

The file written is:

```
Vitamin,Solubility,RDA
A,Fat,800.0
B1,Water,1150.0
B2,Water,1200.0
B3,Water,15000.0
B5,Water,5000.0
B6,Water,1450.0
B7,Water,30.0
B9,Water,400.0
B12,Water,2.4
C,Water,82500.0
D,Fat,15.0
E,Fat,15000.0
K,Fat,115.0
```

---

[6] Full documentation is available at https://pandas.pydata.org/pandas-docs/stable/reference/api/pandas. DataFrame.to_csv.html

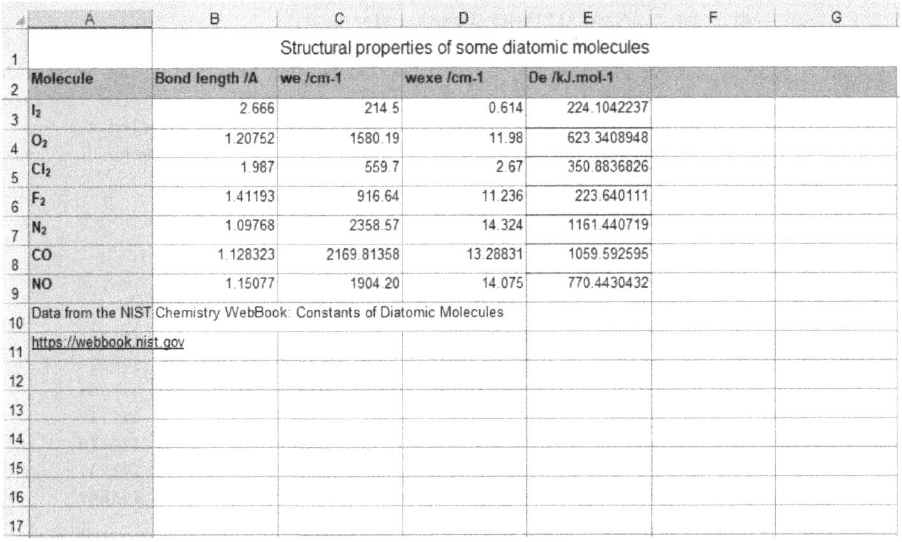

	Structural properties of some diatomic molecules					
Molecule	Bond length /A	we /cm-1	wexe /cm-1	De /kJ.mol-1		
I₂	2.666	214.5	0.614	224.1042237		
O₂	1.20752	1580.19	11.98	623.3408948		
Cl₂	1.987	559.7	2.67	350.8836826		
F₂	1.41193	916.64	11.236	223.640111		
N₂	1.09768	2358.57	14.324	1161.440719		
CO	1.128323	2169.81358	13.28831	1059.592595		
NO	1.15077	1904.20	14.075	770.4430432		
Data from the NIST Chemistry WebBook: Constants of Diatomic Molecules						
https://webbook.nist.gov						

Figure 9.2 An Excel sheet containing data concerning the structural properties of some diatomic molecules.

9.2.3 Microsoft Excel Files

pandas is able to read DataFrames from Excel files with both .xls and .xlsx extensions with the function pd.read_excel. You may need to install the xlrd package[7] separately from your Python package manager or using pip on the command line with, for example:

```
pip install xlrd
```

The file path to the Excel document is passed as the first argument to read_excel. Most of the additional arguments already described for read_csv function in the same way, except that usecols can be passed either a list of columnn indices or a string giving the range of Excel column labels: for example: 'B:K', 'A,D,G:K'.

By default, only the first sheet of the file is used; to read in from a different sheet or more than one sheet, pass one or more indexes or sheet names to the argument sheet_name.

Example E9.7 The Excel file bond-lengths.xlsx, available online at https://scipython.com/eg/bbk, contains data on the bond lengths, vibrational constants and dissociation energies of some diatomic molecules. The single sheet is named 'Diatomics'. Column A contains the molecular formula; the first row is a title, and the second row contains the column names. There is also a footer of two lines, as shown in Figure 9.2.

The following statement can be used to read in a DataFrame containing these data:

---

[7] https://pypi.org/project/xlrd/

```
df = pd.read_excel('bond-lengths.xlsx',
        index_col=0,            # the first column contains the index labels
        skipfooter=2,           # ignore the last two lines of the sheet
        header=1,               # take the column names from the second row
        usecols='A:E',          # use Excel columns labeled A-E
        sheet_name='Diatomics'  # take data from this sheet
        )

print(df)
```

	Bond length /A	we /cm-1	wexe /cm-1	De /kJ.mol-1
Molecule				
I2	2.666000	214.50000	0.61400	224.104224
O2	1.207520	1580.19000	11.98000	623.340895
Cl2	1.987000	559.70000	2.67000	350.883683
F2	1.411930	916.64000	11.23600	223.640111
N2	1.097680	2358.57000	14.32400	1161.440719
CO	1.128323	2169.81358	13.28831	1059.592595
NO	1.150770	1904.20000	14.07500	770.443043

Should you be in the unfortunate position of needing to *write* to an Excel spreadsheet file, use `to_excel`, as in the following example. Again, there may be a dependency to resolve: if the openpyxl module[8] is not available, you can install through your package manager or using pip:

```
pip install openpyxl
```

Example E9.8   To create some data to write to a file, the following program generates a `DataFrame` with the height of a projectile launched at three different angles (in the columns) as a function of time (rows):

Listing 9.2 The height of a projectile as a function of time

```
import numpy as np
import pandas as pd
import matplotlib.pyplot as plt

# Acceleration due to gravity, m.s-2.
g = 9.81

# Time grid, s.
t = np.linspace(0, 5, 500)
# Projectile launch angles, deg.
theta0 = np.array([30, 45, 80])
# Projectile launch speen, m.s-1.
v0 = 20

def z(t, v0, theta0):
    """Return the height of the projectile at time t > 0."""
    return -g/2 * t**2 + v0*t*np.sin(theta0)
```

---

[8] https://pypi.org/project/openpyxl/

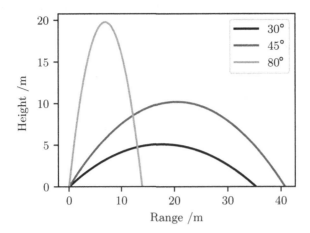

Figure 9.3 Trajectories of a projectile launched with $v_0 = 20$ m s^{-1} at three different angles.

```python
def x(t, v0, theta0):
    """Return the range of the projectile at time t > 0."""
    return v0 * t * np.cos(theta0)

# An empty DataFrame with columns for the different launch angles.
df = pd.DataFrame(columns=theta0, index=t)
# Populate df with the projectile heights as a function of time.
for theta in theta0:
    df[theta] = z(t, v0, np.radians(theta))
# Once the projectile has landed (z <= 0), set the height data as invalid.
df[df<=0] = np.nan

# Create a Matplotlib figure with the trajectories plotted.
fig, ax = plt.subplots()
for theta in theta0:
    ax.plot(x(t, v0, np.radians(theta)), df[theta], label=f'${theta}^\circ$')

# The maximum height obtained by the projectile for each value of theta0.
heights = df.max()
print(heights)
# Set the y-limits with a bit of padding at the top; label the axes.
ax.set_ylim(0, heights.max()*1.05)
ax.set_xlabel('Range /m')
ax.set_ylabel('Height /m')
ax.legend()
plt.show()
```

Figure 9.3 shows the plot of the trajectories that is produced by this code.

To save the DataFrame df to an Excel file in a single sheet, use to_excel:

```python
df.to_excel('projectile.xlsx', sheet_name='Dependence on angle')
```

To write an Excel file with more than one sheet, create a pd.ExcelWriter object and call to_excel for each pandas object to output:

```
with pd.ExcelWriter('projectile2.xlsx') as writer:
    for theta in theta0:
        # Only retain the valid data for each trajectory.
        ser = df[theta].dropna()
        # Change the Series index to be the range instead of time.
        ser.index = x(ser.index, v0, np.radians(theta))
        ser.to_excel(writer, sheet_name=f'{theta} deg')
```

9.2.4   Web Scraping

The pandas function read_html can be used to parse web pages for data contained in HTML tables. A list of DataFrames is returned, and the default arguments for this function do a pretty good job on most well-formed pages. The most useful arguments are listed below.

- io: A URL, filepath or file object from which to parse the HTML
- match: An optional string to search for within the table: only tables containing this string are parsed and returned.[9]
- header: The row index to be used for the column headers; the default, None, uses the HTML <th> header cells, if present.
- index_col: The column(s) to use as the row labels in the DataFrame.
- attrs: A dictionary of HTML attributes to identify the required table; for example, attrs={'id': 'data-table'}.
- thousands: The separator used in grouping the digits of large numbers; defaults to ','.
- decimal: The character used in denoting the decimal point; the default is '.', as used in the United States, United Kingdom, Australia, Japan, China and South Korea; non-British European countries often use ','.
- na_values: String(s) used to denote NaN data, as for read_csv.

Example E9.9   At the time of writing, the first table on the Wikipedia page https://en.wikipedia.org/wiki/List_of_wine-producing_regions contains columns of the rank, country name and wine production for the principal wine-producing countries in the world. To parse it with pandas:

```
In [x]: dfs = pd.read_html(
            'https://en.wikipedia.org/wiki/List_of_wine-producing_regions',
❶           index_col=1, match="Wine production by country")
❷ In [x]: dfs[0].head()
Out[x]:
                                    Rank  Production(tonnes)
Country(with link to wine article)
```

---

[9] match can be a *regular expression*.

Italy	1	4796900
France	2	4607850
Spain	3	4293466
United States	4	3300000
China	5	1700000

❶ In this case, the table is identified by a match to the the text inside the `<caption>` element of the first `<table>` on the page.

❷ `dfs` is a list containing a single item, the `DataFrame` parsed from the matching table.

9.2.5   Exercises

Problems

P9.2.1   The web page at https://scipython.com/ex/bab gives tables for the total ozone column amounts in October in "Dobson units,"[10] and the concentrations of two chlorofluorocarbon (CFC) compounds, "F11" and "F12", in parts per trillion by volume (pptv) for the years 1957 – 1984; see Farman *et al.*, *Nature* **315**, 207 (1985).

Read in and parse these data, and plot them on a suitable chart.

P9.2.2   At   https://en.wikipedia.org/wiki/Abundances_of_the_elements_(data_page) Wikipedia gives a list of element abundances for the Sun and solar system in an HTML table (amongst other, similar data). Use pandas' `read_html` method to read in and parse the Kaye and Laby data (column headed "Y1") and plot a bar chart demonstrating *Oddo–Harkins rule*: that elements with even atomic numbers are more abundant than those with neighboring odd atomic numbers.

P9.2.3   The *Hertzsprung–Russell diagram* classifies stars on a scatter plot: each star is represented as a point with an *x*-coordinate of effective temperature and a *y*-coordinate of *luminosity*, a measure of the star's radiated electromagnetic power. The page at https://scipython.com/ex/bak can be used to obtain a version of the HYG-database,[11] which provides data on 119614 stars. Read in these data with pandas and plot a Hertzsprung–Russell diagram. The luminosity column is identified as `'lum'` in the header and the star temperature can be calculated from its *color index* (also referred to as $(B - V)$ and identified as the column labeled `'ci'`) using the Ballesteros formula:

$$T\ /\mathrm{K} = 4600\left(\frac{1}{0.92(B - V) + 1.7} + \frac{1}{0.92(B - V) + 0.62}\right).$$

---

[10] The Dobson unit is defined as the thickness, in units of 0.01 mm, that a layer of pure gas would form at standard conditions for temperature and pressure from its total column amount in the atmosphere above a region of the Earth's surface.

[11] https://github.com/astronexus/HYG-Database, released under a Creative Commons Attribution-ShareAlike license

Note that the luminosity is best visualized on a logarithmic scale and the temperature axis is usually plotted in reverse (decreasing temperature towards the right-hand side of the diagram).

P9.2.4 Transport for London (TfL) is the UK local government body responsible for the public transport system of Greater London; they make available an Excel document, available from the link at https://scipython.com/ex/bam, which provides statistics about the usage of the underground network (the Tube) in the form of entry and exit passenger numbers for a "typical" day at each station over the years 2007–2017.

Read in this document with pandas, and analyze it to determine: (a) the busiest station on a typical weekday in 2017; (b) the station with the greatest percentage increase in passengers over the period 2007–2017; (c) the station with the largest relative difference in passenger numbers between the working week and a typical Sunday in 2017.

P9.2.5 The HITRAN database (https://hitran.org) provides a list of molecular line intensities for modeling radiative transmission in planetary atmospheres. Its native format consists of 160-character records of fixed-width fields.

Use pandas to read in the file `CO2-transitions.par`, available from https://scipython.com/ex/ban (where a description of the fields can also be found). Plot line intensity against wavelength for these transitions in the infrared region of the spectrum ($\lambda = 10$ mm to 700 nm, corresponding to wavenumber $\tilde{\nu} = 1$ cm^{-1} to about $14\,000$ cm^{-1}), where carbon dioxide (CO_2) is responsible for a significant fraction of the greenhouse effect in Earth's atmosphere.

9.3   More Advanced Indexing

9.3.1   Hierarchical Indexes with `MultiIndex`

A `DataFrame` is an intrinsically two-dimensional array of data: to represent data in higher dimensions, it is common to use hierarchical indexing to represent multiple *levels* within a single index. If the data are sparse or heterogeneous, this is much more efficient than creating a multidimensional NumPy array. For example, consider a data set concerning the mean monthly temperature and rainfall in five European cities. This could be considered three-dimensional, the dimensions being "city," "month" and "data type" (this last meaning either temperature or rainfall). For five cities (Paris, Berlin, Vienna, London, Madrid) and four months (Jan, Apr, Jul, Oct), there would therefore be 40 data points in total.

We *could* create a conventional single-level index consisting of (city, month) tuples, but it wouldn't be very convenient or flexible. Instead, we can create a hierarchical index with two levels from a sequence of two-item tuples using `pd.MultiIndex.from_tuples`:

```
In [x]: cities = ('Paris', 'Berlin', 'Vienna', 'London', 'Madrid')
In [x]: months = ('Jan', 'Apr', 'Jul', 'Oct')
In [x]: index = pd.MultiIndex.from_tuples(
```

```
    ...:        (city, month) for city in cities for month in months)
In [x]: index
Out[x]:
MultiIndex([( 'Paris', 'Jan'),
           ( 'Paris', 'Apr'),
           ( 'Paris', 'Jul'),
           ( 'Paris', 'Oct'),
           ('Berlin', 'Jan'),
           ...
           ('Madrid', 'Jul'),
           ('Madrid', 'Oct')],
          )
```

MultiIndexes of this form (the Cartesian product of two or more sequences) are so common that there is a convenience function, from_product, for their creation:

```
index = pd.MultiIndex.from_product((cities, months))
```

We can create a DataFrame with this index by assigning an array of data in the shape (20, 2):

```
In [x]: index.names = ['City', 'Month']

In [x]: # Mean monthly temperature (degC) for each city in each of Jan, Apr, Jul,
Oct.
In [x]: temps = [[4.9, 11.5, 20.5, 13.0], [0.1, 9.0, 19.1, 9.4],
    ...:         [0.3, 10.7, 20.8, 10.2], [5.2, 9.9, 18.7, 12.0],
    ...:         [6.3, 12.9, 25.6, 15.1]
    ...:        ]
In [x]: # Mean monthly rainfall (mm) for each city in each of Jan, Apr, Jul,
Oct.
In [x]: rainfall = [[51.0, 51.8, 62.3, 61.5], [37.2, 33.7, 52.5, 32.2],
    ...:            [38., 45., 70., 38.], [55.2, 43.7, 44.5, 68.5],
    ...:            [33., 45., 12., 60.]
    ...:           ]

In [x]: arr = np.array((temps, rainfall)).reshape((2, 20)).T
In [x]: df = pd.DataFrame(arr, index=index, columns=['Mean temperature /degC',
    ...:                                              'Mean rainfall /mm'])

In [x]: df
Out[x]:
              Mean temperature /degC  Mean rainfall /mm
City   Month
Paris  Jan                      4.9               51.0
       Apr                     11.5               51.8
       Jul                     20.5               62.3
       Oct                     13.0               61.5
Berlin Jan                      0.1               37.2
       Apr                      9.0               33.7
       Jul                     19.1               52.5
       Oct                      9.4               32.2
Vienna Jan                      0.3               38.0
       Apr                     10.7               45.0
       Jul                     20.8               70.0
       Oct                     10.2               38.0
London Jan                      5.2               55.2
       Apr                      9.9               43.7
       Jul                     18.7               44.5
       Oct                     12.0               68.5
```

```
Madrid Jan                          6.3                    33.0
       Apr                         12.9                    45.0
       Jul                         25.6                    12.0
       Oct                         15.1                    60.0
```

The `loc` method can be used to index into the DataFrame's `MultiIndex`:

```
In [x]: df.loc['Vienna']
Out[x]:
           Mean temperature /degC  Mean rainfall /mm
Month
Jan                         0.3                 38.0
Apr                        10.7                 45.0
Jul                        20.8                 70.0
Oct                        10.2                 38.0
```

```
In [x]: df.loc[('Paris', 'Jul')]
Out[x]:
Mean temperature /degC     20.5
Mean rainfall /mm          62.3
Name: (Paris, Jul), dtype: float64
```

```
In [x]: df.loc[('Paris', 'Jul'), 'Mean rainfall /mm']
Out[x]: 62.3
```

To slice a `MultiIndex`, however, it must first be sorted:

```
In [x]: df['Berlin':'London']
Out[x]: ...
UnsortedIndexError: 'Key length (1) was greater than MultiIndex lexsort depth (0)'
```

This somewhat cryptic error message is a result of the way pandas is optimized to slice only indexes which are in lexicographical order. There are several methods to sort a `MultiIndex`, but the simplest is to use `sort_index` as we did previously:

```
In [x]: df.sort_index(inplace=True)
In [x]: df['Berlin':'London']
Out[x]:
               Mean temperature /degC  Mean rainfall /mm
City   Month
Berlin Apr                       9.0                33.7
       Jan                       0.1                37.2
       Jul                      19.1                52.5
       Oct                       9.4                32.2
London Apr                       9.9                43.7
       Jan                       5.2                55.2
       Jul                      18.7                44.5
       Oct                      12.0                68.5
```

Note that this has sorted the months into alphabetical order as well. To keep them in chronological order, one approach would be to number the months instead by relabeling the index:

```
In [x]: df2 = df.rename({'Jan': 1, 'Apr': 4, 'Jul': 7, 'Oct': 10})
In [x]: df2.sort_index(inplace=True)
In [x]: df2.loc['Vienna', 'Mean temperature /degC']
Out[x]:
Month
```

```
1      0.3
4     10.7
7     20.8
10    10.2
Name: Mean temperature /degC, dtype: float64
```

The useful xs function makes selecting data indexed at different levels of a MultiIndex easier, and does not require the index to be sorted. For example, to retrieve the climate data for January in all cities:

```
In [x]: df.xs('Jan', level=1)    # look in second level of the MultiIndex for 'Jan'
Out[x]:
         Mean temperature /degC  Mean rainfall /mm
City
Berlin                    0.1               37.2
London                    5.2               55.2
Madrid                    6.3               33.0
Paris                     4.9               51.0
Vienna                    0.3               38.0
```

A Series or DataFrame with a hierarchical row index can be reshaped so as to create a MultiIndex on the columns instead by using the unstack() function:

```
In [x]: df.unstack()
Out[x]:
        Mean temperature /degC          ... Mean rainfall /mm
Month                    Apr  Jan  Jul  ...             Jan   Jul   Oct
City                                    ...
Berlin                   9.0  0.1  19.1 ...            37.2  52.5  32.2
London                   9.9  5.2  18.7 ...            55.2  44.5  68.5
Madrid                  12.9  6.3  25.6 ...            33.0  12.0  60.0
Paris                   11.5  4.9  20.5 ...            51.0  62.3  61.5
Vienna                  10.7  0.3  20.8 ...            38.0  70.0  38.0

[5 rows x 8 columns]
```

9.3.2 Timestamps and Time Series

pandas provides a Timestamp object, representing an instant in time to some precision. The to_datetime method provides a powerful and flexible way of parsing a human-readable string into a Timestamp. The following examples all evaluate to a timestamp representing midnight on the 12th of March, 2020: Timestamp('2020-03-12 00:00:00'). Note that where the date is ambiguous, by default it is resolved in favor of the US convention: MM/DD/YY: to force a string to be interpreted as DD/MM/YY set dayfirst=True.

```
pd.to_datetime('2020-03-12')
pd.to_datetime('12/3/20', dayfirst=True)
pd.to_datetime('3/12/20')
pd.to_datetime('12 March, 2020')
pd.to_datetime('12th of March 2020')
pd.to_datetime('Mar 12, 2020')
```

Times are also handled gracefully:

Table 9.1 Some string codes for pandas
time frequencies and offsets

Code	Description
A	Year end
M	Month end
W	Week
D	Calendar day
H	Hour
T	Minute
S	Second
L	Millisecond
U	Microsecond

```
In [x]: pd.to_datetime('9:05 21 August 2017')
Out[x]: Timestamp('2017-08-21 09:05:00')
In [x]: pd.to_datetime('21 August 2017 09:05:23')
Out[x]: Timestamp('2017-08-21 09:05:23')
```

Indexes can be constructed as a range of regularly spaced Timestamps with the
date_range function. Ranges can be specified by passing the start and end date, or
by passing the start date and the number of *periods*. The range interval is one day
by default, but this can be controlled with the freq argument (see Table 9.1 for valid
values).

```
In [x]: pd.date_range('1997-03-12', '1997-03-15')
Out[x]: DatetimeIndex(['1997-03-12', '1997-03-13', '1997-03-14', '1997-03-15'],
               dtype='datetime64[ns]', freq='D')

In [x]: pd.date_range('1997-03-12', periods=4)
Out[x]: DatetimeIndex(['1997-03-12', '1997-03-13', '1997-03-14', '1997-03-15'],
               dtype='datetime64[ns]', freq='D')
```

❶
```
In [x]: pd.date_range('1997-03', periods=4, freq='M')
Out[x]: DatetimeIndex(['1997-03-31', '1997-04-30', '1997-05-31', '1997-06-30'],
               dtype='datetime64[ns]', freq='M')
```

❷
```
In [x]: pd.date_range('1997-03', periods=4, freq='MS')
Out[x]: DatetimeIndex(['1997-03-01', '1997-04-01', '1997-05-01', '1997-06-01'],
               dtype='datetime64[ns]', freq='MS')
```

❶ By defaults, monthly ranges specified with freq='M' are marked at the end of the
month. The same is true for annual ranges (freq='A').

❷ To set timestamps at the start of each month use freq='MS' (and freq='AS' for
annual ranges).

pandas makes a distinction between a timestamp (represented by a Timestamp object)
and a *time period*: an interval of time between two points in time. A time period is
represented by the Period object and its start and end points are accessed through its
attributes start_time and end_time. The syntax for creating time periods is similar to
date ranges:

```
In [x]: p = pd.Period('2020-04', freq='M')
In [x]: t = pd.Timestamp('2020-04-03 14:30')
In [x]: p.start_time < t < p.end_time
Out[x]: True
```

It is often necessary to resample a time series at a different (higher or lower) frequency. The `resample` method assists with this: it returns a `Resampler` object which can be used to aggregate the data in some appropriate way. For example, in *downsampling* (resampling the data to a wider time frame), it may be appropriate to take the mean, minimum, maximum or sum of the values in the resampling interval. The following example should make this clearer.

Example E9.10 The file `river-level.csv`, available at https://scipython.com/eg/bal, lists the height in meters above sea level of Chitterne Brook, a small river in Wiltshire, England. Heights are given as minimum, average, and maximum values for each day between 1 January 2014 and 31 December 2016.

The following code reads in the data and plots the daily river height along with its monthly average, minimum and maximum values.

```
import pandas as pd
import matplotlib.pyplot as plt

df = pd.read_csv('river-level.csv', index_col=0, comment='#', parse_dates=True)

rs_monthly = df.resample('M')

df['avg_level'].plot(label='Daily average')
rs_monthly['avg_level'].mean().plot(label='Monthly average')
rs_monthly['min_level'].min().plot(label='Monthly minimum')
rs_monthly['max_level'].max().plot(label='Monthly maximum')

plt.xlabel('Date')
plt.ylabel('River level /m')
plt.gca().legend()
plt.show()
```

❶ Note that we need to set `parse_dates=True` to force pandas to interpret the first column as a `DatetimeIndex`.

Figure 9.4 shows the resulting plot.

9.3.3   Exercises

Problems

P9.3.1 Use pandas to read in the file, `tb-cases.txt`, available from https://scipython.com/ex/bao, which provides numbers of cases of tuberculosis in the USA, broken down by state for the years 1993–2018. Create a `DataFrame` with a hierarchical index (`MultiIndex`) consisting of the state name and year. Plot these data appropriately and determine the state with the greatest relative decrease in tuberculosis over the time period considered.

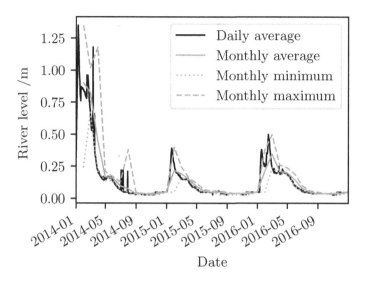

Figure 9.4 The level of Chitterne Brook in meters over the period 2014–2016.

P9.3.2 The populations of each state in the USA over the years 1993–2018 are given in the file US-populations.txt, available from https://scipython.com/ex/bap. Read these data into a pandas DataFrame with a suitable index, and analyze them for any interesting trends. Then combine these data with those of Problem P9.3.1 to determine the states with the greatest and least prevalence of tuberculosis per head of population in 2018.

9.4   Data Cleaning and Exploration

Any scientific research, particularly experimental research, will generate data sets containing invalid or missing values. Data points can be dropped or fall outside the detectable range of the measuring instrument, may get transcribed incorrectly, or are obtained incompletely from various sources. pandas provides a variety of methods for dealing with such missing values, including functions for removing them or replacing them with average or default values.

This book does not attempt to provide a guide to the scientific method, but the reader should be aware that the way in which one deals with missing or invalid data can bias the subsequent analysis towards a particular set of conclusions.

9.4.1   Missing Values

The default *sentinel value* indicating missing data is NaN. In Section 9.1.2 we have already used the methods `isnull()` and `notnull()` to test for the presence or absence of such values, and the method `dropna()`, which returns a new `DataFrame` with rows containing only non-null data:

```
In [x]: df = pd.DataFrame([[   1.1, np.nan, np.nan,   10.3],
        ...:                 [   0.8, np.nan,   3.6,    2.9],
        ...:                 [   1.2,   2.5,   1.6,    2.7],
        ...:                 [np.nan, np.nan, np.nan, np.nan],
        ...:                 [np.nan, np.nan,   3.6,    5.3]],
        ...:                 columns=list('ABCD'))

In [x]: df
     A    B    C     D
0  1.1  NaN  NaN  10.3
1  0.8  NaN  3.6   2.9
2  1.2  2.5  1.6   2.7
3  NaN  NaN  NaN   NaN
4  NaN  NaN  3.6   5.3

In [x]:  df.dropna()
Out[x]:
     A    B    C    D
2  1.2  2.5  1.6  2.7
```

You may wish to drop only rows (or columns) which consist entirely of NaN. In that case, pass the argument `how='all'` instead of using the default, `how='any'`:

```
In [x]: df.dropna(how='all')
Out[x]:
     A    B    C     D
0  1.1  NaN  NaN  10.3
1  0.8  NaN  3.6   2.9
2  1.2  2.5  1.6   2.7
4  NaN  NaN  3.6   5.3
```

It is also possible to specify a threshold number of NaN values to trigger the drop of a column or row:

```
In [x]: df.dropna(thresh=3, axis=1)     # only drop columns with three or more NaNs
Out[x]:
     A    C     D
0  1.1  NaN  10.3
1  0.8  3.6   2.9
2  1.2  1.6   2.7
3  NaN  NaN   NaN
4  NaN  3.6   5.3
```

An alternative to dropping the NaN values is to replace them with valid data according to some process. This is the purpose of the `fillna()` method. Common options are given in the following examples.

Replace all NaN values with a single value:

```
In [x]: df.fillna(3.6)
```

```
Out[x]:
     A    B    C     D
0  1.1  3.6  3.6  10.3
1  0.8  3.6  3.6   2.9
2  1.2  2.5  1.6   2.7
3  3.6  3.6  3.6   3.6
4  3.6  3.6  3.6   5.3
```

Replace NaN values with the last encountered valid value down the columns ("fill forward"):

```
In [x]: df.fillna(method='ffill')
Out[x]:
     A    B    C     D
0  1.1  NaN  NaN  10.3
1  0.8  NaN  3.6   2.9
2  1.2  2.5  1.6   2.7
3  1.2  2.5  1.6   2.7
4  1.2  2.5  3.6   5.3
```

Replace NaN values with the last encountered valid value along the rows:

```
In [x]: df.fillna(method='ffill', axis=1)
Out[x]:
     A    B    C     D
0  1.1  1.1  1.1  10.3
1  0.8  0.8  3.6   2.9
2  1.2  2.5  1.6   2.7
3  NaN  NaN  NaN   NaN
4  NaN  NaN  3.6   5.3
```

Passing a dictionary of column or index names enables close control over the filling values; chaining calls can then give a powerful and flexible way to clean data. For example, to fill in the missing data in columns A and C with their means:

```
In [x]: df.fillna({'A': df['A'].mean(), 'C': df['C'].mean()})
Out[x]:
          A    B         C     D
0  1.100000  NaN  2.933333  10.3
1  0.800000  NaN  3.600000   2.9
2  1.200000  2.5  1.600000   2.7
3  1.033333  NaN  2.933333   NaN
4  1.033333  NaN  3.600000   5.3
```

A further example:

```
In [x]: df.fillna({'A': df['A'].mean(), 'B': df['B'].mean()}).fillna(0)
Out[x]:
          A    B    C     D
0  1.100000  2.5  0.0  10.3
1  0.800000  2.5  3.6   2.9
2  1.200000  2.5  1.6   2.7
3  1.033333  2.5  0.0   0.0
4  1.033333  2.5  3.6   5.3
```

It may be that the data set being used uses a different sentinel value to indicate invalid data, for example -1 or -99. The replace method can canonicalize such data:

```
In [x]: ser = pd.Series([1.2, 3.5, -99, -99, 4.0, -99, -0.5])
In [x]: ser.replace(-99, np.nan)
Out[x]:
0    1.2
1    3.5
2    NaN
3    NaN
4    4.0
5    NaN
6   -0.5
dtype: float64
```

`replace` can also take a dictionary mapping values to their replacements:

```
In [x]: ser.replace({-99: 0, -0.5: np.nan})
Out[x]:
0    1.2
1    3.5
2    0.0
3    0.0
4    4.0
5    0.0
6    NaN
dtype: float64
```

9.4.2   Duplicate Values

The `DataFrame` method `duplicated()` returns a `Series` of boolean values indicating whether each row is a duplicate of a previous row; `drop_duplicates()` drops such rows. By default, both methods consider all columns; to remove rows with duplicate entries in a single column or several columns, pass a column name or a sequence of column names explicitly. A further argument, `keep`, determines whether the first encountered row (`'first'`, the default) or last encountered row (`'last'`) is retained.

```
In [x]: df = pd.DataFrame([['Lithium', 'Li', 3, 6, 0.0759],
   ...:                     ['Lithium', 'Li', 3, 7, 0.9241],
   ...:                     ['Sodium', 'Na', 11, 23, 1],
   ...:                     ['Potassium', 'K', 19, 39, 0.932581],
   ...:                     ['Potassium', 'K', 19, 40, 1.17e-4],
   ...:                     ['Potassium', 'K', 19, 41, 0.067302]],
   ...:                     columns=['Element', 'Symbol', 'Z', 'A', 'Abundance'])

In [x]: df
Out[x]:
     Element Symbol   Z   A  Abundance
0    Lithium     Li   3   6   0.075900
1    Lithium     Li   3   7   0.924100
2     Sodium     Na  11  23   1.000000
3  Potassium      K  19  39   0.932581
4  Potassium      K  19  40   0.000117
5  Potassium      K  19  41   0.067302

In [x]: df.drop_duplicates(['Symbol'])
Out[x]:
     Element Symbol   Z   A  Abundance
0    Lithium     Li   3   6   0.075900
```

```
2      Sodium     Na   11   23    1.000000
3   Potassium     K    19   39    0.932581

In [x]: df.drop_duplicates(['Symbol', 'Z'], keep='last')
Out[x]:
       Element Symbol   Z    A   Abundance
1      Lithium     Li   3    7    0.924100
2       Sodium     Na   11   23   1.000000
5    Potassium     K    19   41   0.067302
```

9.4.3   Binning Data

It is often necessary to bin together large amounts of continuous data, either to reduce it to a manageable size or to categorize it based on value. The pandas function `cut` can be used to do this in a similar way to NumPy's `histogram` function (Section 6.3.3):

```
In [x]: marks = [67, 80, 34, 55, 77, 66, 59, 52, 70, 67, 58, 63, 49, 72]
In [x]: bins = [0, 40, 60, 70, 80, 100]
In [x]: dist = pd.cut(marks, bins)
In [x]: dist
[(60, 70], (70, 80], (0, 40], (40, 60], ..., (60, 70], (40, 60], (70, 80]]
Length: 14
Categories (5, interval): [(0, 40] < (40, 60] < (60, 70] < (70, 80] < (80, 100]]
```

Each mark is placed in a bin with edges defined by the sequence bins. The number of values in each bin is returned by `value_counts`:

```
In [x]: pd.value_counts(dist)
Out[x]:
(60, 70]     5
(40, 60]     5
(70, 80]     3
(0, 40]      1
(80, 100]    0
dtype: int64
```

By default, the right side of each interval is closed (values equal to this side are included in the bin, indicated by ']') and the left side is open (indicated by '('); this can be swapped by setting the argument `right=False`:

```
In [x]: pd.value_counts(pd.cut(marks, bins, right=False))
Out[x]:
[40, 60)     5
[60, 70)     4
[70, 80)     3
[80, 100)    1
[0, 40)      1
dtype: int64
```

The bins can also be named by passing a sequence of strings to the `labels` argument:

```
In [x]: dist = pd.cut(marks, bins, labels=list(reversed('ABCDE')), right=False)
In [x]: dist
[C, A, E, D, B, ..., C, D, C, D, B]
Length: 14
```

```
Categories (5, object): [E < D < C < B < A]

In [x]: pd.value_counts(dist)
Out[x]:
D    5
C    4
B    3
A    1
E    1
dtype: int64
```

Note that the categories do not have any particular order. To put the counts in order of decreasing grade, we can sort the Series index:

```
In [x]: pd.value_counts(dist).sort_index(ascending=False)
Out[x]:
A    1
B    3
C    4
D    5
E    1
dtype: int64
```

9.4.4 Dealing with Outliers

Detecting and filtering outliers is, like dealing with missing values or invalid data, a potentially subtle process and careful thought should be given to the assumptions behind the expected underlying distribution. However, often, outlying values are expected based on detector failure (sticky pixels, cosmic rays, and the like), obvious errors, or well-understood exceptional cases. Filtering them automatically can be achieved with NumPy-like array operations.

For example, consider a simulated village in which the 200 houses have normally-distributed prices ($\mu = \$250\,000$, $\sigma = \$55\,000$), with the exception of a couple of mansions worth many times more than the average home:

```
In [x]: nhouses = 200
In [x]: mu, sigma = 250, 55        # mean, standard deviation in $1000s
❶ In [x]: prices = np.clip(np.random.randn(nhouses)*sigma + mu, 0, None).astype(int)
In [x]: prices[-2] = 1.e3
In [x]: prices[-1] = 2.e3
In [x]: df = pd.DataFrame(prices, columns=['price, $1000s'])
In [x]: df.tail()
Out[x]:
     price, $1000s
195            247
196            218
197            236
198           1000
199           2000
```

❶ np.clip(arr, min, max) constrains the values of arr to fall within min and max, here to prevent negative house prices being produced by the random sampling. This is itself a type of outlier filtering!

These outliers distort the mean and (especially) the standard deviation of the house price distribution:

```
In [x]: df.median()          # the median is a robust measure of central tendency
Out[x]:
price, $1000s    247.8
dtype: float64
```

```
In [x]: df.mean()            # the mean is affected more by the outliers
Out[x]:
price, $1000s    258.775
dtype: float64
```

```
In [x]: df.std()             # the standard deviation is greatly affected
Out[x]:
price, $1000s    145.796907
dtype: float64
```

We may be interested in analyzing the prices of "ordinary" houses in the village, ignoring the mansions. One way to do this is to identify the mansions as deviating from the mean house price by, say, three standard deviations and setting their prices to NaN:

```
In [x]: df[df > 3*df.std()+df.mean()] = np.nan
In [x]: df.tail()
     price, $1000s
195          247.0
196          218.0
197          236.0
198           NaN
199           NaN
```

Now we find values closer to the (non-mansion) house price distribution:

```
In [x]: df.mean()
Out[x]:
price, $1000s    246.237374
dtype: float64
```

```
In [x]: df.std()
Out[x]:
price, $1000s    55.995279
dtype: float64
```

Example E9.11   Robert Millikan's famous oil-drop experiments were carried out at the University of Chicago from 1909 to determine the magnitude of the charge of the electron.[12] In a single experiment, an electrically charged oil droplet was observed to fall a known distance, d, between two uncharged plates at its terminal velocity, v_g: from the time taken, t_g, the droplet's radius, a, can be deduced. Next, a voltage was applied to the plates, inducing an electric field between them. As the droplet rises under the resulting net force, the time taken, t_e, for it to move back up through the same distance, d, can be used to deduce its total charge, q, which is observed to be an integer multiple of the same base value, e, that is: $q = Ne$.

---

[12] Since May 2019, this quantity has been fixed by definition at $1.602176634 \times 10^{-19}$ C.

For the free-fall part of the experiment,[13] when the droplet falls at constant terminal velocity $v_g = -d/t_g$ there is no net force on it: the sum of the gravitational and drag forces is zero:

$$F_g + F_d = 0 \quad \Rightarrow -m'g - 6\pi\eta a v_g = 0,$$

where $m' = \frac{4}{3}\pi a^3 \rho' = \frac{4}{3}\pi a^3 (\rho_{\text{oil}} - \rho_{\text{air}})$ is the effective mass of the droplet (after the mass of air it displaces is taken into account), $g = 9.803$ m s^{-1} is the acceleration due to gravity in Chicago, and $\eta = 1.859 \times 10^{-5}$ kg m^{-1} s^{-1} is the air viscosity under the experimental conditions (temperature, humidity, etc.). Rearranging, we get the following expression for the droplet radius:

$$a = \sqrt{\frac{-9\eta v_g}{2\rho' g}}.$$

When a suitable voltage is applied to the plates and the droplet moves upwards at a constant velocity $v_e = d/t_e$, the force due to the electric field is balanced by gravity and the drag force (at this new velocity):

$$F_e + F_g + F'_d = 0 \quad \Rightarrow qE + 6\pi\eta a v_g - 6\pi\eta a v_e = 0$$

$$\Rightarrow \quad q = \frac{6\pi\eta a(v_e - v_g)}{E} = \frac{6\pi\eta a d}{E}\left(\frac{1}{t_g} + \frac{1}{t_e}\right)$$

Each droplet (labeled A–H) was observed three times for each different charge, q, acquired by exposure to X-rays (up to seven experiments per droplet).

The data at https://scipython.com/eg/bam give the time data for a number of such experiments conducted with an oil of density $\rho_{\text{oil}} = 917.3$ kg m^{-3} on a day for which $\rho_{\text{air}} = 1.17$ kg m^{-3}. The magnitude of the electric field was $E = 322.1$ kN C^{-1} and the distance the drops move, $d = 11.09$ mm. We can use these data to estimate e (assuming it is not fixed by definition) as follows.

drop	expt	tg	te	tg	te	tg	te
A	1	13.102	46.822	12.941	46.896	13.086	46.681
A	2	12.938	86.767	13.032	86.952	13.086	86.746
A	3	13.023	61.082	12.958	60.826	12.998	60.860
A	4	12.943	86.747	12.922	86.840	13.054	86.899
B	1	11.434	56.305	11.350	56.097	11.246	56.282
B	2	11.402	75.823	11.584	75.819	11.487	76.063
B	3	11.591	44.717	11.397	44.851	11.364	44.776
B	4	11.443	75.905	11.368	75.975	11.457	76.041
B	5	11.434	75.939	11.414	75.880	11.444	75.929
B	6	11.559	75.892	11.414	75.924	11.292	75.985
B	7	11.394	44.716	11.589	44.753	11.401	44.794
C	1	16.197	100.458	16.010	100.486	16.329	100.461
C	2	16.241	47.727	16.106	47.714	16.177	47.625
C	3	16.133	37.879	16.267	37.746	16.203	37.709
C	4	16.170	64.765	16.136	64.649	16.229	64.508
D	1	16.176	38.017	16.127	37.910	16.282	38.020

---

[13] In this example we adopt a coordinate system in which the droplet's vertical position, z, increases in the "up" direction.

```
D    2    16.275    38.280    16.092    38.208    16.133    38.092
D    3    16.422    48.327    16.073    48.284    16.212    48.184
D    4    16.134    38.202    16.258    38.270    16.105    38.229
D    5    16.164   102.562    16.217   102.673    16.194   102.696
E    1    12.275    55.020    12.116    54.962    12.307    54.978
E    2    12.157    54.772    12.183    54.967    12.046    55.219
E    3    12.146    55.004    12.118    54.938    12.346    54.869
E    4    12.319    43.635    12.243    43.552    12.073    43.582
F    1    14.172    61.946    14.174    61.970    14.069    61.959
F    2    14.145    90.718    13.955    90.707    14.075    90.866
F    3    14.070    62.147    14.074    61.961    14.247    61.892
F    4    14.017    61.968    14.101    61.921    14.106    62.174
G    1     9.723    50.375     9.527    50.482     9.502    50.508
G    2     9.463    63.755     9.670    63.853     9.509    63.827
G    3     9.448    63.804     9.407    63.899     9.563    63.768
G    4     9.327    63.855     9.518    63.967     9.533    63.824
H    1    13.192    73.375    13.167    73.338    13.316    73.449
H    2    13.042    42.642    13.387    42.428    13.334    42.459
H    3    13.389    42.379    13.244    42.373    13.055    42.610
H    4    13.114    73.161    13.226    73.384    13.207    73.257
H    5    13.030    73.295    13.022    73.419    13.438    73.512
```

First, define the necessary parameters:

```
eta = 1.859e-5                  # air viscosity, kg.m-1.s-1
rho_air = 1.17                  # air density, kg.m-3
rho_oil = 917.3                 # oil density, kg.m-3
rhop = rho_oil - rho_air
g = 9.803                       # acceleration due to gravity, m.s-2
d = 11.09e-3                    # rise/fall distance, m
E = -322.1e3                    # electric field vector (points down!)
```

Next, read in the data, assigning the first two columns to a `MultiIndex`:

```
In [x]: import pandas as pd
In [x]: df = pd.read_csv('eg10-millikan-data.txt', delim_whitespace=True,
                         index_col=[0, 1])
In [x]: df.head()
Out[x]:
                 tg      te     tg.1    te.1     tg.2    te.2
drop expt
A    1        13.102  46.822  12.941  46.896  13.086  46.681
     2        12.938  86.767  13.032  86.952  13.086  86.746
     3        13.023  61.082  12.958  60.826  12.998  60.860
     4        12.943  86.747  12.922  86.840  13.054  86.899
B    1        11.434  56.305  11.350  56.097  11.246  56.282
```

Note that pandas has added a counting integer to the column names to make them distinct.

We will start with just a single droplet, taking the transpose of its data:

```
In [x]: dropA = df.loc['A'].T
In [x]: dropA
Out[x]:
expt       1       2       3       4
tg     13.102  12.938  13.023  12.943
te     46.822  86.767  61.082  86.747
tg.1   12.941  13.032  12.958  12.922
```

```
te.1  46.896  86.952  60.826  86.840
tg.2  13.086  13.086  12.998  13.054
te.2  46.681  86.746  60.860  86.899
```

We would prefer to label each row as simply `'tg'` or `'te'`:

```
In [x]: dropA.index = dropA.index.str.slice(0, 2)
In [x]: dropA
Out[x]:
expt       1       2       3       4
tg     13.102  12.938  13.023  12.943
te     46.822  86.767  61.082  86.747
tg     12.941  13.032  12.958  12.922
te     46.896  86.952  60.826  86.840
tg     13.086  13.086  12.998  13.054
te     46.681  86.746  60.860  86.899
```

We require the average of all of the values of t_g (in the absence of the electric field the droplet takes the same time to fall the distance d) and the average value of t_e for each column (each experiment may have a different droplet charge, but the fall–rise times are measured three times for each experiment):

```
In [x]: tg = dropA.loc['tg'].values.mean()
In [x]: te = dropA.loc['te'].mean()
In [x]: tg
Out[x]: 13.006916666666667
In [x]: te
Out[x]:
expt
1    46.799667
2    86.821667
3    60.922667
4    86.828667
dtype: float64
```

Now use the value of t_g to calculate the droplet's radius:

```
In [x]: a = np.sqrt(9*eta*d/tg/2/rhop/g)
In [x]: a
Out[x]: 2.8181654881967875e-06
```

or about 2.82 μm. The charge we deduce for each experiment is:

```
In [x]: q = 6 * np.pi * eta * a * d / E * (1/tg + 1/te)
In [x]: q
Out[x]:
expt
1    -3.340563e-18
2    -3.005663e-18
3    -3.172143e-18
4    -3.005631e-18
dtype: float64
```

Repeating this for all the droplets, we can add a column, q to the `DataFrame df`:

```
for drop in df.index.levels[0]:
    drop_df = df.loc[drop].T
    drop_df.index = drop_df.index.str.slice(0, 2)
```

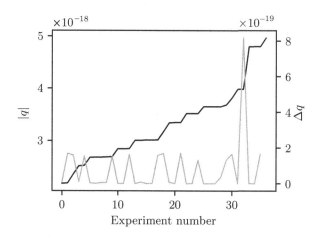

Figure 9.5 Sorted droplet charges, q, and neighbouring differences, Δq.

```
tg = drop_df.loc['tg'].values.mean()
te = drop_df.loc['te'].mean()
a = np.sqrt(9*eta*d/tg/2/rhop/g)
q = 6 * np.pi * eta * a * d / E * (1/tg + 1/te)
df.loc[drop, 'q'] = q.values
```

It is now helpful to sort the droplet charges by magnitude and to plot the sorted array and its differences (Figure 9.5):

```
In [x]: sorted_q = sorted(-df.loc[:, 'q'])
In [x]: plt.plot(sorted_q)
In [x]: plt.ylabel('$|q|$')
In [x]: plt.twinx()
In [x]: dq = np.diff(sorted_q)
In [x]: plt.plot(dq)
In [x]: plt.ylabel(r'$\Delta q$')
In [x]: plt.show()
```

It certainly seems possible that the droplet charge is always a multiple of some value between 1×10^{-19} C and 2×10^{-19} C. We can therefore estimate the value of $|e|$:

```
In [x]: e_estimate = dq[(dq>1.e-19) & (dq<2.e-19)].mean()
In [x]: e_estimate
Out[x]: 1.5697150510604604e-19
```

We can now add a column to df for the number of elementary charges we hypothesise for each experiment:

```
In [x]: df['N'] = (df['q'] / e_estimate).astype(int)
```

Considering all the data then gives us our estimate for the magnitude of the electron charge:

```
In [x]: (df['q']/df['N']).mean()
Out[x]: 1.5923552150386455e-19
```

within 1% of the defined value.

9.4.5 Exercises

Problems

P9.4.1 Use pandas' cut method to classify the stars in the data set of Problem P9.2.3 according to their temperature by placing them into the bins labeled M, K, G, F, A, B, and O with left edges (in K) at 2400, 3700, 5200, 6000, 7500, 10 000, and 30 000.

Hence modify the code in the solution to this problem to plot the stars in a color appropriate to their temperature by establishing the following mapping:

```
color_mapping = {'M': '#FFB56C', 'K': '#FFDAB5', 'G': '#FFEDE3', 'F': '#F9F5FF',
                 'A': '#D5E0FF', 'B': '#A2C0FF', 'O': '#92B5FF'}
```

Hint: pandas provides a map method for mapping input values from an existing column to output values in a new column using a dictionary.

P9.4.2 Reanalyze the data from Example E9.11, concerning Millikan's oil-drop experiment, to use a more accurate approximation for the effective air viscosity:

$$\eta = \frac{\eta_0}{1 + \frac{b}{ap}},$$

where $p = 100.82$ kPa is the air pressure, $\eta_0 = 1.859 \times 10^{-5}$ kg m^{-1} s^{-1}, $b = 7.88 \times 10^{-3}$ Pa m, and a is the droplet radius.

P9.4.3 The Cambridge University Digital Technology Group have been recording the weather from the roof of their department building since 1995 and make the data available to download at www.cl.cam.ac.uk/research/dtg/weather/.

Read in the entire data set and parse it with pandas to determine (a) the most common wind direction; (b) the fastest wind speed measured; (c) the year with the sunniest June; (d) the day with the highest rainfall; (e) the coldest temperature measured. Note that there are occasional missing and invalid data points in the data set.

P9.4.4 The data set at https://scipython.com/ex/baq lists the following quantities, in US dollars over time: (a) the price of gold; (b) the S&P 500 US stock market index; and (c) the price of the cryptocurrency Bitcoin. Compare the performance of these indexes over the period 2010–2020 with respect to the regular investment of $100 per month.

9.5 Data Grouping and Aggregation

9.5.1 `DataFrame` Grouping with `groupby`

The powerful pandas method groupby can be used to analyze data in a Series or DataFrame based on their categorization according to some key row (or column) values. The term *split–apply–combine* describes the process succinctly: first, the data is split according to its categorization; next, the analysis technique or statistical method required (for example, summing values or finding their mean) is applied to the split groups; finally, the results of the analysis are combined into a result object. Figure 9.6 depicts a simple example of the process.

Example E9.12   Consider the following table of the yields of three compounds, A, B and C, attained in a synthesis experiment by three students, Anu, Jenny and Tom.

```
In [x]: data = [['Anu', 'A', 5.4], ['Anu', 'B', 6.7], ['Anu', 'C', 10.1],
   ...:         ['Jenny', 'A', 6.5], ['Jenny', 'B', 5.9], ['Jenny', 'C', 12.2],
   ...:         ['Tom', 'A', 4.0], ['Tom', 'B', None], ['Tom', 'C', 9.5]
   ...:        ]

In [x]: df = pd.DataFrame(data, columns=['Student', 'Compound', 'Yield /g'])

In [x]: print(df)
  Student Compound  Yield /g
0     Anu        A       5.4
1     Anu        B       6.7
2     Anu        C      10.1
3   Jenny        A       6.5
4   Jenny        B       5.9
5   Jenny        C      12.2
6     Tom        A       4.0
7     Tom        B       NaN
8     Tom        C       9.5
```

One way of analyzing these data is to group them by compound ("split" into separate data structures, each with a common value of 'Compound') and then apply some operation (say, finding the mean) to each group, before recombining into a single DataFrame, as illustrated in Figure 9.6.

```
In [x]: grouped = df.groupby('Compound')
```

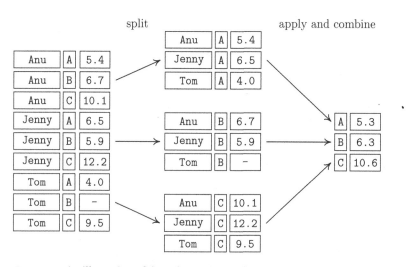

Figure 9.6   An illustration of the *split–apply–combine* paradigm for analyzing data with grouped data in pandas: the DataFrame is split into groups by compound (A, B and C); the mean function is applied to the groups; these values are combined into the returned object.

```
In [x]: grouped.mean()
Out[x]:
          Yield /g
Compound
A            5.3
B            6.3
C           10.6
```

Here, the 'Student' column has been ignored as a so-called "nuisance" column: there is no helpful way to take the mean of a string. The max() and min() functions, however, consider the strings' lexigraphical ordering:

```
In [x]: grouped.max()
Out[x]:
          Student  Yield /g
Compound
A            Tom      6.5
B            Tom      6.7
C            Tom     12.2
```

Note that max() has returned 'Tom' for every row, since this name is lexigraphically last ('greatest') in the 'Student' column. The column 'Yield /g' consists of the maximum yields for each compound, across all students. To apply the function to a subset of the columns only (which may be necessary for very large DataFrames), select them before the function call, for example:

```
In [x]: grouped['Yield /g'].min()
Out[x]:
Compound
A     4.0
B     5.9
C     9.5
Name: Yield /g, dtype: float64
```

The object returned by groupby() can be iterated over:

```
In [x]: for compound, group in grouped:
   ...:         print('Compound:', compound)
   ...:         print(group)
   ...:
Compound: A
   Student Compound  Yield /g
0     Anu        A       5.4
3   Jenny        A       6.5
6     Tom        A       4.0
Compound: B
   Student Compound  Yield /g
1     Anu        B       6.7
4   Jenny        B       5.9
7     Tom        B       NaN
Compound: C
   Student Compound  Yield /g
2     Anu        C      10.1
5   Jenny        C      12.2
8     Tom        C       9.5
```

We can also group by the `'Student'` column:

```
In [x]: grouped = df.groupby('Student')
```

```
In [x]: grouped.mean()
Out[x]:
        Yield /g
Student
Anu        7.40
Jenny      8.20
Tom        6.75
```

Another powerful feature is the ability to group on the basis of a specified mapping, provided, for example, by a dictionary. Suppose each student is undertaking a different degree programme:

```
In [x]: degree_programmes = {'Anu': 'Chemistry',
                             'Jenny': 'Chemistry',
                             'Tom': 'Pharmacology'}
```

First, turn the `'Student'` column into an `Index` and then group, not by the `Index` itself but using the provided mapping:

```
In [x]: df.set_index('Student', inplace=True)
In [x]: df.groupby(degree_programmes).mean()
Out[x]:
              Yield /g
Chemistry        7.80
Pharmacology     6.75
```

That is, the average yield for students of chemistry was 7.8 g, whereas for pharmacology it was only 6.75 g.

9.5.2   Exercises

Problems

P9.5.1 The Organisation for Economic Co-operation and Development (OECD), within its Programme for International Student Assessment (PISA), publishes an evaluation of the educational systems around the world by measuring the performance of 15-year-old school pupils on mathematics, science, and reading. The evaluation is carried out every three years.

Historical PISA data can be downloaded from https://scipython.com/ex/bza. Read these data in to a pandas `DataFrame` and use its grouping functionality to determine and visualize (a) the overall performance of all studied countries over time; (b) the gender disparity (if any) in each of reading, mathematics and science; and (c) the correlation between the performances in each of these areas across all countries.

P9.5.2 Read in the data at https://scipython.com/ex/bar concerning recent Formula One Grands Prix seasons, and rank (a) the drivers by their number of wins; (b) the constructors by their number of wins; and (c) the circuits by their average fastest lap per race.

9.6   Examples

The following examples demonstrate the practical use of pandas in two case studies involving the analysis and visualization of real data.

Example E9.13 The file `nuclear-explosion-data.csv`, available to download at https://scipython.com/eg/ban , contains data on all nuclear explosions between 1945 and 1998.[14] We will use pandas to analyze it in various ways.

Inspection of the file in a text editor shows that it contains a header line naming the columns, so we can load it straight away with `pd.read_csv` and inspect its key features:

```
In [x]: import pandas as pd
In [x]: df = pd.read_csv('nuclear-explosion-data.csv')
In [x]: df.head()

Out[x]:
         date      time     id country  ... yield_upper purpose      name     type
0    19450716  123000.0  45001     USA  ...        21.0      WR   TRINITY    TOWER
1    19450805  231500.0  45002     USA  ...        15.0  COMBAT  LITTLEBOY  AIRDROP
2    19450809   15800.0  45003     USA  ...        21.0  COMBAT    FATMAN  AIRDROP
3    19460630  220100.0  46001     USA  ...        21.0      WE      ABLE  AIRDROP
4    19460724  213500.0  46002     USA  ...        21.0      WE     BAKER       UW

[5 rows x 16 columns]

In [x]: df.index
Out[x]: RangeIndex(start=0, stop=2051, step=1)

In [x]: df.columns
Out[x]:
Index(['date', 'time', 'id', 'country', 'region', 'source', 'lat', 'long',
       'mb', 'Ms', 'depth', 'yield_lower', 'yield_upper', 'purpose', 'name',
       'type'],
      dtype='object')
```

There are 16 columns; here we will be concerned with those described in Table 9.2.

It is natural to assign the date and time of the explosion to the `DataFrame` index. Some helper functions facilitate this:

```
from datetime import datetime
def parse_time(t):
    hour, t = divmod(t, 10000)
    minute, t = divmod(t, 100)
    return int(hour), int(minute), int(t)


def parse_datetime(date, time):
    date_and_time = datetime.strptime(str(date), '%Y%m%d')
    hour, minute, second = parse_time(time)
    return date_and_time.replace(hour=hour, minute=minute, second=second)

df.index = pd.DatetimeIndex([parse_datetime(date, time) for date, time in
                zip(df['date'], df['time'])])
```

---

[14] from N.-O. Bergkvist and R. Ferm, *Nuclear Explosions 1945–1998*, Swedish Defence Research Establishment/SIPRI, Stockholm, July 2000.

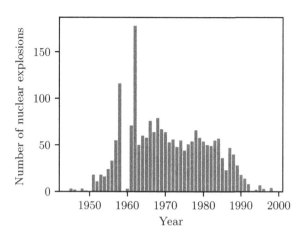

Figure 9.7 Bar chart of the number of nuclear explosions by year between 1945 and 1998.

We can plot the number of explosions in each year by grouping on `index.year` and finding the size of each group; a regular Matplotlib bar chart can then be produced:

```
explosion_number = df.groupby(df.index.year).size()
```

```
import matplotlib.pyplot as plt
fig, ax = plt.subplots()
ax.bar(explosion_number.index, explosion_number.values)
ax.set_xlabel('Year')
ax.set_ylabel('Number of nuclear explosions')
plt.show()
```

Figure 9.7 shows the resulting plot.

Table 9.2 Important columns of nuclear explosion data in file `nuclear-explosion-data.csv`

Column	Description
date	Date of explosion in format YYYYMMDD
time	Time of explosion in format HHMMSS.Z, where Z represents tenths of seconds
country	The state that carried out the explosion
lat	The latitude of the explosion in degrees, relative to the equator
long	The longitude of the explosion in degrees, relative to the prime meridian
yield_lower	Lower estimate of the yield in kilotons (kt) of TNT
yield_upper	Upper estimate of the yield in kilotons (kt) of TNT
type	The method of deployment of the nuclear device

A stacked bar chart can break down the annual count of explosions by country. First, group by both year and country and get the explosion counts for this grouping with `size()`:

```
df2 = df.groupby([df.index.year, df.country])
explosions_by_country = df2.size()
print(explosions_by_country.head(7))
```

```
        country
1945    USA       3
1946    USA       2
1948    USA       3
1949    USSR      1
1951    USA      16
        USSR      2
1952    UK        1
dtype: int64
```

Next, unstack the second index into columns, filling the empty entries with zeros:

```
explosions_by_country = explosions_by_country.unstack().fillna(0)
print(explosions_by_country.head(7))
```

```
country CHINA FRANCE INDIA PAKISTAN   UK   USA  USSR
1945     0.0    0.0   0.0      0.0   0.0   3.0   0.0
1946     0.0    0.0   0.0      0.0   0.0   2.0   0.0
1948     0.0    0.0   0.0      0.0   0.0   3.0   0.0
1949     0.0    0.0   0.0      0.0   0.0   0.0   1.0
1951     0.0    0.0   0.0      0.0   0.0  16.0   2.0
1952     0.0    0.0   0.0      0.0   1.0  10.0   0.0
1953     0.0    0.0   0.0      0.0   2.0  11.0   5.0
```

Each row in this `DataFrame` can then be plotted as stacked bars on a Matplotlib chart:

```
countries = ['USA', 'USSR', 'UK', 'FRANCE', 'CHINA', 'INDIA', 'PAKISTAN']
bottom = np.zeros(len(explosions_by_country))
fig, ax = plt.subplots()
for country in countries:
    ax.bar(explosions_by_country.index, explosions_by_country[country],
           bottom=bottom, label=country)
    bottom += explosions_by_country[country].values

ax.set_xlabel('Year')
ax.set_ylabel('Number of nuclear explosions')
ax.legend()
plt.show()
```

Figure 9.8 shows the resulting stacked bar chart.

The geopandas package provides a convenient way to plot the yield data on a world map. A full description of geographic information systems (GIS) is beyond the scope of this book, but geopandas is relatively self-contained and easy to use. First, read in the `DataFrame` for a low-resolution earth map (included with geopandas), and plot it on a Matplotlib Axes object. We'll accept the default equirectangular projection but customize the borders and fill the land areas in gray:

```
import geopandas
```

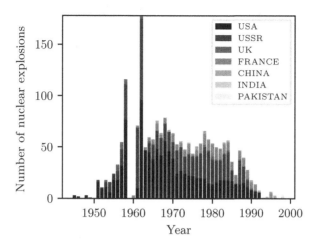

Figure 9.8 Stacked bar chart of the number of nuclear explosions by year caused by different countries between 1945 and 1998.

```
world = geopandas.read_file(geopandas.datasets.get_path('naturalearth_lowres'))
fig, ax = plt.subplots()
world.plot(ax=ax, color="0.8", edgecolor='black', linewidth=0.5)
```

The data provide lower and upper estimates of the explosion yield, so take the average and add circles as a scatter plot at the explosions' latitudes and longitudes. There is quite a large dynamic range from a few kilotons of TNT up to 50 million tons for the Tsar Bomba hydrogen bomb test of 1961, so clip the lower circle size to ensure that the smaller explosions are visible on the map:

```
df['yield_estimate'] = df[['yield_lower','yield_upper']].mean(axis=1)
sizes = (df['yield_estimate'] / 120).clip(10)
ax.scatter(df['long'], df['lat'], s=sizes, fc='r', ec='none', alpha=0.5)
ax.set_ylim(-60, 90)
plt.axis('off')
plt.show()
```

The result is Figure 9.9.

Example E9.14   The file `volcanic-eruptions.csv`, available to download at https://scipython.com/eg/bap , contains data concerning 822 significant volcanic events on Earth between 1750 BCE and 2020 CE from the US National Centres for Environmental Information (NCEI).[15] The information on each event is given in comma-separated fields and includes date, volcano name, location, type, estimated number of human deaths and "Volcanic Explosivity Index" (VEI).

---

[15] https://www.ngdc.noaa.gov/hazard/volcano.shtml.

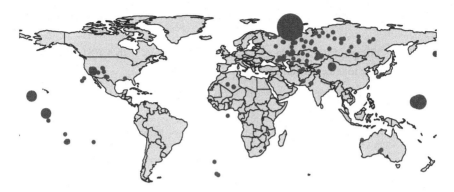

Figure 9.9 A map of nuclear explosions, showing the blast yield, between 1945 and 1998.

The data are readily parsed into a `DataFrame` with:

```
In [x]: df = pd.read_csv('volcanic-eruptions.csv', index_col=0)
```

The most deadly volcanic eruption in the database is that of Ilopango, around the middle of the fifth century CE:

```
In [x]: df.loc[df['Deaths'].idxmax()]
Out[x]:
Year                     450
Month                    NaN
Day                      NaN
Name                Ilopango
Location        El Salvador
Country         El Salvador
Latitude              13.672
Longitude            -89.053
Elevation                450
Type                 Caldera
VEI                        6
Deaths                 30000
Name: 25, dtype: object
```

It would be helpful to have a column with the day, month and year of the explosion parsed into a string. Define a helper function, `get_date`:

```python
def get_date(year, month, day):
    if year < 0:
        s_year = f'{-year} BCE'
    else:
        s_year = str(year)
    if pd.isnull(month):
        return s_year
    s_date = f'{int(month)}/{s_year}'
    if pd.isnull(day):
        return s_date
    return f'{int(day)}/{s_date}'
```

and apply it to the `DataFrame`:

```
In [x]: df['date'] = [get_date(year, month, day) for year, month, day in
                          zip(df['Year'], df['Month'], df['Day'])]
```

Simple filtering can give us a list of eruptions with a VEI of at least 6 since the start of the nineteenth century:

```
In [x]: df[(df['VEI'] >= 6) & (df['Year'] >= 1800)]
Out[x]:
      Year  Month   Day        Name  ...          Type  VEI   Deaths        date
218   1815    4.0  10.0     Tambora  ...  Stratovolcano  7.0  11000.0   10/4/1815
322   1883    8.0  27.0    Krakatau  ...       Caldera  6.0   2000.0   27/8/1883
365   1902   10.0  25.0  Santa Maria  ...  Stratovolcano  6.0   2500.0  25/10/1902
386   1912    9.0   6.0    Novarupta  ...       Caldera  6.0      2.0   6/9/1912
650   1991    6.0  15.0    Pinatubo  ...  Stratovolcano  6.0    350.0   15/6/1991
```

To find the 10 most explosive eruptions, we could filter out those with unknown VEI values before sorting:

```
In [x]: df[pd.notnull(df['VEI'])].sort_values('VEI').tail(10)[
    ...:          ['date', 'Name', 'Type', 'Country', 'VEI']]
Out[x]:
          date            Name           Type         Country  VEI
29         653        Dakataua        Caldera  Papua New Guinea  6.0
25         450        Ilopango        Caldera      El Salvador  6.0
22         240         Ksudach  Stratovolcano           Russia  6.0
21         230           Taupo        Caldera      New Zealand  6.0
18          60  Bona-Churchill  Stratovolcano    United States  6.0
99   19/2/1600    Huaynaputina  Stratovolcano             Peru  6.0
1     1750 BCE     Veniaminof  Stratovolcano    United States  6.0
40        1000   Changbaishan  Stratovolcano      North Korea  7.0
218  10/4/1815        Tambora  Stratovolcano        Indonesia  7.0
3     1610 BCE       Santorini  Shield volcano           Greece  7.0
```

However, there are many entries with a VEI of 6 and their ordering here is not clear. A better approach might be to sort first by VEI and next by deaths, setting na_position='first' to ensure that the null values are placed before numerical values (and therefore effectively rank lowest):

```
In [x]: df.sort_values(['VEI', 'Deaths'], na_position='first').tail(10)[
    ...:          ['date', 'Name', 'Type', 'Country', 'VEI', 'Deaths']]
Out[x]:
          date            Name            Type          Country  VEI   Deaths
386   6/9/1912        Novarupta         Caldera    United States  6.0      2.0
650  15/6/1991        Pinatubo   Stratovolcano      Philippines  6.0    350.0
99   19/2/1600    Huaynaputina   Stratovolcano             Peru  6.0   1500.0
120       1660     Long Island  Complex volcano  Papua New Guinea  6.0   2000.0
322  27/8/1883        Krakatau         Caldera        Indonesia  6.0   2000.0
365 25/10/1902     Santa Maria   Stratovolcano        Guatemala  6.0   2500.0
25         450        Ilopango         Caldera      El Salvador  6.0  30000.0
3     1610 BCE       Santorini  Shield volcano           Greece  7.0      NaN
40        1000   Changbaishan   Stratovolcano      North Korea  7.0      NaN
218  10/4/1815        Tambora   Stratovolcano        Indonesia  7.0  11000.0
```

We can also plot some histograms summarizing the data (Figure 9.10):

```
fig, axes = plt.subplots(nrows=2, ncols=2)
df['Day'].hist(bins=31, ax=axes[0][0], grid=False)
```

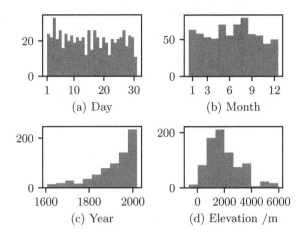

Figure 9.10 Histograms summarizing some of the columns of volcanic event data: (a) day of month; (b) month of year; (c) frequency by year since 1600 – hopefully, volcanic events have been progressively better recorded since 1600 and have not actually increased in frequency; (d) volcano elevation.

```
axes[0][0].set_xlabel('(a) Day')
df['Month'].hist(bins=np.arange(1, 14) - 0.5, ax=axes[0][1], grid=False)
axes[0][1].set_xticks(range(1, 13))
axes[0][1].set_xlabel('(b) Month')
df[df['Year']>1600]['Year'].hist(ax=axes[1][0], grid=False)
axes[1][0].set_xlabel('(c) Year')
df['Elevation'].hist(ax=axes[1][1], grid=False)
axes[1][1].set_xlabel('(d) Elevation /m')
plt.tight_layout()
plt.show()
```

10 General Scientific Programming

10.1   Floating-Point Arithmetic

10.1.1   The Representation of Real Numbers

The real numbers, such as 1.2, −0.36, π, 4 and 13256.625 may be thought of as points on a continuous, infinite *number line*.[1] Some real numbers (including the integers themselves) can be expressed as a ratio of two integers, for example, $\frac{5}{8}$ and $\frac{1}{3}$. Such numbers are called *rational*. Others, such as π, e and $\sqrt{2}$ cannot and are called *irrational*.

There can therefore be several ways of writing a real number, depending on which category it falls into, and not all of these ways can express the number precisely (using a finite amount of ink!). For example, the rational real number $\frac{5}{8}$ may be written exactly as a *decimal expansion* as 0.625:

$$\frac{5}{8} = \frac{6}{10} + \frac{2}{100} + \frac{5}{1000},$$

but the number $\frac{1}{3}$ cannot be written in a finite number of terms of a decimal expansion:

$$\frac{1}{3} = \frac{3}{10} + \frac{3}{100} + \frac{3}{1000} + \ldots = 0.333\ldots$$

In writing $\frac{1}{3}$ as a decimal expansion we must truncate the infinite sequence of 3s somewhere.

The irrational numbers can be *described* exactly (given some presumed geometrical or other knowledge), for example, π is the ratio of a circle's circumference to its diameter, $\sqrt{2}$ is the length of the hypotenuse of a right-angled triangle whose other sides have length 1. To *represent* or store such a number numerically, however, some level of approximation is necessary. For example, $\frac{355}{113}$ is a famous rational approximation to π. A (better) decimal approximation is 3.14159265358979. But, as a decimal expansion,[2] an infinite number of digits are needed to express the value of π precisely, just as an infinite number of 3s are needed in the decimal expansion of $\frac{1}{3}$.

Computers store numbers in binary, and the same considerations that apply to the limits of the decimal representation of a real number apply to its binary representation.

---

[1] Obviously, an integer such as 4 is just a special sort of real number.

[2] Note that a decimal expansion is simply a rational number with a power of 10 in the denominator, $3.14159265358979 = \frac{314159265358979}{100000000000000}$.

For example, $\frac{5}{8}$ has an exact binary representation in a finite number of bits:

$$\frac{5}{8} = \frac{1}{2} + \frac{0}{4} + \frac{1}{8} = 0.101_2$$

but

$$\frac{1}{10} = 0.000110011001100110011\ldots_2$$

is an infinitely repeating sequence. Only a finite number of these digits can be stored, and the truncated series of bits converted back to decimal is

$$\frac{1}{10} \approx 0.10000000000000006$$

using the so-called *double-precision* standard common to most computer languages on most operating systems. This is the *nearest representable number* to $\frac{1}{10}$.

The format of the double-precision floating-point representation of numbers is dictated by the IEEE-754 standard. There are three parts to the representation, stored in a total of 64 bits (8 bytes): the single *sign bit*, an 11-bit *exponent* and a 52-bit *significand* (also called the *mantissa*). This is best demonstrated by an example in decimal: the number 13256.625 can be written in scientific notation as:

$$13256.625 = +1.3256625 \times 10^4$$

and stored with the sign bit corresponding to +, a significand equal to 13256625 (where the decimal point is implicitly to be placed after the first digit) and the exponent 4. This notation is called "floating point" because the decimal point[3] is moved by the number of places indicated by the exponent.

The floating-point representation of numbers in binary works in the same way, except that each digit can only be 0 or 1, of course. This allows for a neat trick: when the number's binary point (equivalent to the decimal point in base-10) is shifted so that its significand has no leading zeros, then it will start with 1. Because all significands *normalized* in this way will start with 1, there is no need to store it, and effectively 53 bits of precision can be stored in a 52-bit significand.[4] The omitted bit is sometimes called the *hidden bit*.

In our example, 13256.625 can in fact be represented exactly in binary as

$$13256.625_{10} \equiv 11001111001000.101_2.$$

The normalized form of the significand is therefore 11001111001000101 and the exponent is 13, since:

$$11001111001000.101_2 = 1.1001111001000101 \times 2^{13}.$$

Now, as discussed, the first digit of the normalized signifcand will always be 1, so it is omitted and the significand is stored as

---

[3] More generally known as the *radix* point in bases other than base-10.
[4] Note that this trick works only in the binary sysetm.

```
1001111001000101000000000000000000000000000000000000000000
```

In order to allow for negative exponents (numbers with magnitudes less than 1), the exponent is stored with a *bias*: 1023 is added to the actual exponent. That is, actual exponents in the range -1022 to $+1023$ are stored as numbers in the range 1 to 2046. In this case, the 11-bit exponent field is $13 + 1023 = 1036$:

```
10000001100
```

Finally, the sign bit is 0, indicating a positive number. The full, 64-bit floating-point representation of 13256.625 (with spaces for clarity) is

```
0 10000001100 1001111001000101000000000000000000000000000000000000
```

and is exact. However, 0.1 is

```
0 01111111011 1001100110011001100110011001100110011001100110011010
```

and is not exact (note the truncation and rounding of the infinitely repeating sequence 0011...) – in decimal, this number is

```
0.1000000000000000055511151231257
```

In general, the 53 bits (including the hidden bit) of the significand give about 15 decimal digits of precision: $\log_{10}(2^{53}) = 15.95$. Any calculation resulting in more significant digits is subject to *rounding error*. The upper bound of the relative error due to rounding is called the *machine epsilon*, ϵ. In Python,

```
In [x]: import sys
In [x]: eps = sys.float_info.epsilon
In [x]: eps
Out[x]: 2.220446049250313e-16
```

It can be shown that the maximum spacing between two normalized floating-point numbers is 2ϵ. That is, `x + 2*eps == x` is guaranteed always to be `False`.

10.1.2   Comparing Floating-Point Numbers

Because of the finite precision of the floating-point representation of (most) real numbers it is extremely risky to compare two `floats` for equality. For example, consider squaring 0.1:

```
In [x]: (0.1)**2
Out[x]: 0.010000000000000002
```

As we have come to expect, this is not exactly 0.01, but it is also *not even the nearest representable number to 0.01*, since the number squared was, in fact, 0.100000000000000006. The unfortunate consequence of this is

```
In [x]: (0.1)**2 == 0.01
Out[x]: False
```

NumPy provides the methods `isclose` and `allclose` (see Section 6.1.12) for comparing two floating-point numbers or arrays to within a specified or default tolerance:

```
In [x]: np.isclose(0.1**2, 0.01)
Out[x]: True
```

Note also that floating-point addition is not necessarily *associative*:

```
In [x]: a, b, c = 1e14, 25.44, 0.74
In [x]: (a + b) + c
Out[x]: 100000000000026.17

In [x]: a + (b + c)
Out[x]: 100000000000026.19
```

Nor, in general, is floating-point multiplication *distributive* over addition:

```
In [x]: a, b, c = 100, 0.1, 0.2

In [x]: a*b + a*c
Out[x]: 30.0

In [x]: a * (b + c)
Out[x]: 30.000000000000004
```

10.1.3 Loss of Significance

Most floating-point operations (such as addition and subtraction) result in a loss of significance. That is, the number of significant digits in the result can be smaller than in the original numbers (operands) used in the calculation. To illustrate this, consider a hypothetical floating-point representation working in decimal with a six-digit significand and perform the following calculation, written in its exact form:

$$1.2345432 - 1.23451 = 0.0000332.$$

Our hypothetical system cannot store the first operand to its full precision but can only get as close as 1.23454. The floating-point subtraction then yields

$$1.23454 - 1.23451 = 0.00003.$$

The original numbers were accurate in the most significant six digits, but the result is only accurate in its first significant digit. Note that it isn't the case that the exact result cannot be *represented* in all its digits by our floating-point architecture: $0.0000332 \equiv 3.32 \times 10^{-5}$ only has three significant digits, well within the six available to us. The drastic loss of significance occurred because there was only a very small difference between the two numbers. This effect is sometimes called *catastrophic cancellation* and should always be a consideration when subtracting two numbers with similar values.

A similar loss of significance can occur when a small number is subtracted from (or added to) a much larger one:

$$12345.6 + 0.123456 = 12345.72345 \quad \text{(exactly)},$$
$$12345.6 + 0.123456 = 12345.7 \quad \text{(six-digit decimal significand).}$$

Even though the 15 or so significant digits of a double-precision floating-point number may seem like sufficient accuracy for a single calculation, be aware that repeatedly carrying out such calculations can increase this rounding error dramatically if the numbers involved cannot be represented exactly. For example, consider the following:

```
In [x]: for i in range(10000000):
   ....:     a += 0.1
   ....:
```

```
In [x]: a
Out[x]: 999999.9998389754
```

The difference between this approximate value and the exact answer, 1 000 000, is over 1.61×10^{-4}.

Python's math module has a function, fsum, which uses a technique called the *Shewchuk algorithm* to compensate for rounding errors and loss of significance. Compare these two implementations of the previous sum using a generator expression:

```
In [x]: sum((0.1 for i in range(10000000)))
Out[x]: 999999.9998389754
In [x]: math.fsum((0.1 for i in range(10000000)))
Out[x]: 1000000.0
```

10.1.4   Underflow and Overflow

Another consequence of the way that floating-point numbers are handled is that there is a minimum and maximum magnitude to the numbers that can be stored. For example, Bayesian calculations frequently require small probabilities to be multiplied together, with each probability a number between 0 and 1. For a large number of such probabilities this product can reach a value that is too small to represent, resulting in *underflow* to zero:

```
In [x]: P = 1
In [x]: for i in range(101):
   ....:     print(P)
   ....:     P *= 5.e-4
```

```
1
0.0005
2.5e-07
1.25e-10
6.250000000000001e-14
...
1.0097419586828971e-307
5.0487097934146e-311        # denormalization starts
2.5243548965e-314
1.2621776e-317
6.31e-321
5e-324
0.0                         # underflow
0.0
```

Below this value, Python begins to sacrifice some of the precision and maintains a modified representation of the number (a denormal, or subnormal number), a process called *gradual underflow*. Eventually, however, the number underflows its representation totally and becomes indistinguishable from zero. The minimum number that can be represented at full IEEE-754 double precision is

```
In [x]: import sys
In [x]: sys.float_info.min
Out[x]: 2.2250738585072014e-308
```

There are several possible tactics for dealing with underflow (beyond using higher-precision numbers such as np.float128, if available). In the earlier example, it is common to take the sum of the logarithms of the probabilities, which has a much more modest magnitude, instead of taking the product directly. Alternatively, one could start the earlier code with P = 1.e100 and manipulate the resulting numbers on the understanding that they are larger than they should be by this constant factor.

Floating-point *overflow* is the problem at the other end of the number scale: the largest double-precision number that can be represented is

```
In [x]: sys.float_info.max
Out[x]: 1.7976931348623157e+308
```

In NumPy, numbers that overflow are set to the special values inf or -inf depending on sign:

```
In [x]: f = 1
In [x]: for x in range(1, 40, 4):
   ...:        print('exp({}) = {}'.format(x**2, np.exp(x**2)))
   ...:
exp(1) = 2.718281828459045
exp(25) = 72004899337.38588
exp(81) = 1.5060973145850306e+35
exp(169) = 2.487524928317743e+73
exp(289) = 3.2441824460394912e+125
exp(441) = 3.340923407659982e+191
exp(625) = 2.7167594696637367e+271
exp(841) = inf
exp(1089) = inf
exp(1369) = inf
```

This leads to some curious relations between numbers that are too big to represent:

```
In [x]: a, b = 1.e500, 1.e1000
In [x]: a == b
Out[x]: True
In [x]: a, b
Out[x]: (inf, inf)
```

There is another special value, nan ("Not a Number," NaN), which is returned by some operations involving overflowed numbers:

```
In [x]: a / b
Out[x]: nan
```

(NumPy also implements its own values, numpy.nan and numpy.inf, see Section 6.1.4.) Never check if an object is nan with the == operator: nan is not even equal to itself(!)[5]:

---

[5] This means that the == operator is not an *equivalence relation* for floating-point numbers as it is not reflexive.

```
In [x]: c = a / b
In [x]: c == c
Out[x]: False
```

Python int objects are not subject to overflow, as Python will automatically allocate memory to hold them to full precision (within the limitations of available machine memory). However, NumPy integer arrays, which map to the underlying C data structures, are stored in a fixed number of bytes (see Table 6.2) and may overflow. For example,

```
In [x]: a = np.zeros(3, dtype=np.int16)
In [x]: a[:] = -30000, 30000, 40000
In [x]: a
Out[x]: array([-30000,  30000, -25536], dtype=int16)

In [x]: b = np.zeros(3, dtype=np.uint16)
In [x]: b[:] = -30000, 40000, 70000
In [x]: b
Out[x]: array([35536, 40000,  4464], dtype=uint16)
```

Signed 16-bit integers have the range -32768 to 32767, i.e. -2^{15} to $(2^{15} - 1)$. Due to the way they are stored, an attempted assignment to the number 40000 has resulted instead in the assignment of $40000 - 2^{16} = -25536$ to a[2] above. Similarly, *unsigned* 16-bit integers are limited to values in the range 0 to 65535, i.e. 0 to $(2^{16} - 1)$. Negative numbers cannot be represented at all and b[0] = -30000 gets converted to -30000 mod $2^{16} = 35536$; b[2] = 70000 overflows and ends up as 70000 mod $2^{16} = 4464$.

10.1.5   Further Reading

- From the Python documentation: *Floating-Point Arithmetic: Issues and Limitations*, available at https://docs.python.org/tutorial/floatingpoint.html.
- The article "What Every Computer Scientist Should Know About Floating-Point Arithmetic" by David Goldberg (*Computing Surveys*, March 1991) has become something of a classic and for a rigorous approach to the topic of floating-point arithmetic is highly recommended. It is available at https://docs.oracle.com/cd/E19957-01/806-3568/ncg_goldberg.html.
- S. Oliveira and D. Stewart, *Writing Scientific Software: A Guide to Good Style*, Cambridge University Press, Cambridge (2006).
- N. J. Higham, *Accuracy and Stability of Numerical Algorithms*, 2nd edn., Society for Industrial and Applied Mathematics, Philadelphia, PA (2002).
- The Standard Library module, decimal, supports decimal fixed-point and correctly rounded decimal floating-point arithmetic, but its calculations are generally much slower than those of the native float datatype. See https://docs.python.org/3/library/decimal.html for more details.

10.1.6 Exercises

Questions

Q10.1.1 The *decimal representation* of some real numbers is not unique. For example, prove mathematically that $0.\dot{9} \equiv 0.9999\ldots \equiv 1$.

Q10.1.2 $\sqrt{\tan(\pi)} = 0$ is mathematically well defined, so why does the folllowing calculation fail with a math domain error?

```
In [x]: math.sqrt(math.tan(math.pi))
--------------------------------------------------------------------
ValueError                        Traceback (most recent call last)
<ipython-input-135-7bfdceeef434> in <module>()
----> 1 math.sqrt(math.tan(math.pi))
```

Q10.1.3 Fermat's last theorem states that no three positive integers x, y and z can satisfy the equation $x^n + y^n - z^n = 0$ for any integer $n > 2$. Explain this apparent counter-example to the theorem:

```
In [x]: 844487.**5 + 1288439.**5 - 1318202.**5
Out[x]: 0.0
```

Q10.1.4 The functions $f(x) = (1 - \cos^2 x)/x^2$ and $g(x) = \sin^2 x/x^2$ are mathematically indistinguishable, but plotted using Python in the region $-0.001 \leq x \leq 0.001$ show a significant difference. Explain the origin of this difference.

Q10.1.5 How can you establish whether a floating-point number is nan or not without using `math.isnan` or `numpy.isnan`?

Q10.1.6 Predict and explain the outcome of the following:

(a)   `1e1001 > 1e1000`
(b)   `1e350/1.e100 == 1e250`
(c)   `1e250 * 1.e-250 == 1e150 * 1.e-150`
(d)   `1e350 * 1.e-350 == 1e450 * 1.e-450`
(e)   `1 / 1e250 == 1e-250`
(f)   `1 / 1e350 == 1e-350`
(g)   `1e450/1e350 != 1e450 * 1e-350`
(h)   `1e250/1e375 == 1e-125`
(i)   `1e35 / (1e1000 - 1e1000) == 1 / (1e1000 - 1e1000)`
(j)   `1e1001 > 1e1000 or 1e1001 < 1e1000`
(k)   `1e1001 > 1e1000 or 1e1001 <= 1e1000`

Problems

P10.1.1 *Heron's formula* for the area of a triangle (as used in Example E2.3),

$$A = \sqrt{s(s-a)(s-b)(s-c)} \text{ where } s = \tfrac{1}{2}(a+b+c),$$

is inaccurate if one side is very much smaller than the other two ("needle-shaped" triangles). Why? Demonstrate that the following reformulation gives a more accurate result in this case by considering the triangle with sides $(10^{-13}, 1, 1)$, which has the area 5×10^{-14}.[6]

$$A = \frac{1}{4}\sqrt{(a + (b + c))(c - (a - b))(c + (a - b))(a + (b - c))},$$

where the sides have been relabeled so that $a \geq b \geq c$.

What happens if you rewrite the factors in this equation to remove their inner parentheses? Why?

P10.1.2 Write a function to determine the machine epsilon of a numerical data type (`float`, `np.float128`, `int`, etc.).

10.2   Stability and Conditioning

10.2.1   The Stability of an Algorithm

The stability of an algorithm may be thought of in relation to how it handles approximation errors that occur in its operation or its input data. These errors typically arise from experimental uncertainties (imperfect measurements providing the input data) or from the sort of floating-point approximations involved in the calculations of the algorithm discussed in the previous section. Another common source of error is in the approximations made in "discretizing" a problem: the need to represent the values of a continuous function, $y = f(x)$ say, on a discrete "grid" of points: $y_i = f(x_i)$. An algorithm is said to be numerically stable if it does not magnify these errors and unstable if it causes them to grow.

Example E10.1   Consider the differential equation

$$\frac{dy}{dx} = -\alpha y$$

for $\alpha > 0$ subject to the boundary condition $y(0) = 1$. This simple problem can be solved analytically:

$$y = e^{-\alpha x},$$

but suppose we want to solve it numerically. The simplest approach is the *forward* (or *explicit*) *Euler* method: choose a step size, h, defining a grid of x values, $x_i = x_{i-1} + h$, and approximate the corresponding y values through:

$$y_i = y_{i-1} + h\left.\frac{dy}{dx}\right|_{x_{i-1}} = y_{i-1} - h\alpha y_{i-1} = y_{i-1}(1 - \alpha h).$$

The question arises: what value should be chosen for h? A small h minimizes the error introduced by the approximation above, which basically joins y values by straight-line

---

[6] This formula is due to William Kahan, one of the designers of the IEEE-754 floating-point standard.

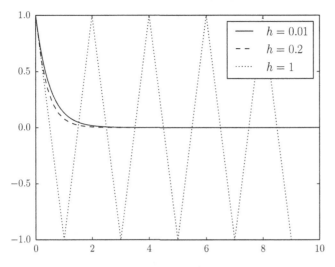

Figure 10.1 Instability of the forward Euler solution to $dy/dx = -\alpha y$ for large step size, h.

segments,[7] but if h is too small there will be cancellation errors due to the finite precision used in representing the numbers involved.[8]

The following code implements the forward Euler algorithm to solve the earlier differential equation. The largest value of h (here, $h = \alpha/2 = 1$) clearly makes the algorithm unstable (see Figure 10.1).

Listing 10.1 Comparison of different step sizes, h, in the numerical solution of $y' = -\alpha y$ by the forward Euler algorithm

```
import numpy as np
import matplotlib.pyplot as plt

alpha, y0, xmax = 2, 1, 10

def euler_solve(h, n):
    """ Solve dy/dx = -alpha.y by forward Euler method for step size h."""
    y = np.zeros(n)
    y[0] = y0
    for i in range(1, n):
        y[i] = (1 - alpha * h) * y[i-1]
    return y

def plot_solution(h):
    x = np.arange(0, xmax, h)
    y = euler_solve(h, len(x))
    plt.plot(x, y, label='$h={}$'.format(h))

for h in (0.01, 0.2, 1):
```

---

[7] That is, the Taylor series about y_{i-1} has been truncated at the linear term in h.

[8] In the extreme case that h is chosen to be smaller than the *machine epsilon*, typically about 2×10^{-16}, then we have $x_i = x_{i-1}$ and there is no grid of points at all.

```
    plot_solution(h)

plt.legend()
plt.show()
```

Example E10.2 The integral

$$I_n = \int_0^1 x^n e^x \, dx \quad n = 0, 1, 2, \dots$$

suggests a recursion relation obtained by integration by parts:

$$I_n = [x^n e^x]_0^1 - n \int_0^1 x^{n-1} e^x \, dx = e - nI_{n-1},$$

terminating with $I_0 = e - 1$. However, this algorithm, applied "forward" for increasing n is numerically unstable since small errors (such as floating-point rounding errors) are magnified at each step: if the error in I_n is ϵ_n such that the estimated value of $I'_n = I_n + \epsilon_n$ then

$$\epsilon_n = I'_n - I_n = (e - nI'_{n-1}) - (e - nI_{n-1}) = n(I_{n-1} - I'_{n-1}) = -n\epsilon_{n-1},$$

and hence $|\epsilon_n| = n! \epsilon_0$. Even if the error in ϵ_0 is small, that in ϵ_n is larger by a factor $n!$, which can be huge.

The numerically stable solution, in this case, is to apply the recursion backward for decreasing n:

$$I_{n-1} = \frac{1}{n}(e - I_n) \quad \Rightarrow \epsilon_{n-1} = -\frac{\epsilon_n}{n}.$$

That is, errors in I_n are *reduced* on each step of the recursion. One can even start the algorithm at $I'_N = 0$ and, providing enough steps are taken between N and the desired n, it will converge on the correct I_n.

Listing 10.2 Comparison of algorithm stability in the calculation of $I(n) = \int_0^1 x^n e^x \, dx$

```
# eg9-integral-stability.py
import numpy as np
import matplotlib.pyplot as plt

def Iforward(n):
    if n == 0:
        return np.e - 1
    return np.e - n * Iforward(n-1)

def Ibackward(n):
    if n >= 99:
        return 0
    return (np.e - Ibackward(n+1)) / (n+1)


N = 35
Iforward = [np.e - 1]
for n in range(1, N+1):
```

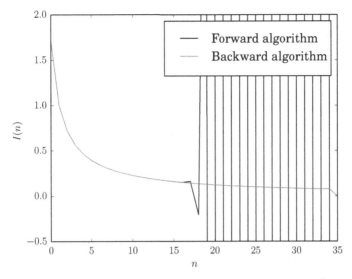

Figure 10.2 Instability of the forward recursion relation for $I_n = \int_0^1 x^n e^x \, dx$.

```
    Iforward.append(np.e - n * Iforward[n-1])

Ibackward = [0] * (N+1)
for n in range(N-1,-1,-1):
    Ibackward[n] = (np.e - Ibackward[n+1]) / (n+1)

n = range(N+1)
plt.plot(n, Iforward, label='Forward algorithm')
plt.plot(n, Ibackward, label='Backward algorithm')
plt.ylim(-0.5, 2)
plt.xlabel('$n$')
plt.ylabel('$I(n)$')
plt.legend()
plt.show()
```

Figure 10.2 shows the forward algorithm becoming extremely unstable for $n > 16$ and fluctuating between very large positive and negative values; conversely, the backward algorithm is well behaved.

10.2.2   Well-Conditioned and Ill-Conditioned Problems

In numerical analysis, a further distinction is made between problems which are well- or ill-conditioned. A *well-conditioned problem* is one for which small relative errors in the input data lead to small relative errors in the solution; an *ill-conditioned problem* is one for which small input errors lead to large errors in the solution. Conditioning is a property of the problem, not the algorithm, and is distinct from the issue of stability: it is perfectly possible to use an unstable algorithm on a well-conditioned problem and end up with erroneous results.

Example E10.3   Consider the two lines given by the equations:

$$y = x$$
$$y = mx + c$$

These lines intersect at $(x_\star, y_\star) = (c/(1 - m), c/(1 - m))$. Finding the intersection point is an ill-conditioned problem when $m \approx 1$ (lines nearly parallel).

For example, the lines $y = x$ and $y = (1.01)x + 2$ intersect at $(x_\star, y_\star) = (-200, -200)$. If we perturb m slightly by $\delta m = 0.001$, to $m' = m + \delta m = 1.011$, the intersection point becomes $(x'_\star, y'_\star) = (-181.8182, -181.8182)$. That is, a relative error of $\delta m / m \approx 0.001$ in m has created a relative error of $|(x'_\star - x_\star)/x_\star| \approx 0.091$, almost 100 times larger.

Conversely, if the lines have very different gradients, the problem is well-conditioned. Take, for example, $m = -1$ (perpendicular lines): the intersection $(1, 1)$ becomes $(1.0005, 1.0005)$ under the same perturbation to $m' = m + \delta m = -0.999$, leading to a relative error of 0.0005, which is actually *smaller* than the relative error in m.

Example E10.4   The conditioning of polynomial root-finding is notoriously bad. One famous example is *Wilkinson's polynomial*:

$$P(x) = \prod_{i=1}^{20}(x - i) = (x - 1)(x - 2)\cdots(x - 20)$$
$$= x^{20} - 210x^{19} + 20615x^{18} + \ldots + 2432902008176640000,$$

By inspection, the roots are simply $1, 2, \ldots, 20$. However, Wilkinson showed that decreasing the coefficient of x^{19} from -210 to $-210 - 2^{-23} \approx -210.000000119209$ had a drastic effect on many of the roots, some of which become complex. For example, the root at $x = 20$ moves to $x = 20.8$, a change of 4% on a perturbation of one coefficient by less than one part in a billion (see also Problem P10.2.2).

10.2.3   Exercises

Problems

P10.2.1   The simplest (and least accurate) way to calculate the first derivative of a function is to simply use the definition:

$$f'(x) = \lim_{h \to 0} \frac{f(x + h) - f(x)}{h}.$$

Fixing h at some small value, our approximation is

$$f'(x) \approx \frac{f(x + h) - f(x)}{h}.$$

Using the function $f(x) = e^x$, which value of h (in double-precision arithmetic, to the nearest power of 10) gives the most accurate approximation to $f'(1) = e$?

P10.2.2   Use NumPy's `Polynomial` class (see Section 6.4) to generate an object representing Wilkinson's polynomial from its roots to the available numerical precision; then find the roots of this representation of the polynomial.

10.3   Programming Techniques and Software Development

10.3.1   General Remarks

Commenting Code

Throughout this book we have tried to comment the code examples and exercise solutions helpfully. This is a good practice, even for short scripts, but the effective use of comments is not an entirely trivial activity. Here is some general advice:

- Generally, place comments on their own lines rather than "inline" with code (that is, after but on the same line as the code they describe):

```
# Volume of a dodecahedron of side length a.
V = (15 + 7 * np.sqrt(5)) / 4 * a**3
```

rather than

```
V = (15 + 7 * np.sqrt(5)) / 4 * a**3 # volume of a dodecahedron of side a
```

Longer comments should be complete sentences, with the first word capitalized (if appropriate, i.e. unless it is an identfier that begins with a lower-case letter) and ending in a period.

This book uses inline comments to clarify points of syntax; most of these would not be necessary in fluent "production" Python code.

- Explain *why* your code does what it does, don't simply explain *what* it does. Assume that the person reading your code knows the syntax of the language already. Thus,

```
# Increase i by 10:
i += 10
```

is a terrible comment which adds nothing to the line of code it purports to explain. On the other hand,

```
# Skip the next 10 data points.
i += 10
```

at least gives some indication of the reason for the statement.

- Keep comments up to date with the code they explain. It is all too easy to change code without synchronizing the corresponding comments. This can lead to a situation that is worse than having no comment at all:

```
# Skip the next 10 data points.
i += 20
```

Which is correct? Is the comment correct in explaining the programmer's intention but the line of code buggy, or has the line of code been updated for some reason without changing the comment? If your code is likely to be subject to such changes, consider defining a separate variable to hold the change in `i`:

```
DATA_SKIP = 10
...
# Skip the next DATA_SKIP data points.
i += DATA_SKIP
```

In fact, some programmers advocate aiming to minimize the number of comments by carefully choosing meaningful identifier names. For example, if we rename our index, we might even do away with the comment altogether:

```
data_index += DATA_SKIP
```

- Explain functions carefully using docstrings. In Python, all functions have an attribute `__doc__` which is set to the docstring provided in the function definition (see Section 2.7.1). A docstring is usually a multiline, triple-quoted string providing an explanation of what the function does, the arguments it takes and the nature of its return value(s), if any. From an interactive shell, typing `help(`*function_name*`)` provides more detailed information concerning the function, including this docstring.

Example E10.5   An example of a well-commented function (to calculate the volume of a tetrahedron) is given here.

Listing 10.3 A function to calculate the volume of a tetrahedron

```
# eg9-tetrahedron.py
import numpy as np

def tetrahedron_volume(vertexes=None, sides=None):
    """
    Return the volume of the tetrahedron with given vertexes or side lengths.
    If vertexes are given they must be in an array with shape (4, 3): the
    position vectors of the four vertexes in three dimensions; if the six sides are
    given, they must be an array of length 6. If both are given, the sides
    will be used in the calculation.

    Raises a ValueError if the vertexes do not form a tetrahedron (e.g.
    because they are coplanar, colinear or coincident).

    """

    # This method implements Tartaglia's formula using the Cayley-Menger
    # determinant:
    #                |0    1    1    1    1  |
    #                |1    0   s1^2 s2^2 s3^2|
    # 288 V^2 =  |1   s1^2   0   s4^2 s5^2|
    #                |1   s2^2 s4^2  0   s6^2|
    #                |1   s3^2 s5^2 s6^2  0  |
    # where s1, s2, ..., s6 are the tetrahedron side lengths.
```

```
# Warning: this algorithm has not been tested for numerical stability.

# The indexes of rows in the vertexes array corresponding to all
# possible pairs of vertexes.
vertex_pair_indexes = np.array(((0, 1), (0, 2), (0, 3),
                                (1, 2), (1, 3), (2, 3)))
if sides is None:
    # If no sides were provided, work them out from the vertexes.
    vertexes = np.asarray(vertexes)
    if vertexes.shape != (4, 3):
        raise TypeError('vertexes must be a numpy array with shape (4, 3)')
    # Get all the squares of all side lengths from the differences between
    # the six different pairs of vertex positions.
    vertex1, vertex2 = vertex_pair_indexes.T
    sides_squared = np.sum((vertexes[vertex1] - vertexes[vertex2])**2,
                           axis=-1)
else:
    # Check that sides has been provided as a valid array and square it.
    sides = np.asarray(sides)
    if sides.shape != (6,):
        raise TypeError('sides must be an array with shape (6,)')
    sides_squared = sides**2

# Set up the Cayley-Menger determinant.
M = np.zeros((5, 5))
# Fill in the upper triangle of the matrix.
M[0, 1:] = 1
# The squared-side length elements can be indexed using the vertex
# pair indexes (compare with the determinant illustrated above).
M[tuple(zip(*(vertex_pair_indexes + 1)))] = sides_squared

# The matrix is symmetric, so we can fill in the lower triangle by
# adding the transpose.
M = M + M.T

# Calculate the determinant and check it is positive (negative or zero
# values indicate the vertexes to not form a tetrahedron).
det = np.linalg.det(M)
if det <= 0:
    raise ValueError('Provided vertexes do not form a tetrahedron')
return np.sqrt(det / 288)
```

❶ Using np.asarray to convert vertexes into a NumPy array if it isn't one already enables the function to work with any compatible object (such as a list of lists).

Magic Numbers

In the context of programming style, a *magic number* is any constant numerical value used directly in a program, instead of being assigned to a meaningful variable name. Avoiding magic numbers generally leads to more readable, flexible and maintainable code, despite the increased verbosity. Wherever a number appears without an explanation (except, perhaps, for trivial cases such as initializing a variable to zero, or incrementing a quantity by one), it is worth considering whether that number can be assigned to a variable. If it appears more than once, it is almost always a good idea to do so, as illustrated in the following example.

Example E10.6 The following program estimates the probability of obtaining different totals on rolling two dice:

Listing 10.4 Code to simulate rolling two dice containing magic numbers

```python
import random

# Initialize the rolls dictionary to a count of zero for each possible outcome.
rolls = dict.fromkeys(range(2, 13), 0)

# The simulation: roll two dice 100000 times.
for j in range(100000):
    roll_total = random.randint(1, 6) + random.randint(1, 6)
    rolls[roll_total] += 1

# Report the simulation results.
for i in range(2, 13):
    P = rolls[i] / 100000
    print(f'P({i}) = {P:.5f}')
```

Several magic numbers appear without explanation in the program above: the number of pips showing on each die is selected randomly from the integers 1–6; the total number of each of the possible outcomes 2–12 are stored in a dictionary, `rolls`, whose keys are generated by the function call `range(2, 13)`; and the number of simulated rolls is hard-coded as 100 000.

Note that if we wanted, for example, to change the number of rolls we would have to edit the code in three places: the simulation loop, the comment above the simulation loop, and in the probability calculation. Maintaining and adapting code like this in a longer program is likely to be time-consuming and error-prone.

A little thought about how to assign these magic-number constants to variables also suggests a way to make the code more flexible, as shown below. As with other languages, it is common, but not necessary, to signal the definition of such a constant by defining its name in capitals:

Listing 10.5 Code to simulate rolling two dice refactored to use named constants

```python
import random

NDICE = 2
NFACES_PER_DIE = 6
NROLLS = 100000

# Calculate all the possible roll totals.
min_roll, max_roll = NDICE, NDICE * NFACES_PER_DIE
roll_total_range = range(min_roll, max_roll+1)

# Initialize the rolls dictionary to a count of zero for each possible outcome.
rolls = dict.fromkeys(roll_total_range, 0)

# The simulation: roll NDICE dice NROLLS times.
```

```
for j in range(NROLLS):
    roll_total = 0
    for i in range(NDICE):
        roll_total += random.randint(1, NFACES_PER_DIE)
    rolls[roll_total] += 1

# Report the simulation results.
for i in roll_total_range:
    P = rolls[i] / NROLLS
    print(f'P({i}) = {P:.5f}')
```

In this program, we can simulate the rolling of any number of dice with any number of sides any number of times by changing, in a single code location, the variables NDICE, NFACES_PER_DIE and NROLLS.

Style Guide for Python Code

The officially recommended coding conventions for Python are provided by a document known as PEP8 (available at www.python.org/dev/peps/pep-0008/). While it is acknowledged that it isn't always appropriate to follow these conventions all the time, Python programmers generally agree that they maximize the comprehensibility and maintainability of code. The focus is on consistency, readability and in minimizing the probability of hard-to-find typographical errors. Some of the highlights are

- Use *four spaces* per indentation level (and never tabs).[9]
- In assignments, put spaces around the = sign; for example, a = 10, not a=10.
- Use a maximum of 79 characters per line, where you need to split a line of code over more than one line:

    - favor implicit line continuation inside parentheses over the explicit use of the character, \ (see Section 2.3.1);
    - in arithmetic expressions, break around binary operators so that the new line is *after* the operator;
    - as far as possible, line up code so that expressions within parentheses line up.

    For example, the following is considered poor style:

    ```
    lengthy_calculation = margin*margin_px + (border*border_px\
                            + padding*padding_px)
    ```

    and might be better written as

    ```
    lengthy_calculation = (margin*margin_px + (border*border_px +
                                        padding*padding_px))
    ```

- Separate top-level function and class definitions by two blank lines; within a class, separate them by one blank line.

---

[9] A good text editor can be configured to automatically expand tabs to a fixed number of spaces.

- Use UTF-8 encoding for your source code (in Python 3 this is the default encoding anyway).
- Avoid wildcard imports (`from foo import *`), as they introduce (and potentially over-write) the imported module's names into the local namespace (recall Question Q2.2.5).
- Separate operators from their operands with single spaces unless operations with different priorities are being combined; for example, write `x = x + 5` but `r2 = x**2 + y**2`.
- Don't use spaces around the `=` in keyword arguments; for example, in function calls use `foo(b=4.5)` not `foo(b = 4.5)`.
- Avoid putting more than one statement on the same line separated by semicolons; for example, instead of `a = 1; b = 2`, write `a, b = 1, 2` (see Section 4.3.1).
- *Functions*, *modules* and *packages* should have short, all-lowercase names. Use underscores in function and module names if necessary, but avoid them in package names.
- *Class names* should be in (upper) CamelCase, also known as CapWords; for example, `AminoAcid`, not `amino_acid` (see Section 4.6.2).
- Define *constants*[10] in all-capitals with underscores separating words; for example, `MAX_LINE_LENGTH`.

10.3.2   Editors

While, to some extent, the choice of text editor for writing code is a personal one, most programmers favor one with syntax highlighting and the possibility to define macros to speed up repetitive tasks. Popular choices include:

- Visual Studio Code, a popular, free and open-source editor developed by Microsoft for Windows, Linux and macOS;
- Sublime Text, a commercial editor with per-user licensing and a free-evaluation option;
- Vim, a widely used, cross-platform keyboard-based editor with a steep learning curve but powerful features; the more basic vi editor is installed on almost all Linux and Unix operating systems;
- Emacs, a popular alternative to Vim;
- Notepad++, a free Windows-only editor;
- SciTE, a fast, lightweight source code editor;
- Atom, another free, open-source, cross-platform editor.

Beyond simple editors, there are fully featured integrated development environments (IDEs) that also provide debugging, code-execution, intelligent code-completion and access to operating-system services. Here are some of the options available:

- Eclipse with the PyDev plugin, a popular free IDE (www.eclipse.org/ide/);

---

[10] Note that Python doesn't really have constants in the same way that, for example, C does.

- JupyterLab, an open-source browser-based IDE for data science and other applications in Python (https://jupyter.org/);
- PyCharm, a cross-platform IDE with commercial and free editions (www.jetbrains.com/pycharm/);
- PythonAnywhere, an online Python environment with free and paid-for options (www.pythonanywhere.com/);
- Spyder, an open-source IDE for scientific programming in Python, which integrates NumPy, SciPy, Matplotlib and IPython (www.spyder-ide.org/).

10.3.3   Version Control

Unless properly managed, larger software projects (in practice, anything consisting of more than a single file of code) often rapidly descend into a tangle with modified versions, experimental code, ad hoc features and temporary files. The management of changes to the files comprising a software project is called *version control* (or *revision control*).

At its simplest, version control can involve simply keeping code in a number of parallel directories (folders), numbered chronologically as the software evolves. This approach can work, but if a small change in a large amount of code leads to a new version, it is inefficient (a lot of unchanged code is copied across to the new directory). If a new version is created only when the code changes a lot, then there is scope for a lot of tangled code to be generated between versions.

To solve these problems, there are several version-control software packages available, some of which are listed here. Most of these run as stand-alone applications on an operating system and can be invoked from the command line or used through a graphical interface. Some advantages are as follows:

- many developers can collaborate on one project;
- *branching*: the parallel development of two versions of the software at the same time, for example, to test out new features;
- *tagging* (or *labeling*): a way of referring to a snapshot of the project in a particular state;
- roll-back of a file in the project to a previous version;
- *cloning*: a means of distributing a software project along with its history of changes;
- some version-control systems integrate with online repositories for storing and sharing code; the most famous of these is GitHub (https://github.com/).

The working of version-control systems is not described in detail here (the syntax varies between systems and there are extensive tutorials, documentation and even entire books written about each one). Some recommended options are:

- Git: the most widely adopted version-control system, Git works on a *distributed* (or *decentralized*) basis, allowing developers to work on a project without sharing a common network or central reference code repository; open-source projects can

be hosted for free at online services such as GitHub and Bitbucket:
https://git-scm.com/

- Mercurial: another distributed version-control system:
https://www.mercurial-scm.org/

- Subversion (SVN): a centralized option with free (for open-source projects) hosting at SourceForge (https://sourceforge.net/). As Git has gained in popularity, SVN is not as widely used as it once was:
https://subversion.apache.org/.

10.3.4　Unit Tests

Unit testing is a way of validating software by focusing on individual units of source code. As an object-oriented programming language, for Python this usually means that individual classes (and sometimes even individual functions) are tested against a set of trial data (some of which may be deliberately incorrect or malformed). The aim is to catch any bugs which lead to the faulty interpretation of data. The set of unit tests also serves as a documented and verifiable assertion that the code does what it is supposed to. In some paradigms of code development, unit tests are written before the code itself.[11]

An important aspect of unit testing is "regression testing": it provides a means of ensuring that subsequent changes to the code (perhaps the addition of some functionality) do not break it: the upgraded code should pass the same unit tests that the original code did.

Unit testing your own code for a small project takes discipline. The tests are, themselves, computer code (and, perhaps, associated data) and need careful thought to write. The devising of suitable unit tests often prompts the programmer to think more deeply about the implementation of their code and can catch possible bugs before it is written.

Python's native unit testing framework is based around the `unittest` module: a simple application is given in the example below. The external `pytest` framework is a popular and well-supported alternative.

Example E10.7　Suppose we want to write a function to convert a temperature between the units degrees Fahrenheit, degrees Celsius and kelvins (identified by the characters `'F'`, `'C'` and `'K'`, respectively). The six formulas involved are not difficult to code, but we might wish to handle gracefully a couple of conditions that could arise in the use of this function: a physically unrealizable temperature (< 0 K) or a unit other than `'F'`, `'C'` or `'K'`.

Our function will first convert to kelvins and then to the units requested; if the from-units and the to-units are the same for some reason, we want to return the original value unchanged. The function `convert_temperature` is defined in the file `temperature_utils.py`.

---

[11] In particular, so-called "extreme" programming.

Listing 10.6 A function for converting between different temperature units

```
# temperature_utils.py

def convert_temperature(value, from_unit, to_unit):
    """ Convert and return the temperature value from from_unit to to_unit. """

    # A dictionary of conversion functions from different units *to* K.
    toK = {'K': lambda val: val,
           'C': lambda val: val + 273.15,
           'F': lambda val: (val + 459.67)*5/9,
          }
    # A dictionary of conversion functions *from* K to different units.
    fromK = {'K': lambda val: val,
             'C': lambda val: val - 273.15,
             'F': lambda val: val*9/5 - 459.67,
            }

    # First convert the temperature from from_unit to K.
    try:
        T = toK[from_unit](value)
    except KeyError:
        raise ValueError('Unrecognized temperature unit: {}'.format(from_unit))

    if T < 0:
        raise ValueError('Invalid temperature: {} {} is less than 0 K'
                         .format(value, from_unit))

    if from_unit == to_unit:
        # No conversion needed!
        return value

    # Now convert it from K to to_unit and return its value.
    try:
        return fromK[to_unit](T)
    except KeyError:
        raise ValueError('Unrecognized temperature unit: {}'.format(to_unit))
```

To use the unittest module to conduct unit tests on the convert_temperature, we write a new Python script defining a class, TestTemperatureConversion, derived from the base unittest.TestCase class. This class defines methods that act as tests of the convert_temperature function. These test methods should call one of the base class's *assertion functions* to validate that the return value of convert_temperature is as expected. For example,

```
self.assertEqual(<returned value>, <expected value>)
```

returns True if the two values are exactly equal and False otherwise. Other assertion functions exist to check that a specific exception is raised (e.g. by invalid arguments) or that a returned value is True, False, None, and so on. The unit test code for our convert_temperature function is here.

Listing 10.7 Unit tests for the temperature conversion function

```
from temperature_utils import convert_temperature
import unittest
```

```
class TestTemperatureConversion(unittest.TestCase):

    def test_invalid(self):
        """
        There's no such temperature as -280 C, so convert_temperature should
        raise a ValueError.
        """
```
❶
```
        self.assertRaises(ValueError, convert_temperature, -280, 'C', 'F')

    def test_valid(self):
        """ A series of valid temperature conversions to test. """

        test_cases = [((273.16, 'K',), (0.01, 'C')),
                      ((-40, 'C'), (-40, 'F')),
                      ((450, 'F'), (505.3722222222222, 'K'))]

        for test_case in test_cases:
            ((from_val, from_unit), (to_val, to_unit)) = test_case
            result = convert_temperature(from_val, from_unit, to_unit)
```
❷
```
            self.assertAlmostEqual(to_val, result)

    def test_no_conversion(self):
        """
        Ensure that if the from-units and to-units are the same the
        temperature is returned exactly as it was passed and not converted
        to and from kelvins, which may cause loss of precision.

        """
        T = 56.67
        result = convert_temperature(T, 'C', 'C')
```
❸
```
        self.assertTrue(result is T)

    def test_bad_units(self):
        """ Check that ValueError is raised if invalid units are passed. """
        self.assertRaises(ValueError, convert_temperature, 0, 'C', 'R')
        self.assertRaises(ValueError, convert_temperature, 0, 'N', 'K')

unittest.main()
```

❶ assertRaises verifies that a specified exception is raised by the method convert_
temperature. The necessary arguments to this method are passed after the method
object itself.

❷ We need assertAlmostEqual here because the floating-point arithmetic is likely to
cause a loss of precision due to rounding errors.

❸ We use assertTrue here to ensure that the temperature value is returned as the same
object that was passed and not converted to and from kelvins.

Running this script shows that our function passes its unit tests:

```
$ python eg9-temperature-conversion-unittest.py
...
--------------------------------------------------------------------
Ran 4 tests in 0.000s

OK
```

10.3.5   Further Reading

- F. Brooks, *The Mythical Man-Month*, Addison-Wesley, Boston, MA (1975, 1995). Near-legendary monograph on software development explaining why "adding manpower to a late software project makes it later."
- J. Loeliger and M. McCullough, *Version Control with Git*, O'Reilly, Sebastopol, CA (2012).
- S. McConnell, *Code Complete: A Practical Handbook of Software Construction*, Microsoft Press, Redmond, WA (2004).
- A. Hunt and D. Thomas, *The Pragmatic Programmer*, Addison-Wesley, Boston, MA (1999).

Appendix A Solutions

Answers to selected questions are given here. For further exercises and solutions, see
https://scipython.com.

Q2.2.5 This question illustrates the danger of "wildcard" imports: the value of the
variable e = 2 is replaced by the definition of e in the math module. The expression d
** e therefore raises 8 to the power of $e = 2.71828\ldots$ instead of squaring it.

Q2.2.7 Using Python's operators:

```
>>> a = 2
>>> b = 6
>>> 3 * (a**3*b - a*b**3) % 7
3
>>> a = 3
>>> b = 5
>>> 3 * (a**3*b - a*b**3) % 7
1
```

Q2.2.8 The thickness of the paper on the nth fold is $2^n t$, so we require $2^n t \geq d \Rightarrow$
$n_{min} = \lceil \log_2(d/t) \rceil$:

```
>>> d = 384_400 * 1.e3      # distance to Moon, m
>>> t = 1.e-4               # paper thickness, m
>>> math.log(d / t, 2)      # base-2 logarithm
41.805745474760016
```

Hence the paper must be folded 42 times to reach to the Moon ($\lceil x \rceil$ denotes the *ceiling*
of x: the smallest integer not less than x).

Q2.2.10 The ^ operator does not raise a number to another power (that is the **
operator). It is the *bitwise xor* operator, and in binary 10^2 is 1010 xor 0010 = 1000,
which is 8 in decimal.

Q2.3.1 Slice the string s = 'seehemewe' as follows (other solutions are possible in
some cases):

(a)   s[:3]
(b)   s[3:5]
(c)   s[5:7]
(d)   s[7:]
(e)   s[3:6]
(f)   s[5:2:-1]

(g)   s[-2::-3]

Q2.3.2 Simply slice the string backward and compare with the original:

```
>>> s = 'banana'
>>> s == s[::-1]
False
>>> s = 'deified'
>>> s == s[::-1]
True
```

Q2.3.5 This is not the correct way to test if the string s is equal to either 'ham' or 'eggs'. The expression ('eggs' or 'ham') is a boolean one in which both arguments, being nonempty strings, evaluate to True. The expression short-circuits at the first True equivalent and this operand is returned (see Section 2.2.4): that is, ('eggs' or 'ham') returns 'eggs'. Because s is, indeed, the string 'eggs' the equality comparison returns True. However, if the order of the operands is swapped, the boolean or again short-circuits at the first True equivalent, which is now 'ham' and returns it. The equality comparison with s fails, and the result is False.

There are two correct ways to test if s is one of two or more strings:

```
>>> s = 'eggs'
>>> s == 'ham' or s == 'eggs'
True
>>> s in ('ham', 'eggs')
True
```

(See Section 2.4.2 for more information about the syntax of the second statement.)

Q2.4.2 The problem is that enumerate, by default, returns the indexes and items of the array passed to it with the indexes starting at 0. The array passed to it is the slice P[1:] = [5, 0, 2] and so enumerate generates, in turn, the tuples (0, 5), (1, 0) and (2, 2). However, for our derivative we need the indexes into the original list, P, giving (1, 5), (2, 0) and (3, 2). There are two alternatives: pass the optional argument start=1 to enumerate or add 1 to the default index:

```
>>> P = [4, 5, 0, 2]
>>> dPdx = []
>>> for i, c in enumerate(P[1:], start=1):
...     dPdx.append(i*c)
>>> dPdx
[5, 0, 6]
```

```
>>> P = [4, 5, 0, 2]
>>> dPdx = []
>>> for i, c in enumerate(P[1:]):
...     dPdx.append((i+1)*c)
>>> dPdx
[5, 0, 6]
```

Q2.4.3 Here is one solution:

```
>>> scores = [87, 75, 75, 50, 32, 32]
>>> ranks = []
```

```
>>> for score in scores:
...     ranks.append(scores.index(score) + 1)
...
>>> ranks
[1, 2, 2, 4, 5, 5]
```

Q2.4.4 The following calculates π to 10 decimal places.

```
>>> import math

>>> pi = 0
>>> for k in range(20):
...     pi += pow(-3, -k) / (2*k+1)
...
>>> pi *= math.sqrt(12)
>>> print('pi = ', pi)
pi =  3.1415926535714034
>>> print('error = ', abs(pi - math.pi))
error =  1.8389734179891093e-11
```

❶ ... (annotation for line marked ❶)

❶ The built-in pow(x, j) is equivalent to (x)**j.

Q2.4.5 any(x) and not all(x) is True if at least one item in x is equivalent to True but not all of them:

```
>>> x1, x2, x3 = [False, False], [1, 2, 3, 4], [1, 2, 3, 0]
>>> any(x1) and not all(x1)
False
>>> any(x2) and not all(x2)
False
>>> any(x3) and not all(x3)
True
```

Q2.4.6 Recall that the * operator unpacks a tuple into a positional argument list to a function. So if z = zip(a, b) is the (iterator) sequence: (a0, b0), (a1, b1), (a2, b2), ... Unpacking this sequence in the call zip(*z) is equivalent to calling zip with these tuples as arguments:

zip((a0, b0), (a1, b1), (a2, b2), ...)

zip takes the first and second items from each tuple in turn, reproducing the original sequences:

(a0, a1, a2, ...), (b0, b1, b2, ...)

Q2.4.7 Simply zip the lists of sunshine hours and month names together and reverse-sort the resulting list of tuples:

```
>>> months = ['Jan', 'Feb', 'Mar', 'Apr', 'May', 'Jun',
...           'Jul', 'Aug', 'Sep', 'Oct', 'Nov', 'Dec']
>>> sun = [44.7, 65.4, 101.7, 148.3, 170.9, 171.4,
...        176.7, 186.1, 133.9, 105.4, 59.6, 45.8]
>>> for s, m in sorted(zip(sun, months), reverse=True):
...     print('{}: {:.1f} hrs'.format(m, s))
...
Aug: 186.1 hrs
Jul: 176.7 hrs
Jun: 171.4 hrs
```

```
May: 170.9 hrs
Apr: 148.3 hrs
Sep: 133.9 hrs
Oct: 105.4 hrs
Mar: 101.7 hrs
Feb: 65.4 hrs
Nov: 59.6 hrs
Dec: 45.8 hrs
Jan: 44.7 hrs
```

Q2.5.1   To normalize a list:

```
>>> a = [2, 4, 10, 6, 8, 4]
>>> amin, amax = min(a), max(a)
>>> for i, val in enumerate(a):
...     a[i] = (val - amin) / (amax - amin)
...
>>> a
[0.0, 0.25, 1.0, 0.5, 0.75, 0.25]
```

Q2.5.2   The following code calculates Gauss's constant to 14 decimal places.

```
>>> import math
>>> tol = 1.e-14
>>> an, bn = 1., math.sqrt(2)
>>> while abs(an - bn) > tol:
...     an, bn = (an + bn) / 2, math.sqrt(an * bn)
...
>>> print('G = {:.14f}'.format(1/an))
G = 0.83462684167407
```

Q2.5.3   The following code produces the first 100 "Fizzbuzz" numbers.

```
nmax = 100
for n in range(1, nmax + 1):
    message = ''
    if not n % 3:
        message = 'fizz'
    if not n % 5:
        message += 'buzz'
❶   print(message or n)
```

❶   Note that if n is not divisible by either 3 or 5, message will be the empty string, which evaluates to False in this logical expression, so n is printed instead.

Q2.5.4   Here's one solution, using stoich = 'C8H18' as an example:

Listing A.1 The structural formula of a straight-chain alkane

```
# qn2-5-c-alkane-a.py

stoich = 'C8H18'

fragments = stoich.split('H')
nC = int(fragments[0][1:])
nH = int(fragments[1])
if nH != 2*nC + 2:
    print('{} is not an alkane!'.format(stoich))
```

```
else:
    print('H3C', end='')
    for i in range(nC-2):
        print('-CH2', end='')
    print('-CH3')
```

The output is:

```
H3C-CH2-CH2-CH2-CH2-CH2-CH2-CH3
```

Q2.7.1   Only (b) and (f) behave as intended:

(a)   In the absence of an explicit return statement, the line function returns None. Because None cannot be joined into a string, an error occurs:

```
my_sum = '\n'.join(['   56', '  +44', line, '  100', line])
...
TypeError: sequence item 2: expected str instance, NoneType found
```

(b)   This code works as intended.

(c)   The function line returns a string, as required, but is not called as line(): without the parentheses, line refers to the function object itself, which cannot be joined in a string, so an error occurs:

```
my_sum = '\n'.join(['   56', '  +44', line, '  100', line])
...
TypeError: sequence item 2: expected str instance, function found
```

(d)   This code does not cause an error, but outputs a string representation of the function instead of the string returned when the function is called:

```
  56
 +44
<function line at 0x103d9e9e0>
 100
<function line at 0x103d9e9e0>
```

(e)   This code generates unwanted None output:

```
  56
 +44
-----
None
 100
-----
None
```

This happens because the statement print(line()) calls the function line, which prints a line of hyphens but also prints its return value (which is None since it doesn't return anything else explicitly).

(f)   This code works as intended.

Q2.7.2   The problem is within the add_interest function:

```
def add_interest(balance, rate):
    balance += balance * rate / 100
```

This creates a new float object, balance, *local* to the function, which is independent of the original balance object. When the function exits, the local balance is destroyed and the original balance never updated. One fix would be to return the updated balance value from the function:

```
>>> balance = 100
>>> def add_interest(balance, rate):
...     balance += balance * rate / 100
...     return balance
...
>>> for year in range(4):
...     balance = add_interest(balance, 5)
...     print('Balance after year {}: ${:.2f}'.format(year + 1, balance))
...
Balance after year 1: $105.00
Balance after year 2: $110.25
Balance after year 3: $115.76
Balance after year 4: $121.55
```

Q2.7.3   The problem is that the function digit_sum does not return the sum of the digits of n that it has calculated. In the absence of an explicit return statement, a Python function returns None, but None isn't an acceptable object to use in a modulus calculation and so a TypeError is raised.

The fix is simply to add return dsum:

```
def digit_sum(n):
    """ Find and return the sum of the digits of integer n. """

    s_digits = list(str(n))
    dsum = 0
    for s_digit in s_digits:
        dsum += int(s_digit)
    return dsum

def is_harshad(n):
    return not n % digit_sum(n)
```

Now, as expected:

```
>>> is_harshad(21)
True
```

Q2.7.4   The code outputs:

```
[1, 2, 'a']
[1, 2, 'a']
```

because a new list is created *once* when the function is defined, and it is this list that is appended to and returned with each call. Therefore, lst1 and lst2 are the same object, as you can confirm with:

```
print(lst1 is lst2)
True
```

Q4.1.1   It is a good idea to keep the try block as small as possible to prevent exceptions that you do not want to catch being caught instead of the one you do. For instance, in Example E4.5, suppose we read the file after opening it within the same `try` block:

```
try:
    fi = open(filename, 'r')
    lines = fi.readlines()
except IOError:
    ...
```

Now there are two errors that could give rise to an `IOError` exception being raised: failure to open the file and failure to read its lines. The `except` clause is intended to handle the first case, but it will also be executed in the second case when it would be more appropriate to handle it differently (or leave it unhandled and stop program execution).

Q4.1.2 The point of `finally` in Example E4.5 is that statements in this block get executed *before* the function returns. If the line

```
print('   Done with file {}'.format(filename))
```

were moved to after the `try` block, it would not be executed if an `IOError` exception is raised (because the function would have returned to its caller before this `print` statement is encountered.

Q4.2.1 This can easily be achieved with a `set`. Given the string, s:

```
set(s.lower()) >= set('abcdefghijklmnopqrstuvwxyz')
```

is True if it is a pangram. For example,

```
>>> s = 'The quick brown fox jumps over the lazy dog'
>>> set(s.lower()) >= set('abcdefghijklmnopqrstuvwxyz')
True
>>> s = 'The quick brown fox jumped over the lazy dog'
>>> set(s.lower()) >= set('abcdefghijklmnopqrstuvwxyz')
False
```

Q4.2.2 This function can be used to remove duplicates from an ordered list.

```
>>> def remove_dupes(l):
...     return sorted(set(l))
...
>>> remove_dupes([1, 1, 2, 3, 4, 4, 4, 5, 7, 8, 8, 9])
[1, 2, 3, 4, 5, 7, 8, 9]
```

Note that although sets don't have an order, they are iterable and can be passed to the `sorted()` built-in method (which returns a `list`).

Q4.2.3 From within the Python interpreter:

```
>>> set('hellohellohello')
{'h', 'o', 'l', 'e'}
>>> set(['hellohellohello'])
{'hellohellohello'}
>>> set(('hellohellohello'))
{'h', 'o', 'l', 'e'}
>>> set(('hellohellohello',))
```

```
{'hellohellohello'}
>>> set(('hello', 'hello', 'hello'))
{'hello'}
>>> set(('hello', ('hello', 'hello')))
{'hello', ('hello', 'hello')}
>>> set(('hello', ['hello', 'hello']))
Traceback (most recent call last):
  File "<stdin>", line 1, in <module>
TypeError: unhashable type: 'list'
```

Note the difference between initializing a set with a list of objects and attempting to add a list as an object in a set.

Q4.2.4 Note that the statement

```
>> a |= {2, 3, 4, 5}
```

does not change the frozenset but rather creates a new one from the union of the old one and the set {2, 3, 4, 5}. (In the same way, we have seen that for int object i, the assignment i = i + 1 rebinds the label i to a new integer object with value i + 1 rather than changing the value of the immutable int object previously bound to i.)

Q4.2.5 The following code snippet should be added after the definition of text: don't forget to import defaultdict from the collections module.

```
words_by_length = defaultdict(list)
for word in text.split():
    words_by_length[len(word)].append(word)

for length in sorted(words_by_length.keys()):
    print(f'{length}: {words_by_length[length]}')
```

Output:

```
1: ['a']
2: ['on', 'in', 'to']
3: ['and', 'ago', 'our', 'new', 'and', 'the', 'all', 'men', 'are']
4: ['four', 'this', 'that']
5: ['score', 'seven', 'years', 'forth', 'equal']
6: ['nation']
7: ['fathers', 'brought', 'liberty', 'created']
9: ['continent', 'conceived', 'dedicated']
11: ['proposition']
```

Q4.3.1 The list comprehension

```
>>> flist = [lambda x, i=i: x**i for i in range(4)]
```

creates the same list of anonymous functions as that in Example E4.11.

Note that we need to pass each i into the lambda function explicitly or else Python's closure rules will lead to every lambda function being equivalent to x**3 (3 being the final value of i in the loop).

Q4.3.2 The code snippet outputs the first nmax+1 rows of Pascal's triangle:

```
[1]
[1, 1]
```

```
[1, 2, 1]
[1, 3, 3, 1]
[1, 4, 6, 4, 1]
[1, 5, 10, 10, 5, 1]
```

In the list comprehension assignment,

```
x = [([0] + x)[i] + (x + [0])[i] for i in range(n+1)]
```

the elements of two lists are added. The two lists are formed from the list representing the previous row by, in the first case, adding a 0 to the beginning of the list, and in the second case, by adding a 0 to the end of the list. In this way, the sum is taken over by neighboring pairs of numbers, with the end numbers unchanged. For example, if x is [1, 3, 3, 1], the next row is formed by summing the elements in the lists

```
[0, 1, 3, 3, 1]
[1, 3, 3, 1, 0]
```

which yields the required [1, 4, 6, 4, 1].

Q4.3.3

(a)   Index the items of a using the elements of b:

```
>>> [a[x] for x in b]
['E', 'C', 'G', 'B', 'F', 'A', 'D']
```

(b)   Index the items of a using the sorted elements of b. In this case, the returned list is just (a copy of) a:

```
>>> [a[x] for x in sorted(b)]
['A', 'B', 'C', 'D', 'E', 'F', 'G']
```

(c)   Index the items of a using the elements of b indexed at the elements of b(!)

```
>>> [a[b[x]] for x in b]
['F', 'G', 'D', 'C', 'A', 'E', 'B']
```

(d)   Associate each element of b with the corresponding element of a in a sequence of tuples: [(4, 'A'), (2, 'B'), (6, 'C'), ...], which is then sorted – this method is used to return the elements of a corresponding to the ordered elements of b.

```
>>> [x for (y,x) in sorted(zip(b,a))]
['F', 'D', 'B', 'G', 'A', 'E', 'C']
```

Q4.3.4   To return a sorted list of (*key*, *value*) pairs from a dictionary:

```
>>> d = {'five': 5, 'one': 1, 'four': 4, 'two': 2, 'three': 3}
>>> d
{'four': 4, 'one': 1, 'five': 5, 'two': 2, 'three': 3}
>>> sorted([(k, v) for k, v in d.items()])
[('five', 5), ('four', 4), ('one', 1), ('three', 3), ('two', 2)]
```

Note that sorting the list of (*key*, *value*) tuples requires that the keys all have data types that can be meaningfully ordered. This approach will not work, for example, if the keys are a mixture of integers and strings since (in Python 3) there is no defined order to sort these types into: a `TypeError: unorderable types: int() < str()` exception will be raised.

To sort by *value* we could sort a list of (*value*, *key*) tuples, but to keep the returned list as (*key*, *value*) pairs, use

```
>>> sorted([(k, v) for k, v in d.items()], key=lambda item: item[1])
[('one', 1), ('two', 2), ('three', 3), ('four', 4), ('five', 5)]
```

The `key` argument to `sorted` specifies how to interpret each item in the list for ordering: here we want to order by the second entry (`item[1]`) in each (`k`, `v`) tuple to order by value.

Q4.3.5 The following code encrypts (and decrypts) a telephone number held as a string using the "jump the 5" method.

```
''.join(['5987604321'[int(i)] if i.isdigit() else '-' for i in '555-867-5309'])
```

Q4.3.6 One solution is to construct a tuple with two elements for each item in the list to be sorted: first, a boolean value indicating whether the item is `None` or not; second, the value itself. When these tuples are sorted, the first element is `False` for all the numbers (which can be compared as the second element) and `True` for all the `None` values. Since `False` always compares as "less than" `True`, there is no need to compare different types:

```
In [x]: lst = [4, 2, 6, None, 0, 8, None, 3]
In [x]: lst.sort(key=lambda e: (e is None, e))
In [x]: lst
Out[x]: [0, 2, 3, 4, 6, 8, None, None]
```

Q4.3.7 Here are two suggested solutions:

(a) Using an assignment expression with a `tuple`:

```
>>> t = 1, 1
>>> while (t := (t[0] + t[1], t[0])) < (5000, 0):
...     continue
...
>>> t[0]
6765
```

(b) With an assignment expression involving `input` in a `while` loop:

```
>>> while (s := input("> ").lower()) != "exit":
...     print(s)
...
> hello
hello
> bye
bye
> quit
quit
> :q
:q
> exit
>>>
```

Q6.1.1   An np.ndarray is a NumPy class for representing multidimensional arrays in Python; in this book, we often refer to instances of this class simply as array objects. np.array is a function that constructs such objects from its arguments (usually a sequence).

Q6.1.2   To create a two-dimensional array, array() must be passed a *sequence of sequences* as a single argument: this call passes three sequence arguments instead. The correct call is

```
>>> np.array( ((1, 0, 0), (0, 1, 0), (0, 0, 1)) , dtype=float)
```

Q6.1.3   np.array([0, 0, 0]) creates a one-dimensional array with three elements; a = np.array([[0, 0, 0]]) creates a 1×3 two-dimensional array (i.e. a[0] is the one-dimensional array created in the first example).

Q6.1.4   Changing an array's type by setting dtype directly does not alter the data at the byte level, only how the data are interpreted as a number, string and so on. As it happens, the byte-representations of zero are the same for integers (int64) and floats (float64), so the result of setting dtype is as expected. However, the 8 bytes representing 1.0 translate to the integer 4602678819172646912. To convert the data type properly, use astype(), which returns a new array (with its own data):

```
In [x]: a = np.ones((3,))
In [x]: a
Out[x]: array([ 1.,   1.,   1.])

In [x]: a.astype('int')
In [x]: a
Out[x]: array([1, 1, 1])
```

Q6.1.5   Indexing and slicing a NumPy array:

(a)   a[1, 0, 3]

(b)   a[0, 2, :] (or just a[0, 2])

(c)   a[2, ...] (or a[2, :, :] or a[2])

(d)   a[:, 1, :2]

(e)   a[2, :, :1:-1] ("in the third block, for each row take (backwards) the items in all but the first column").

(f)   a[:, ::-1, 0] ("for each block, traverse the rows backward and take the item in the first column of each").

(g)   Defining the three 2 × 2 index arrays for the blocks, rows and columns locating our elements as follows:

```
ia = np.array([[0, 0], [2, 2]])
ja = np.array([[0, 0], [3, 3]])
ka = np.array([[0, 3], [0, 3]])
```

a[ia, ja, ka] returns the desired result.

Q6.1.6   For example,

```
In [x]: a = np.array([0, -1, 4.5, 0.5, -0.2, 1.1])
In [x]: a[abs(a) <= 1]
Out[x]: array([ 0. ,  -1. ,   0.5, -0.2])
```

Q6.1.7   In the following code:

```
In [x]: a, b = -2.00231930436153, -2.0023193043615
In [x]: np.isclose(a, b, atol=1.e-14)
Out[x]: True
```

np.isclose() returns True because although the absolute difference between the two numbers is greater than 10^{-14}, it is (significantly) less than rtol * abs(b), the contribution from the default *relative* difference. To obtain the expected behavior, set rtol to 0:

```
In [x]: np.isclose(-2.00231930436153, -2.0023193043615, atol=1.e-14, rtol=0)
Out[x]: False
```

Q6.1.8   The different behavior here is due to the finite precision with which real numbers are stored: double-precision floating-point numbers are only represented to the equivalent of about 15 decimal places and so the two numbers being compared here are the same to within this precision:

```
In [x]: 3.1415926535897932 - 3.141592653589793
Out[x]: 0.0
```

Q6.1.9   For example,

```
In [x]: N = 5
In [x]: Nsq = N**2
In [x]: np.allclose(np.sort(magic_square.flatten()),
                    np.linspace(1, Nsq, Nsq).astype(int))
Out[x]: True

In [x]: Nsum = N * (N**2 + 1) // 2
In [x]: np.allclose(np.sum(magic_square, axis=0), Nsum)
Out[x]: True

In [x]: np.allclose(np.sum(magic_square, axis=1), Nsum)
Out[x]: True

In [x]: n.allclose(np.diag(magic_square), Nsum)
Out[x]: True
```

❶ ```
In [x]: n.allclose(np.diag(np.fliplr(magic_square)), Nsum)
Out[x]: True
```

❶ np.fliplr flips the array in the left/right direction. An alternative way to get this "other" diagonal is with a.ravel()[N-1:-N+1:N-1].

**Q6.1.10**   The following statement will determine if a sequence a is increasing or not:

```
np.all(np.diff(a) > 0)
```

Q6.1.11   In the first case, a single object is created of the requested dtype and multiplied by a scalar (regular Python int). Python "upcasts" to return the result in dtype that can hold it:
```

```
In [x]: x = np.uint8(250)
In [x]: type(x*2)
Out[x]: numpy.int64
```

However, an ndarray, because it has a fixed byte size, cannot be upcast in the same way: its own dtype takes precedence over that of the scalar multiplying it, and so the multiplication is carried out modulo 256.

Compare this with the result of multiplying two scalars with the same dtype:

```
In [x]: np.uint8(250) * np.uint8(2)
Out[x]: 244              # (of type np.uint8)
```

(You may also see a warning: RuntimeWarning: overflow encountered in ubyte_ scalars.)

Q6.4.1   The Polynomial deriv method returns a Polynomial object (in this case with a single term, the coefficient of x^0, equal to 18). This object is not equal to the integer object with value 18.

Q6.4.2   Using numpy.polynomial.Polynomial,

```
In [x]: p1 = Polynomial([-11, 1, 1])
In [x]: p2 = Polynomial([-7, 1, 1])
In [x]: p = p1**2 + p2**2
In [x]: dp = p.deriv()        # first derivative
In [x]: stationary_points = dp.roots()
In [x]: ddp = dp.deriv()      # second derivative
In [x]: minima = stationary_points[ddp(stationary_points) > 0]
In [x]: maxima = stationary_points[ddp(stationary_points) < 0]
In [x]: inflections = stationary_points[np.isclose(ddp(stationary_points),0)]
In [x]: print(np.array((minima, p(minima))).T)
[[-3.54138127  8.         ]
 [ 2.54138127  8.         ]]
In [x]: print(np.array((maxima, p(maxima))).T)
[[  -0.5  ,   179.125]]
In [x]: print(np.array((inflections, p(inflections))).T)
[]
```

That is, the function has two minima,

$$f(-3.54138127) = 8,$$
$$f(2.54138127) = 8,$$

one maximum,

$$f(-0.5) = 179.125,$$

and no points of inflection/undulation.

Q6.5.1   Without overcomplicating things,

```
In [x]: pauli_matrices = np.array((
                          ((0, 1), (1, 0)),
                          ((0, -1j), (1j, 0)),
                          ((1, 0), (0, -1))
                         ))
In [x]: I2 = np.eye(2)
In [x]: for sigma in pauli_matrices:
```

```
    ...:        print(np.allclose(sigma.T.conj().dot(sigma), I2))
True
True
True
```

Q6.5.2 The following code fits the coefficients to the required quadratic equation. Note that this is a *linear* least-squares fit even though the function is nonlinear in time because it is linear with respect to the coefficients.

Listing A.2 Least-squares fit to the function $x = x_0 + v_0 t + \frac{1}{2} g t^2$

```
# qn6-9-b-quadratic-fit-a.py
import numpy as np
import matplotlib.pyplot as plt
Polynomial = np.polynomial.Polynomial

x = np.array([1.3, 6.0, 20.2, 43.9, 77.0, 119.6, 171.7, 233.2, 304.2,
              384.7, 474.7, 574.1, 683.0, 801.3, 929.2, 1066.4, 1213.2,
              1369.4, 1535.1, 1710.3, 1894.9])
dt, n = 0.1, len(x)
tmax = dt * (n-1)
t = np.linspace(0, tmax, n)

A = np.vstack((np.ones(n), t, t**2)).T
coefs, resid, _, _ = np.linalg.lstsq(A, x)

# Initial position (cm) and speed (cm.s-1), acceleration due to gravity (m.s-2).
x0, v0, g = coefs[0], coefs[1], coefs[2] * 2 / 100

print('x0 = {:.2f} cm, v0 = {:.2f} cm.s-1, g = {:.2f} m.s-2'.format(x0, v0, g))

xfit = Polynomial(coefs)(t)
plt.plot(t, x, 'ko')
plt.plot(t, xfit, 'r')
plt.xlabel('Time (sec)')
plt.ylabel('Distance (cm)')
plt.show()
```

The fitted function is shown in Figure A.1.

Q6.6.1 The first case,

```
In [x]: a = np.array([6, 6, 6, 7, 7, 7, 7, 7, 7])
In [x]: a[np.random.randint(len(a), size=5)]
array([7, 7, 7, 6, 7])        # (for example)
```

takes random samples from the array a with replacement: for each item selected the probability of a 6 is $\frac{1}{3}$ and the probability of a 7 is $\frac{2}{3}$.

In the second case,

```
In [x]: np.random.randint(6, 8, 5)
array([6, 6, 7, 7, 7])        # (for example)
```

the numbers are drawn from [6, 7] uniformly, so the probabilities of each number being selected is $\frac{1}{2}$.

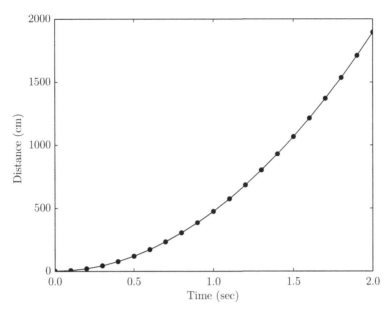

Figure A.1 Least squares fit to the function $x = x_0 + v_0 t + \frac{1}{2}gt^2$.

Q6.6.2   The function np.random.randint samples uniformly from the half-open interval, [low, high), so to get the equivalent behavior to np.random.random_integers in Example E6.18 we need:

```
In [x]: a, b, n = 0.5, 3.5, 4
In [x]: a + (b - a) * (np.random.randint(1, n + 1, size=10) - 1) / (n - 1)
Out[x]: array([ 0.5,   1.5,   0.5,   3.5,   1.5,   3.5,   2.5,   0.5,   1.5,   1.5])
```

Q6.6.3   The probability of winning is one in

$$\binom{70}{5}\binom{25}{1} = \frac{70 \cdot 69 \cdot 68 \cdot 67 \cdot 66}{1 \cdot 2 \cdot 3 \cdot 4 \cdot 5} \cdot 25 = 302575350.$$

To pick five random numbers from 1 to 70 and one from 1 to 25:

```
In [x]: (sorted(np.random.choice(np.arange(1, 71), 5, replace=False)),
            np.random.randint(25) + 1)
([23, 45, 51, 52, 67], 11)
```

Q6.6.4   Here is a more general solution to the problem. Draw the distribution of misprints across the book from the binomial distribution using np.random.binomial and count up how many pages have more than q misprints on them. To compare with the Poisson distribution, for the number of misprints on a page, X, we must calculate $\Pr(X >= q) = 1 - \Pr(X < q) = 1 - (\Pr(X = 0) + \Pr(X = 1) + \ldots + \Pr(X = q - 1))$:

Listing A.3 Calculating the probability of q or more misprints on a given page of a book.

```
# qn6-7-d-misprints-a.py
import numpy as np
```

```
n, m = 500, 400
q = 2
ntrials = 100
errors_per_page = np.random.binomial(m, 1/n, (ntrials, n))
av_ge_q = np.sum(errors_per_page>=q) / n / ntrials
print('Probability of {} or more misprints on a given page'.format(q))
print('Result from {} trials using binomial distribution: {:.6f}'
            .format(ntrials, av_ge_q))

# Now calculate the same quantity using the Poisson approximation,
# Pr(X>=q) = 1 - exp(-lam)[1 + lam + lam^2/2! + ... + lam^(q-1)/(q-1)!]
lam = m/n
poisson = 1
term = 1
for k in range(1, q):
    term *= lam/k
    poisson += term
poisson = 1 - np.exp(-lam) * poisson
print('Result from Poisson distribution: {:.6f}'.format(poisson))
```

A sample output is

```
Probability of 2 or more misprints on a given page
Result from 100 trials using binomial distribution: 0.190200
Result from Poisson distribution: 0.191208
```

Q6.7.1 The two methods for calculating the DFT can be timed using the IPython %timeit magic function

```
In [x]: import numpy as np
In [x]: n = 512
In [x]: # Our input function is just random numbers.
In [x]: f = np.random.rand(n)

In [x]: # Time the NumPy (Cooley-Tukey) DFT algorithm.
In [x]: %timeit np.fft.fft(f)
100000 loops, best of 3: 13.1 us per loop

In [x]: # Now calculate the DFT by direct summation.
In [x]: k = np.arange(n)
In [x]: m = k.reshape((n, 1))
In [x]: w = np.exp(-2j * np.pi * m * k / n)
In [x]: %timeit np.dot(w, f)
1000 loops, best of 3: 354 us per loop

In [x]: # Check the two methods produce the same result.
In [x]: ftfast = np.fft.fft(f)
In [x]: ftslow = np.dot(w, f)
In [x]: np.allclose(ftfast, ftslow)
Out[x]: True
```

The Cooley–Tukey algorithm is found to be almost 30 times faster than the direct method. In fact, this algorithm can be shown to scale as $O(n \log n)$ compared with $O(n^2)$ for direct summation.

Q8.1.1 Simply change the line:

```
for rec in constants[-10:]:
```

to:

```
for rec in constants[constants['rel_unc'] > 0][:10]:
```

and the format specifier in the output line to ':g' (since the uncertainties are less than 1 ppm). The most accurately known constant is the electron g-factor.

```
1.74797e-07 ppm: electron g factor = -2.00232
1.79792e-07 ppm: electron mag. mom. to Bohr magneton ratio = -1.00116
1.90811e-06 ppm: hertz-hartree relationship = 1.51983e-16 E_h
1.91096e-06 ppm: Rydberg constant times hc in eV = 13.6057 eV
...
```

Q8.1.2   The calculation $N/V = p/k_B T$ for the stated conditions can be done entirely with constants from `scipy.constants`:

```
In [x]: scipy.constants.atm / scipy.constants.k / scipy.constants.zero_Celsius
Out[x]: 2.686780501003883e+25
```

This is the *Loschmidt constant*, which is defined by the 2010 CODATA standards and included in `scipy.constants` (see the documentation for details):

```
In [x]: from scipy import constants
In [x]: constants.value('Loschmidt constant (273.15 K, 101.325 kPa)')
Out[x]: 2.6867805e+25
```

Q8.2.1   By numerical integration, the result is seen to be 3:

```
In [x]: from scipy.integrate import quad
In [x]: import numpy as np
In [x]: func = lambda x: np.floor(x) - 2*np.floor(x/2)
In [x]: quad(func, 0, 6)
Out[x]: (2.999964948683555,0.00095207666146406472)
```

Q8.2.2   In the following we assume the following imports:

```
In [x]: import numpy as np
In [x]: from scipy.integrate import quad
```

(a)   ```
In [x]: f1 = lambda x: x**4 * (1 - x)**4/(1 + x**2)
```

```
In [x]: quad(f1, 0, 1)
Out[x]: (0.0012644892673496185, 1.1126990906558069e-14)
```

```
In [x]: 22/7 - np.pi
Out[x]: 0.0012644892673496777
```

(b)   ```
In [x]: f2 = lambda x: x**3/(np.exp(x) - 1)
```

```
In [x]: quad(f2, 0, np.inf)
Out[x]: (6.49393940226683, 2.628470028924825e-09)
```

```
In [x]: np.pi**4 / 15
Out[x]: 6.493939402266828
```

(c)   In [x]: f3 = **lambda** x: x**-x

    In [x]: quad(f3, 0, 1)
    Out[x]: (1.2912859970626633, 3.668398917966442e-11)

    In [x]: np.sum(n**-n **for** n **in** range(1, 20))
    Out[x]: 1.2912859970626636

(d)   In [x]: **from** scipy.misc **import** factorial
    In [x]: f4 = **lambda** x, p: np.log(1/x)**p
    In [x]: **for** p **in** range(10):
       ...:   print(quad(f4, 0, 1, args=(p,))[0], factorial(p))
       ...:
    1.0 1.0
    0.9999999999999999 1.0
    1.9999999999999991 2.0
    6.000000000000064 6.0
    24.000000000000014 24.0
    119.9999999999327 120.0
    719.9999999989705 720.0
    5039.99999945767 5040.0
    40320.00000363255 40320.0
    362880.00027390465 362880.0

(e)   In [x]: **from** scipy.special **import** i0
    In [x]: z = np.linspace(0, 2, 100)
    In [x]: y1 = 2 * np.pi * i0(z)
    In [x]: f5 = **lambda** theta, z: np.exp(z*np.cos(theta))
    In [x]: y2 = np.array([quad(f5, 0, 2*np.pi, args=(zz,))[0] **for** zz **in** z])
    In [x]: np.max(abs(y2-y1))
    Out[x]: 2.1863399979338283e-11

Q8.2.3   To estimate π by integration of the constant function $f(x,y) = 4$ over the quarter circle with unit radius in the quadrant $x > 0$, $y > 0$:

```
In [x]: from scipy.integrate import dblquad
In [x]: dblquad(lambda y, x: 4, 0, 1, lambda x: 0, lambda x: np.sqrt(1 - x**2))
Out[x]: (3.1415926535897922, 3.533564552071766e-10)
```

Q8.2.4   The integral to be calculated is

$$\int_0^1 \int_0^{2\pi} r \, d\theta \, dr = \pi.$$

Note that the inner integral is over θ and the outer is over r. Therefore, the call to dblquad should call the function $f(r, \theta) = r$ as lambda theta, r: r (note the order of the arguments).

```
In [x]: dblquad(lambda theta, r: r, 0, 1, lambda r: 0, lambda r: 2*np.pi)
Out[x]: (3.141592653589793, 3.487868498008632e-14)
```

Alternatively, swap the order of the integration:

```
dblquad(lambda r, theta: r, 0, 2*np.pi, lambda theta: 0, lambda theta: 1)
(3.141592653589793, 3.487868498008632e-14)
```

Q8.4.1 Rewrite the equation as

$$f(x) = x + 1 + (x - 3)^{-3} = 0.$$

This function is readily plotted and the roots may be bracketed in $(-2, -0.5)$ and $(0, 2.99)$ (avoiding the singularity at $x = 3$).

```
In [x]: f = lambda x: x + 1 + (x-3)**-3
In [x]: brentq(f, -2, -0.5), brentq(f, 0, 2.99)
Out[x]: (-0.984188231211512, 2.3303684533047426)
```

Q8.4.2 Some examples of root-finding for which the Newton–Raphson algorithm fails and how to solve this.

(a)
```
In [x]: newton(lambda x: x**3 - 5*x, 1, lambda x: 3*x**2 - 5)
...
RuntimeError: Failed to converge after 50 iterations, value is 1.0
```

The Newton–Raphson algorithm enters an endless cycle of values for x:

$$x_0 = 1: \quad x_1 = x_0 - f(x_0)/f'(x_0) = -1,$$
$$x_1 = -1: \quad x_2 = x_1 - f(x_1)/f'(x_1) = 1,$$
$$x_2 = 1: \quad x_3 = x_2 - f(x_2)/f'(x_2) = -1,$$
$$\cdots$$

Alternative starting points converge correctly on a root. Even a very small displacement from $x = 0$ ensures convergence:

```
In [x]: newton(lambda x: x**3 - 5*x, 1.0001, lambda x: 3*x**2 - 5)
Out[x]: 2.23606797749979
In [x]: newton(lambda x: x**3 - 5*x, 1.1, lambda x: 3*x**2 - 5)
Out[x]: -2.23606797749979
In [x]: newton(lambda x: x**3 - 5*x, 0.5, lambda x: 3*x**2 - 5)
Out[x]: 0.0
```

(b)
```
In [x]: f, fp = lambda x: x**3 - 3*x + 1, lambda x: 3*x**2 - 3
In [x]: newton(f, 1, fp)
Out[x]: 1.0
In [x]: f(1.0)
Out[x]: -1
```

The algorithm converged, but not on a root! Unfortunately, the gradient of the function is zero at the chosen starting point and because of round-off error this has not led to a ZeroDivisionError. To find the roots, choose different starting points such that $f'(x_0) \neq 0$, or use a different method after bracketing the roots by inspection of a plot of the function:

```
In [x]: brentq(f, -0.5, 0.5), brentq(f, -2, -1.5), brentq(f, 1, 2)
Out[x]: (0.34729635533386066, -1.879385241571423, 1.532088886237956)
```

(c)   The function $f(x) = 2 - x^5$ has a flat plateau around $f(0) = 2$ and the small gradient here leads to slow convergence on the root:

```
In [x]: newton(f, 0.01, fp)
...
RuntimeError: Failed to converge after 50 iterations, value is ...
```

To find it using `newton`, either move the starting point closer to the root, or increase the maximum number of iterations:

```
In [x]: newton(f, 0.01, fp, maxiter=100)
Out[x]: 1.148698354997035
```

(d)   This is another example of a function that generates an endless cycle of values from the Newton–Raphson method:

```
In [x]: f = lambda x: x**4 - 4.29 * x**2 - 5.29
In [x]: fp = lambda x: 4*x**3 - 8.58 * x
In [x]: newton(f, 0.8, fp)
...
RuntimeError: Failed to converge after 50 iterations, value is ...
```

Unlike the function in (a), the region $0.6 \leq x_0 \leq 1.1$ *attracts* this cyclic behavior, so one needs to initialize the algorithm outside this range to obtain the roots ± 2.3. For example,

```
In [x]: newton(f, 1.2, fp)
Out[x]: -2.3
```

Q8.4.3   In general, there are two (physically distinct) possible angles θ_0 corresponding to the projectile passing through the specified point, $(x_1, y_1) = (5, 15)$, on the way up or on the way down. These values are the roots in $(0, \pi/2)$ of the function

$$f(\theta_0; x_1, z_1) = x_1 \tan \theta_0 - \frac{g x_1^2}{2 v_0^2 \cos^2 \theta_0} - z_1.$$

After bracketing the roots with a rough plot of $f(\theta_0)$, we can use `brentq`:

```
In [x]: g = 9.81
In [x]: v0, x1, z1 = 25, 5, 15
In [x]: f = lambda theta0, x1, z1: x1 * np.tan(theta0) - g / 2\
                    * (x1 / v0 / np.cos(theta0))**2 - z1
In [x]: th1 = brentq(f, 1, 1.4, args=(x1,z1))
In [x]: th2 = brentq(f, 1.5, 1.6, args=(x1,z1))
In [x]: np.degrees(th1), np.degrees(th2)
Out[x]: (74.172740936822834, 87.392310240255171)
```

That is, $\theta_0 = 74.2°$ or $\theta_0 = 87.4°$.

Q10.1.1   Let $x = 0.9999\ldots$. Then,

$$10x = 9.9999\ldots = 9 + x \implies 9x = 9 \implies x = 1.$$

Q10.1.2   This occurs because `math.pi` is only a (double-precision floating-point) approximation to π, and the tangent of this approximate value happens to be negative:

```
In [x]: math.tan(math.pi)
Out[x]: -1.2246467991473532e-16
```

Taking the square root leads to the math domain error.

Q10.1.3   The problem, of course, is that the expression has been written using double-precision floating-point numbers and the difference between the sum of the first two terms and the third is smaller than the precision of this representation. Using the exact representation in integer arithmetic,

```
In [x]: 844487**5 + 1288439**5
Out[x]: 3980245235185639013055619497406
In [x]: 1288439**5
Out[x]: 3980245235185639013290924656032
```

giving a difference of

```
In [x]: 844487**5 + 1288439**5 - 1318202**5
Out[x]: -235305158626
```

The finite precision of the floating-point representation used, however, truncates the decimal places before this difference is apparent:

```
In [x]: 844487.**5 + 1288439.**5
Out[x]: 3.980245235185639e+30
In [x]: 1318202.**5
Out[x]: 3.980245235185639e+30
```

This is an example of *catastrophic cancellation.*

Q10.1.4   The expression 1 - np.cos(x)**2 suffers from catastrophic cancellation close to x = 0 resulting in a dramatic loss of precision and wild oscillations in the plot of $f(x)$ (Figure A.2). Consider, for example, x = 1.e-9: in this case, the *difference* 1 - np.cos(x)**2 is indistinguishable from zero (at double precision) so f(x) returns 0. Conversely, np.sin(x)**2 is indistinguishable from x**2 and g(x) returns 1.0 correctly.

Listing A.4 A comparison of the numerical behavior of $f(x) = (1 - \cos^2 x)/x^2$ and $g(x) = \sin^2 x/x^2$, close to $x = 0$.

```
# qn9-1-c-cos-sin-a.py

import numpy as np
import matplotlib.pyplot as plt


f = lambda x: (1 - np.cos(x)**2)/x**2
g = lambda x: (np.sin(x)/x)**2


x = np.linspace(-0.0001, 0.0001, 10000)

plt.plot(x, f(x))
plt.plot(x, g(x))
plt.ylim(0.99995, 1.00005)
plt.show()
```

Q10.1.5   We cannot compare with == because nan is not equal to itself. However, it is the *only* floating-point number that is not equal to itself, so use != instead:

```
In [x]: c = 0 * 1.e1000     # 0 * inf is nan
In [x]: c != c
Out[x]: True                # c isn't equal to itself, so must be nan
```

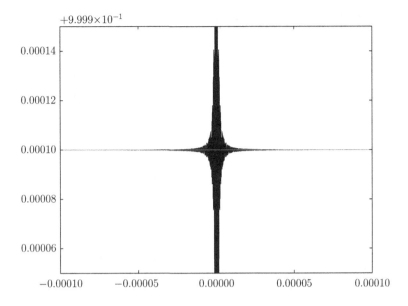

Figure A.2 A comparison of the numerical behavior of $f(x) = (1 - \cos^2 x)/x^2$ and $g(x) = \sin^2 x/x^2$, close to $x = 0$.

Appendix B Differences Between Python Versions 2 and 3

B.0.1   Integers and Arithmetic

In Python 2, there were two kinds of integer: "simple" integers (system-dependent, but usually stored in either 32 or 64 bits) and "long" integers (of any size), indicated with the suffix L. Python 3 is simpler: there is only one integer type, which can be of any magnitude (subject to the availability of memory on your computer).

Python 2's division operator, /, would always perform integer (floor) division on its arguments if they are both integers. Conversely, in Python 3, this operation always returns a float. In both versions, the // operator can be used to force floor division. In summary,

Python 2:

```
>>> 8 / 4
2              # Python 2 only: an int
>>> 8 // 4
2
>>> 7.7 / 2
3.85
>>> 7.7 // 2
3.0            # the largest integer not greater than 3.85
>>> -7.7 // 2
-4.0           # the largest integer not greater than -3.85
```

Python 3:

```
>>> 8 / 4
2.0            # a float, even though both 8 and 4 are ints
>>> 8 // 4
2
>>> 7.7 / 2
3.85
>>> 7.7 // 2
3.0
>>> -7.7 // 2
-4.0
```

The round() built-in function acts a little differently between the two versions. In rounding to zero decimal places, Python 3 employs *bankers' rounding*: if a number is midway between two integers, then the even integer is returned; the return type is int. In Python 2, round() rounded *away from zero* and always returned a float.

Python 2:

```
>>> round(-4.5)
-5.0
>>> round(6.5)
7.0
```

Python 3:

```
>>> round(-4.5)
-4
>>> round(6.5)
6
```

B.0.2   Comparisons

Python 2 allowed comparisons between objects of different types. When comparing a numeric and non-numeric type, the numeric type was always less than the non-numeric one; other objects of different types were ordered consistently but arbitrarily (that is, different Python interpreters may produce a different ordering; CPython ordered by the type name, e.g. all dicts were "less than" strs):

```
>>> '2' > 5
True
>>> {} > 'a'              # (CPython)
False
```

Python 3 does not allow objects of different types to be compared:

```
>>> '2' > 5
TypeError                                Traceback (most recent call last)
...
----> 1 '2' > 5
TypeError: unorderable types: str() > int()
```

B.0.3   Keywords

The reserved keywords in Python, those that cannot be used as variable names (identifiers), were given in Table 2.4. The corresponding list for Python 2.7 excluded async, await, nonlocal, True, False, None but included exec and print.

One of the most obvious differences to new users between Python 2 and Python 3 is that in Python 2 print was a *statement* whereas in Python 3 print() is a *function*:

Python 2:

```
>>> print '2 + 2 =', 4
2 + 2 = 4
>>> print >>fo, 'A string to write to file with open handle fo'
```

Python 3:

```
>>> print('2 + 2 =', 4)
2 + 2 = 4
>>> print('A string to write to file with open handle fo', file=fo)
```

It was also the case that since `True` and `False` were not reservered keywords in Python 2, they were valid identifier names, allowing problematic usage such as:

Python 2:

```
>>> True = False
>>> True
False
```

B.0.4   Strings and Unicode

In Python 2, a distinction was made between Unicode strings (whose string literals were preceded with the character u, for example `u'El Niño'` and 8-bit strings (which only contain 7-bit ASCII bytes). This was the cause of many headaches, including the dreaded `UnicodeEncodeError`, which occurred when strings of different types were mixed. For example, attempting to interpolate a Unicode string into an 8-bit string:

```
>>> s1 = 'I live in'
>>> s2 = u'Saint Étienne'
>>> '{} {}'.format(s1, s2)
----> 1 '{} {}'.format(s1, s2)

UnicodeEncodeError: 'ascii' codec can't encode characters in position 0-6:
ordinal not in range(128)
```

All strings of text in Python 3 are Unicode strings and have the type `str`:

```
>>> s1 = 'I live in'
>>> s2 = 'Saint Étienne'
>>> '{} {}'.format(s1, s2)
'I live in Saint Étienne'
```

If you really do have a string of 8-bit data values, then that's a bytestring of type `bytes`. Binary data literals may be defined with the `b'...'` syntax, or cast from other compatible objects as follows:

```
>>> b1 = b'ABC'
>>> b2 = b'ABC\xff'
>>> b3 = bytes([65, 66, 67, 255])
>>> b4 = bytes([65, 66, 67, 0xff])
```

The last three generate the same `bytes` object, with the final byte being defined in different ways. It is an error to try to assign a value greater than 255 in a `byte` object, since bytes have 8 bits by definition:

```
>>> b5 = bytes([65, 66, 67, 256])
----> 1 b5 = bytes([65, 66, 67, 256])

ValueError: bytes must be in range(0, 256)
```

B.0.5   Iterators and Lists

In Python 2, the `range` built-in returned a list, and memory was allocated for every element:

```
>>> range(10)
[0, 1, 2, 3, 4, 5, 6, 7, 8, 9]
```

This created obvious problems for very large sequences, and so there was a separate built-in, xrange that returned an iterable object which would yield each value of an arithmetic progression in turn without storing all the elements.

In Python 3, range returns an iterable object similar to the old xrange, and xrange no longer exists. Python 3's range (but not Python 2's xrange) is clever enough to work out whether an integer is a member of the sequence without iterating over it, so the following is extremely fast:

```
>>> 999999999998 in range(0, 10**12, 2)   # all even numbers less than 1 trillion
True

>>> 999999999997 in range(0, 10**12, 2)
False
```

Similarly, the Python 2 dictionary methods, keys(), values() and items(), returned lists of these values:

```
>>> d = {'a': 1, 'b': 2, 'c': 3}
>>> d.keys()
['a', 'c', 'b']
>>> d.items()
[('a', 1), ('c', 3), ('b', 2)]
```

Python 3 returns special iterable objects. If a list is required, it must be cast explicitly:

```
>>> d = {'a': 1, 'b': 2, 'c': 3}
>>> d.keys()
dict_keys(['a', 'b', 'c'])
>>> list(d.keys())
['a', 'b', 'c']     # NB Python 3.6+: dictionaries are in order of insertion
```

This is, of course, more efficient for large dictionaries.

The zip built-in provides a further example. Python 2:

```
>>> zip(['a', 'b', 'c'], [1, 2, 3])
[('a', 1), ('b', 2), ('c', 3)]     # a list of tuples
```

Python 3:

```
>>> zip(['a','b','c'], [1, 2, 3])
<zip at 0x102bcf5c8>               # a special iterable zip object ...
>>> list(zip(['a','b','c'], [1, 2, 3]))
[('a', 1), ('b', 2), ('c', 3)]   # that can be cast into a list if you want.
```

Appendix C SciPy's odeint Ordinary Differential Equation Solver

Section 8.2.3 described SciPy's solve_ivp method for solving ordinary differential equations (ODEs): this function provides access to a suite of ODE solvers through a unified interface and is currently the recommended approach for most situations likely to be encountered by SciPy users. The older function, scipy.integrate.odeint,[1] is still available as of SciPy version 1.4, and because much legacy code is likely to depend on it, its use is described in this appendix. odeint is based on the well-tested Fortran LSODA routine, which can automatically switch between stiff and nonstiff algorithms.

A Single First-Order ODE

In its simplest use for the solution of a single first-order ODE,

$$\frac{dy}{dt} = f(y, t),$$

odeint takes three arguments: a function object returning dy/dt, an initial condition, y_0, and a sequence of t values at which to calculate the solution, $y(t)$.

Returning to the example of a first-order chemical reaction, $A \rightarrow P$, in terms of the reactant concentration, $y = [A]$, which satisfies

$$\frac{dy}{dt} = -ky,$$

we would define

```
def dydt(y, t):
    return -k * y
```

Note that the order of the arguments is the opposite to that required for solve_ivp. Using odeint, the equivalent solution to the one given in Section 8.2.3 would be

```
from scipy.integrate import odeint
# First-order reaction rate constant, s-1.
k = 0.2
# Initial condition on y: 100% of reactant is present at t=0.
y0 = 100
# A suitable grid of time points for the reaction.
t = np.linspace(0, 20, 21)
```

---

[1] The odeint function is a simplified interface to the more advanced scipy.integrate.ode method, which provides a range of different numerical integrators, including Runge–Kutta algorithms and support for complex-valued variables.

```
y = odeint(dydt, y0, t)
```

Note that, by default, odeint returns just the solution, *y*, as a function of the specified grid of time points.[2]

A Single Second-Order ODE

Solving a second-order ODE such as that of the simple harmonic oscillator problem in Section 8.2.3 again requires that it be decomposed into first-order equations:

```
# Harmonic oscillator frequency (s-1).
omega = 0.9
# Initial conditions on x1 = x and x2 = dx/dt at t = 0.
x0 = 3, 0          # cm, cm.s-1

def dxdt(x, t, omega):
    """ Return dx/dt = f(x,t) at time t for the harmonic oscillator."""
    x1, x2 = x
    dx1dt = x2
    dx2dt = -omega**2 * x1
    return dx1dt, dx2dt

# Integrate the differential equations with odeint.
x1, x2 = odeint(dxdt, x0, t, args=(omega,)).T
```

odeint returns a two-dimensional array with the values of each dependent variable in *rows*: if we want to unpack this array into separate one-dimensional arrays of position and velocity, we need to transpose the returned array.

---

[2] An optional argument full_output can be set to True to additionally output a dictionary, infodict, with statistics about how well (or badly) the integration went.

Glossary

argument A value passed to a function when calling it. *Positional arguments* are identified by their order in the sequence specified by the function definition (e.g. `complex(1, -2)`); *keyword arguments* are associated with an identifier when calling the function (e.g. `complex(real=1, imag=-2)`).

attribute In Python, an object (item of data, function, etc.) associated with ("belonging to") another object and accessed through the dot notation, `object.attribute`. For example, the `upper` function is an attribute of the `str` object: given an instance of a string, such as `'python'`, it is accessed as `'python'.upper` and called as `'python'.upper()`.

built-in Any of a number of functions and types that are predefined by the Python interpreter and always available without explicitly importing them from another module. Examples are `print()`, `float` and `range()`.

bytecode An "intermediate" language into which source code is compiled by the Python interpreter, to be executed by the Python "virtual machine" as machine code understood by the computer processor.

catastrophic cancellation A drastic loss of significance experienced in floating-point arithmetic when a quantity is subtracted from another with a very similar value.

class A template for creating (*instantiating*) objects, which defines their attributes, interfaces and behaviors. Classes may be related to each other through *inheritance* (q.v.).

conditioning (of a numerical problem) A problem whose solution is relatively insensitive to errors and uncertainties in its input data is called *well-conditioned*; a problem for which small relative errors in the input result in large differences in the output is said to be *ill-conditioned*.

docstring A string literal written as the first statement in a function, class or module definition, which becomes the `__doc__` attribute of that object and documents what it does.

double-underscore methods Special methods ("dunder" methods) are denoted in Python by names starting and ending with two underscore characters. These methods usually implement an operation that is invoked by some special syntax or built-in function. For example, indexing a `list`, `x[i]` is basically equivalent to calling `x.__getitem__(i)`. Redefining these methods is how operator overloading (q.v.) is achieved in Python.

duck-typing The determination of the suitability of an object with respect to some operation from the methods and properties it possesses rather than its formally declared data type. Python is a duck-typed language.

dynamic type-checking The determination of the suitability of an object for an operation at runtime rather than before the program is executed (static type-checking). Python is a dynamically typed language.

EAFP *"Easier to Ask Forgiveness than to seek Permission"*: the idiomatic Python coding style that attempts to carry out an operation on data with the intention of handling any errors (exceptions) that occur gracefully. This is in contrast to the *"Look Before You Leap"* paradigm) adopted in other languages, in which the data are carefully checked for their type and suitability before attempting the operation.

endianness The platform-dependent order of the bytes representing a number. Big-endian systems place the most-significant byte first (in the smallest memory address); little-endian systems place it last. The name derives from the sectarian division between Lilliputs over which end of a boiled egg should be broken in Jonathan Swift's 1726 novel, *Gulliver's Travels*.

exception An error occuring during the execution of a Python program; unless dealt with, an exception leads to termination of the program. It is common to anticipate and catch (*handle*) specific exceptions in a `try ... except` clause (q.v. EAFP). User-defined exceptions can also be defined and triggered (*raised*).

expression A syntactically correct combination of valid Python identifiers, literals and operators, which evaluates to some value. For example, (assuming x is defined), `x + 2` is a valid Python expression evaluating to the sum of the value of x and the integer 2.

floating-point number The representation of a real number used in computer arithmetic, in which the number is stored in separate parts representing its significant digits and exponent within a fixed number of bits to a generally finite precision.

function A set of statements, grouped into a unit of the program to perform a task when called (executed). Functions may be passed arguments (data in the form of Python object references) and may return one or more values. Functions that belong to an object are called its *methods*; since everything in Python is an object, there is little difference between these terms.

hash A Python object is *hashable* if it can return a hash value from a special method, `__hash__`, which never changes during its lifetime, and if it can be compared to other objects for equality. This is important for data structures such as dictionaries, which map hash values to objects for high-performance storage and access.

identifier name The symbolic name associated with an object, by which it is referred to in a program's source code.

inheritance The realization of a hierarachy between classes in an object-oriented program. A class that is defined as being based on ("sub-classed" from) another is said to inherit its attributes, which it may modify and add to. Inheritance encourages code-reuse and the hierarchy it represents is often inherent in the underlying concepts the classes represent.

instance A specific occurence of an object, created ("instantiated") from a class (q.v.) with its own particular data and identifier name. In addition to attribute values belonging to a specific instance of an object, there may be data shared by all instances of a given class (class attributes).

introspection The ability of a program or interactive shell session to provide information about an object at runtime. For example, the expression `type(x)` returns the type of the object identified by the name `x`.

IPython A command-line, interactive computing environment for Python, providing a variety of convenient functionalities including introspection, command history, interactive visualization and tab completion.

iterable An iterable object is one containing or yielding a sequence of values, one after another, when looped over. Examples include `lists`, `tuples` and `strs`.

Jupyter Notebook A browser-based interactive computing environment for creating rich-media, shareable documents, which mix code, code output and explanatory text in an ordered sequence of editable "cells."

lambda An *anonymous function*: one not (necessarily) bound to an identifier name. In Python, lambda functions are typically defined inline with other code and can only contain a single expression.

least-squares fitting In the analysis of an overdetermined system (one with more data points than unknown parameters to its model), the process of obtaining the set of parameters that minimizes the sum of the squares of the (observed – modeled) residuals. In *linear* least-squares fitting, the model function depends linearly on its parameters.

list comprehension The Python syntax for creating a `list` in a concise, readable line of code without the explicit use of the append method; for example, `[x**2 for x in range(5)]`.

literal A Python literal is the direct specification of an object when parsed by the interpreter (as opposed to its reference by variable name). For example, `'parrot'` is a string literal.

magic number A constant numerical value used as a literal (q.v.) in a program instead of being assigned to a meaningul variable name. Generally avoided in clear, maintainable code.

Matplotlib A free, open-source plotting library for producing a variety of two- and three-dimensional plots and charts suitable for data visualization, which can be used in print and online.

method A function bound to an object.

module A collection of Python code (object definitions, including classes and functions, and executable code) in one or more files, that may be *imported* for reuse by other programs.

mutable and immutable objects A immutable object cannot be changed after it is defined: objects of this type are assigned a region of memory (associated with an *identity*, which may be thought of as its "address" in memory) and the contents of this region cannot be altered whilst the object exists (though the name associated with the object may be rebound to a different object). A mutable object may be changed "in-place": its properties can be altered without changing its identity. Some immutable object types are: `int`, `float`, `str` and `tuple`; the object types `list`, `dict` and `set` are mutable.

namespace A mapping (dictionary) between identifier names and the objects they are assigned to; which namespaces are used to resolve a name to an object is determined by the name's *scope* (q.v.).

Not a Number (NaN) The unique member of the floating-point arithmetic data type that represents undefined or unrepresentable numbers.

NumPy A free, open-source library supporting fast numerical operations on large, multidimensional arrays of a single data type.

object The abstract conceptualization of an entity realized in a computer program as a data structure which may possess and manipulate its own data (attributes), including methods and other objects.

object-oriented programming (OOP) A programming paradigm in which the components of a system are identified as discrete entities which are defined in code by *objects* described by templates ("blueprints") called classes. This conceptualization can help to solve problems by breaking them into small, related parts with simple individual behaviors, which interact in a defined and controllable way.

operator overloading The implementation of an operator used in one class on another class: for example, the + operator is used to add integers (in, for example, the expression 2 + 3) but also to concatenate `list` objects: `[1, 2] + [3, 4, 5]` returns `[1, 2, 3, 4, 5]`. This is an example of polymorphism (q.v.). User-defined classes can overload an operator by defining its corresponding *double-underscore method* (q.v.).

optimization (mathematics) The process of obtaining the best (according to some criteria) parameterization of a problem (possibly subject to some constraints). For example, finding the values of x and y which give the minimum value of a function $f(x, y)$ is an optimization problem. Finding its maximum value is essentially the same problem, since it is equivalent to minimizing $-f(x, y)$.

overflow and underflow The condition of the result of a calculation being larger or smaller than can be represented in memory for the data type used. For example, `math.exp(1000)` overflows in double-precision floating-point arithmetic because it is bigger than the largest representable float using 64 bits (about 1.8×10^{308}).

package A hierarchically structured collection of modules (q.v.) stored in a directory, which provides them with a namespace (q.v.) and allows larger projects to define how they are imported and used.

pandas A free, open-source library for data manipulation and analysis, with support for tabular data with heterogeneous data types and high-level functionality for grouping, aggregating and cleaning data sets.

PEP Python Enhancement Proposals, a set of usage recommendations, standards descriptions and governance procedures for developers of the Python language. New language features are proposed and described in short documents, which are reviewed by a steering committee and placed in the repository at www.python.org/dev/peps/.

pip The Python package manager for installing packages (q.v.) from the Python Package Index (https://pypi.org/), a repository of free software libraries to extend the language's functionality.

polymorphism The property of a function (method) being able to act meaningfully on data of different types. For example, the Python * operator is polymorphic in that it can be used to multiply together two numbers (2 * 1.2 returns 2.4) but also to create a list with repeated elements ([0] * 4 returns [0, 0, 0, 0].

pseudorandom-number generator (PRNG) An algorithm that generates a sequence of numbers that approximates the properties of random numbers. A PRNG may be initialized (*seeded*) with a fixed value to reproduce a sequence (it is deterministic), which will repeat eventually (though with a long period).

SciPy A free, open-source library used for scientific computing, including algorithms for integration, interpolation, optimization and differential equation solving.

scope The availability of a variable (name) within a block of code. Since the same identifier name may be bound to different objects in different parts of a program, the rules about scope resolution determine which object is resolved when that identifier is used. For example, if a variable is referred to within a function definition, the Python interpreter will first see if it has been assigned to an object within that function (the *local namespace*), and if this fails, look within the block of code enclosing the function, and so on, ending with the program (or module) *global namespace* and finally Python's own *built-in namespace* (which predefines names for objects such as the print function).

shell A user interface to a computing system, particularly one with an interactive command-line interface at which commands are entered and executed.

stability (of an algorithm) An algorithm is said to be *stable* if it is relatively unaffected by approximation errors of various sorts that can occur in its execution or input data; if these errors are magnified (typically leading to the catastrophic failure of the algorithm to obtain a meaningful result), it is said to be unstable.

stack (software) A set of software components which work together to create a platform (or environment) on which a class of computing tasks can be carried out. For example, the "SciPy stack" is sometimes thought of as composed of a Python interpreter installed with the NumPy, SciPy and Matplotlib libraries.

Standard Library A large set of modules containing methods for carrying out common tasks in, for example, mathematics, file I/O and debugging. The Standard

Library is installed automatically with most Python distributions; further packages and modules are available from the Python Package Index using pip (q.v.).

statement One or more lines of Python code, composed of *expressions* (q.v.), which generally have an effect on the state of the executing program.

strongly and weakly typed A language is described as weakly typed if it will silently convert objects to a suitable type in order to allow a function to operate on them; strongly typed languages only allow operations on objects of a prescribed set of types (in the case of statically typed languages) or on those which possess compatible properties (in the case of dynamically typed languages). Python is fairly strongly typed: for example, the expression `'hello'` `+ 4` raises an exception (`TypeError`). A language like JavaScript is relatively weakly typed and coerces the integer 4 into a string in order to return `'hello4'`.

syntactic sugar Programming syntax and constructs which, whilst not strictly necessary to make the language function, make code simpler, clearer, more pleasant to program, and (sometimes) faster. For example, Python supports *augmented assignment* (e.g. `a += 1` as a synonym for `a = a + 1`) and list comprehension (q.v.).

unicode An international standard for the encoding and representation of text expressed in almost all of the world's writing systems. The Unicode standard assigns a code point number to each character and defines how this number should be expressed in bytes, along with rules concerning the directionality of text, comparison (collation) of characters, and so on.

variable The symbolic name associated with an object, by which it is referred to in a program's source code. In Python, a variable in this sense is sometimes more formally called an *identifier name*: a single object may have more than one such identifier.

vectorization The batch operation of a single operation on an entire array, without the need for an explicit Python loop – this improves both speed and readability. NumPy supports vectorization on its `ndarray` object type by implemeting many of its operations in precompiled C code.

version control The management of changes to software as it is developed, often using a tool which allows collaboration, tagging of release versions and branching of parallel code versions for development.

Index